# FUTURE OF FRACTALS

Proceedings of the International Conference

# FUTURE OF FRACTALS

*Nagoya, Japan* *25 – 27 July 1995*

*editor*

**S Miyazima**

*Chubu University*
*Japan*

**World Scientific**
*Singapore • New Jersey • London • Hong Kong*

*Published by*

World Scientific Publishing Co. Pte. Ltd.
P O Box 128, Farrer Road, Singapore 912805
*USA office:* Suite 1B, 1060 Main Street, River Edge, NJ 07661
*UK office:* 57 Shelton Street, Covent Garden, London WC2H 9HE

**British Library Cataloguing-in-Publication Data**
A catalogue record for this book is available from the British Library.

**FUTURE OF FRACTALS**



ISBN 981-02-2570-9

Printed in Singapore.

# Contents

## Vol. 4, Issue Nos. 2

**Mathematics and Physics**

Abstracts of the papers in this issue can be obtained on the World-Wide Web, http://www.singnet.com.sg/~wspclib/Journals/Fractals/Fractals.html

**Subscriptions**, changes of address, single-copy orders should be addressed to Journal Department, World Scientific Publishing Co. Pte. Ltd., Farrer Road, P. O. Box 128, Singapore 912805, or 1060 Main Street, River Edge, NJ 07661, U.S.A., or 57 Shelton Street, London WC2H 9HE, England, or 4911, 9th Floor, High Point IV, 45 Palace Road, Bangalore 560 011, India.



**FRACTALS** (ISSN 0218-348X) is published quarterly by World Scientific Publishing Co. Pte. Ltd., Farrer Road, P. O. Box 128, Singapore 912805. Subscription rates are available upon request. Periodicals postal is pending at Jamaica, N.Y. 11431.

US POST MASTER: Please send change of address to FRACTALS, c/o Publications Expediting, 200 Meacham Avenue, Elmont, N.Y. 11003. Air freight & mailing in the US by Publications Expediting, 200 Meacham Avenue, Elmont, N.Y. 11003 (Tel. 516-352-7300).

Printed in Singapore.

# Acknowledgments

Twenty years have passed since fractals were shown to represent the common aspects of many complex processes occurring in a diverse range of fields, including geology, chemistry, biology, medical science, and so on. Many scientists have applied the concept of fractals to their own subjects and developed various new techniques for more calculations with a wealth of results. It is opportune to discuss and anticipate the future of fractals. At the same time as we were preparing for this International Conference on Fractals in Japan, we were also planning another big conference on Statistical Physics in China. This plan was supported by the Physical Society of Japan, the Japan Society of Applied Physics, the Mathematical Society of Japan, and the Biological Society of Japan. Financial support was given by Cgubu University, Nishina Memorial Foundation, Inoue Foundation for Science and Daiko Foundation.

**S. Miyazima, 1996**

Fractals, Vol. 4, No. 2 (1996) 113–122


# DYNAMICAL MODEL FOR VIRUS SPREAD

G. CAMELO-NETO and S. COUTINHO
*Departamento de Física, Universidade Federal de Pernambuco,*
*CEP 50670-901, Recife, Brazil*

**Abstract**

The steady state properties of the mean density population of infected cells in a viral spread is simulated by a general forest fire like cellular automaton model with two distinct populations of cells (permissive and resistant ones) and studied in the framework of the mean field approximation. Stochastic dynamical ingredients are introduced into this model to mimic cells regeneration (with probability $p$) and to consider infection processes by other means than contiguity (with probability $f$). Simulations are carried out on a $L \times L$ square lattice taking into consideration the eighth first neighbors. The mean density population of infected cells ($D_i$) is measured as a function of the regeneration probability $p$, and analyzed for small values of the ratio $f/p$ and for distinct degrees of cell resistance. The results obtained by a mean field like approach recovers the simulations results. The role of the resistant parameter $R$ ($R \geq 2$) on the steady state properties, is investigated and discussed in comparison with the $R = 1$ monocell case which corresponds to the *self organized critical* forest fire model. The fractal dimension of the dead cells ulcers contours was also estimated and analyzed as a function of the model parameters.

## 1. INTRODUCTION

In this work we study the steady state properties of a simple discrete dynamical stochastic cellular automaton model describing the virus spread. An earlier deterministic version of the present model was proposed by Landini et al.[1] in an attempt to study the morphology of Herpes Simplex Virus (HSV) corneal ulcers. In this latter model, two distinct populations of epithelial cells were considered according to the degree of permissivity to infection by HSV. The susceptibility to viral infection by contiguity was distinguished by assuming that the permissive cells became infected by the presence of one or more infected neighbors while for the resistant ones at least $R$ ($R \geq 2$) infected neighbors cells were required to spread de virus. Landini et at.[1] considering a square lattice monolayer cellular automaton model (taking into account the first eight nearest neighbors), showed that the HSV ulcers can be quantitatively well characterized by the fractal dimension of their contour. Furthermore, they shed some light on our understanding that the complex shape and evolution of the HSV ulcers may depend on intrinsic characteristics (susceptibility to infection) and topological distribution of the ephithelial cells. For instance, when the resistance parameter $R = 5$, Landini's model exhibit a dramatic change from dendritic to amoeboid morphology when the concentration of permissive cells reaches the percolation threshold.

The aim of this work is to investigate the dynamical features of the HSV ulcers evolution, that were well described by the *static* model proposed by Landini et al.[1] We are mainly concerned with the overall distribution of ulcers in the steady state that should occur after a primary infection. Nowadays, it is well known that many viruses have evolved mechanisms to avoid the immune system, e.g., adenovirus, murine and HSV. For instance Hill et al.[2] describe a new mechanism by which HSV may evade the immune control. For those viruses after a primary infection, the virus enters a latent phase and can reactivate later on at any time due to other factors. Therefore we introduce stochastic dynamical ingredients to the model proposed by Landini et al.[1] to mimic regeneration of the ephithelial tissue and to consider reinfection processes due to a latent virus phase. With these ingredients, our model can also describe a forest fire phenomena with two distinct species of trees with different degrees of resistance to burning. Actually, the particular case with $R = 1$ (one species of tree) recovers the Self-Organized Critical forest fire model with lightning probability introduced by Drossel and Schwabl.[3]

In the present work, two methods are employed to study the role of the resistance parameter $R$ on the steady state properties of the dead cell population (destroyed trees or empty sites in the counterpart forest fire model): (a) an analytical mean field approach; and (b) a numerical simulation. In Sec. 2, we describe our cellular automaton model and develop the mean field approach. Section 3 is devoted to discussing the numeric simulation procedure. The results and discussions are given on Sec. 4.

## 2. THE MODEL AND THE MEAN FIELD APPROACH

Our stochastic discrete cellular automaton is defined on a square lattice with $L^2$ sites, the sites being occupied by a permissive cell, a resistant cell, an infected cell or a dead cell. The system is parallel updated and governed by the following rules during one time step:

(1) A permissive cell becomes infected by contiguity if one or more infected neighbor cells exist in its environment.
(2) A resistant cell becomes infected by contiguity if $R$ or more infected neighbors cells exist in its environment.
(3) An infected cell becomes a dead cell.
(4) A living cell (permissive or resistant) may become infected with probability $f$ if there are not enough infected neighbors in its environment.
(5) A dead cell may regenerate with probability $p$, a fraction $q$ being permissive cells and $(1-q)$ being resistant cells.
(6) The cell environment is defined by its eight nearest neighbors. Therefore $R$ ranges from 2 to 8.

To study the mean field behavior of the system, we will assume that the system size $L$ is large enough to prevent finite-size effects, and that the system reaches a steady state after a transient period from arbitrary initial conditions. We also assume that this steady state depends only on the model parameters and is characterized by mean values of the densities of the permissive cells $D_p$, of the resistant cells $D_r$, of the infected cells $D_i$ and that of the dead cells $D_d$. These densities are constrained by normalization condition:

$$D_p + D_r + D_i + D_d = 1\,. \tag{1}$$

Now let us consider the variation of these densities within one time step. The rate equation for dead cells is clearly given by:

$$\Delta D_d = D_i - pD_d\,, \tag{2}$$

while for the permissive cells we have:

$$\Delta D_p = qpD_d - [\Theta + f(1-\Theta)]D_p\,. \tag{3}$$

In Eq. (3), the first term gives the mean number of new regenerate permissive cells and the second reads for the permissive cells becoming infected on the next time step [rules (a) and (d)], $\Theta = 1 - (1 - D_i)^8$ being the probability for a given permissive cell to have one or more infected neighbors. By analogy, the rate of resistant cells is given by:

$$\Delta D_r = (1-q)pD_d - [\Phi_R + f(1-\Phi_R)]D_r\,, \tag{4}$$

where $\Phi_R$ is the probability that a given resistant cell is surrounded by $R$ or more infected cells, i.e.,

$$\Phi_R = \sum_{n}^{8} \binom{8}{n} D_i^n (1-D_i)^{8-n}\,. \tag{5}$$

Finally, the rate of the infected cells is given by:

$$\Delta D_i = (\Theta D_p + \Phi_R D_r) + [(1-\Theta)D_p + (1-\Phi_R)D_r]f - D_i\,. \tag{6}$$

In the above equation, the first term describes the number of cells infected by contiguity in one time step, the second gives the ones infected by other means and the third clearly stands for the number of infected cells that will become dead in the next time step.

The steady state is characterized by constant mean densities of all populations of cells, i.e.,

$$\Delta D_r = \Delta D_p = \Delta D_i = \Delta D_d = 0\,. \tag{7}$$

In that condition, we obtain from Eqs. (2)–(4) that,

$$D_d = \frac{1}{p} D_i \tag{8}$$

$$D_p = \frac{qD_i}{\Theta + f(1-\Theta)} \tag{9}$$

$$D_r = \frac{(1-q)D_i}{\Phi_R + f(1-\Phi_R)}\,. \tag{10}$$

Now substituting these expressions into the normalization conditions [Eq. (1)], we finally obtain an equation for the infected cells populations as a function of the model parameters (provide $f$, $p \neq 0$) given by:

$$D_i\left[1+\frac{1}{p}+\frac{q}{(1-f)\Theta+f}+\frac{1-q}{(1-f)\Phi_R+f}\right]=1\,. \tag{11}$$

This equation, which expresses the mean field behavior of the system, can be solved numerically, and the figures for $D_i$ can be used to obtain the other steady state populations given by Eqs. (8)–(10).

## 3. NUMERICAL SIMULATION PROCEDURE

We carried out numerical simulations of the present model by considering a $L \times L$ square lattice with periodic boundary conditions. The initial state was prepared by randomly distributing permissive cells with probability 0.4 (arbitrary) and resistant cells with probability 0.6. An infected cell was triggered on a central site of the lattice. After a transient time interval (time steps) of the order of $L$, we measure all kinds of cells densities and then average these values over the next $L$ consecutive configurations. During all processes, infected cells were triggered randomly at $1/f$ time steps and regenerated cells were randomly allowed at $1/p$ time steps. As far as we are interested to investigate the possibility of occurrence of *self-organized critical* steady states, we fixed the model parameters such that $1/f \gg 1/p \approx L$. This is a necessary condition to allow that all living cells of an infected cluster become infected before new regenerated cells appear on the clusters ends. Note that $L/2$ is the maximum time interval for infection to spread in a large living cells cluster. The *fractal dimension* of the contours of dead cells clusters was also estimated by the box counting method and averaged over all simulations.

## 4. RESULTS AND DISCUSSION

Since we are mainly interested in the properties of the dynamic steady state of the virus spread, we direct our attention to the mean density of the population of infected cells $D_i$, or mean density of fire in the counterpart forest fire model. In Fig. 1, we show the dependence of $D_i$ as a function of the regeneration probability $p$ obtained by simulation of the model on a $L = 100$ square lattice with the degree of resistance varying from $R = 1$ to 8, for the ratio $f/p = 0.1$ and $q = 0.6$. We notice that the curve for $R = 2$ looks qualitatively similar to the one obtained for $R = 1$ (SOC *forest fire model*) and is very distinct from the ones obtained for $R \geq 4$. Those latter curves show a universal like dependence for lower values of $p$. Furthermore, the plot for $R = 3$ exhibits a cross-over behavior between the $R \leq 2$ and $R \geq 4$ curves as $p$ varies from lower to higher values. Now we make comparison with the mean field like results obtained from Eq. (11). In Fig. 2, we display the results obtained by the mean field procedure recovering the ones obtained by simulations. For $R = 1$ and 2, the mean field curves are in qualitative agreement with the one obtained by simulations, while for $R = 3$ one also observes the cross-over between $R \leq 2$ and $R \geq 4$ behaviors, with very sharp changes at small but definite value of $p$. For $R \geq 4$, the agreement between the two approaches is quantitatively established.

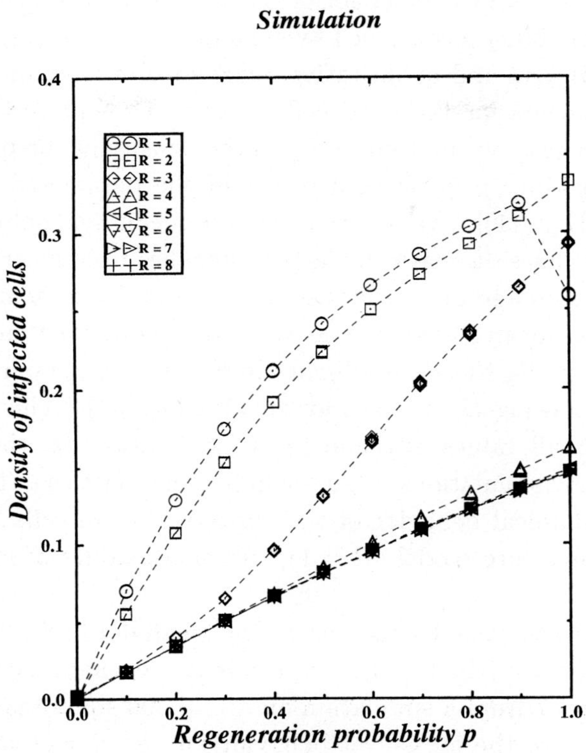


**Fig. 1** Mean density of infected cells $D_i$ obtained by simulation for a $L = 100$ square lattice as a function of the regeneration probability $p$. $R = 1 - 8$, $f/p = 0.1$, $q = 0.6$. Dashed lines are guide for the eyes.

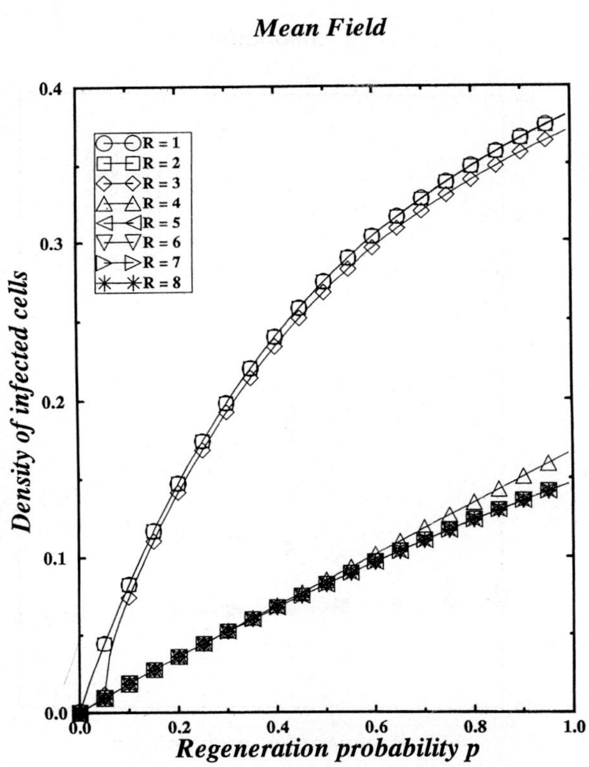


**Fig. 2** Mean density of infected cells $D_i$ obtained by mean field approximation, as function of the regeneration probability $p$. Same parameters as in Fig. 1.

In Figs. 3 and 4, we present the behavior of the density of dead cells (ulcers) as a function of the regeneration probability as obtained by simulation and mean field approaches, respectively. The same qualitative and quantitative behavior observed for the density of infected cells (Figs. 1 and 2) occurs for the dead cell density. To explore the cross-over behavior observed for the $R = 3$ case, we plotted in Fig. 5 the density of the infected cells $D_i$ against the regeneration probability $p$ for several values of the parameter $(f/p)$ obtained by the mean field approach [Eq. (11)]. We observe that the mean field solution (i.e., independent of finite size effects) changes dramatically as $p$ approaches from a certain value which is $f/p$ dependent. As $p$ decreases and approaches zero, the condition $f/p \ll 1$ is no longer fulfilled and the SOC state disappears.[3] This means that clusters of resistant cells remain alive for long time intervals stopping the virus spread. In Fig. 6, we compare $R = 3$ case obtained by both methods by showing $D_i \times p$ plot for small values of $p$. This plot indicates that for $R = 3$ and for very small values of $p$ and for a fixed ratio $f/p$, the mean field predicted behavior is confirmed by simulations, even for finite small lattices. Therefore, the expected SOC state for the dynamical behavior of the clusters of dead cells (ulcers), similar to the one observed in the forest fire model ($R = 1$),[3] for small values of $p$, should occur only for the $R = 2$ case.

We have also investigated the dependence of the mean density of infected cells $D_i$ on the variation of the parameter $q$ (the fraction of permissive regenerate cells) for a fixed value of $p$. This is shown in Fig. 7 from a simulation on a $L = 200$ square lattice and for $f/p = 0.1$ and $p = 0.6$. One observe the cross-over behavior for the $R = 3$ case as $q$ increases from lower to higher values.

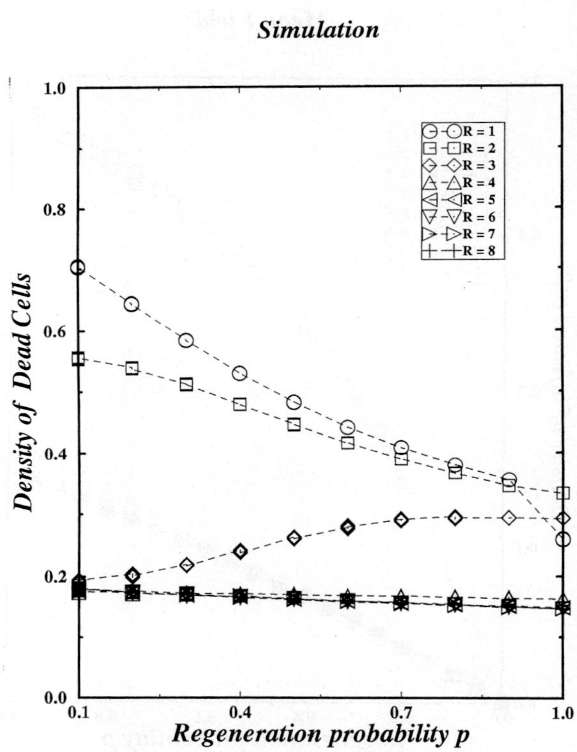


**Fig. 3** Mean density of dead cells $D_d$ obtained by simulation on a $L = 100$ square lattice as a function of the regeneration probability $p$. Same parameters as in Fig. 1.

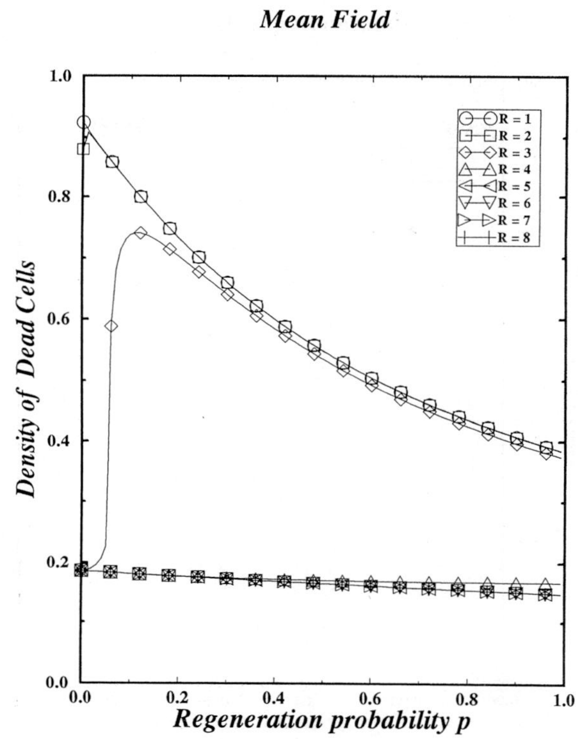


**Fig. 4** Mean density of dead cells obtained by mean field approximation, as function of the regeneration probability. Same parameters as in Fig. 1.

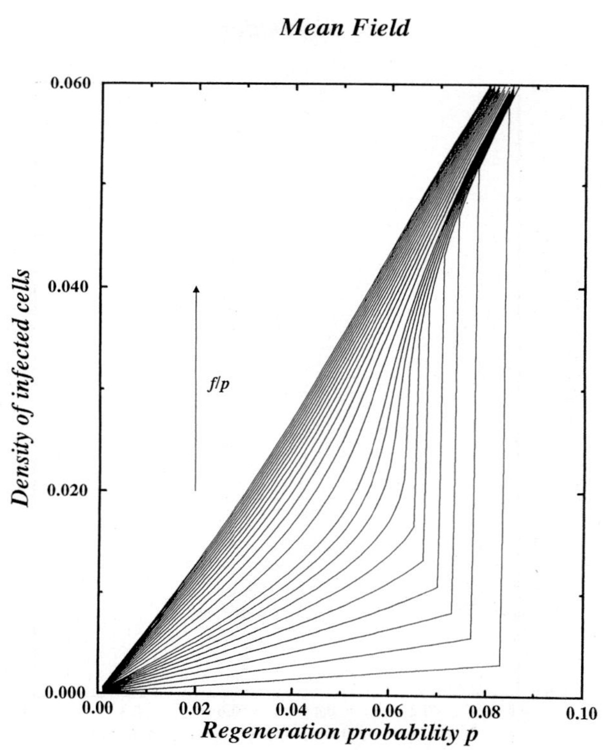


**Fig. 5** Mean density of infected cells $D_i$ obtained by mean field approach as function of the regeneration probability $p$ for $R = 3$ for several values from the ratio $f/p = 0.0001$ to 1.0, with $q = 0.6$.

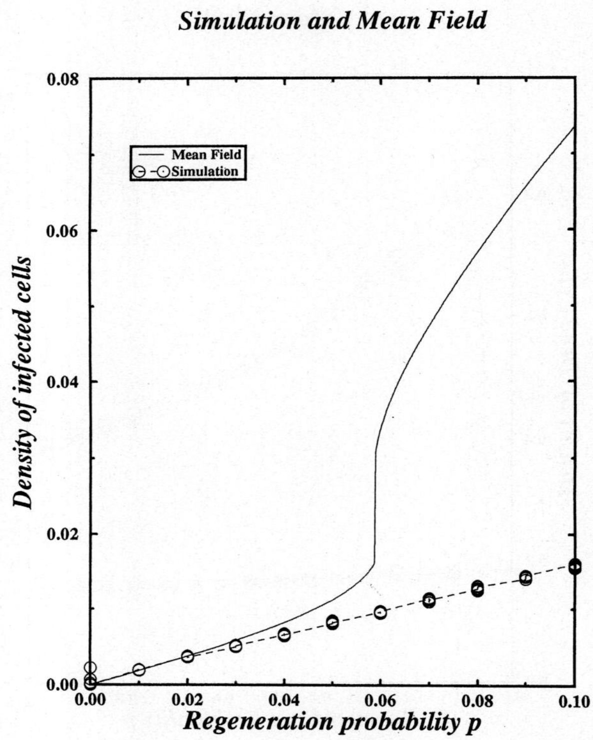


**Fig. 6** Mean density of infected cells $D_i$ as function of the regeneration probability $p$ for resistance parameter $R = 3$. Full line is obtained from mean field approach and open circles indicates figures from simulations. Same parameters as in Fig. 1.

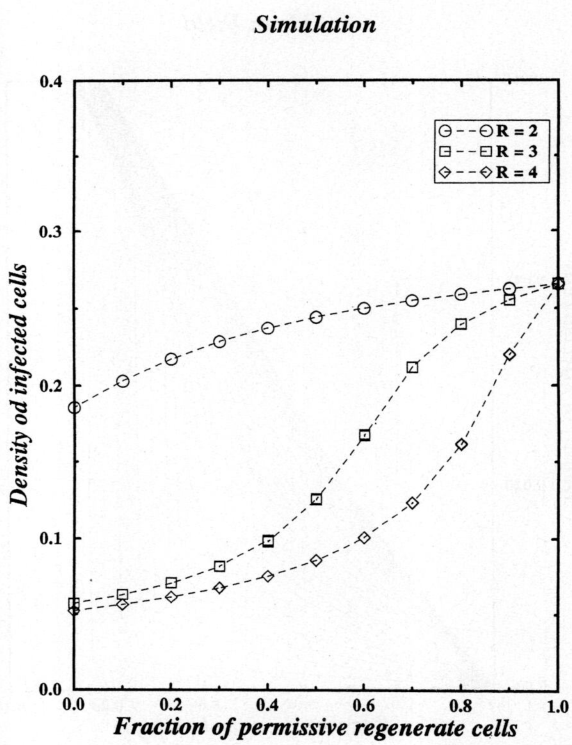


**Fig. 7** Mean density of infected cells $D_i$ obtained by simulation on a $L = 200$ square lattice as function of the fraction of regeneration of permissive cells $q$ for $f/p = 0.1$ and $p = 0.6$.

Finally, we have investigated the dependence of the *fractal dimension* $D_F$ of the contour of the dead cells clusters as a function of the model parameters. The clusters were generated by simulations and after the transient time interval (of the order of $L$), the *fractal dimension* of ulcers contours was calculated by using the *box counting* method and averaged over the next $L$ equilibrium configuration.

In Fig. 8, we show $D_F$ as a function of small values of $(f/p)$ obtained for $p = 0.1$ and for $R = 2$, 3 and 4. The figures for $R > 4$ are quite similar to the ones for $R = 4$. We notice that for the models with resistant cells with $R \geq 3$, the fractal dimension of the ulcers is very sensitive to small values of the $(f/p)$ parameter. Under these conditions, the densities of dead cells is of order of 0.2 (see Fig. 3), indicating that a steady state regime is achieved where small ulcers of very rough contour remain dynamically stable. For $(f/p)$ close to 0.01, the ulcers frontiers are less rough for $R > 2$ with $D_F \approx 1.4$. This figure is close to the one estimated by box counting digitalized ulcers images of dentritic small real ulcers (Feret's diameter ranging from 1.6 to 3.2 mm).[1] On the other hand, if the cell resistance is low ($R \leq 2$), this latter picture does not hold and the ulcers is allowed to spread over the whole tissue with high density and with $D_F \sim 2$.

In conclusion, we show that our present dynamical model indicates that the degree of resistance of the cell to infection by contiguity should play an important role on the ulcer evolution.

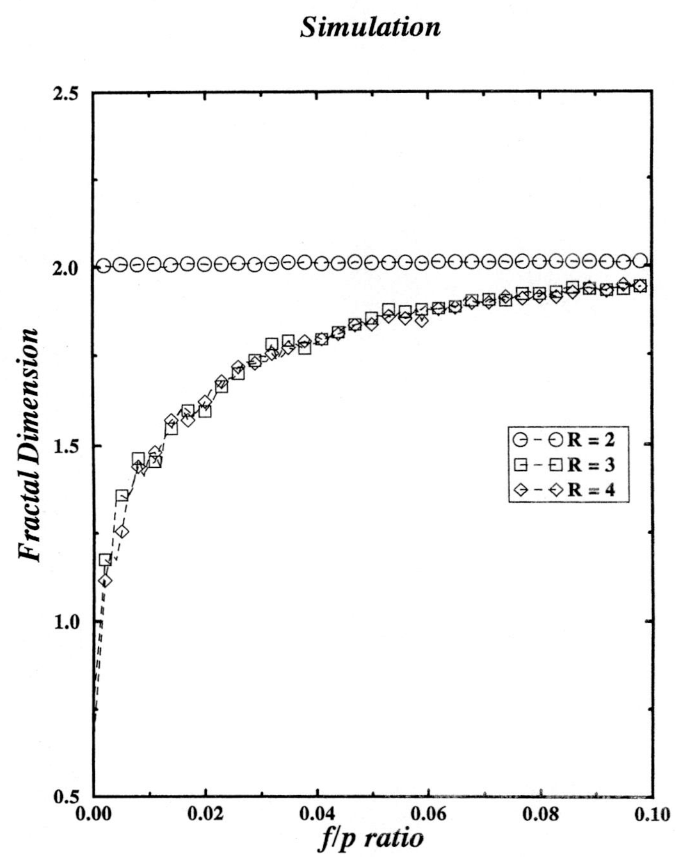


**Fig. 8** Fractal dimension of ulcers contours $D_F$ obtained by simulation as function of the $f/p$ ratio for $p = 0.01$ and $q = 0.6$ on a $L = 100$ square lattice, for $R = 2$, 3 and 4.

## ACKNOWLEDGMENTS

The authors acknowledge the financial support received from CNPq, FINEP, CAPES and FACEPE (Brazilian granting agencies). One of us (S.C) is also grateful for the kind hospitality and financial support received from Sasuke Miyazima during "Future of Fractals" (Aichi, Japan, 1995).

## REFERENCES

1. G. Landini, G. P. Mission and P. I. Murray, in *Fractals in the Natural and Applied Sciences*, ed. M. M. Novak (Elsevier Science B. V., North Holland, 1994).
2. A. Hill, P. Jugovic, I. York, G. Russ, J. Bennink, J. Yewdell, H. Ploegh and D. Johnson, *Nature* **375**, 411 (1995).
3. B. Drossel and F. Schwabl, *Phys. Rev. Lett.* **69**, 1629 (1992).
4. B. Drossel, S. Clar and F. Schwabl, *Phys. Rev.* **E50**, 1009 (1994).

Fractals, Vol. 4, No. 2 (1996) 123–132


# COMPARISON OF THE SURFACE DIMENSION AND THE MASS DIMENSION OF BLOOD VESSEL SYSTEMS

F. JESTCZEMSKI
*Institute for Biochemistry and Endocrinology, Justus-Liebig-University, Frankfurter Str. 100, D-35392 Giessen, Germany*

H. BOLTERAUER
*Institute for Theoretical Physics, Justus-Liebig-University, Heinrich-Buff-Ring 16, D-35392 Giessen, Germany*

M. SERNETZ
*Institute for Biochemistry and Endocrinology, Justus-Liebig-University, Frankfurter Str. 100, D-35392 Giessen, Germany*

### Abstract

The arterial blood vessel system of kidneys is known to be fractal with a global mass dimension of 2.2 to 2.3. The global mass dimensions were determined by the mass-radius method.[1] We support these results of the mass correlation analysis with an examination of the surface, which is in good approximation with the "true mass" of the border between the tissue (parenchyma) and the blood. We applied the Minkowski cover to the vessel surfaces and determined their scaling behavior, which yielded the surface global dimension. The examinations on kidney arteries[1] are supplemented with new data of placentoma arteries[2] and a comparison is made between their surface and mass dimensions.

## 1. INTRODUCTION

The arterial blood vessel system of kidney or placentoma is a typical example of a hierarchically structured, fractal object. If a mass is attached to the hollow arterial branching system, it is best described as a fat fractal with a fractal mass dimension on a "mesoscopic" scale and a fractal surface dimension. The blood vessel system has an almost space-filling character and we are therefore interested in its surface. This approximates the tissue (parenchyma) under certain conditions. We applied the Minkowski cover to the surface of the vessels and calculated its scaling behavior.

At the very beginning of this century, Minkowski considered the length and area of curves and surfaces.[3] He defined the cover of a set $O$ to be the set of all points $U(r)$ defined as $U(r) = \bigcup_{\mathbf{y}\in O} B_r(\mathbf{y}) = \{\mathbf{x} : \|\mathbf{x}-\mathbf{y}\| \leq r, \forall \mathbf{y} \in O\}$, which is constructed by replacing every point of the set $\mathbf{y} \in O$ with a ball $B_r(\mathbf{y})$ of radius $r$. The volume $V(r)$ of this cover can be used to introduce the length or the surface area for curves or surfaces, respectively. The length of a curve is then obtained as the limit of $V(r)/\pi r^2$ for $r \to 0$, whereas the area of a surface results from the limit of $V(r)/2r$. Bouligand[4–6] extended these ideas to introduce a continuous dimension $D$. For an object of dimension $D$ in $\mathbf{R}^d$, the volume turns out to scale with the codimension $d - D$ of the object like $V(r) \propto r^{d-D}$ for small $r$. Mandelbrot reconsidered the formalism in terms of fractal geometry and called $D$ the Minkowski–Bouligand dimension.[7] The construction of the Minkowski covers is a usual method implemented in two-dimensional image processing systems (dilation). Fractal dimensions of two-dimensional images have already been determined. Very refined methods have been developed, like fast algorithms, which use distance transform functions[8,9] and simple shift and logical OR operations.[10] Fat fractals in $\mathbf{R}^3$, like the blood vessel system, have a non-negligible volume, so the dimension of the object itself is 3, but the boundary will have a fractal dimension $D'$. Grebogi et al.[11] suggested to subtract the volume $V(0)$ of the object from the volume $V(r)$, which yields an exterior dimension $D = D_x$ of the boundary. Unfortunately, $D_x$ is less than or equal to $D'$.[11,erratum]

Another way to find the fractal surface dimension of fat fractals is to isolate the boundary by an excavation of all points inside the object, and then to apply the Minkowski method to this (thin) surface. The Minkowski method is known to be valid for thin (skinny) fractals. The application to fat fractals is difficult due to the finite volume. Karlinger et al.[12] have investigated the relation between thin and fat-fractal scaling of river networks, which are the two-dimensional pendant to three-dimensional blood vessel systems. Whereas the thin-fractal scaling is described by a fractal dimension $D$, derived from the length properties of the river channels, the fat-fractal scaling $\beta$ is derived from the behavior of river-channel area. They consider the area $f(\varepsilon)$ of river channels whose widths are less than, or equal to, the scale resolution $\varepsilon$ and define $\beta$ by $f(\varepsilon) \propto \varepsilon^{\beta}$. The linkage between $D$ and $\beta$ is given in terms of the bifurcation and length ratios, and a diameter exponent $\Delta$, which describes the downstream channel-widening process. Less generally, it is given as $\beta = 1 - \Delta(1 - 1/D)$. We explain the relation between thin and fat-fractal scaling from a different point of view by mass and surface dimensions.

An experimental pendant of the Minkowski method is porosimetry. Mercury porosimetry is the classical method for the study of pore size distributions $dV(R)/dR$. Pfeifer and Avnir[13] or recently Ehrburger-Dolle et al.[14] describe the space-filling fractal surfaces of porous solids. If the object is covered with spheres of radius $R$, the total volume of voids of sizes larger than $R$ to the remaining volume of space is $V(R) \propto$ constant $- R^{3-D}$ or $-dV(R)/dR \propto R^{2-D}$. $D$ is the surface fractal dimension, the dimension of the boundary between the solid object and its pore volume.

## 2. MATHEMATICAL BACKGROUND

### 2.1 Skinny Fractals

At first, let the object $O$ under investigation be a mathematical set of points in $\mathbf{R}^3$ without volume. The Minkowski cover $U(r) = \bigcup_{\mathbf{y}\in O} B_r(\mathbf{y})$ is constructed by replacing every point of the set $\mathbf{y} \in O$ with a ball $B_r(\mathbf{y})$ of radius $r$ (dilation). If $O$ is a point, then the volume of $U(r)$ is $V(r) = (4/3)\pi r^{3-0} \propto r^3$; if $O$ is a segment of length $l$, then its volume is $V(r) \propto \pi l r^{3-1} \propto r^2$; and if $O$ is a flat set of area $A$, then the volume is $V(r) \propto 2Ar^{3-2} \propto r$. Generally, for an object of dimension $D$ in $\mathbf{R}^d$, the volume turns out to scale with the codimension $d-D$ of the object like $V(r) \propto r^{d-D}$ for small $r$.[15] $D$ is a kind of box dimension: if an object of dimension $D$ is covered with $N(r) \propto r^{-D}$ non-overlapping balls of radius $r$ instead of boxes, then the volume of the union of balls is $V(r) \propto N(r)r^d \propto r^{d-D}$. $D$ is calculated by a log-log plot of volume vs. radius as usually:

$$D = d - \lim_{r\to 0} \frac{\log V(r)}{\log r}. \tag{1}$$

A closer look at the volume of a segment of length $l$ shows $V(r) = \pi l^3 (r/l)^2 + (4/3)\pi l^3 (r/l)^3$. In the limit of small $r$, the volume is proportional to $r^2$ as stated above and the dimension is 1. If the mathematical limit in Eq. (1) is ignored, $D(r) = d - (\log V(r)/\log r)$ can be regarded as a dimension of the object observed from a distance proportional to $r$. For the segment, the second term proportional to $r^3$ predominates for large $r$ and the object appears as a point of dimension 0. So, $D(r)$ is a kind of "effective dimension". If the segment is given on a grid, the log-log plot of the volume vs. radius will for small grid sizes result in a concave curve instead of a straight line due to the second term of the volume proportional to $r^3$, and adulterates the true dimension value.[16] Similarly, fat fractals exhibit the same concavity due to their finite volume. The surface will then scale, but mass will scale at most in an intermediate range where thicknesses are no longer resolved. This is discussed in the next section.

### 2.2 Fat Fractals

From a practical point of view, solid objects can be classified with respect to the scaling behavior of their mass, their surface, or their pore space.[17] Consider an object in $\mathbf{R}^3$. If its mass and surface scale alike with some (fractal) dimension $D$, it is called a mass fractal. But if its mass scales with $d = 3$ and its surface with $D > 2$, it is called a surface fractal. Objects exhibiting a fractal pore space can be regarded as a reverse mass fractal. Pore space and surface then scale alike and hence are called pore fractals. Natural objects are voluminous and have natural thicknesses. Consider an object with a single thickness $R_T$. Let $R$ denote the resolution yardstick which is used to examine the object and assume that the surface is fractal in a limited range $[R_{\min}, R_{\max}]$ due to the finest detail of the object and its natural diameter. If $R < R_T$, the object will appear as a surface fractal. For resolutions $R > R_T$, the thickness will be no longer resolved, but only the skeleton of the object. Then, mass and surface scale in an identical way indicating a mass fractal (see Fig. 1). Obviously, for thicknesses $R_T \approx R_{\max}$, the object will never appear as a mass fractal but as a surface fractal. This behavior is typical of practical analyses like the mass-radius analysis or the Minkowski analysis. The volume of an object which was dilated by $r$ is $V(r) = V(0) + \Delta V(r)$. For a skinny fractal with $V(0) = 0$, there is no difference in taking $V(r)$ or $\Delta V(r)$ and the volume scales like $V(r) = \Delta V(r) \propto r^{3-D}$. If the object is a fat fractal with $V(0) \neq 0$, then $V(r) \approx V(0)$ for small $r$ holds and $D = 3$. The log-log plot of $V(r)$ vs. $r$ leads to a concave curve instead of a straight line because of:

$$\frac{d \log V(r)}{d \log r} = \left(\frac{\Delta V(r)}{V(0) + \Delta V(r)}\right) \frac{d \log \Delta V(r)}{d \log r}.$$

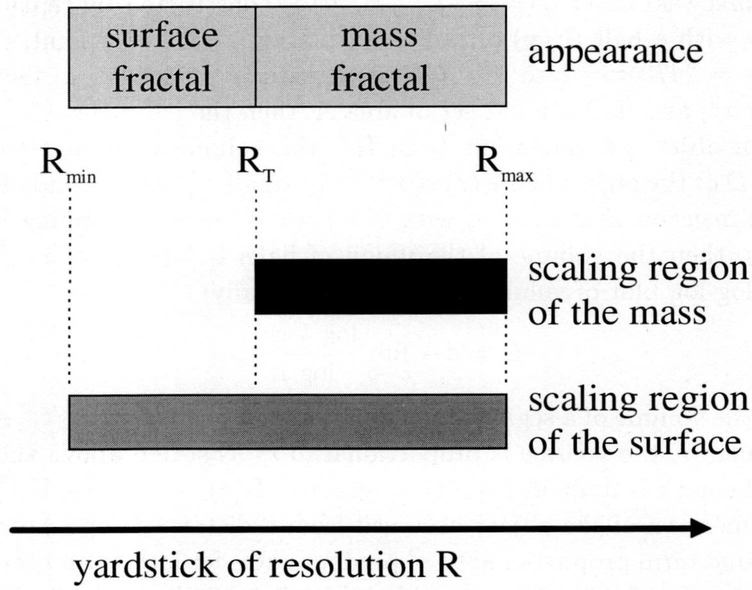


**Fig. 1** Natural objects appear with changing resolution $R$ either as a surface fractal or a mass fractal. If $R_T \approx R_{\max}$ then it will appear as a surface fractal only.

Grebogi et al.[11] tried to catch the surface dimension and distinguished between:

the "capacity dimension" $$D_c = d - \lim_{r\to 0} \frac{\log V(r)}{\log r}$$

and the "exterior capacity dimension" $$D_x = d - \lim_{r\to 0} \frac{\log(V(r) - V(0))}{\log r} .$$

The capacity dimension is identical to the Minkowski-Bouligand dimension. For skinny fractals (mass fractals) $D_c = D_x$ holds. If the object is a solid, $d$-dimensional volume whose boundary is a skinny fractal, then the dimension of the object itself is $D_c = d$. The boundary has a fractal dimension which, unfortunately, is not always equal to $D_x$. The definition of $D_x$ is a single-sided (exterior) limit of the volume and $D_x$ is less than or equal to the true surface dimension.

## 3. EXPERIMENTAL

### 3.1 Samples

We had spatial datasets of arterial vessel systems at our disposal, especially of kidney and placentoma, which were obtained by $\mu$-NMR tomography or serial sectioning technique.[1,2] These datasets are given on a grid with intensity 1 denoting the object, and intensity 0 otherwise. Typically, the edge length of the grids is some hundreds to thousand voxel with a resolution of about 100 $\mu$m edge length per voxel. The casts therefore reach down to the region of the arterioles. The objects are represented with up to 2.4 million occupied sites.

### 3.2 Technique

The Minkowski covers $U(r)$ are produced by a dilation of the object. The method we used to obtain dilations is known as distance transform method (Fig. 2). Distance transforms take a binary image and convert it into a minimal distance array. Every site of the array is labeled with the minimal distance to the object for a given particular metric. The minimal distance array is a series of contours radiating around the object. This method has advantages in relation to conventional sequential methods, which use small matrices of $3 \otimes 3$ or $5 \otimes 5$ pixels and produce square, cross, or octagonal dilations and therefore distort the object. Distance transforms are fast and yield circular shapes when the Euclidean distance is used.[8,9] Such circular shapes are important for reconstruction of the hull of the object approximately. We used the Euclidean distance for all examinations and wrote down the integer part of the floating-point distance. All sites with the same distance $\rho$ to the object form an isosurface. From the minimum distance array, a frequency distribution histogram $N(\rho)$ is obtained. The volume $V(r)$ of the Minkowski cover of radius $r$ is calculated as the cumulative frequency histogram $N(\rho \leq r)$. The double-logarithmic plot of $V(r)$ vs. $r$ yields the codimension $3 - D$. The blood vessel system has non-zero volume and thus the log-log plot results in a concave curve instead of a straight line. Therefore, we excavated the vessel systems and constructed their first isosurface, which is one pixel ($\approx 100\ \mu$m) apart from the vessels. This surface sufficiently approximates the surface of the vessels and is more likely to be a skinny fractal as well. We calculated the minimal distance array for this surface and determined the scaling behavior as usual.

**Table 1 Results of the Surface Analysis for the Menger Sponge**

| Object | Grid Size [voxel] | Surface Dimension $D_{MB} \equiv D_c$ |
|---|---|---|
| Menger sponge | $81 \otimes 81 \otimes 81$ | $2.57 \pm 0.01$ |
| Menger sponge | $243 \otimes 243 \otimes 243$ | $2.66 \pm 0.01$ |

<table>
<tr><td>2</td><td>1</td><td>1</td><td>1</td><td>2</td><td>2</td><td>3</td><td>4</td><td>5</td><td>6</td></tr>
<tr><td>2</td><td>1</td><td>0</td><td>1</td><td>1</td><td>2</td><td>3</td><td>4</td><td>5</td><td>6</td></tr>
<tr><td>2</td><td>1</td><td>1</td><td>0</td><td>1</td><td>2</td><td>3</td><td>4</td><td>5</td><td>6</td></tr>
<tr><td>2</td><td>2</td><td>1</td><td>0</td><td>1</td><td>2</td><td>3</td><td>4</td><td>5</td><td>5</td></tr>
<tr><td>3</td><td>2</td><td>1</td><td>1</td><td>1</td><td>2</td><td>3</td><td>4</td><td>4</td><td>4</td></tr>
<tr><td>3</td><td>2</td><td>2</td><td>2</td><td>2</td><td>2</td><td>3</td><td>3</td><td>3</td><td>3</td></tr>
<tr><td>4</td><td>3</td><td>3</td><td>3</td><td>3</td><td>3</td><td>2</td><td>2</td><td>2</td><td>2</td></tr>
<tr><td>5</td><td>4</td><td>4</td><td>4</td><td>4</td><td>3</td><td>2</td><td>1</td><td>1</td><td>1</td></tr>
<tr><td>5</td><td>5</td><td>5</td><td>5</td><td>4</td><td>3</td><td>2</td><td>1</td><td>0</td><td>1</td></tr>
<tr><td>6</td><td>6</td><td>6</td><td>5</td><td>4</td><td>3</td><td>2</td><td>1</td><td>1</td><td>1</td></tr>
</table>

**Fig. 2** Distance transforms take a binary image and convert it into a minimal distance array. Every site of the array is labeled with the minimal distance to the object.

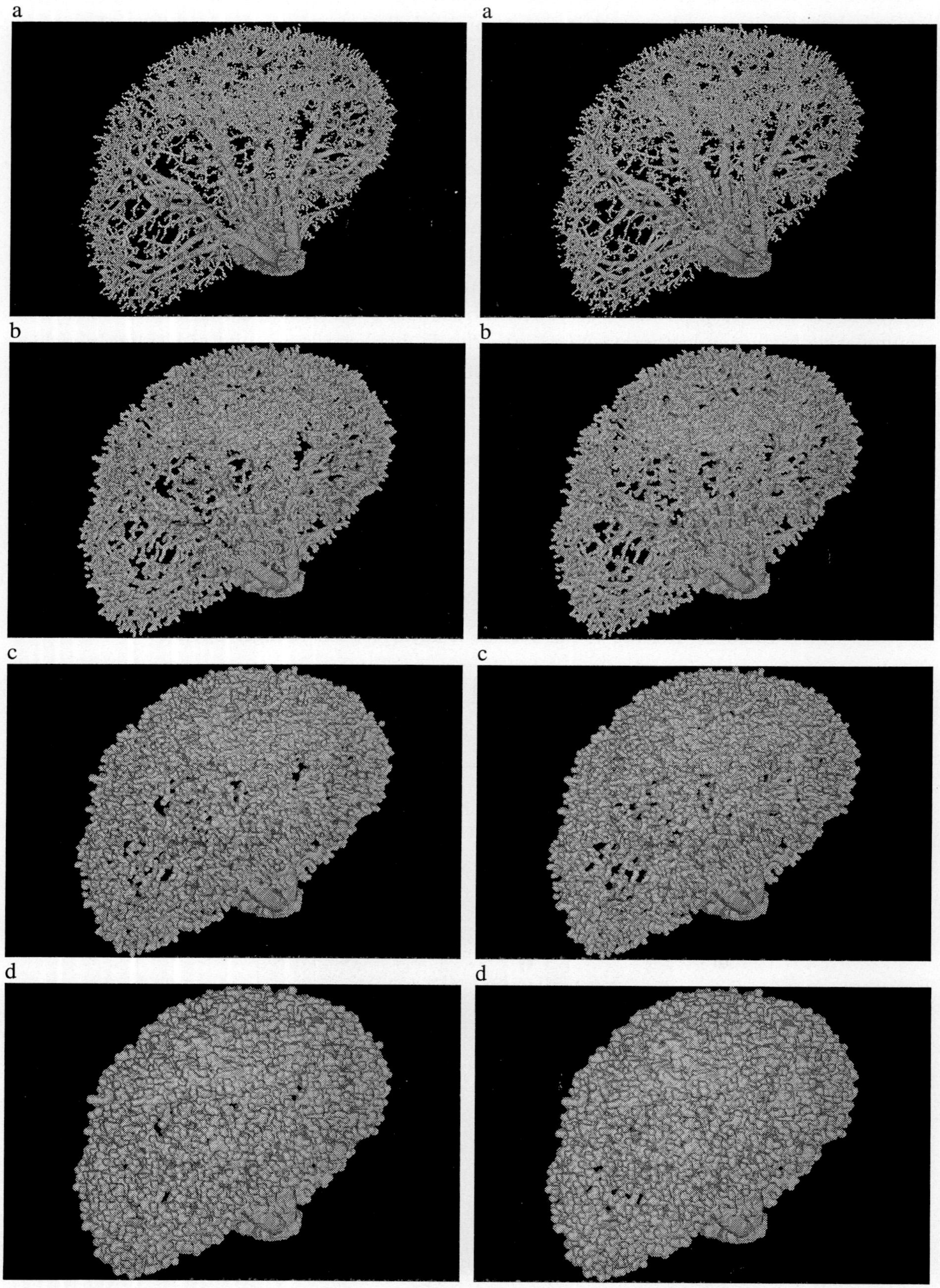

**Fig. 3** Stereoscopic view of the **first three steps of dilation** (b–d) for the rabbit kidney #3(a).

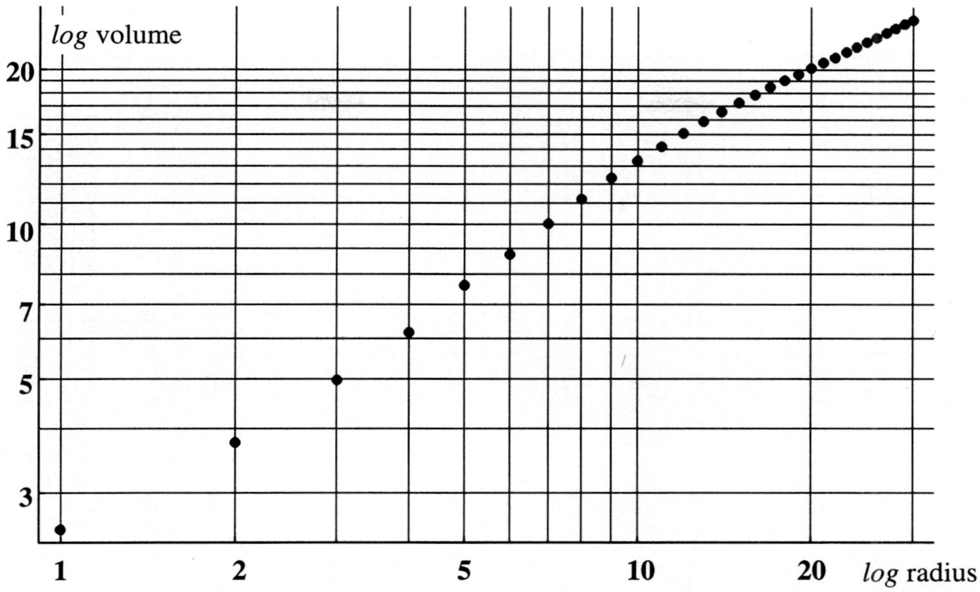


**Fig. 4** Double-logarithmic plot of the cumulative volume of isosurfaces vs. radius for cow placentoma #3. The volume is normalized to the volume of the object.

### 3.3 Quality

We applied the distance transform algorithm to the Menger sponge to verify our implementation. Its true dimension is $(\log 20/\log 3) \approx 2.73$. We used two samples of the Menger sponge with grid sizes of $81^3$ and $243^3$ voxels, respectively. For an edge length of 81 voxels, the Minkowski analysis yielded a dimension of 2.57, and for an edge length of 243 voxels a dimension of 2.66 (see Table 1), which is in good correspondence with the true value. We conclude that grids with an edge length of some hundred voxels as in our case guarantee good dimension estimates.

## 4. RESULTS AND DISCUSSION

The minimal distance array was calculated with up to 30 dilation radii for each vessel system. Figure 3 shows the first three steps of dilation for the rabbit kidney #3. From the minimal distance array, we obtained the frequency distribution histograms and the cumulative frequency histograms. The double-logarithmic plot of volume vs. radius yielded the codimension of the surface. Figure 4 shows a log-log plot of the cow placentoma #3. The histograms contain frequency values for equidistant radii. In a linear regression, this can overemphasize the values of large radii. A correction is given by generating a new histogram with over 100 classes (class $\propto$ log radius). We give the values with this correction, although we found it to be small.

Table 2 summarizes our results of the surface analysis of rabbit and pig kidneys and of cow and sheep placentoma. The given mass dimensions originate from preceding works[1,2] and result from a mass-radius analysis of randomly distributed centers on the vessel system. Although the vessel system was found to be non-homogeneous with a broad distribution of local dimensions, it is of significance to give a mean global dimension on a mesoscopic

**Table 2 Results of the surface analysis for the different vessel systems**

| object | | object size [voxel] | resolution [voxel/mm] | volume [voxel] | surface [voxel] | surface/volume ratio | mass dimension | surface dimension $D_{MB} \equiv D_c$ | exterior capacity dimension $D_x$ |
|---|---|---|---|---|---|---|---|---|---|
| rabbit kidney | 2 | 199 ⊗ 227 ⊗ 177 | 0,098 | 182.723 | 299.664 | 1,64 | **2,08 ± 0,15** | **2,32 ± 0,01** | 2,14 ± 0,02 |
| | 3 | 327 ⊗ 194 ⊗ 248 | 0,098 | 206.379 | 616.441 | 2,99 | **2,05 ± 0,23** | **2,36 ± 0,01** | 2,25 ± 0,01 |
| pig kidney | 1 | 207 ⊗ 135 ⊗ 416 | 0,25 | 150.471 | 246.603 | 1,64 | **1,89 ± 0,15** | **2,30 ± 0,01** | 2,13 ± 0,01 |
| | 2 | 348 ⊗ 247 ⊗ 749 | 0,15 | 1.756.261 | 2.768.379 | 1,58 | **2,23 ± 0,15** | **2,36 ± 0,01** | 2,20 ± 0,02 |
| | 3 | 402 ⊗ 220 ⊗ 794 | 0,15 | 1.400.046 | 2.964.945 | 2,12 | **2,19 ± 0,14** | **2,40 ± 0,01** | 2,23 ± 0,02 |
| | 4 | 371 ⊗ 349 ⊗ 842 | 0,125 | 2.400.292 | 6.283.354 | 2,62 | **2,31 ± 0,14** | **2,56 ± 0,02** | 2,52 ± 0,02 |
| | 5 | 303 ⊗ 499 ⊗ 1003 | 0,1 | 2.102.201 | 4.482.503 | 2,13 | **2,17 ± 0,13** | **2,24 ± 0,01** | 2,05 ± 0,02 |
| | 6 | 332 ⊗ 501 ⊗ 860 | 0,125 | 1.865.511 | 6.829.586 | 3,66 | **2,28 ± 0,16** | **2,47 ± 0,01** | 2,31 ± 0,02 |
| cow placentoma | 1 | 331 ⊗ 550 ⊗ 519 | 0,099 | 1.779.086 | 3.007.502 | 1,69 | **2,10 ± 0,11** | **2,31 ± 0,01** | 2,14 ± 0,01 |
| | 2 | 185 ⊗ 513 ⊗ 489 | 0,099 | 598.167 | 1.384.494 | 2,31 | **2,08 ± 0,14** | **2,36 ± 0,01** | 2,21 ± 0,01 |
| | 3 | 268 ⊗ 415 ⊗ 410 | 0,099 | 463.921 | 1.350.494 | 2,91 | **2,05 ± 0,15** | **2,22 ± 0,01** | 2,11 ± 0,01 |
| | 4 | 299 ⊗ 626 ⊗ 586 | 0,099 | 1.957.975 | 2.982.557 | 1,52 | **2,15 ± 0,09** | **2,36 ± 0,01** | 2,21 ± 0,01 |
| | 5 | 111 ⊗ 473 ⊗ 386 | 0,099 | 412.163 | 748.582 | 1,82 | **2,02 ± 0,15** | **2,37 ± 0,01** | 2,22 ± 0,01 |
| sheep placentoma | 1 | 139 ⊗ 169 ⊗ 184 | 0,099 | 71.056 | 188.402 | 2,65 | **2,06 ± 0,17** | **2,43 ± 0,01** | 2,26 ± 0,01 |
| | 2 | 62 ⊗ 86 ⊗ 61 | 0,219 | 10.205 | 42.000 | 4,12 | **2,06 ± 0,2** | **2,45 ± 0,01** | 2,26 ± 0,02 |
| | 3 | 141 ⊗ 239 ⊗ 100 | 0,097 | 48.230 | 133.572 | 2,77 | **1,95 ± 0,16** | **2,31 ± 0,01** | 2,14 ± 0,01 |

scale. On a microscopic scale ($\approx$ some voxels), voluminous vessels with dimension 3 or linear vessels with dimension 1 (arterioles) are resolved. With an increasing yardstick of resolution the thickness of the vessels is "lost", and the mass dimension describes the skeleton of the object. On the other hand, if a skinny fractal is fattened (dilated), then the structure is flattened and increasingly lost. Dilation is an irreversible process and the mass dimension of its skeleton will be less than the dimension of the original skinny fractal. This mass dimension is compared with the surface dimension which is the aim of this paper.

The surface dimension describes the border between the vessel system (blood) and the tissue (parenchyma). We usually find it to have larger values (between 2.3 and 2.5) than the mass dimension. It also describes the dimension of the underlying skeleton. The dilation method as compared with the mass-radius method shows no marginal errors (but weak grid effects, which are reduced for increasing grid sizes) and is therefore a very valuable tool. We have calculated the exterior capacity dimension as well, which is a lower bound of the true surface dimension.

The results confirm and extend our previous results of the blood vessel characterization.[1] It has been revealed that the mass-radius analysis is more sensitive to marginal or grid effects than the Minkowski analysis. The blood vessel system of kidneys is fractal with a global mass dimension between 2.2 and 2.3 and a surface dimension between 2.3 and 2.5. Placentomae have an arterial vessel system whose global mass dimension is between 2.1 and 2.2, but whose surface dimension is between 2.2 and 2.4.

## NOTES ADDED

In two other papers we describe the origin of geometrical multifractality of arterial systems[18] and more generally[19] the multifractal approach to inhomogeneous fractals.

## ACKNOWLEDGMENT

Supported by grant Se 315/13-4 and 19-1 of the Deutsche Forschungsgemeinschaft.

## REFERENCES

1. M. Sernetz, M. Justen and F. Jestczemski, "Dispersive fractal characterization of kidney arteries by three-dimensional mass-radius analysis," *Fractals* **3**(4), 879–891 (1996).
2. M. Eibenstein, Thesis Giessen (Justus-Liebig-University Giessen, 1995).
3. H. Minkowski, "Über die Begriffe Länge, Oberfläche und Volumen," *Jahresbericht der Deutschen Mathematikervereinigung* **9**, 115–121 (1901).
4. G. Bouligand, "Dimension, étendue, densité," *C. R. Acad. Sci. Paris* **180**(2), 245–248 (1925).
5. G. Bouligand, "Ensembles Impropres Et Nombre Dimensionnel," *Bulletin des Sciences Mathématiques* **II-52**, 320–334, 361–376 (1928).
6. G. Bouligand, "Sur La Notion D'Ordre De Mesure D'Un Ensemble Fermé," *Bulletin des Sciences Mathématiques* **II-53**, 185–192 (1929).
7. B. B. Mandelbrot, *The Fractal Geometry of Nature* (Freeman, San Francisco, 1977).
8. J. Adler and C. D. Melia, "Using a distance transform function in image processing," *Euro. Microscopy Analy.* **29**, 29–31 (1994).
9. J. Adler and D. Hancock, "Advantages of using a distance transform function in the measurement of fractal dimensions by the dilation method," *Powder Technology* **78**(3), 191–196 (1994).
10. R. Creutzburg, A. Mathias and E. Ivanov, "Fast algorithm for computing the fractal dimension of binary images," *Physica* **A185**, 56–60 (1992).

11. C. Grebogi, S. W. McDonald, E. Ott and J. A. Yorke, "Exterior dimension of fat fractals," *Phys. Lett.* **A110**(1), 1–4 (1985) [Erratum in **A113**(9), 495 (1986)].
12. M. R. Karlinger, T. M. Over and B. M. Troutman, "Relating thin and fat-fractal scaling of river-network models," *Fractals* **2**(4), 557–565 (1994).
13. P. Pfeifer and D. Avnir, *J. Chem. Phys.* **79**, 3558 (1983).
14. F. Ehrburger-Dolle, A. Lavanchy and F. Stoeckli, "Determination of the surface fractal dimension of active carbons by mercury porosimetry," *J. Colloid Interface Sci.* **166**, 451–461 (1994).
15. K. Falconer, *Fractal Geometry* (John Wiley & Sons Ltd., 1990).
16. B. Dubuc, J. F. Quiniou, C. Roques-Carmes, C. Tricot and S. W. Zucker, "Evaluating the fractal dimension of profiles," *Phys. Rev.* **A39**(3), 1500–1512 (1989).
17. P. Pfeifer and M. Obert, "Fractals: Basic concepts and Terminology," in *The Fractal Approach to Heterogeneous Chemistry*, ed. D. Avnir, pp. 11–43 (1989).
18. F. Jestczemski and M. Sernetz, "Mass multifractal characterization of blood vessel systems," *Fractals* (1996) in press.
19. F. Jestczemski and M. Sernetz, "Multifractal approach to inhomogeneous fractals," *Physica* **A223**, 275–282 (1996).

Fractals, Vol. 4, No. 2 (1996) 133–138


# MASS MULTIFRACTAL CHARACTERIZATION OF BLOOD VESSEL SYSTEMS

F. JESTCZEMSKI and M. SERNETZ
*Institute for Biochemistry and Endocrinology, Justus-Liebig-University*
*Frankfurter Str. 100, D-35392 Giessen, Germany*

**Abstract**

The blood vessel system as measured on kidney[1] and placenta arteries[2,3] is known to be a non-homogeneous fractal with a distribution of local dimensions. We interpret this distribution as a mass multifractal property and we have therefore examined the average of the masses $M_i(r)$ and their $q$th moments within boxes of increasing size $r$. The centers $i$ of the boxes are randomly distributed on the vessels. The generalized dimensions $D_q$ are introduced by taking the average of $(M_i(r)/M_0)^{q-1}$ over the centers $i$, according to the probability distribution $M_i(r)/M_0$ ($M_0$: total mass of the cluster). Thus, we have determined $D_q$ by calculating $\langle (M_i(r)/M_0)^{q-1}\rangle_{\langle \text{centers } i\rangle} \propto (r/L)^{(q-1)D_q}$ ($L$: diameter of the cluster).

## 1. INTRODUCTION

In a preceding work,[1] it was shown that the arterial blood vessel system of kidneys is a non-homogeneous fractal. Though it is possible to determine a global dimension, its non-homogeneity is more appropriately described by a broad spectrum of local dimensions. This has been done by means of a mass-radius analysis of randomly chosen points of the structure. The expectation value of the frequency distribution histogram yields the global dimension. By encoding the local dimensions as colors and by spatial representation of the centers of the mass-radius analysis as colored points, the three-dimensional distribution of local dimensions within the object could be demonstrated likewise.[1]

The mass-radius analysis is, however, perturbed by the non-negligible thicknesses of the vessels. These induce a three-dimensional scaling of mass which vanishes only at scales with a yardstick $R$ of resolution equal or larger than the thickness $R_{\mathrm{T}}$ of the vessels. Thus as a function of the range of regression, the vessel system can appear as a surface fractal ($R < R_{\mathrm{T}}$) or as a mass fractal ($R \geq R_{\mathrm{T}}$).[3] According to different ranges of scale, a moving-average analysis shows bimodal frequency distribution histograms in the lowest range (microscopic scale), which change to monomodal distributions in a medium range (mesoscopic scale). Only on this mesoscopic scale does it make sense to determine an expectation value as a global dimension of mass.

Both the frequency distribution histogram and the spatial distribution show a broad distribution of the local dimensions. In the present work, this dispersion of local dimension is expressed in terms of multifractal analysis, by definition of a proper measure on the fractal support. The mass distribution of the vessel system turns out to be multifractal and is described by a "dispersion relation" $f(\alpha)$. Though the vessel system can be described by a single, global dimension $D$, corresponding to the surface and the underlying skeleton,[3] it is more appropriately described by an infinite set of exponents $f(\alpha)$.

The present work applies a mass multifractal analysis on arterial vessel systems both of kidneys and fetal arteries of placenta, which have already been analyzed by the usual mass-radius analysis.[2,3] We follow the formalism of the generalized sand box method[4] described in Refs. 5 and 6 for a multifractal analysis of the mass distribution of DLA.

## 2. MULTIFRACTAL ANALYSIS

The usual way to determine a multifractal spectrum is to lay a grid of mesh size $r$ over a given structure which supports a measure $M(\mathbf{x})$. The total measure $M_i$ of box $i$ is $M_i = \int_{\text{box}\, i} M(\mathbf{x}) d\mathbf{x}$. We define the measure $M_i$ to be the mass of the structure within box $i$. A set of general dimensions $D_q$ of order $q$ is introduced by the Ansatz:

$$\sum_{\text{boxes}\, i} \mu_i^q \propto \varepsilon^{(q-1)D_q} , \tag{1}$$

with the normalized mass $\mu_i = M_i/M_0$ relative to the total mass $M_0$ of the structure, and with the normalized mesh size $\varepsilon = r/L$ relative to the diameter $L$ of the structure. $D_0$ is the dimension of the support. If the structure is a monofractal, then $D_q = D_0$ for any $q$. Multifractal measures exhibit:

$$\mu_i \propto \varepsilon^{\alpha} , \tag{2}$$

with $\alpha$ depending on the locus $\mathbf{x}$. The number of boxes belonging to the same $\alpha$ is supposed to scale with an exponent $f(\alpha)$ like $\varepsilon^{-f(\alpha)}$. If $D_q$ has been determined by:

$$D_q = \frac{1}{q-1} \lim_{\varepsilon \to 0} \frac{\log \sum_i \mu_i^q}{\log \varepsilon} , \tag{3}$$

for various $q$, then $f(\alpha)$ can be calculated parametrically as graph $(\alpha_q, f_q)$ with $\alpha_q = \frac{d}{dq}((q-1)D_q)$ and $f_q = q\alpha_q - (q-1)D_q$.

Boxes with a small number of particles give rise to anomalously large contributions to the left-hand side of Eq. (1), and this gives unreliable results for negative $q$. To avoid this, the formalism can be changed from a box-counting to a mass-radius formalism,[4–6] known as generalized sand box method. The sum in Eq. (1) is interpreted as taking the average of $\mu_i^{q-1}$ of boxes, whose centers are randomly distributed on the structure according to the probability distribution $\mu_i$. Therefore, Eq. (1) is rewritten as:

$$\langle \mu_i^{q-1} \rangle_{\text{centers}\, i} \propto \varepsilon^{(q-1)D_q} \,. \tag{1a}$$

Equation (2) now describes the mass-radius analysis we have done so far.[1,2] It follows that $\alpha$ must not exceed 3 in the formalism of mass (geometrical) multifractals. $f(\alpha)$ is the dimension of the subset belonging to $\alpha$. Therefore, the structure is classified in subsets which are commingled according to $f(\alpha)$.

## 3. EXPERIMENTAL

We used datasets of arterial vessel systems of kidney and placentoma, which we have analyzed so far by ordinary mass-radius analysis.[1–3] These datasets are given on a grid with a typical edge length of some hundred voxels. According to Eq. (2), a mass-radius analysis was made with boxes whose centers were randomly distributed on the structure. Typically, 1% of all the structure points or more were evaluated. For every given center, the maximum radius of a box was chosen to be the shortest distance to the margin of the grid. We calculated $\mu_i^{q-1}$ for every box size and $q = -20, \ldots, +20$, and averaged over all centers with respect to the probability distribution $\mu_i$. According to Eq. (3), we calculated the generalized dimensions $D_q$ from a double-logarithmic plot of $\langle \mu_i^{q-1} \rangle_{\text{centers}\, i}$ vs. the box size.

We tested our implementation with two different datasets, namely:

(1) With a random mass distribution on a grid of $300 \otimes 300 \otimes 300$ sites. We found for box sizes between 20 and 200 voxels, $D_q = 2.99 \pm 0.01$ for any $q$ (Fig. 1).

(2) We tested our implementation with a Menger sponge with $243 \otimes 243 \otimes 243$ sites. The Menger sponge is a monofractal due to the similarity relation:

$$\sum_{j=1}^{20} r_j^{(q-1)D_q} \mu_j^q = 20 \cdot 3^{(q-1)D_q} \left(\frac{1}{20}\right)^q = 1 \,.$$

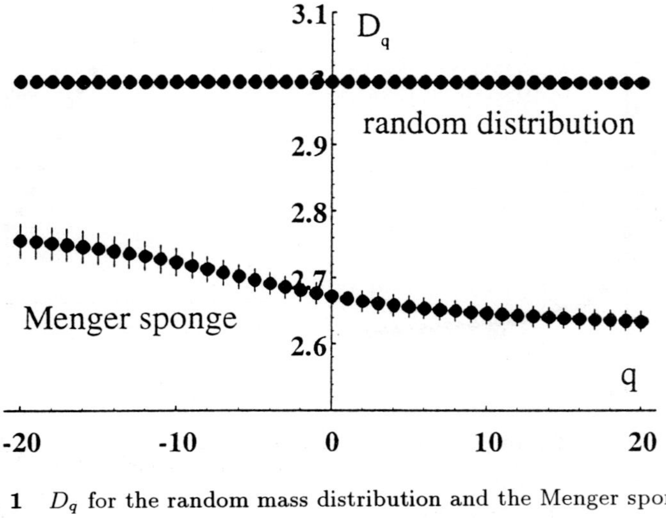


**Fig. 1** $D_q$ for the random mass distribution and the Menger sponge.

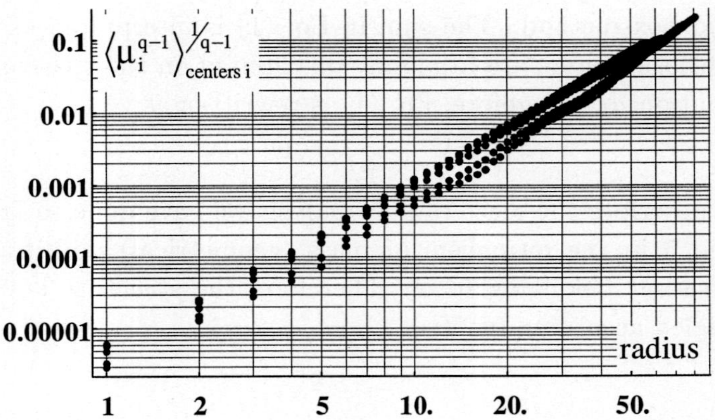


**Fig. 2** Double-logarithmic plot of $\langle \mu_i^{q-1} \rangle^{1/q-1}_{\text{centers } i}$ of the Menger sponge vs. radius for $q = -20, -10, 0, +10, +20$.

It is recursively built of 20 identical parts with $r_j = 3$ and $\mu_j = 1/20$, which yields $D_q = \log 20/\log 3 \approx 2.73$ for any $q$. For radii between 5 and 40 voxels, $D_q$ slightly changes from $2.73 \pm 0.03$ ($q = -20$) to $2.69 \pm 0.02$ ($q = 0$) to $2.64 \pm 0.02$ ($q = +20$) (Figs. 1 and 2). We attribute the weak dependence on $q$ to the influence of a marginal error of the mass-radius method and will discuss it later (see Discussion).

## 4. RESULTS

The results are summarized in Figs. 3 and 4, and Table 1. According to Eq. (1a), we have determined the generalized dimensions on a mesoscopic scale of box sizes. Although the dimension of the subsets with identical $\alpha$ is given by $f(\alpha)$, we show only the generalized dimensions $D_q$ for $q = -20$ to $q = +20$. The generalized dimensions together with the error bars are drawn for four typical examples (Figs. 3 and 4). For negative $q$, their values exceed

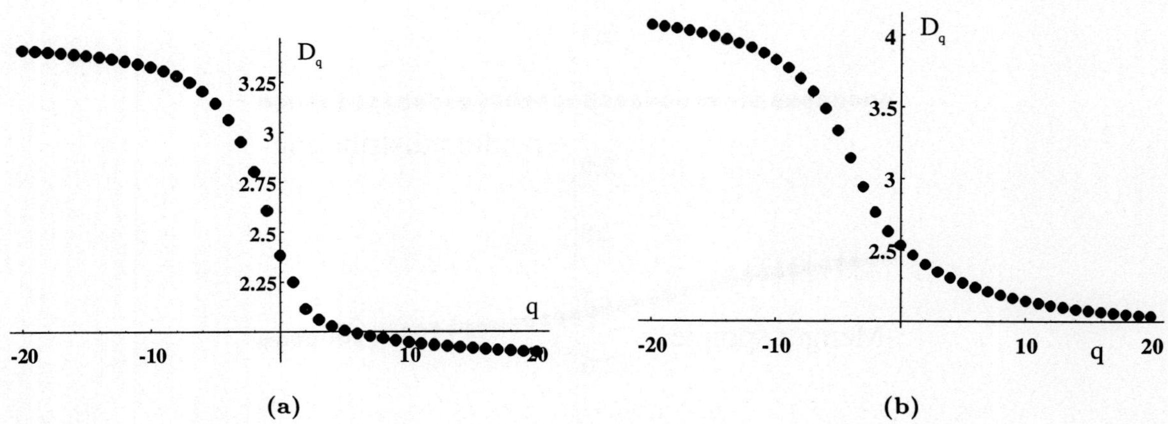


**Fig. 3** $D_q$ for pig kidneys. **(a)** kidney #2; and **(b)** kidney #4.

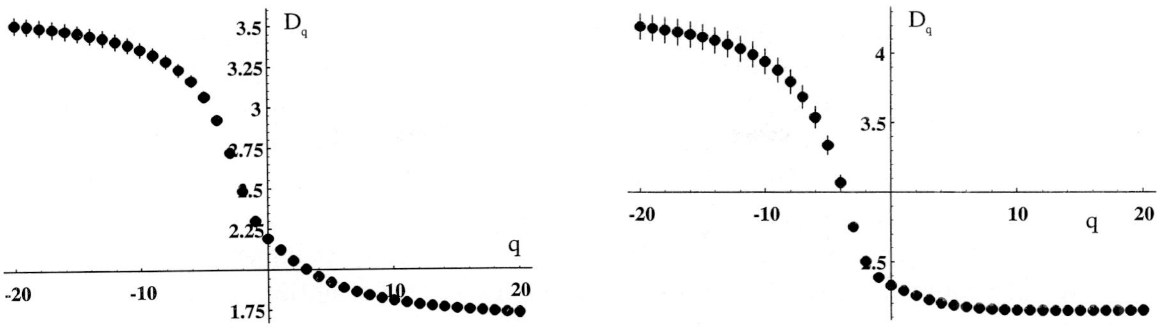

(a) (b)

**Fig. 4** $D_q$ for cow placentoma. **(a)** placentoma #1; and **(b)** placentoma #4.

**Table 1 The Generalized Dimensions $D_0$ and $D_2$ for the Different Vessel Systems**

| object (numbering as in [1]) | | object size [voxel] | resolution [voxel/mm] | volume [voxel] | global mass dimension (Euclidean metric) | $D_0$ (box metric) | $D_2$ (box metric) |
|---|---|---|---|---|---|---|---|
| pig kidney | 2 | 348 ⊗ 247 ⊗ 749 | 0,15 | 1.756.261 | **2,23 ± 0,15** | **2,38 ± 0,01** | **2,11 ± 0,01** |
| | 3 | 402 ⊗ 220 ⊗ 794 | 0,15 | 1.400.046 | **2,19 ± 0,14** | **2,31 ± 0,01** | **2,04 ± 0,01** |
| | 4 | 371 ⊗ 349 ⊗ 842 | 0,125 | 2.400.292 | **2,31 ± 0,14** | **2,51 ± 0,02** | **2,42 ± 0,02** |
| | 5 | 303 ⊗ 499 ⊗ 1003 | 0,1 | 2.102.201 | **2,17 ± 0,13** | **2,23 ± 0,01** | **1,82 ± 0,01** |
| | 6 | 332 ⊗ 501 ⊗ 860 | 0,125 | 1.865.511 | **2,28 ± 0,16** | **2,41 ± 0,02** | **1,94 ± 0,03** |
| cow placentoma | 1 | 331 ⊗ 550 ⊗ 519 | 0,099 | 1.779.086 | **2,10 ± 0,11** | **2,19 ± 0,03** | **2,05 ± 0,02** |
| | 2 | 185 ⊗ 513 ⊗ 489 | 0,099 | 598.167 | **2,08 ± 0,14** | **2,48 ± 0,01** | **2,40 ± 0,01** |
| | 3 | 268 ⊗ 415 ⊗ 410 | 0,099 | 463.921 | **2,05 ± 0,15** | **2,57 ± 0,01** | **2,46 ± 0,01** |
| | 4 | 299 ⊗ 626 ⊗ 586 | 0,099 | 1.957.975 | **2,15 ± 0,09** | **2,33 ± 0,01** | **2,26 ± 0,01** |

the theoretical maximum limit of 3. We suggest that this behavior is due to the fat-fractal structure of the vessels.

Table 1 shows the evaluated generalized dimensions $D_0$ and $D_2$. $D_0$ is the dimension of the support, while $D_2 < D_0$ is the correlation dimension. Standard methods including the mass-radius method yield $D_2$. We compare these dimensions with our previous results of the mass-radius analysis,[1,2] where the Euclidean metric was used. In the present work we have used a box metric.

## 5. DISCUSSION

The random mass distribution has identical generalized dimensions $D_q = 2.99 \pm 0.01$ for any $q$ (Fig. 1). This means that randomness does not give rise to multifractality. The generalized dimensions of the Menger sponge show a weak dependence on $q$, although it should be a monofractal. We argue that this behavior is due to an inherent marginal error of the mass-radius method. The central region of the Menger sponge with a width of one third of the edge length is empty. Thus only mass centers within one third of the diameter (here, e.g., 81 voxels) contribute to the analysis. Centers within a distance of 1/6 (here, e.g., about 40 voxels) contribute with boxes which fit entirely into the structure. The empty central region effectively induces a marginal error, such as boxes which would pass into the external empty environment of the object. Thus centers near to the central region contribute to the analysis with nearly one half of the box being empty, changing the constant of proportionality. In the regression analysis, this situation leads to a different dimension and pretends a different mass distribution.

As has already been discussed,[1] the fat-fractal structure of the vessel system can lead to a shift of dimension by some amount $\pm\delta$. Passing from a thin-vessel region to a fat-fractal region, the scaling of mass changes from $\mu_i = \rho_t \varepsilon^\alpha$ to $\mu_i = \rho_f \varepsilon^\alpha$ with $\rho_t < \rho_f$. A regression analysis including these regions will therefore result in $\mu_i \propto \varepsilon^{\alpha+\delta}$. We suspect this behavior to be relevant for the large values of $D_q$ obtained for negative $q$, which theoretically must not exceed 3.

In summary, the analysis of the mass distribution of arterial blood vessel systems of kidney and placentoma shows their multifractality, irrespective of some irregularities for negative $q$. While the Menger sponge is an example of a monofractal with a single dimension, the blood vessel system is a non-homogeneous multifractal due to its changing branching behavior from the big vessels down to the capillaries.

## NOTES ADDED

Meanwhile, in another paper about the multifractal approach to inhomogeneous fractals[7] we have examined the origin of multifractality of thin and fat fractals and given a new classification of thin and fat monofractals and multifractals.


## ACKNOWLEDGMENT

Supported by grant Se 315/13–4 and 19–1 of the Deutsche Forschungsgemeinschaft.


## REFERENCES


1. M. Sernetz, M. Justen and F. Jestczemski, "Dispersive fractal characterization of kidney arteries by three-dimensional mass-radius analysis," *Fractals* **3**(4), 879–891 (1996).
2. M. Eibenstein, Thesis, Giessen (Justus-Liebig-University Giessen, 1995).
3. F. Jestczemski, H. Bolterauer and M. Sernetz, "Comparison of the surface dimension and the mass dimension of blood vessel systems," *Fractals* **4**(2), 123–132 (1996).
4. T. Tél, A. Fülöp and T. Vicsek, *Physica* **A159**, 155 (1989) [erratum: *Physica* **A163**, 932 (1990)].
5. T. Vicsek, "Mass multifractals," *Physica* **A168**, 490–497 (1990).
6. T. Vicsek, F. Family and P. Meakin, "Multifractal geometry of diffusion-limited aggregates," *Europhys. Lett.* **12**(3), 217–222 (1990).
7. F. Jestczemski and M. Sernetz, "Multifractal approach to inhomogeneous fractals," *Physica* **A223**, 275–282 (1996).

Fractals, Vol. 4, No. 2 (1996) 139–147


# A FRACTAL-LIKE STRUCTURE IN THE SKIN

HISAO HONDA*§, SHUHEI IMAYAMA† and MASAHARU TANEMURA‡
*Kanebo Institute for Cancer Research,
Tomobuchicho 1-5-90, Miyakojimaku, Osaka 534 Japan
and
Yoshizato MorphoMatrix Project, ERATO, JRDC,
in Hiroshima-Techno-Plaza, Kagamiyama 3-chome 13-26, HigashiHiroshima 739 Japan
†Department of Dermatology, Faculty of Medicine, Kyushu University,
Fukuoka 812 Japan
‡Institute of Statistical Mathematics,
Minami-Azabu 4-6-7, Minatoku, Tokyo 106 Japan

## Abstract

Objects in a biological body consist of structures which themselves have finer structures, and such a relation repeats several orders. We show that the interface between the dermis and the epidermis in human skin is one of them and is considered as a fractal-*like* structure. To determine the degree of complexity of the structure, we estimated the fractal dimension. Its fractal dimension was 2.4. Folding of the interface increases the area that occupies a three-dimensional volume, enhancing functionality and promoting strong adhesion between the dermis and the epidermis. The fractal dimension quantifies this space-filling; it was found to increase from 2 to 3 as the surface folds more extensively. We discuss the functional significance of the fractal-like structure of the human skin and the other structures in a biological body.

§Author for correspondence at his present address: Hyogo University, Kakogawa, Hyogo 675-01 Japan.

## 1. INTRODUCTION

When observing a biological body with increasing magnification, we find that it consists of structural elements which themselves have finer structures, and the relation repeats several orders,[1] e.g., the inner surface of the small intestine,[2] the fibrous structures in the chromosomes,[2–5] and so on. Such a hierarchical structure is characteristic in biological substances, and reminds us of fractal objects.[6] Strictly speaking, each of the new details of a biological body observed after magnification is not similar to those details observed on a smaller scale. We have to distinguish hierarchical structures of biological bodies from fractal objects because of the imperfect self-similarity. However, the biological hierarchical structures are considered to be fractal-*like* structures, and we can try to apply a method of analysis used in fractal objects to biological hierarchical structures to grasp the total structures.

Here, we will show that an epidermal-dermal interface in the skin has a fractal-like structure, and its fractal dimension is estimated to determine the degree of complexity of the structure.

## 2. MATERIALS AND METHODS

### 2.1 Materials

The specimens used in this study were normal skin obtained from the finger of a 17-year-old female, and from the palm of a 20-year-old female, and from plantar skin of a 49-year-old female. The protocol of this study was approved by the Kyushu university board for human research. The operations in the three cases were a reconstructive surgery procedure for a traffic accident case, excision of a congenital pigmented nevus, and excision of an acquired pigmented nevus.

### 2.2 Processing of the Specimens

Each specimen was cut into three pieces and processed as follows: one piece was fixed with formaldehyde and stained with hematoxylin and eosin for light microscopy; the second piece was processed for Transmission Electron Microscopy (TEM) by fixing with 3% glutaraldehyde and 1% osmium tetroxide and examined with TEM (Hitachi 7000); and the last piece was processed for Scanning Electron Microscopic (SEM) observation of the interface between the epidermis and the dermis according to the technique described elsewhere.[7–9] Briefly, a skin specimen was incubated with 20 mM ethylenediaminetetra-acetic acid in Hanks' balanced salt solution at 0°C for 6–12 hours, to separate the epidermis from the dermal connective tissue. The epidermal sheet and the dermal connective tissue, thus separated, were fixed with 3% glutaraldehyde for 2 hours, immersed for 1 hour in 1% tannic acid for conduction staining, and then dehydrated using the *t*-butyl alcohol freeze-drying method. Preparations were mounted on specimen holders so as to observe the epidermal-dermal interface, and were examined with SEM (JEOL T300 and T880).

### 2.3 Estimation of Fractal Dimension

The shapes of the actual surface of tissues or interface of organs are approximated by a parallel array of ridges [Fig. 1(a)] or a hexagonal lattice of protrusions [Fig. 1(b)], and their size is specified by width, height and distance as defined in Fig. 1(c). Possibilities of four types of cross sectional shape as shown in Figs. 2(a)–2(d) are considered for cross section of ridges or protrusions. After determination of width, height and distance, we can estimate surface area using simple geometries. The surface area ($S$) varies with the

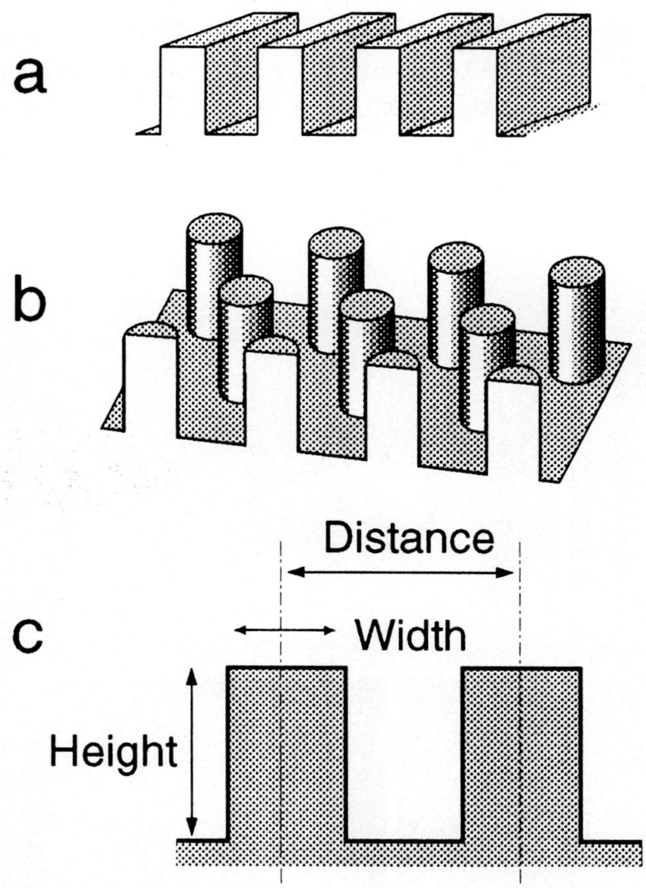


**Fig. 1** **(a)** Parallel arrays of ridges; **(b)** Hexagonal lattice distribution of protrusions; **(c)** The definition of width, height and distance of ridges or protrusions.

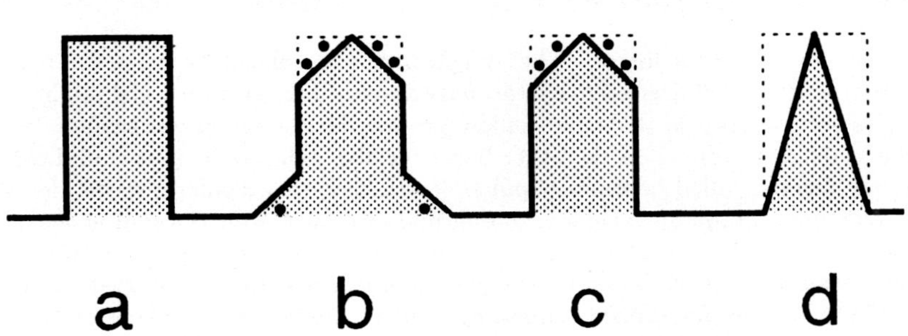


**Fig. 2** Four types **(a)**–**(d)** of cross sectional shape. Dot means 45 degrees.

scale of observation, and was estimated at several different scales, where the scale was characteristically represented by the width ($w$) of ridge or protrusion. Then the fractal dimension ($D_F$) is:

$$D_F = d - \log\, S(w)/\log\, w\,,$$

where $d = 2$, because the surface or interface is two-dimensional. $D_F$ can be calculated as 2 minus the slope of the regression of $\log S(w)$ vs. $\log w$.

Let us add some words about the definition of fractal dimension $D_F$. In the book by Mandelbrot,[6] $D_F$ is introduced as $-\log N(w)/\log w$, where $w$ is the size of scale of observation as above and $N(w)$ is the total number of units of the specimen measured by using the scale $w$. Then it turns out that $S(w)/w^d = N(w)$. This relation gives us our formula as above for $d = 2$.

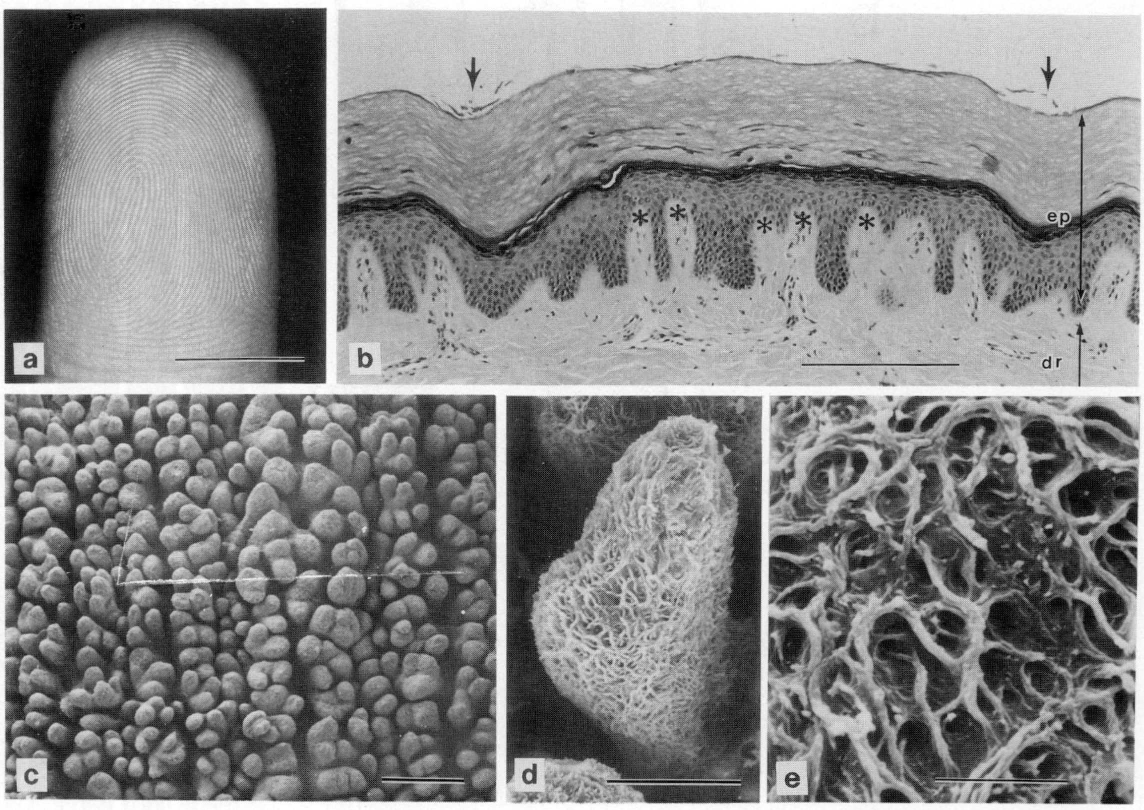


**Fig. 3** **(a)** A close-up view of a finger. The "fingerprint", produced by an ordered rows of ridges and grooves of the surface of the skin, is seen by the naked eye. Bar, 10 mm. **(b)** Light micrograph of the finger skin. A cross section reveals that the shallow grooves of the epidermal surface (arrows) correspond to the "fingerprint". The epidermis, on the other basal surface, exhibits further complicated interdigitation with the underlying dermis, called "dermal papillae" (asterisks). ep, epidermis. dr, dermis. Bar, 100 $\mu$m. **(c)** Scanning electron micrograph of the epidermal-dermal interface, after removal of the covering epidermis. The procedure applied in the present study allows us to observe relatively large areas of the epidermal-dermal interface in detail. At a surface view, there are shallow grooves with the interval of every 6 to 8 dermal papillae, corresponding to the fingerprint, superimposed on the separated plane of the epidermal-dermal interface. Bar, 200 $\mu$m. d, e, SEM micrographs at higher magnifications. We can see further folding at the surface of each dermal papilla. Bars, 50 $\mu$m **(d)** and 10 $\mu$m **(e)**.

## 3. RESULTS

We see an ordered undulation of the surface of the skin: the "fingerprint" on the palmar seen by the naked eye [Fig. 3(a)]. Light microscopy reveals that the "fingerprint" is produced by the quasi-periodic shallow grooves (or wrinkles) of the surface of the epidermal sheet [Fig. 3(b), arrows]. The epidermis exhibits another fine undulation at the other basal surface, where it faces the underlying dermal connective tissue, thus producing an interdigitation of the different tissues which is called as "dermal papillae" [Fig. 3(b), asterisks].

Such nesting structure can be directly seen after separation of the epidermis from the dermis [Figs. 3(c)–3(e)]. SEM shows that the dermal papillae are arranged in an orderly manner and with similar height and diameter. At an interval of every 6 to 8 dermal papillae, shallow grooves are superimposed, thus producing periodic parallel rows, on the plane of the epidermal-dermal interface [Fig. 3(c)]. This corresponds to the "fingerprint"

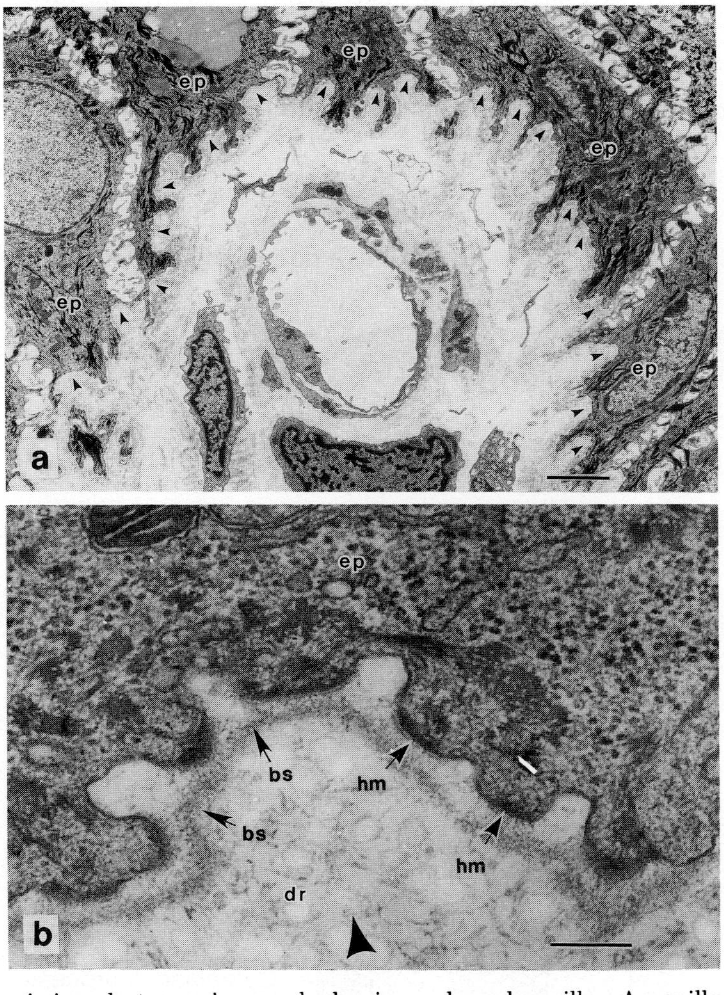


**Fig. 4** **(a)** Transmission electron micrograph showing a dermal papilla. A papilla has numerous small fold (arrow heads) at the interface with epidermal basal cells. Bar, 2 $\mu$m. **(b)** TEM micrograph showing the interface between the dermal papilla and the epidermal basal cell. Those two tissues interdigitate with numerous cytoplasmic processes whereby epidermal cells attach to the dermal connective tissue with hemidesmosomes. ep, epidermis. bs, basement membrane. hm, hemidesmosome. Large arrow head, fold of papilla surface. Bar, 0.2 $\mu$m.

seen with the naked eye [Fig. 3(a)] and with light microscopy [Fig. 3(b)], respectively. At higher magnifications, the surface of the individual papillae are shown to be not smooth but further folded [Figs. 3(d) and 3(e)].

TEM confirms the fine nesting structure revealed by the SEM study. Each dermal papilla is provided with folding structures at the ultrastructural level [Fig. 4(a), arrow heads]. Furthermore, at higher magnifications, another structure, the cytoplasmic process becomes evident. The epidermal cells covering the dermal papilla, have numerous cytoplasmic processes whereby the cells attach to the dermal connective tissue with hemidesmosomes [Fig. 4(b)].

The interface between the epidermis and the dermis thereby has a hierarchical structure: the parallel row of grooves of the fingerprint at the level of naked eye, the dermal papilla at the level of light microscopy, the fold of the papilla surface at the level of SEM and TEM, and the cytoplasmic process at the level of TEM (Fig. 5). These structures are ridge-type, protrusion-type, ridge-type, and protrusion-type, respectively. The width, height and distance of these structures are roughly determined from the photographs (Figs. 3 and 4). Starting with a piece of skin (1 $cm^2$), we successively estimate the total area $S(w)$ of the epidermal-dermal interface with varying scale of observation, and the result is shown in Table 1 and Fig. 6. The slope of the regression of $\log S(w)$ vs. $\log w$ in Fig. 6 was $-0.37$. The fractal dimension $D_F$ of the epidermal-dermal interface was 2.37.

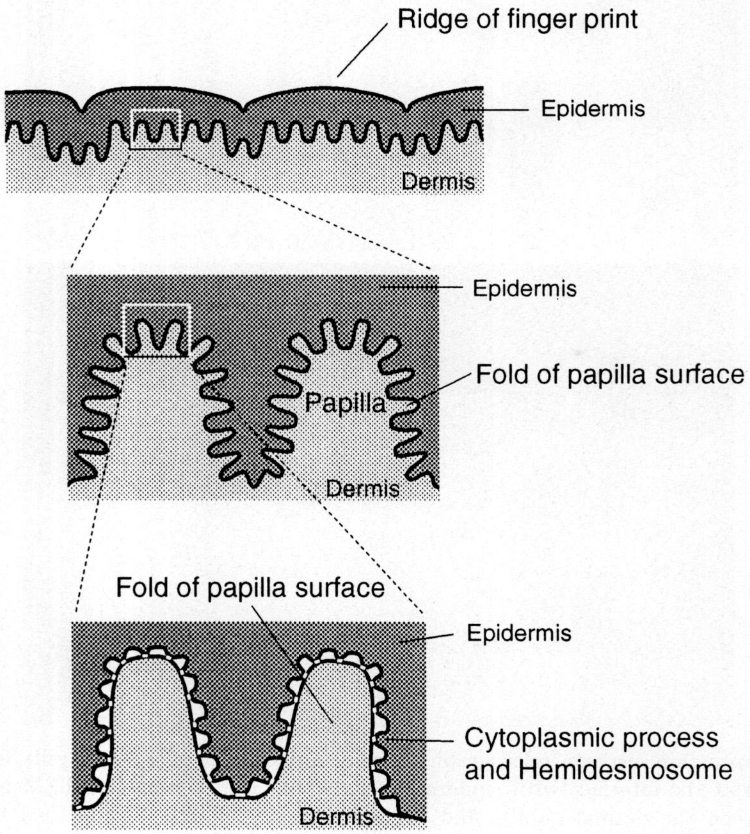


**Fig. 5** Schematic drawing of fractal-like structures of the boundary interface between the dermis and the epidermis of human skin.

Table 1 Estimation of the Interface Area between the Dermis and the Epidermis of Human Skin

| Interface Structure | Width ($\mu$m) | Height ($\mu$m) | Distance ($\mu$m) | Shape+ | Interface Area of a Piece of Skin (1 $cm^2$) ($cm^2$) | Average ($cm^2$) |
|---|---|---|---|---|---|---|
| 1. Ridge of fingerprint | 500 | 100 | 500 | r | 1.1(d*) – 1.4(a*) | 1.3 |
| 2. Dermal papilla | 70 | 150 | 100 | p | 4.7(b*) – 6.0(a*) | 5.4 |
| 3. Fold of papilla surface | 0.5 | 1.5 | 1 | r | 18(b*) – 22(a*) | 20 |
| 4. Cytoplasmic process with hemidesmosomes | 0.2 | 0.1 | 0.3 | p | 23(d*) – 36(a*) | 30 |

+r, ridges in parallel arrays; p, protrusions in hexagonal lattice
*a, b, d, type of cross sectional shape. See Fig. 2.

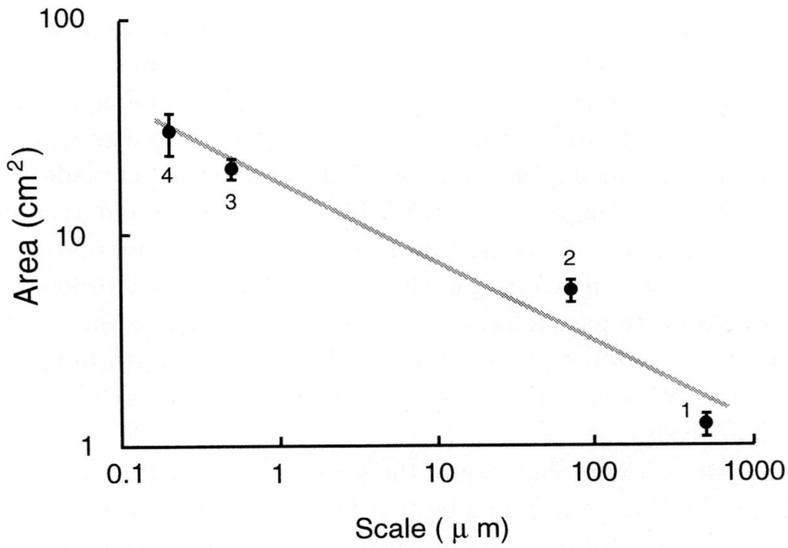


**Fig. 6** Fractal analysis of the interface between the dermis and the epidermis of human skin. Vertical bars show the possibility range according to the types of cross sectional shape in Fig. 2. See Table 1.

## 4. DISCUSSION

Each detail of a biological body has nested within it more details that become evident with further magnification, and each of these details has even finer details. Although each detail is not strictly self-similar, there is often a weak resemblance between details at different scales, for example, between a ridge and a protrusion, or between a helix and a chain of loops. We therefore consider that biological objects are often fractal-like, and we applied a method of analysis used in fractal objects to the epidermal-dermal interface in the skin to obtain the degree of complexity of the structure.

The epidermal-dermal interface whose fractal dimension is 2.4 gives rise to the complexity of the structure with a tendency to fill the three-dimensional space. The increase of the interface drives the fractal dimension from 2 toward 3 by 0.4. The skin consists of two tissues, dermis and epidermis, each having different properties. The dermis, a rough but tenacious tissue containing collagen fibers, must adhere strongly to the smooth, waterproof and protective epidermis. Fractal-like folding of the interface between them increases its area and allows more hemidesmosomes to mediate the adhesion. The fractal-like properties of the epidermal-dermal interface is adaptive for the animal life.

We know of other hierarchical structures in the biological body. Two of them are described here. One is an inner surface of the small intestine. In the small intestine, several structural devices increase the total area of the inner surface.[2] The plicae circulares (valves of Kerckring) extend as crescentic folds around the lumen of the small intestine. The intestinal villi are finger-like projections and cover the entire surface of the plicae circulares. The surface of the villi is a monolayer sheet of columnar epithelial cells. The luminal surface of the cells has a prominent brush border which consists of closely packed microvilli. The inner surface of the small intestine thereby has a hierarchical structure, the plicae circulares, the villi, and the microvilli. These structures are ridge-type, protrusion-type, and protrusion-type, respectively. The small intestine thus augments the efficiency of the absorption of nutrients by a number of structural devices that increase the total area of the inner surface.

The other is the fibrous structures in the chromosome. At the time of cell division, the genetic material in the cell becomes tightly packed, and individual chromosomes become recognizable. Each chromosome is believed to contain a single very long molecule of DNA.[3–5] DNA is coiled around the histone octamer to form a string of nucleosomes which are helically coiled to form a 30-nm chromatin fiber. The chromosome is made up of the 30-nm chromatin fiber. The total length of the DNA in the human diploid genome is estimated to be 1.8 m, which is distributed among the chromosomes.[2] The fractal-like structures, the several orders of coiling and supercoiling of the fibers, achieve about 10 000-fold reduction in length, making it possible to pack a large amount of genetic information in a nucleus. These objects of restricted size produce great extension of surface or length in the construction of fractal-like structures. We estimate the degree of complexity of these objects by using the method of fractal analysis.

Finally, we will discuss the mechanism of the formation of the fractal-like structures. We interpret the fractal-like structures as adaptive for the function of the organs, the strong adhesiveness, the effective absorption of nutrients and the compact packing. However, the mechanisms of the formation of fractal-like structures remain to be unsolved. Detailed observation of the developmental process will provide clues to the mechanisms. For example, during the developmental process, the epidermal-dermal interface generates primary ridges on the upper surface of the dermis. The ridges are separated from one another to produce the secondary ridges. Then conical dermal papillae project upward from these ridges. These observations suggest that a fractal-like set of cascades generates the fractal-like structures.[2,10]

## ACKNOWLEDGMENT

The authors thank Dr. Jay E. Mittenthal for critical reading of the manuscript.

## REFERENCES

1. J. B. Bassingthwaighte, L. S. Liebovitch and B. J. West, *Fractal Physiology* (Oxford University Press, London, 1994).

2. Don W. Fawcett, *Bloom and Fawcett, A Textbook of Histology*, 12th ed. (Chapman & Hall, London, 1994).
3. B. Alberts, D. Bray, J. Lewis, M. Raff, K. Roberts and J. D. Watson, *Molecular Biology of the Cell* (Garland Publishing, Inc., New York, 1983).
4. A. R. Rees and M. J. E. Sternberg, *From Cells to Atoms — An Illustrated Introduction to Molecular Biology* (Blackwell Scientific Pub., 1984).
5. B. Lewin, *Genes V.* (Oxford University Press, Oxford, 1994).
6. B. B. Mandelbrot, *The Fractal Geometry of Nature* (Freeman, San Francisco, 1982).
7. S. Imayama, Y. Yashima, R. Higuchi and H. Urabe, *Am. J. Pathol.* **128**, 497–504 (1987).
8. S. Imayama, Y. Yashima and Y. Hori, *J. Histochem. Cytochem.* **40**, 1191–1196 (1992).
9. S. Imayama, M. Furumura and Y. Hori, *Pigment Cell Res.* **7**, 170–174 (1994).
10. M. Okajima and L. Newell-Morris, *Am. J. Anatomy* **183**, 323–337 (1988).

Fractals, Vol. 4, No. 2 (1996) 149–159


# PATTERN FORMATION IN AN ETCHED RADIAL HELE SHAW CELL

M. T. BATCHELOR
*Department of Mathematics, School of Mathematical Sciences, Australian National University, Canberra ACT 0200, Australia*

B. I. HENRY
*Department of Applied Mathematics, University of New South Wales, Sydney NSW 2052, Australia*

**Abstract**

We have introduced numerical schemes for simulating pattern formation at the interface between two imiscible fluids in an etched radial Hele Shaw cell. Our simulations, which have been carried out over a range of two model parameters, are compared with experimental results that have been reported for a range of plate spacings and applied pressures. The simulations reproduce the complete range of morphology transitions in the experiments.

## 1. INTRODUCTION

Pattern formation at the interface between two imiscible fluids (with different viscosities) confined between narrowly spaced glass plates (a Hele Shaw cell) is one of the simplest problems in fluid mechanics.[1–4] In typical experiments, a fluid such as air is forced under pressure at the centre of the cell into a more viscous fluid such as glycerol which is confined between the glass plates. If the applied pressure is low then the less viscous fluid is stabilized by surface tension and the interface pattern shows stable fingers of the less viscous fluid penetrating the more viscous fluid. This pattern becomes increasingly unstable and ramified as the applied pressure is increased.

Ben-Jacob and co-workers[5–7] have carried out viscous fingering experiments using a modified Hele Shaw cell with etched square lattice grooves (in one experiment) and triangular lattice grooves (in another experiment) on one of the glass plates. The etched grooves, which result in competition between surface tension and anisotropy, lead to a complex range of interfacial patterns as the plate spacing and the applied pressure are varied. The range of patterns has been comprehensively described for experiments with triangular lattice grooves.[6] As the applied pressure is increased the morphology changes in the sequence: faceted growth (a hexagon with the vertices directed along the lattice axes) → tip-splitting → surface tension dendrite (a stable six-armed dendrite with the arms aligned at 30° to the lattice axes) → tip-splitting → kinetic dendrite (a stable six-armed dendrite with the arms aligned along the direction of the lattice axes) → kinetic dendrite decorated with side branches.

The interface patterns in etched radial Hele Shaw cell experiments also depend on the spacing between the plates. Essentially the cut-off applied pressures for the different pattern regimes are increased as the plate spacing is decreased (e.g., the faceted growth region persists for larger applied pressures when the plate spacing is decreased).

Approximate theoretical equations that govern the motion of the interface $\Gamma$ between two imiscible fluids, when the fluids are confined between closely spaced parallel plates and the less viscous fluid $a$ is injected under pressure $P_0$ into the more viscous fluid, are given by[1–6]:

$$\frac{\partial P}{\partial t} = \nabla^2 P\,, \qquad (1)$$

$$\nu_n = -k\nabla P \cdot \boldsymbol{n}\,, \qquad (2)$$

$$P_\infty = 0\,, \qquad (3)$$

$$P_a = P_0\,. \qquad (4)$$

In these equations $\nu_n = \boldsymbol{\nu} \cdot \boldsymbol{n}$ is the velocity normal to the interface and $k$ is a constant whose value is proportional to the square of the plate spacing. The equation for the velocity of the interface [Eq. (2)], is known as Darcy's law. In the quasi-stationary approximation, Eq. (1) is replaced by Laplace's equation. An additional boundary condition is required to represent the pressure jump across the interface. Under equilibrium conditions of no flow, the pressure on the interface is related to the applied pressure by $P_\Gamma = P_0 - d\kappa$, where $d$ is the capillary length (related to surface tension) and $\kappa$ is the curvature of the interface. This boundary condition is not clearly relevant for the non-equilibrium flow; however it has been found to be a useful simplest assumption.[1] Ben-Jacob and co-workers[6] have suggested that a kinetic term should be introduced into the boundary conditions to account for enhanced growth in the direction of the grooves when modelling viscous fingering in etched Hele Shaw cells. With these two approximations, the final boundary condition may be written as:

$$P_\Gamma = P_0 - d\kappa - \beta\nu_n\,. \qquad (5)$$

There have been numerous numerical schemes proposed for simulating viscous fingering in fluids. These schemes can be separated into two broad classes:

(1) *Direct schemes* — schemes that attempt a direct numerical solution of Eqs. (1)–(5) to simulate the moving interface. This includes the so called Boundary Layer Models that utilize Green's functions to convert the continuum equations into an integro-differential equation.[4,8,9]

(2) *Indirect schemes* — schemes that capture the essential physics of Eqs. (1)–(5) and incorporate this in a computer growth model for the moving interface. This includes variants of the Diffusion-Limited Aggregation (DLA) model[10] modified to include surface tension.[4,7,11–14]

A major difference between the two approaches is that the direct schemes are deterministic whereas the indirect schemes are stochastic. It has been pointed out that the direct schemes and limiting noise-reduced variants of the indirect schemes[15] are essentially equivalent.[11,15]

The purpose of this paper is to introduce numerical schemes, intermediate between the two approaches, that simulate the essential physical features of Eqs. (1)–(5) and to compare the results of our numerical simulations with the experimental results of Ben-Jacob and co-workers.[5–7] The numerical schemes that we propose are equivalent to the zero-noise limit of modified DLA models. In the zero-noise limit, the cluster grows in levels and the growth probability matches the velocity of the interface precisely. We have described (and implemented) the basic algorithm for simulating the zero-noise limit of modified DLA models in previous publications[16–18] [see also Refs. (19 and 20)]. At a given level $l$, surface sites are associated with a weight:

$$w^{(l)} = \frac{1 - r^{(l-1)}}{p^{(l)}}, \tag{6}$$

where $p^{(l)}$ denotes the normalized *a priori* growth probability and $r^{(l-1)}$ denotes the accrued residue from growth in previous levels. Each level of growth corresponds to the conversion of all minimum weight surface sites into aggregate sites and the remaining surface sites accrue an increased residue:

$$r^{(l)} = p^{(l)} w^{(l)}_{\min} + r^{(l-1)} . \tag{7}$$

The numerical schemes that we describe in this paper provide simple approximations to the underlying physics. Advantages of the schemes are:

(1) They are amenable to theoretical analysis.
(2) They can be readily implemented over a large parameter space.
(3) The simulations can be compared directly with experiments on radial viscous fingering in etched Hele Shaw cells.

## 2. NUMERICAL SCHEME

### 2.1 Lattice Growth by Darcy's Law

Starting with the continuum [Eqs. (1)–(5)], we introduce a new field variable:

$$\phi = 1 - \frac{P}{P_0} . \tag{8}$$

The Laplacian approximation to the continuum equations now becomes:

$$\begin{aligned}\nabla^2\phi &= 0\,, &(9)\\ \nu &= \frac{k}{P_0}\nabla\phi\,, &(10)\\ \phi_\Gamma &= \frac{d}{P_0}\kappa + \frac{\beta}{P_0}\nu_n\,, &(11)\\ \phi_\infty &= 1\,, &(12)\\ \phi_a &= 0\,. &(13)\end{aligned}$$

The kinetic growth term in the boundary condition, Eq. (11), can be approximated by solving Laplace's equation on a discrete lattice. For example, to simulate pattern formation in a radial Hele Shaw cell in which one of the plates is etched with triangular lattice grooves, we solve Laplace's equation on a triangular lattice. Thus we replace Eq. (9) by[18]:

$$\phi_{i,j} = \frac{1}{6}(\phi_{i+1,j} + \phi_{i,j+1} + \phi_{i+1,j+1} + \phi_{i-1,j-1} + \phi_{i,j-1} + \phi_{i-1,j})\,. \tag{14}$$

The discrete Laplacian is most applicable if the spacing between the plates is sufficiently narrow that the fluids experience a strong lattice effect. For example, in this regime the interface pattern at low pressures is a hexagon rather than a circle. Darcy's law still applies in this case and the lattice version of the velocity equation for a site on the interface is now:

$$\nu_{i,j} = k_P\left[\frac{1}{6}(\phi_{i+1,j} + \phi_{i,j+1} + \phi_{i+1,j+1} + \phi_{i-1,j-1} + \phi_{i,j-1} + \phi_{i-1,j}) - \phi_{i,j}\right], \tag{15}$$

where $k_P = k/P_0$. The field on the interface, Eq. (11), in this approximation is reduced to:

$$\phi_{i,j} = d_P\kappa\,, \tag{16}$$

where $d_P = d/P_0$.

The curvature of the interface on the triangular lattice can be approximated by the local approximation:

$$\kappa = 2 - B\,, \tag{17}$$

where $B$ is the number of (nearest) aggregate neighbors. This is the simplest approximation to Vicsek's estimate of surface curvature based on counting the number of aggregate sites within a circle of diameter $L$ centred on the site in question.[12] Equations (12)–(17) are the governing equations for viscous fingering between very narrow plates. The plate separation dependence $k$ can be scaled out of these equations and the velocity at a point on the interface may be negative. We have solved Eqs. (12)–(17) exactly using a modification of our weights and residues scheme, Eqs. (6)–(7), to accomodate the negative velocities. The strategy here was to determine the surface site with the smallest absolute value of the weight at each level. If the selected site had positive weight, it was converted to an aggregate site and new surface sites were added in the usual fashion, whereas if it had negative weight the neighboring aggregate sites were converted back to surface sites and the selected site became a background site. Using the modified weights and residues scheme, we found a smooth transition from faceted growth to a kinetic dendrite as the effective pressure was increased (see Fig. 1). There was no intermediate tip-splitting and no surface tension dendrite in this model.

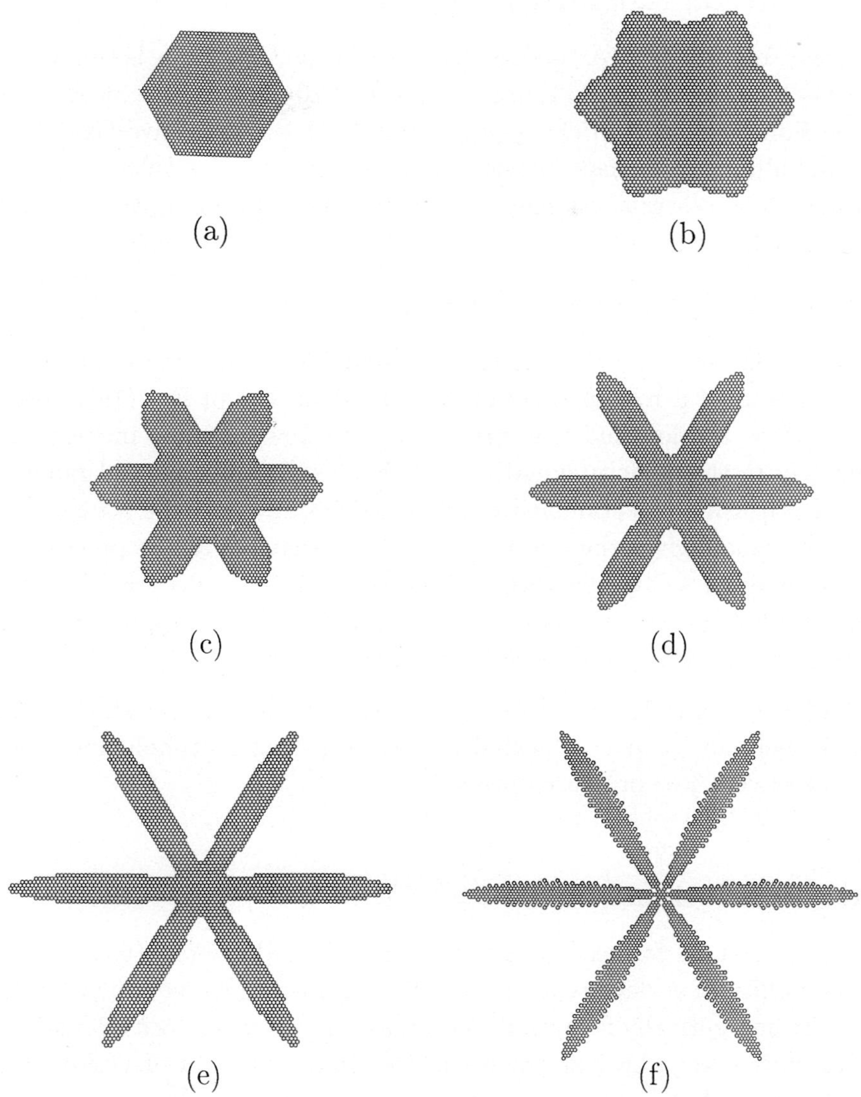

**Fig. 1** The initial 200 levels for triangular lattice growth by Darcy's law [Eqs. (12)–(17)]. The parameters are: **(a)** $d_P = 0.2$; **(b)** $d_P = 0.05$; **(c)** $d_P = 0.03$; **(d)** $d_P = 0.02$; **(e)** $d_P = 0.01$; **(f)** $d_P = 0.0001$. All clusters are shown at the same scale.

We attempted several refinements of the surface curvature term including next nearest neighbors ($\kappa = 7 - B$), next-next nearest neighbors ($\kappa = 15 - B$), and next-next-next nearest neighbors ($\kappa = 26 - B$) in the aggregate count $B$; however, this did not make any qualitative difference to the overall trend. As an aside, we note that care must be exercised in going beyond nearest neighbors since the calculated surface curvature might change sign as more neighbors are included in the count for $B$. To avoid this problem, we calculated the surface curvature by counting as many neighbors as possible (up to next-next-next nearest neighbors) without a change in sign for the curvature. The overall effect of including more neighbors in the surface curvature calculation was to obtain more rounded growth patterns (particularly in the low pressure limit), but we still only observed the same limited range of morphologies.

## 2.2 Surface Tension Modified Darcy's Law

In surface tension modified DLA models, the surface tension term is not incorporated as a boundary condition as in Eq. (16) but is included directly in the probability equation as a surface tension dependent sticking probability.[12–14] We have investigated the patterns obtained by including the surface tension term directly in the velocity equation in our model equations. The essential change is to replace Eq. (16) with $\phi_\Gamma = 0$ and to replace the velocity equation with:

$$\nu_{i,j} = k_P(1 - d_P\,\kappa)\frac{1}{6}(\phi_{i+1,j} + \phi_{i,j+1} + \phi_{i+1,j+1} + \phi_{i-1,j-1} + \phi_{i,j-1} + \phi_{i-1,j})\,. \tag{18}$$

Here again the scale factor $k_P/6$ is redundant and hence the plate spacing dependence through $k$ does not play a role in this model. An advantage of Eq. (18), however, is that the effects of surface tension and the directed lattice flow are now included on the same footing — directly in the velocity equation. We have solved the revised model [Eqs. (12)–(14), (17) and (18)] using the standard weights and residues scheme for $d_P$ in the interval $[0,1]$. The revised model gives the following range of behaviors as the pressure is increased; surface tension dendrite → tip-splitting → kinetic dendrite → decorated kinetic dendrite. The range of morphologies from this model as a function of the applied pressure is shown in Fig. 2. There are two clear shortcomings with this model:

(1) The model has one parameter whereas there are two distinct physical parameters.
(2) The model does not yield the faceted and tip-splitting morphologies observed in the experiments at very low pressures ($d_P = 1$).

## 2.3 Surface Relaxations

The continuum equations for viscous fingering in fluids [Eqs. (1)–(5)], do not take into account the possibility of surface relaxations in which surface molecules are microscopically reshuffled to more energetically favourable locations. However, surface relaxations have been identified as an important physical process in the fluid dynamics of viscous fingering.[11,12] The simplest way to include surface relaxations is to homogeneously re-distribute a fraction of the interface particles over the entire interface.[21] Here we consider a more realistic surface relaxation, whereby a fraction of interface particles are re-distributed over the interface according to the probability distribution:

$$p = \frac{B}{\sum_{\mathcal{N}} B}\,, \tag{19}$$

where $B$ is the number of nearest neighbor aggregate particles for a given surface site and the sum is over the total number of surface sites $\mathcal{N}$. This surface relaxation effect thus re-distributes interface particles with a bias for filling in holes and rounding off bumps. This is the simplest deterministic approximation to the sort of surface relaxations suggested by Kadanoff.[11]

A revised interface velocity for the model including surface relaxations is now given by:

$$\begin{aligned}\nu_{i,j} &= (1 - d_P\kappa)(1 - k_P)\frac{1}{6}[\phi_{i+1,j} + \phi_{i,j+1} + \phi_{i+1,j+1} + \phi_{i-1,j-1} + \phi_{i,j-1} + \phi_{i-1,j}] \\ &\quad + \frac{k_P B_{i,j}}{\sum_{\mathcal{N}} B_{i,j}}\,. \end{aligned} \tag{20}$$

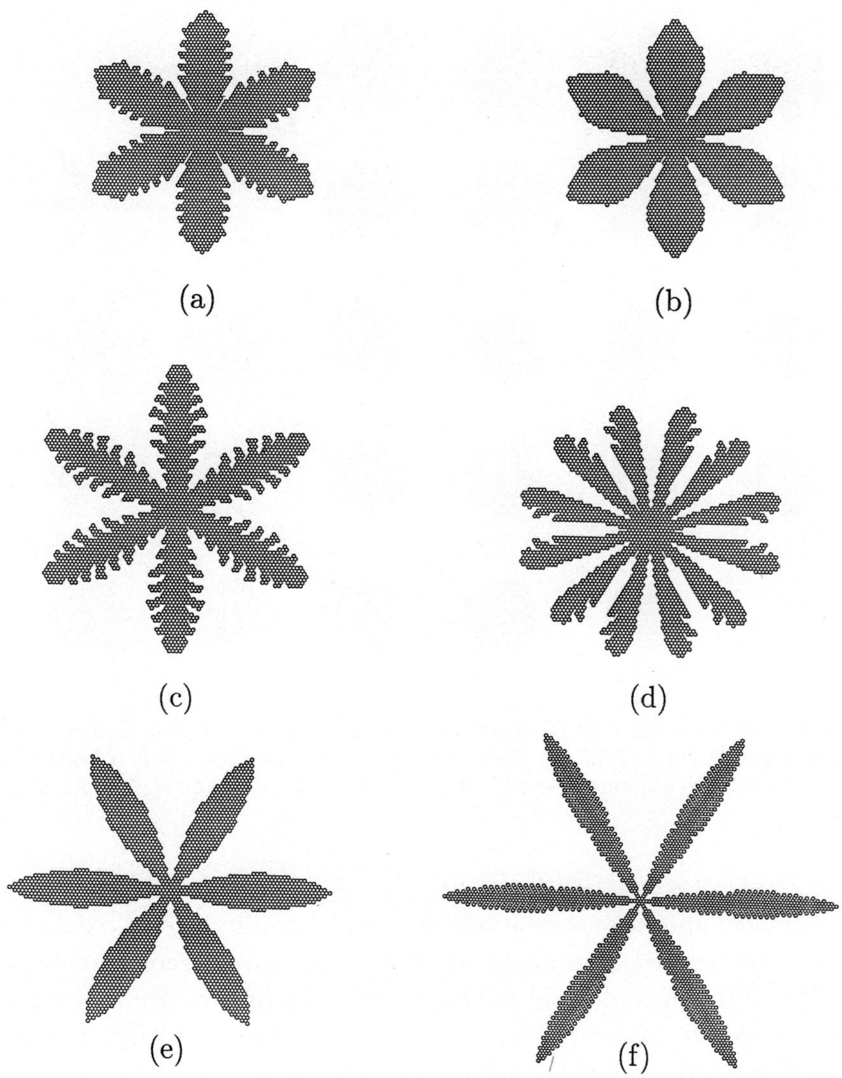

**Fig. 2** The initial 200 levels for triangular growth by surface tension modified Darcy's law [Eqs. (12)–(14), (17) and (18)]. The parameters are: **(a)** $d_P = 1.0$; **(b)** $d_P = 0.8$; **(c)** $d_P = 0.5$; **(d)** $d_P = 0.45$; **(e)** $d_P = 0.4$; **(f)** $d_P = 0.01$. All clusters are shown at the same scale.

At first sight, Eq. (20) may seem inappropriate since the pre-factor of the gradient of the field now decreases with increasing plate separation, whereas the fluid mobility should increase with increasing plate separation. We have compensated the reduction in the directed lattice mobility (via the gradient of the field) with an increase in surface mobility (via surface relaxations). Thus the factor $k_P$ has been incorporated in a manner that is consistent with the physical expectation that more surface relaxations will occur, relative to the outward motion of the interface:

(1) As the plate separation is increased.
(2) As the applied pressure is decreased.

The parameter $k_P$ is no longer redundant in this scheme. Our numerical scheme is now reduced to a deterministic solution of [Eqs. (12)–(14), (19) and (20)].

**Fig. 3** The initial 200 levels for triangular lattice growth by surface tension modified Darcy's law with surface relaxation [Eqs. (12)–(14), (17) and (18)]. The values of $\hat{P}$ used were (left to right) 1.0, 1.5, 2.0, 2.3, 2.5, 3.0, 3.5, 4.5. The values of $1/\hat{k}$ used were (bottom to top) 0.5, 0.7, 1.0, 1.5, 5.0. All clusters are shown at the same scale.

Although the model above was essentially derived by modifying the continuum [Eqs. (1)–(5)], it could also be derived as a modified DLA model in the zero-noise limit. The fully stochastic version of this model could be implemented as follows. Release a random walker from a distant boundary and terminate the walker at the first surface site it contacts. Implement one of the following three outcomes:

(1) Convert no surface sites to aggregate sites. The probability for implementing this outcome is $d_P\kappa(1 - k_P)$.
(2) Convert the termination surface site to an aggregate site — probability $(1 - d_P\kappa)$ $(1 - k_P)$.
(3) Select a surface site at random using a biased probability distribution with a bias for choosing sites with greater numbers of aggregate neighbors and convert this surface site to an aggregate site. The probability for this scenario is $k_P$.

We have used the weights and residues scheme to solve the above zero-noise model [Eqs. (12)–(14), (19) and (20)] for $d_P$ and $k_P$ both in the interval $[0, 1]$. It is convenient to introduce normalized variables $\hat{k} = k/d$ and $\hat{P} = P/d$. In Fig. 3, we have shown representative morphologies as a function of $1/\hat{k}$ vs. $\hat{P}$. At intermediate values of $\hat{k}$, as $\hat{P}$ is increased, we now get the full range of behaviors: faceted growth → tip-splitting → surface tension dendrite → tip-splitting → kinetic dendrite → decorated kinetic dendrite. From Fig. 3, it can also be seen that the cut-off values of $\hat{P}$ for the different morphology regimes are increased as $\hat{k}$ is increased. These results are consistent with the experimental results if we identify increasing $\hat{k}$ with increasing plate separation.

**Fig. 4** The initial 300 levels for square lattice growth by surface tension modified Darcy's law with surface relaxation. The values of $\hat{P}$ used were (left to right) 1.0, 1.5, 2.0, 2.5, 3.0, 3.5, 4.5, 5.5. The values of $1/\hat{k}$ used were (bottom to top) 0.5, 0.7, 1.0, 1.5, 5.0. All clusters are shown at the same scale.

## 3. DISCUSSION

In this paper we introduced and implemented new numerical schemes for simulating viscous fingering in an etched radial Hele Shaw cell. The numerical schemes represent the zero-noise limit of modified Diffusion-Limited-Aggregation processes on the one hand, and a discrete numerical solution of the governing continuum equations of motion on the other. In each of the schemes the quasi-stationary diffusion equation (Laplace's equation) was solved on a triangular lattice. This feature was not purely a computational convenience, but mirrored the effect of the triangular lattice grooves etched on one of the plates of the Hele Shaw cell. The variety of patterns in the etched Hele Shaw cell experiments is known to arise from a competition between the lattice anisotropy and surface tension. We investigated numerical schemes in which a surface tension term based on the curvature of the interface between the two fluids was introduced:

(1) As a boundary condition.
(2) As a velocity scale factor.

The second approach was found to give better agreement with experimental results, although this one parameter model failed to reproduce the full range of morphologies exhibited in the two parameter experiments. We incorporated a second parameter in our numerical schemes by building surface relaxations into the model. This two parameter model reproduced the full range of morphologies reported in the experiments.

The simulations that we reported above solved Laplace's equation on a triangular lattice. By solving Laplace's equation on a square lattice, we can implement the same schemes to model viscous fingering in a radial Hele Shaw cell in which one of the plates is etched with square lattice grooves. The results of such simulations from the two parameter model are summarized in Fig. 4. These results compare favorably with experiments[7]; however, we have explored a much larger range of parameters in our numerical studies than has been reported for the experiments. Features of note in the numerical simulations on the square lattice are:

(1) Surface tension dendrites (a stable four-armed star with the arms aligned at 45° to the lattice axes) and kinetic dendrites (a stable four-armed star with the arms aligned along the direction of the lattice axes) can co-exist.
(2) The faceted growth pattern is a square. Note that if the surface relaxation obeyed a uniform distribution, then the faceted growth pattern would be a diamond.[21] Physical experiments have not yet revealed the shape of the faceted growth pattern.
(3) As the parameter representing the applied pressure is increased, there is a morphology transition: faceted growth → tip-splitting → co-existing surface tension dendrite plus kinetic dendrite → tip-splitting → kinetic dendrite.
(4) The cut-off applied pressures for the different morphology regimes are increased as the parameter representing the plate separation $\hat{k}$ is increased.

**Fig. 5** Representative snowflakes. The initial 250 levels for triangular lattice growth by surface tension modified Darcy's law with surface relaxation [Eqs. (12)–(14), (17) and (18)]. The parameters $k_P$, $d_P$ are varied (with $k_P/d_P$ constant) after each 50 levels of growth. The different parameter regimes are represented by different colors. All clusters are shown at the same scale.

Finally, we note that much has been written about the similarities between the model equations for viscous fingering in etched Hele Shaw cells and for solidification in anisotropic crystals (see, for example, Refs. 4 and 7). Based on these similarities, we would expect that our two parameter numerical scheme on the triangular lattice should also provide a good model for the formation of snowflakes by the accretion of diffusing supersaturated water vapor molecules, if the applied pressure (now interpreted as the difference between the partial pressure of the supersaturated water vapor and the equilibrium vapor pressure of ice) is varied during the course of the simulation (see also Ref. 14). This can be achieved in our simulations by varying $d_P$ and $k_P$ but keeping the ratio $d_P/k_P$ fixed. Figure 5 shows the results of such simulations; the different colors represent different parameter regimes. The striking diversity of patterns that can be achieved in this fashion do indeed bear a close similarity with the diverse array of morphologies exhibited by real snowflakes. Martian snowflakes[7] could be produced in a similar fashion from the square lattice code.

## ACKNOWLEDGMENTS

This work has been supported by the Australian Research Council and the ANU Faculties Research Fund. The calculations reported here were carried out on the Fujitsu VP2200 at the Australian National University Supercomputer Facility.

## REFERENCES

1. P. G. Saffman and G. I. Taylor, *Proc. R. Soc. London Ser.* **A245**, 312 (1958).
2. L. Paterson, *J. Fluid Mech.* **113**, 513 (1981).
3. D. Bensimon, L. P. Kadanoff, S. Liang, B. I. Shraiman and C. Tang, *Rev. Mod. Phys.* **58**, 977 (1986).
4. T. Vicsek, *Fractal Growth Phenomena*, 2nd ed. (World Scientific, Singapore, 1992).
5. E. Ben-Jacob, R. Godbey, N. D. Goldenfeld, J. Koplik, H. Levine, T. Mueller and L. M. Sander, *Phys. Rev. Lett.* **55**, 1315 (1985).
6. E. Ben-Jacob, P. Garik, T. Mueller and D. Grier, *Phys. Rev.* **A38**, 1370 (1988).
7. E. Ben-Jacob, *Contemp. Phys.* **34**, 247 (1993).
8. D. A. Kessler, J. Koplik and H. Levine, *Phys. Rev.* **A30**, 2820 (1984).
9. L. M. Sander, P. Ramanlal and E. Ben-Jacob, *Phys. Rev.* **A32**, 3160 (1985).
10. T. A. Witten and L. M. Sander, *Phys. Rev. Lett.* **47**, 1400 (1981); *Phys. Rev.* **B27**, 5686 (1983).
11. L. P. Kadanoff, *J. Stat. Phys.* **39**, 267 (1985).
12. T. Vicsek, *Phys. Rev.* **A32**, 3084 (1985).
13. P. Meakin, F. Family and T. Vicsek, *J. Colloid. Int. Sci.* **117**, 394 (1987).
14. J. Nittmann and H. E. Stanley, *J. Phys.* **A20**, L1185 (1987).
15. C. Tang, *Phys. Rev.* **A31**, 1977 (1985).
16. M. T. Batchelor and B. I. Henry, *Phys. Rev.* **A45**, 4180 (1992).
17. M. T. Batchelor and B. I. Henry, *Physica* **A187**, 551 (1992).
18. M. T. Batchelor, C. R. Dun and B. I. Henry, *Physica* **A193**, 553 (1993).
19. C. Moukarzel, *Physica* **A188**, 469 (1992).
20. K. Kassner, *Fractals* **1**, 205 (1993).
21. M. T. Batchelor and B. I. Henry, *Physica* **A203**, 566 (1994).

Fractals, Vol. 4, No. 2 (1996) 161–168


# SPREADING OF *N* DIFFUSING SPECIES WITH DEATH AND BIRTH FEATURES

S. HAVLIN[*], A. BUNDE[*,†], H. LARRALDE[‡], Y. LEREAH[§], M. MEYER[†], P. TRUNFIO[¶] and H. E. STANLEY[¶]
[*] *Minerva Center and Department of Physics, Bar-Ilan University, Ramat-Gan 52900, Israel*
[†] *Institut für Theoretische Physik III, University of Giessen, Germany*
[‡] *Cavendish Laboratory, Cambridge University, Cambridge CB3 OHE, United Kingdom*
[§] *Department of Physical Electronics, Tel-Aviv University, Ramat-Aviv 69978, Israel*
[¶] *Center for Polymer Studies and Department of Physics, Boston University, Boston, MA 02215 USA*

## Abstract

The number of distinct sites visited by a random walker after $t$ steps is of great interest, as it provides a direct measure of the territory covered by a diffusing particle. We review the analytical solution to the problem of calculating $S_N(t)$, the mean number of distinct sites visited by $N$ random walkers on a $d$-dimensional lattice, for $d = 1, 2, 3$ in the limit of large $N$. There are three distinct time regimes for $S_N(t)$. A remarkable transition, for dimension $\geq 2$, in the geometry of the set of visited sites is found. This set initially grows as a disk with a relatively smooth surface until it reaches a certain size, after which the surface becomes increasingly rough. We also review the results for a model for migration and spreading of populations and diseases. The model is based on $N$ diffusing species, where each species has a probability $\alpha_-$ of dying (or recovery from a disease) and a probability $\alpha_+$ to give birth (or to infect another species). It is found analytically that when $\alpha_+ \approx \alpha_- \neq 0$, after a crossover time $t_\times \sim N/2\alpha_-$, the territory covered by the population is *localized* around its center of mass while the center of mass diffuses regularly. When $\alpha_+ > \alpha_-$, the localization breaks down after a second crossover time and the species diffuse and spread around their center of mass. These results may explain the phenomena of migration and spreading of diseases and population appearing in nature.

## 1. NUMBER OF DISTINCT SITES VISITED BY *N* RANDOM WALKERS

One of the most important properties of a discrete-time lattice random walk is the number of distinct sites visited by a $t$-step walk.[1–19] This quantity enters into the description of phenomena ranging from relaxation processes[3–6] and diffusion limited reactions[3,20,21] such as defect annealing[2,7] and exciton trapping,[8,9] to the spread of populations in ecology.[22–25]

The analyses in the literature to date refer to the calculation of properties of $S_1(t)$, the distinct number of sites visited by a *single* random walker. For this case, the asymptotic forms for the mean number of sites visited, $\langle S_1(t)\rangle$, in any number of dimensions $d$, are well known. Knowledge of $\langle S_1(t)\rangle$ enables one to find a lowest order approximation to the survival probability for the "trapping problem", in which a random walker moves in the presence of randomly distributed static traps.[3] This problem can be regarded as the simplest generalization of the Smoluchowski model for the rate of chemical reactions of the form $A+B\to B$, taking into account the possibility of a concentration of $B$'s rather than the single $B$ envisaged by Smoluchowski.[3,20,21] $\langle S_1(t)\rangle$ also appears in the solution of the "target" problem,[3] in which a single target is static within a concentration of diffusing traps. The survival probability in this case is given exactly by $\exp(-k\langle S_1(t)\rangle)$ (where $k$ depends on the lattice and the concentration of traps), if the traps are initially Poisson distributed.

The properties of the number of distinct sites visited in the more general situation where there are $N$ random walkers in the system, are not related simply to the single walker case. Here we calculate asymptotic properties of $S_N(t)$, the number of distinct sites visited by $N\gg 1$ independent random walkers.

The quantity $S_N(t)$ has direct application to any situation that might be conceptualized as a "multiple scavenger problem", in which mobile traps (or scavengers) react with stationary particles initially distributed homogeneously on the lattice, and a particle disappears when any one of the traps reaches it. It can also be useful for the analysis of simultaneous survival of several particles diffusing in the presence of randomly distributed static traps, which is the generalization of the trapping problem described above. The probability for the simultaneous survival of $M$ of the initial $N$ particles will be given by a combinatorial factor times $\langle\exp(-kS_M(t))\rangle$, where $k$ again depends on the concentration of traps and on the lattice. For short times or low concentration of traps, this survival probability can be approximated by $\exp(-k\langle S_M(t)\rangle)$.

We consider the case for which all $N$ random walkers are assumed to be initially at the origin. The single-step displacements are characterized by a finite variance, so that the $t$-step transition probabilities (in the absence of boundaries) tend toward the Gaussian form predicted by the central-limit theorem. The random walkers move independently of one another with the consequence that multiple occupancy of a single site is allowed.

Reference 26 finds that the mean number of distinct sites $\langle S_N(t)\rangle$, passes through several distinct growth regimes in time. At very short times, the simple expression is found:

$$\langle S_N(t)\rangle \sim At^d \quad [t\ll t_\times] \quad [\text{Regime I}]\,, \tag{1.1}$$

where $A$ depends on the lattice. Equation (1.1) simply states that every accessible site is occupied by a walker.

Regime I holds so long as there are many walkers at every accessible site, i.e., so long as $NP_{\min}(t)\gg 1$, where $P_{\min}$ is the smallest non-zero occupation probability on the lattice at time $t$. Then $P_{\min}(t) = z^{-t}$, where $z$ is the number of nearest neighbors of a site, so Regime I must terminate at a crossover time $t_\times$ which scales logarithmically with $N$,

$$t_\times \sim \ln N .$$

To discuss times greater than $t_\times$, we calculate $\langle S_N(t)\rangle$ using generating function techniques. This analysis leads to a compact scaling expression for $\langle S_N(t)\rangle$:

$$\langle S_N(t)\rangle \sim t^{d/2} f(x) \quad [t \gg t_\times] , \tag{1.2}$$

where the tilde denotes the fact that Eq. (1.2) holds for $N$ and $t$ both large. The scaled variable $x$ is given by:

$$x \equiv \begin{cases} N & [d=1] \\ N/\ln t & [d=2] \\ N/\sqrt{t} & [d=3] , \end{cases}$$

and the scaling function $f(x)$ by:

$$f(x) = \begin{cases} (\ln x)^{d/2} & t_\times \ll t \ll t'_\times \quad \text{[Regime II]} \\ x & t \gg t'_\times \quad \text{[Regime III]} . \end{cases}$$

Here the second crossover time $t'_x$ is:

$$t'_x \sim \begin{cases} \infty & [d=1] \\ e^N & [d=2] \\ N^2 & [d=3] . \end{cases} \tag{1.3}$$

The appearance of Regime III (for $d \geq 2$) can be understood from the following heuristic argument. In Regime II, all but an exponentially small fraction of the walkers are contained within a $d$-dimensional sphere of radius $\xi \sim t^{1/2}$. Hence $\langle S_N(t)\rangle$ must be bounded from above by the volume of this sphere, $V(t) \sim t^{d/2}$. A second upper bound on $\langle S_N(t)\rangle$ is $N\langle S_1(t)\rangle$, where:

$$\langle S_1(t)\rangle \sim \begin{cases} t^{1/2} & [d=1] \\ t/\ln t & [d=2] \\ t & [d=3] , \end{cases} \tag{1.4}$$

is the number of distinct sites visited by one random walker. A crossover in $\langle S_N(t)\rangle$ will occur if the system passes from one constraint to the other. For $d=1$, $V(t) < N\langle S_1(t)\rangle$ for all $t$, so no crossover occurs — Regime II holds for arbitrarily large $t$, confirming the result [Eq. (1.3)] above. For $d = 2$, 3, we find $V(t) < N\langle S_1(t)\rangle$ initially, but for sufficiently large $t$, $V(t) > N\langle S_1(t)\rangle$. Thus $t'_\times$ is obtained from the condition:

$$V(t'_\times) \sim N\langle S_1(t'_\times)\rangle . \tag{1.5}$$

For $d = 2$, Eqs. (1.4) and (1.5) lead to $t'_\times \sim N t'_\times/\ln t'_\times$, so that $t'_\times \sim e^N$; this confirms the result [Eq. (1.3)] above. Similarly, for $d = 3$, $(t'_\times)^{3/2} \sim N t'_\times$ implies $t'_\times \sim N^2$, confirming the result [Eq. (1.3)]. One can interpret $t'_\times$ as the time up to which the walkers visit the same places very frequently. For times longer than $t'_\times$, the walkers "almost" do not see each other, and can be treated independently. Thus one would expect the form $S_N(t) \sim N S_1(t)$ under these conditions.

Following the same kind of reasoning, we can generalize the above argument to any spatial dimension $d$. The crossover time to the final regime will be given by:

$$t'_\times \sim N^{2/(d-2)} \quad [d-2]\,. \tag{1.6}$$

This result is a consequence of the fact that $\langle S_1(t)\rangle \sim t$ for any dimension larger than 2. Equation (1.6) shows the effect of the space dimension on $t'_\times$; it shows that when the dimension increases, the walkers become "independent" at shorter times $t'_\times$.

The remarkable feature is the appearance of Regime II. The behavior in Regime I corresponds to the limit [$N \to \infty$, $t$ fixed]; the interface of the set of visited sites is smooth and $S_N(t)$ is easy to understand ($S_N \sim t^d$). The behavior in Regime III corresponds to the opposite limit [$t \to \infty$, $N$ fixed]; the interface of the set of visited sites is extremely rough and $S_N$ is also easy to understand ($S_N \sim NS_1$). In Regime II, the function $S_N$ takes on an unexpected and nontrivial form. The walkers are largely confined to a sphere of radius $\sqrt{t}$ (in contrast to Regime I, where they populate a sphere of radius $t$); the interface of the set of visited sites undergoes a progressive roughening, which is readily apparent on visual inspection of the set of visited sites.

We also carried out numerical calculations for $\langle S_N(t)\rangle$ using both the methods of Monte Carlo and exact enumeration. In particular, we confirmed the scaling form [Eq. (1.2)].[26]

## 2. MODEL FOR SPREADING AND MIGRATION OF POPULATIONS AND EPIDEMICS

For populations and diseases, birth (infection) and death (recovery) are important features that affect their spatial dynamics. To include these features, "branching diffusion" models have been developed earlier and quantities related to the mean number of (infected) species $\langle N(r, t)\rangle$ at position $r$ and time $t$ have been calculated.[27–30] Since, however, the mean number of species gives no information about spatial correlations between two species $\langle N(r, t)\rangle$ cannot describe effects like clustering or segregation that are observed in field experiments.[31–33] Here we review the results[34] for a Fluctuation Reduced Branching Diffusion (FRBD) model, where the correlations can be calculated analytically.

Consider $N_0$ species, initially located at the origin of a $d$-dimensional coordinate system. At $t = 0$, the species start to diffuse independently and death and birth events occur at certain instances of time $t_k$. The increments $\Delta t$ between two successive time instances are exponentially distributed with mean $\langle \Delta t\rangle = 1/\alpha_+ N(t)$. At every instance of time $t_k$, one randomly chosen species is removed with probability $p_d$, and another randomly chosen species is replaced by two species with probability 1. By definition, the birth rate is $\alpha_+$. The death rate is $\alpha_- = p_d\alpha_+ \le \alpha_+$, and the mean number of species increases exponentially $\langle N(t)\rangle = N_0 \exp(\Delta\alpha t)$, where $\Delta\alpha = \alpha_+ - \alpha_-$ is the growth rate.

By construction, the number $N(t)$ of species shows only weak statistical fluctuations around $\langle N(t)\rangle$ (that vanish in the "critical" case, $\alpha_+ = \alpha_-$), which prevents the community from dying out. In contrast, in the usual branching diffusion mode,[27] where death and birth events occur *independently* with rate $\alpha_+$ and $\alpha_-$, there is a nonzero dying-out probability (which is one in the critical case). For the FRBD model, the spatial dynamics of the community can be treated analytically and the correlations between the species can be calculated, at and above, critically.

Spatial correlations between the species show up most clearly in the mean-squared pair distance $\langle r_2^2(t)\rangle \equiv (N(N-1))^{-1}\sum_{i,j}\langle (r_i - r_j)^2\rangle$. The detailed analytical treatment[34] shows that in the critical case ($\Delta\alpha = 0$), $\langle r_2^2(t)\rangle$ can be written as:

$$\langle r_2^2(t)\rangle = \rho_0^2\left[1 - \exp\left(-\frac{t}{\tau_0}\right)\right], \tag{2.1}$$

with a localization radius $\rho_0 \equiv \sqrt{(N_0-1)/\alpha_+}$ and $\tau_0 \equiv (N_0-1)/2\alpha_+$. Equation (2.1) shows that $\langle r_2^2(t)\rangle$ is linear in $t$ for short times $t \ll \tau_0$, while for large times, $t \gg \tau_0$, $\langle r_2^2(t)\rangle$ saturates and becomes identical to $\rho_0^2$. Accordingly, after an initial spreading period, the species stay together in one community, that typically covers a region of radius $\rho_0$ The localization behavior for large times reflects strong correlations that are caused by the natural asymmetry between birth/infection and death/recovery processes (birth and infection occur only next to a species while death and recovery occur everywhere).

These correlations also affect the dynamics of the community as a whole, which can be measured by the mean-squared displacement of the center of mass $\langle r_{\rm cm}^2(t)\rangle \equiv \langle (N^{-1}\sum_i r_i)^2\rangle$. This quantity is simply related to $\langle r_2^2(t)\rangle$ and the mean-squared displacement $\langle r^2(t)\rangle \equiv N^{-1}\sum_i\langle r_i^2\rangle = 1$,

$$\langle r_{\rm cm}^2(t)\rangle = \langle r^2(t)\rangle - \frac{N_0-1}{2N_0}\langle r_2^2(t)\rangle\,. \tag{2.2}$$

Equation (2.2) follows directly from the definitions of the three quantities. By substituting Eq. (2.1) into Eq. (2.2) and expanding the exponential function up to the second order, one can identify three different time regimes in $\langle r_{\rm cm}^2(t)\rangle$ ,

$$\langle r_{\rm cm}^2(t)\rangle = \begin{cases} \dfrac{t}{N_0} & \text{for} \quad t \ll \dfrac{2\tau_0}{N_0-1} \\ \dfrac{N_0-1}{2N_0\tau_0}t^2 & \text{for} \quad \dfrac{2\tau_0}{N_0-1} \ll t \ll \tau_0 \\ t & \text{for} \quad t \gg \tau_0\,. \end{cases} \tag{2.3}$$

Only for very short times, $t \ll \frac{2\tau_0}{N_0-1}$, the center of mass moves as in the case of independently diffusing particles. In the intermediate time regime $\frac{2}{N-1} \ll \frac{t}{\tau_0} \ll 1$, $\langle r_{\rm cm}^2(t)\rangle \propto t^2$ shows ballistic behavior while for long times $\langle r_{\rm cm}^2(t)\rangle \approx \langle r^2(t)\rangle = t$, independent of $N_0$.

Thus we have found that in the critical case ($\Delta\alpha = 0$), the community shows a kind of *migration* behavior for $t > \tau_0$: the mean-squared pair distance becomes constant $\langle r_2^2(t)\rangle = \rho_0^2$, and hence the species are located around their center of mass within a "ball" of radius of $\rho_0$. In this state of "dynamical localization", the community as a whole moves like a single diffusing entity.

For a growing community with $\Delta\alpha > 0$, the mean number of species is not constant, and the localization radius $\rho_0$ in Eq. (2.1) depends on time, $\rho_0^2 \to \rho^2(t) = (\langle N(t)\rangle - 1)/\alpha_+ \simeq (N_0-1)/\alpha_+ \exp(\Delta\alpha t)$. On the other hand, Eq. (2.2) shows that $\langle r_2^2(t)\rangle$ is bounded from above by $2N_0\langle r^2(t)\rangle/(N_0-1) \simeq 2t$ and accordingly $\langle r_2^2(t)\rangle$ shows a second crossover at time $t'_\times$, when $\rho^2(t)$ becomes larger than $2t$,

$$\langle r_2^2(t)\rangle = \begin{cases} 2t & t \ll \tau_0 \\ \rho^2(t) & \tau_0 \ll t \ll t'_\times \\ 2t & t \gg t'_\times\,. \end{cases} \tag{2.4}$$

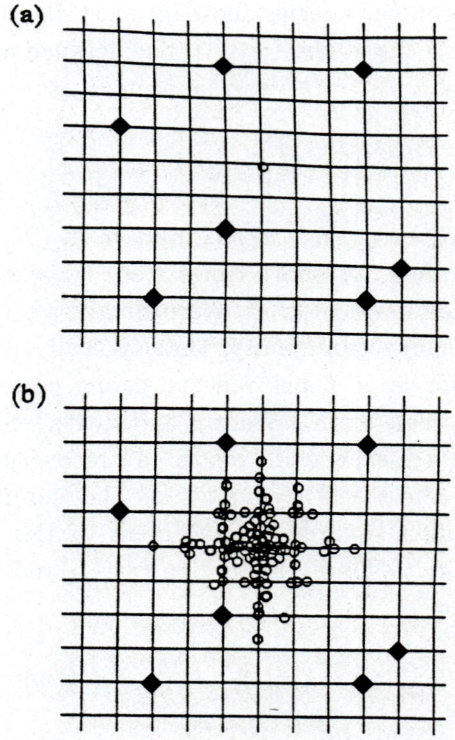


**Fig. 1** Schematic illustration of the trapping problem in which the number of distinct sites enters explicitly into the expression for the survival probability $P_{surv}$. **(a)** A single walker, starting in the center of an infinite square lattice, can be trapped by fixed traps distributed randomly in the plane. **(b)** $N$ walkers starting at the origin in the presence of the same set of randomly distributed traps. In both cases, for short times, $P_{surv} \sim \exp(-\langle S_N \rangle)$.

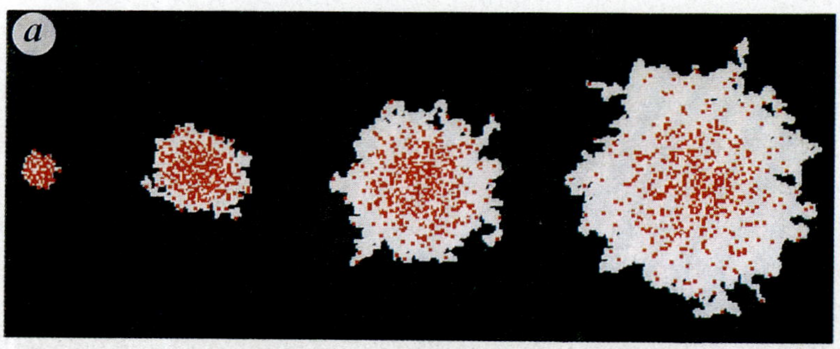


**Fig. 2** Contours of the surface obtained from snapshots at successive times of the territory covered by $N$ random walkers for the case $N = 200$ for a sequence of times in Regime II. Note the roughening of the disc surface as time increases.

Thus for a growing community, the localization occurs only up to times $t'_{\times}$, which is in the order of $\Delta\alpha^{-1}$, and the community starts to spread for $t \gg \Delta\alpha^{-1}$. We therefore have two crossover times $\tau_0$ and $\Delta\alpha^{-1}$ in the non-critical case $\Delta\alpha > 0$.

In summary, we have found two different types of phenomena, *migration* and *spreading*. In the case of *migration*, the mean distance between the species saturates after some time, i.e., the population territory (epidemics area) does not increase, but the territory as a whole moves with time. In the case of *spreading*, the mean distance between the species increases all the time, but the center of the growing territory almost does not move. Both, *migration* and *spreading* phenomena, occur in the spatial behavior of population dynamics and epidemics.[22,35–38]

## ACKNOWLEDGMENTS

We thank the Deutsche Forschungsgemeinschaft and the Alexander von Humboldt-Stiftung for financial support.

## REFERENCES


1. A. Dvoretzky and P. Erdós, in *Proc. 2nd Berkeley Symposium* (University of California, Berkeley, 1951), p. 33.
2. G. H. Vineyard, *J. Math. Phys.* **4**, 1191 (1963).
3. A. Blumen, J. Klafter and G. Zumofen, in *Optical Spectroscopy of Glasses*, ed. I. Zschokke (D. Reidel, New York, 1986), pp. 199–265.
4. R. Czech, *J. Chem. Phys.* **91**, 2498–2504 (1989).
5. P. Bordewijk, *Chem. Phys. Lett.* **32**, 592 (1975).
6. C. A. Condat, *Phys. Rev.* **A41**, 3365 (1990).
7. R. J. Beeler and J. A. Delaney, *Phys. Rev.* **A130**, 926 (1963).
8. H. B. Rosenstock, *Phys. Rev.* **187**, 1166 (1969).
9. R. D. Wieting, M. D. Fayer and D. D. Dlott, *J. Chem. Phys.* **69**, 1996 (1978).
10. E. W. Montroll, in *Stochastic Processes in Applied Mathematics* XVI (American Mathematical Society, Providence RI, 1964) p. 193.
11. E. W. Montroll and G. H. Weiss, *J. Math. Phys.* **6**, 167 (1965).
12. N. C. Jain and S. Orey, *Israel J. Math.* **6**, 373 (1968).
13. N. C. Jain and W. E. Pruitt, *Proc. 6th Berkeley Symposium* **III**, 31 (1971); *J. d'Analyze Math.* **27**, 94 (1974).
14. F. S. Henyey and V. Seshadri, *J. Chem. Phys.* **76**, 5330 (1982).
15. D. C. Torney, *J. Stat. Phys.* **44**, 49 (1986).
16. S. Alexander, J. Bernasconi and R. Orbach, *Phys. Rev.* **B17**, 4311 (1978).
17. K. Heinrichs, *Phys. Rev.* **B22**, 3093 (1982).
18. G. H. Weiss and S. Havlin, *J. Stat. Phys.* **37**, 17 (1984).
19. J. W. Haus and K. W. Kehr, *Physics Reports* **150**, 263–416 (1987).
20. M. V. Smoluchowski, *Z. Phys. Chem.* **29**, 129 (1917).
21. S. A. Rice, *Diffusion-Controlled Reactions* (Elsevier, Amsterdam, 1985).
22. J. G. Skellam, *Biometrika* **38**, 196–218 (1951).
23. J. G. Skellam, *Biometrika* **39**, 346–362 (1952).
24. E. C. Pielou, *An Introduction to Mathematical Ecology* (Wiley-Interscience, New York, 1969).
25. L. Edelstein-Keshet, *Mathematical Models in Biology* (Random House, New York, 1988).
26. H. Larralde, P. Trunfio, S. Havlin, H. E. Stanley and G. H. Weiss, *Nature* **355**, 423 (1992); *Phys. Rev.* **45**, 7128 (1992).
27. K. B. Althreya and P. E. Ney, *Branching Processes* (Springer, Berlin, 1972).
28. S. Asmussen and N. Kaplan, *Stoch. Proc. Appl.* **4**, 1–13 (1976); S. Asmussen and N. Kaplan, *Stoch. Proc. Appl.* **4**, 15–31 (1976).
29. J. D. Biggins, in *Spatial Spread in Branching Processes* (Springer, Heidelberg, 1980).
30. Y. Iwasa and E. Teramoto, *J. Math. Biol.* **19**, 109–124 (1984).
31. C. E. Pielou, *Mathematical Ecology* (Wiley and Sons, New York, 1977).
32. C. E. van de Lande, *Neth. J. Pl. Path.* **99** (Suppl.), **3**, 129–138 (1993).
33. R. Lecoustre, D. Fargette, C. Fauguet and P. de Reffye, *Phytopatology* **79**, 913–920 (1989).
34. M. Meyer, A. Bunde and S. Havlin, preprint.
35. J. B. Ristaino, R. P. Larkin and C. L. Campbell, *Phytopatology* **84**, 1015–1024 (1994).

36. C. L. Bos, H. J. Mandersloot, F. Vader and B. Steenbergen, *J. Pl. Path.* **95**, 225–240 (1989).
37. H. Riehl, *Die Vólkerwanderung* (VMA-Vertiebsgesellschaft, Wiesbaden, 1991); F. Lot, *Les invasions germaniques* (Payot, Paris 1945).
38. R. M. May, *Stability and Complexity in Model Ecosystems* (Princeton University Press, Princeton, New York, 1973).

Fractals, Vol. 4, No. 2 (1996) 169–174


# GIBBS-THOMSON AND OTHER QUANTITATIVE RELATIONS IN DENDRITIC CRYSTAL GROWTH

A. TANAKA and T. YOKOYAMA
*Department of Electrical and Inf. Engineering,*
*Yamagata University, 4-3-16 Jonan, Yonezawa, 992, Japan*

**Abstract**

The Gibbs-Thomson relation is well-known in crystal growth. That implies that the temperature of interface is not equal to the melting temperature due to its surface energy. It is, however, a qualitative relation near equilibrium, not clear not only qualitatively but also quantitatively far equilibrium. One of the authors has measured quantitative relation in experimental dendritic, i.e., far equilibrium, crystal growth. We also carried out numerical simulations using Phase Field Model, and obtained new relation. We will discuss this in detail here.

## 1. INTRODUCTION

There are many geometrical patterns formed in various physical systems. Their processes of formation are attractive problems in physical science. In particular, snow crystals have fascinated us since long ago. We have been able to treat the subject experimentally during the last tens of years. The fast development of computers and numerical techniques enabled us to simulate the phenomena. However, the problem of dendritic growth is one of free boundary problems, which are analytically intractable and difficult to solve even with numerical techniques. Hence several kinds of numerical models have been proposed in recent years.

Since it is well known that fully non-local calculation takes huge time (see Ref. 1 for example), most of the models are local ones; e.g., the Geometric model[2] and the Boundary layer model.[3] These models have succeeded in simulating the formation of needle and dendritic crystals, and clarified many important effects. Nevertheless, they are all local models, and they lack global information of the system. As we know, a developed dendrite often has many sidebranches and complicated shapes, and local models do not work in these cases as they cannot treat the interaction between the branches to be concrete. Therefore we need a non-local model in those situations, and here we use a simplified non-local model called "Phase Field Model".

## 2. DEFINITION OF THE MODEL

Phase Field Model of crystal growth has previously been proposed by R. Kobayashi, who presented splendid pictures of dendrites.[4–6] His results can be summarized as follows:

- If there is no anisotropy, the growth processes isotropically and each branch sometimes splits (tip-splitting).
- If we introduce anisotropy, the branches grow selectively in the direction of anisotropy (4-fold or 6-fold) and the growth in other directions is suppressed. This effect becomes more remarkable as the degree of anisotropy increases.

The whole results give us the importance of anisotropy in the mechanism of the dendritic growth. Moreover his scheme has several advantages as compared with other numerical schemes.

In Phase Field Model of 2-dimensional crystal growth, two field variables are introduced; namely, a phase field $p(x,\, t)$ and a temperature field $T(x,\, t)$. The former variable $p$ is an ordering parameter, and $p = 0$ and 1 means liquid and solid states, respectively. The model is derived from the conservation law of enthalpy with the following Ginzburg-Landau type free energy including $m$ as a parameter,

$$\Phi[p;\, m] = \int \frac{1}{2}\, \varepsilon^2 |\nabla p|^2 + F(p;\, m) dx \tag{1}$$

$$F(p;\, m) = \frac{1}{4} p^4 - \left(\frac{1}{2} - \frac{m}{3}\right) p^3 + \left(\frac{1}{4} - \frac{m}{2}\right) p^2 \tag{2}$$

where $\varepsilon$ is a small parameter representing the thickness of interface and $|m| < 1/2$. The resulting equations are written as follows:

$$\tau \frac{\partial p}{\partial t} = -\frac{\partial}{\partial x}\left(\varepsilon \frac{\partial \varepsilon}{\partial \theta} \frac{\partial p}{\partial y}\right) + \frac{\partial}{\partial y}\left(\varepsilon \frac{\partial \varepsilon}{\partial \theta} \frac{\partial p}{\partial x}\right) + \nabla \cdot (\varepsilon^2 \nabla p) + p(1-p)\left(p - \frac{1}{2} + m\right) \tag{3}$$

$$\frac{\partial T}{\partial t} = \nabla^2 T + K \frac{\partial p}{\partial t} \tag{4}$$

where $K$ is the dimensionless latent heat. The anisotropy is included in $\varepsilon$ as follows:

$$\varepsilon = 1 + \delta \cos j\theta \tag{5}$$

where $j$ is a mode number of anisotropy.

One of us has already obtained experimental results using $NH_4Cl$ aqueous solution. $NH_4Cl$ has 4-fold symmetric anisotropy and we denote the direction of anisotropy $\langle 100 \rangle$. As is well-known, it also has an anisotropy in the $\langle 110 \rangle$ direction. For consistency with the experiment, the anistropy in that direction should be taken into account, and thus we introduced the anisotropy as follows:

$$\varepsilon = 1 + \delta_1 \cos 4\theta + \delta_2 \cos 8\theta \,. \tag{6}$$

We varied 4-fold anisotropy parameter $\delta_1$ and 8-fold one $\delta_2$.

## 3. NUMERICAL RESULTS AND DISCUSSION

As has been previous reported,[7] we can obtain parabolic and oscillatory dendrites by varying the parameters, $\delta_1$, $\delta_2$ and $K$.

The tip velocities and the shapes of dendrite vary according to change of supercooling $\Delta T$. In the experiments of $NH_4Cl$, $NH_4Br$ and others, it has been confirmed that the tip velocity is proportional to the square of the tip curvature.[8] We show our numerical results in Fig. 1 and find that the tip radius is proportional to the 5th power of tip velocity. It is quite different from the experimental results.

We also investigated the temperature field around dendrite. It is important because a dendrite grows in the temperature field. In Fig. 2, we show dendrite with isothermal lines. The spacing between the isothermal lines becomes narrow near the tip. We can obviously see one isothermal line cross the interface, so the temperature at the tip is not equal to the melting temperature.

How does the difference depend on the parameters? This reminds us of the Gibbs-Thomson relation, which says that the temperature of the tip is not equal to the melting

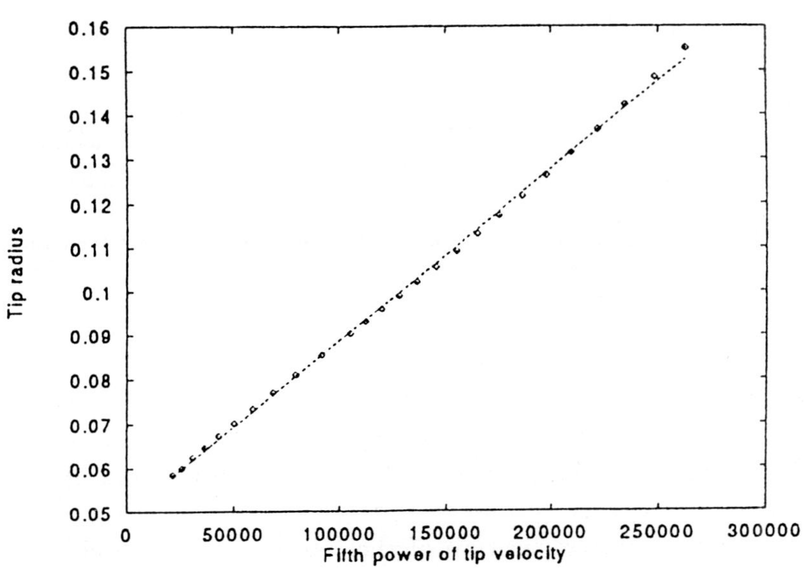


**Fig. 1** The relation between tip radius and velocity.

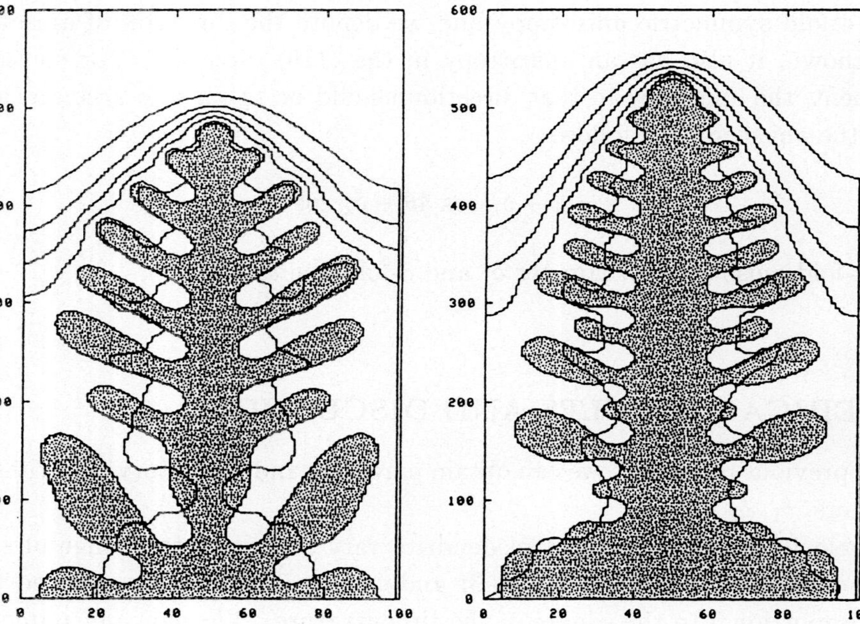


(a) Oscillating dendrite (b) Parabolic dendrite

**Fig. 2** Dendrite with isothermal lines.

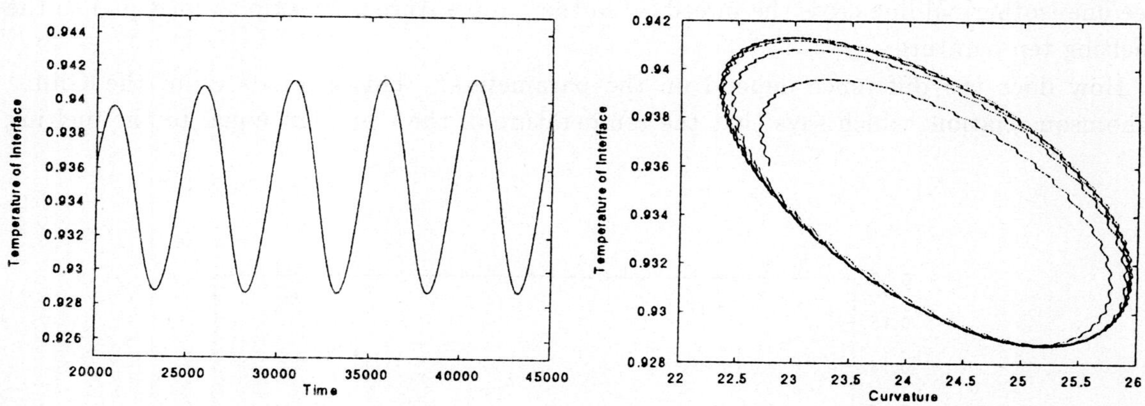


(a) (b)

**Fig. 3** The change of temperature of the interface vs. time **(a)** and tip curvature **(b)**.

temperature because of its surface energy. Since it is a qualitative relation near equilibrium originally, it has never confirmed for far equilibrium dendrite. One of us has already measured the concentration field quantitatively in experiments[9] and obtained the Gibbs-Thomson relation.

In oscillatory growth, the temperature of the interface is also oscillating (Fig. 3). In parabolic growth in our simulation, we examined the quantitative relation as shown in Fig. 4. In this figure, the difference between the tip and melting temperatures $(T_M - T_i)$

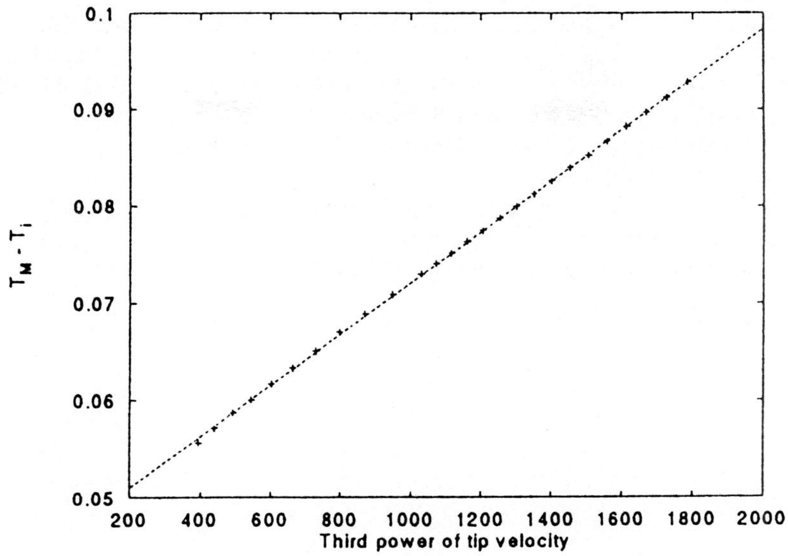


**Fig. 4** The change of the difference between temperature of the tip and melting temperature.

is proportional to the 3rd power of tip velocity. We can see that it is not consistent with experiments and the Gibbs-Thomson relation does not hold in our simulation.

Our quantitative numerical results are both inconsistent with the experiments. To investigate the reason, we now think about the processes of crystal growth. In melting growth, the dominant process is the elimination of latent heat, i.e., how fast it is possible to exclude latent heat around the interface. On the other hand, in solution growth the solute is rare compared with melting growth, and for the solute there is some time to migrate on the interface. Thus we also have to take into account the diffusion of solute and interfacial kinetics. Since the original Phase Field Model is a 2-field model, temperature field and phase field, we have to consider the diffusion of solute or interfacial kinetics for modification. In concrete terms, one possible idea is a 3-field model including the concentration field, and the other is to modify the free energy of Eq. (2) suitable for experiments. We do not know which is better or whether other ideas are needed. We will propose a modified model and clarify that point in the near future.

## 4. SUMMARY

In summary, we carried out numerical simulations of dendritic growth using a model called Phase Field Model, and obtained quantitative relations. In our simulation, the tip radius of the dendrite is proportional to the 5th power of the tip velocity. Secondarily we could not confirm the Gibbs-Thomson relation at far equilibrium numerically. The above numerical results are both inconsistent with experiments. So we need to modify our model.

## ACKNOWLEDGMENTS

We thank R. Kobayashi, Y. Sawada and M. Sano for helpful discussions and suggestions.

## REFERENCES

1. Y. Saito, G. Goldbeck-Wood and H. Müller-Krumbhaar, *Phys. Rev. Lett.* **58**, 1541 (1987).
2. R. C. Brower, D. A. Kessler, J. Koplik and H. Levine, *Phys. Rev. Lett.* **51**, 1111 (1983).
3. E. Ben-Jacob, N. Goldenfeld, J. S. Langer and G. Schon, *Phys. Rev. Lett.* **51**, 1930 (1983).
4. R. Kobayashi, *RIMS kokyuroku* **614**, 39 (in Japanese, 1987).
5. R. Kobayashi, *Simulations on Dendrite Growth*, videotape (1990).
6. R. Kobayashi, *Physica* **D63**, 410 (1993).
7. A. Tanaka and R. Kobayashi, *Sup. Int. J. Appl. Electromag. Mat.* 521 (1992).
8. A. Tanaka, Doctoral thesis, Tohoku University (in Japanese, 1991).
9. A. Tanaka and M. Sano, *J. Crystal Growth* **125**, 59 (1993).

Fractals, Vol. 4, No. 2 (1996) 175–180


# THE ELECTRIC PROPERTIES OF SINTERED CERMET AS A CONTINUUM PERCOLATION MEDIUM

KATSUKUNI YOSHIDA*
*Institute of Atomic Energy, Kyoto University, Uji, Kyoto 611, Japan*

**Abstract**

The percolation process in the sintered Nb-$Al_2O_3$ cermet has been investigated by measuring the statistical behavior of the metallic cluster and the critical exponent of the conductivity. The present continuum system has been found to exhibit a percolation character similar to that of the conventional discrete-lattice percolation systems.

## 1. INTRODUCTION

Many computer simulations for the conventional discrete-lattice percolation systems have revealed that the electric properties such as the conductivity $\sigma$ and the dielectric constant $\varepsilon$ obey the power-law with exponents $t$ and $s$:

$$\sigma \propto (p - p_c)^t \text{ for } p > p_c \quad \text{and} \quad \varepsilon \propto (p_c - p)^{-s} \text{ for } p < p_c\,, \tag{1}$$

where $p$ is the concentration of conducting elements (occupied sites or open bonds on the lattice) and $p_c$ is the percolation threshold. The exponents have been found to have values $t = 2.0$ and $s = 0.7 \sim 0.73$ for the three-dimensional space (3D) and $t = s = 1 \sim 1.3$ for 2D. These exponents are believed to possess universality. It has been an interesting controversy as to what extent the percolation process in the continuum systems made of

*Present address: Faculty of Energy Science, Kyoto University, Uji, Kyoto 611, Japan.

realistic materials resembles that in the computer-generated discrete systems. However, there seems to be far less relevant information for the continuum percolation systems than for the discrete percolation systems. This situation is largely due to technical problems in fabricating well-defined samples and also to difficulty in observing the realistic clusters in three-dimensional (3D) space.

With this problem in mind, we have investigated the percolation process in a metal-ceramics sintered mixture (cermet) as a continuum percolation system different from the Swiss-cheese type,[1,2] in particular reference to the critical behavior of the conductivity and the behavior of the cluster distribution.

## 2. EXPERIMENTS

The starting materials in fabricating samples were powders of Nb and $\alpha$-$Al_2O_3$ of particle size of several tens of micrometers, and the powders were mixed thoroughly in a V-shaped mixer. Both the particles of Nb and $Al_2O_3$ were almost the same in size and shape. By virtue of this, they were mixed so well that no special structure was formed other than a homogeneously random structure. The mixture was then hot-pressed under a uniaxial pressure of $4.2 \times 10^7$ Pa at 1200°C in a vacuum of $1.3 \times 10^{-3}$ Pa. The resultant ingots were cut into the shape of a bar or slab. The distribution of the Nb clusters was measured with the aid of an image processing system consisting of an image processor, video camera, microscope and a host computer.

## 3. RESULTS AND DISCUSSION

Figure 1 exhibits a statistical behavior of the cluster distribution of the Nb clusters when $f$ (volume fraction) increases; the cluster stated here is a cross section of the real cluster in 3D.[3] The ordinate represents the relative probability that a particle of unit size belongs to a cluster with a perimeter ratio $l/l_o$, where $l$ is the perimeter of a chosen cluster and $l_o$ is the perimeter of the corresponding circle with the same area as that of the cluster. This ratio $l/l_o$ is a simple but efficient representation of the morphological complexity of the clusters, and a larger $l/l_o$ indicates a more complex geometry. The growth process of the clusters is achieved by interlink between adjacent precursory clusters, and this interlinking makes the shape of the resultant cluster more complex. Therefore, the distribution in Fig. 1 well depicts the growth process. By using the image processor system we measured the above statistics over 1000 $\sim$ 3000 clusters. At low volume fraction below 0.15, the distribution is narrow and simple in shape. However, a trivial increase in $f$ from 0.15 to 0.166 causes a considerable change in the distribution, giving a sudden broadening. This change means that the cluster growth by interlinking becomes remarkable when $f$ exceeds 0.15, and implies the occurrence of the percolation transition (in 3D space) between $0.15 < f < 0.166$. This observation shows that the 3D percolation transition can be detected even from the 2D profiles of the real clusters.

This percolation transition can clearly be observed by measuring the electric conductivity, as shown in Fig. 2, where the samples are in the shape of a bar with typical dimensions $5 \times 5 \times 25$ mm$^3$, and the current was passed in a direction perpendicular ($\perp$) to the uniaxial

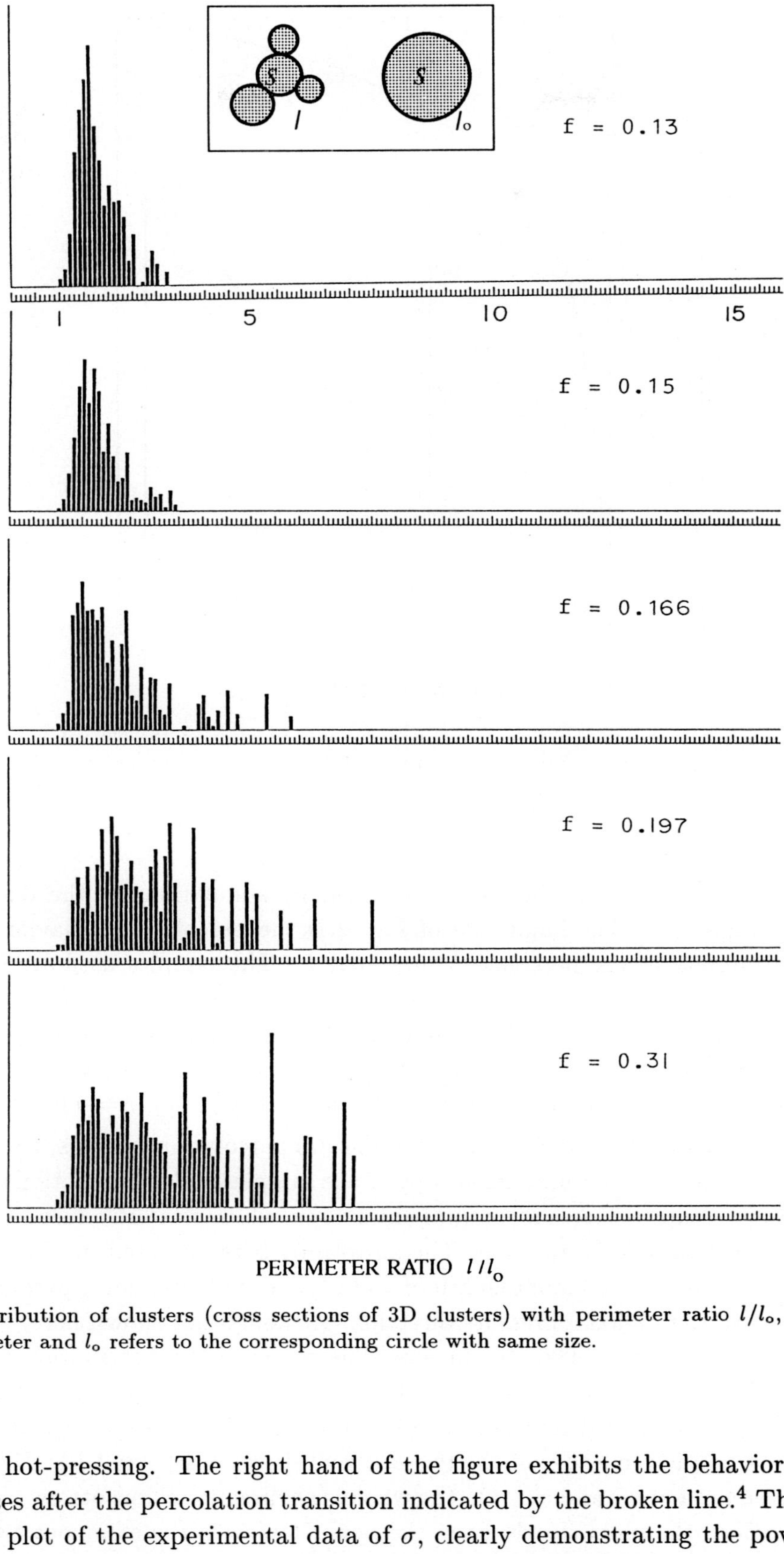


**Fig. 1** Distribution of clusters (cross sections of 3D clusters) with perimeter ratio $l/l_o$, where $l$ is the cluster perimeter and $l_o$ refers to the corresponding circle with same size.

pressure in hot-pressing. The right hand of the figure exhibits the behavior of $\sigma$, and it abruptly rises after the percolation transition indicated by the broken line.[4] The inset is the logarithmic plot of the experimental data of $\sigma$, clearly demonstrating the power-law. The exponent $t$ and the threshold $f_c$ (volume fraction corresponding to $p_c$) were determined simultaneously by the least squares fitting,[4] as $t = 2.06$ and $f_c = 0.154$. This result

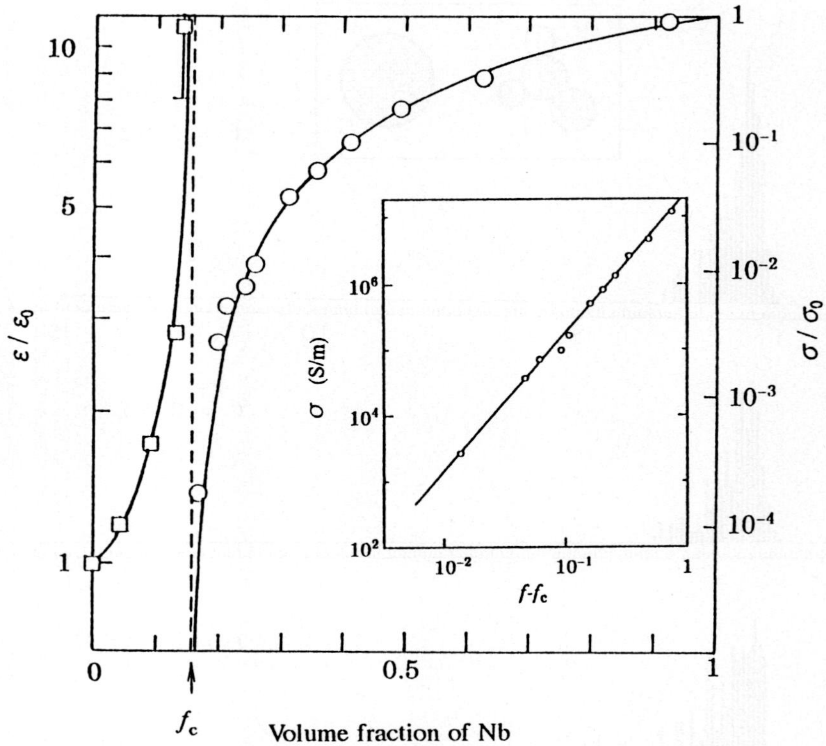


**Fig. 2** The conductivity $\sigma$ and the dielectric constant $\varepsilon$ as a function of the Nb volume fraction $f$; the broken line indicates the percolation transition, and the inset shows the power-law $\sigma \propto (f - f_c)^t$.

is in excellent agreement with the values obtained by simulation for the discrete-lattice percolation systems. The left hand side of Fig. 2 is the percolative behavior of $\varepsilon$, rising steeply toward infinity at the percolation threshold $f_c$. The samples used are in the shape of a slab of size $9 \times 9 \times 3.5$ mm$^3$. The metallic phase has infinite dielectric constant, because the external electric field is completely shielded by the free electron inside the cluster. Therefore, the growth of metallic cluster leads to increase in $\varepsilon$. Although not shown here, the logarithmic plot of $\varepsilon$ also illustrates the power-law giving $s = 0.6 \sim 0.7$.

To investigate further the similarity between the two percolation systems, we consider the relationship between the structural anisotropy of the samples and the directional anisotropy of $\sigma$. The present samples have a structural anisotropy, which was introduced by a uniaxial press applied in the sample fabrication. The anisotropy is represented by the ratio $d_\perp/d_{//}$, where $d_\perp$ and $d_{//}$ are the maximum length of a cluster in two directions; perpendicular ($\perp$) and parallel ($//$) to the uniaxial pressure direction. Although not shown here, we measured the ratio $d_\perp/d_{//}$ over $1000 \sim 3000$ Nb clusters. The center of the distribution was found to be located at about $d_\perp/d_{//} = 1.5$, almost independently of $f$. To measure the directional anisotropy of $\sigma$, we prepared another set of samples in the shape of a slab, where the supplied current was fed in either of the two different directions ($\perp$ and $//$). Figure 3 shows the experimental data of $\sigma_\perp$ and $\sigma_{//}$. On the whole, the directional anisotropy is found to be $\sigma_\perp/\sigma_{//} \approx 2.5$. Both the plots clearly obey the power-law with the exponent $t \approx 2$, though exhibiting a difference between the percolation thresholds for two directions; $f_c = 0.154$ for $\sigma_\perp$ and 0.185 for $\sigma_{//}$.

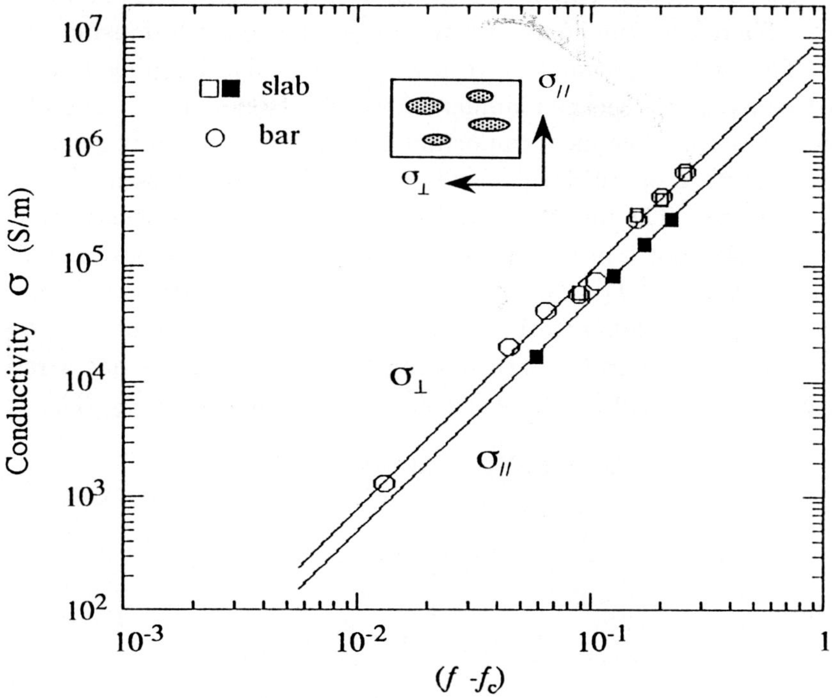


**Fig. 3** Logarithmic plot of the anisotropic conductivities depicting the power-law.

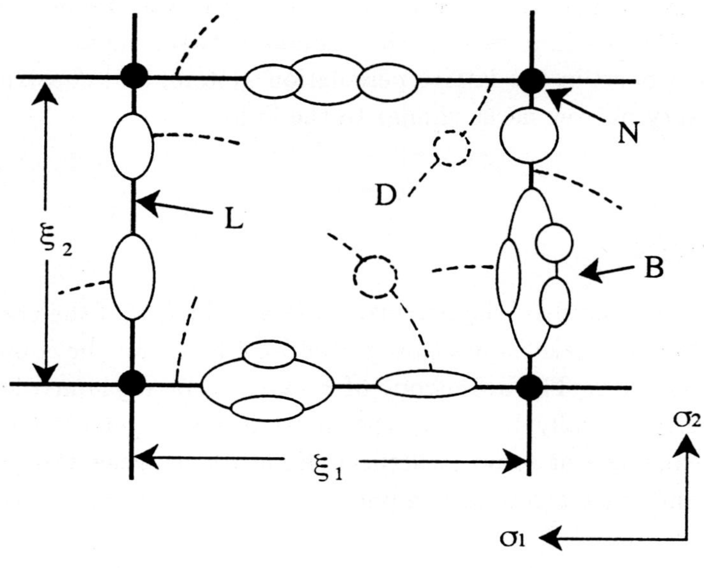


**Fig. 4** An anisotropic network (3D) consisting of links (L), blobs (B), dangling branches (D) and nodes (N), with dimensions $\xi_1 \times \xi_1 \times \xi_2$.

In the discrete-lattice system, the percolation cluster is found to be reduced to a simplified network structure, as illustrated in Fig. 4, which is known as the links-nodes-blobs model.[5–7] The network consists of cells with a diameter of order of the correlation length $\xi$, and each cell consists of the nodes (N), links (L), blobs (B) and the dangling branches (D).

The dangling branches do not contribute to the conductivity, since the local current does not pass there. Therefore, the conductivity has no straight relationship with the fractal dimension $d_f$. This will be evident if we consider that the dangling branch is the most important part constructing the self-similar structure of the cluster. The blob is a bundle of ramified current-paths. The most important constituent is the link consisting of a single open-bond, since cutting off this will result in an overall breakdown of the current path between adjacent nodes. The number of links per length of $\xi$ has been proved[7] to be given by $N \propto (f - f_c)^{-1}$. As shown in Fig. 4, we consider an anisotropic network ($3D$) consisting of unit cells with dimensions $\xi_1 \times \xi_1 \times \xi_2$, and further approximate as $(f - f_{c1}) \approx (f - f_{c2})$. The conductivity is proportional to the density of the nodes, so we have $\sigma_1 \propto (\xi_1\xi_2)^{-1}$ and $\sigma_2 \propto (\xi_1\xi_1)^{-1}$. The single current-path passing through the nodes has a resistance $r$ per unit length, which is governed by the links: $r_1 \propto N/\xi_1$ and $r_2 \propto N/\xi_2$. Therefore we have:

$$\sigma_1 \propto 1/(\xi_2 N) \quad \text{and} \quad \sigma_2 \propto \xi_2/(\xi_1^2 N)\,. \tag{2}$$

These give

$$\sigma_1/\sigma_2 \approx (\xi_1/\xi_2)^2 \approx (d_1/d_2)^2\,. \tag{3}$$

Thus we obtain

$$\sigma_\perp/\sigma_{//} \approx (d_\perp/d_{//})^2 \approx (1.5)^2 \approx 2.3\,, \tag{4}$$

which is in good agreement with the experimental observation ($\sigma_\perp/\sigma_{//} \approx 2.5$) stated before. Therefore, our continuum percolation medium of the sintered cermet is well described by the links-blobs-nodes model, though it is based on the lattice percolation systems. This result together with the experimental observation of $f_c$, $t$ and $s$ demonstrates the strong resemblance between our sintered percolation system and the discrete-lattice percolation systems. This resemblance implies that the contact between metallic grains plays a role like that of the open bond in the lattice percolation system, and constructs a percolation cluster containing very narrow necks similar to the links.

## 4. CONCLUSION

In the Nb-$Al_2O_3$ sintered cermet, the percolation threshold $f_c$ and the critical exponents of $\sigma$ and $\varepsilon$ have been found to take values close to those obtained for the conventional discrete-lattice percolation systems. The anisotropy of $\sigma$ can also be explained by the links-blobs-nodes model, which is initially based on the discrete-lattice percolation systems. These results imply the existence of some structural similarity between the present continuum percolation media and the discrete-lattice ones.

## REFERENCES


1. B. I. Halperin, S. Feng and P. N. Sen, *Phys. Rev. Lett.* **54**, 2391 (1985).
2. S. Miyazima, K. Maruyama and K. Okumura, *J. Phys. Soc. Jpn.* **60**, 2805 (1991).
3. K. Yoshida, Y. Tomii and S. Ueda, *Jpn. J. Appl. Phys.* **27**, 2224 (1988).
4. K. Yoshida, *J. Phys. Soc. Jpn.* **59**, 4087 (1990).
5. A. Skal and B. I. Shklovskii, *Soviet Phys. Semicond.* **8**, 1029 (1975).
6. H. E. Stanley, *J. Phys.* **A10**, L211 (1977).
7. A. Coniglio, *Phys. Rev. Lett.* **46**, 250 (1981).

Fractals, Vol. 4, No. 2 (1996) 181–185


# VISCOUS FINGERING PATTERNS IN HIGH MOLECULAR WEIGHT POLYMER SOLUTIONS

MASAMI KAWAGUCHI*, KYOKO MAKINO and TADAYA KATO
*Department of Chemistry for Materials, Faculty of Engineering,*
*Mie University, 1515 Kamihama, Tsu, Mie 514 Japan*

**Abstract**

Characteristics of viscous fingering patterns in polymer solutions were investigated by introducing air into a radial Hele-Shaw cell, in which high molecular weight Hydroxypropyl Methyl Cellulose (HPMC) solution was placed. HPMC chains form a chain entanglement easily, which leads to an increase in elasticity. The resulting patterns showed a drastic morphological change from highly branched patterns to tip-splitting patterns to skewering patterns, with an increase in injection pressure. Such a crossover in the fingering patterns could be well correlated with the characteristic quantities of pattern growth, such as the finger width and tip velocity.

## 1. INTRODUCTION

During the past decade, great attention has been focused viscous fingering patterns in the Hele-Shaw cell from both the experimental and theoretical aspects.[1–16] Presently, while the viscous fingering problem for simple Newtonian fluids is well understood, our understanding of non-Newtonian fluids are far from complete. The reason for our limited knowledge could be attributed to richer pattern dynamics, a variety of patterns, and complex morphological changes.

*To whom correspondence should be addressed.

Recently, experiments using polymer solutions have been carried out by several research groups.[3,12–14,16] Among them, Zhao and Maher[13,14] studied the fingering experiments for aqueous solutions of Polyethylene Oxide (PEO) with the hydrophobic end caps and hydroxyl groups in a radial Hele-Shaw cell injected with water. A transition from a fingering to fracturing pattern for the hydrophobically terminated PEO solution was obtained with an increase in the injection rate of water, whereas the fingering-fracturing transition was not observed for homopolymers.

More recently, Makino, Kawaguchi, Aoyama, and Kato[16] reported a morphological change from a highly branched to a tip-splitting pattern in aqueous Hydroxypropyl Methyl Cellulose (HPMC) solutions pushed by air in a radial Hele-Shaw cell, with an increase in isopropyl alcohol, accompanied by a decrease in surface tension and an increase in viscosity. In this paper, in order to investigate the injection pressure effect on the fingering patterns, we report fingering experiments using aqueous solutions of high molecular weight HPMC in a radial Hele-Shaw cell.

## 2. EXPERIMENTAL SECTION

### 2.1 Materials

A HPMC-1 sample (kindly supplied by Shin-Etsu Chemical company) was purified by the method described previously.[17] Its molecular weight was determined to be $738 \times 10^3$ by intrinsic viscosity measurements in a 0.1 M aqueous NaCl solution at 25° C.[17] HPMC-1 was dissolved in water at a fixed concentration of 0.5 g/100 mL. In order to obtain the high-contrast patterns, the HPMC solutions were colored by adding methylene blue.

### 2.2 Rheometer

Rheological measurements of the HPMC-1 solutions were performed using a Rheometrics fluid spectrometer RFS-II with a cone-plate geometry (plate diameter: 50 mm; cone angle: 0.04 rad; gap: 0.05 mm). The temperature of the sample chamber was maintained at $25 \pm 0.5°$ C.

### 2.3 Hele-Shaw Cell

The fingering experiments were performed by using the same Hele-Shaw cell as reported before.[16] Air was injected through a small hole (of 0.2 cm diameter) at the center of the top plate under different pressures of 1.0–5.0 kPa, which are applied by the hydrostatic pressure generated using a water filled cylinder with the air in a cup. Measuring temperature was fixed at $25 \pm 0.5°$ C. Records of the generating patterns were taken by a CCD video camera/recorder. The images of the recorded patterns were printed by an image printer and analyzed by a Himawari-60 digital image analyzer (Library Co. Tokyo).

## 3. RESULTS AND DISCUSSION

Figure 1(a) shows the fingering patterns of the HPMC-1 solutions as a function of injection pressure. The resulting finger patterns change drastically with an increase in the injection

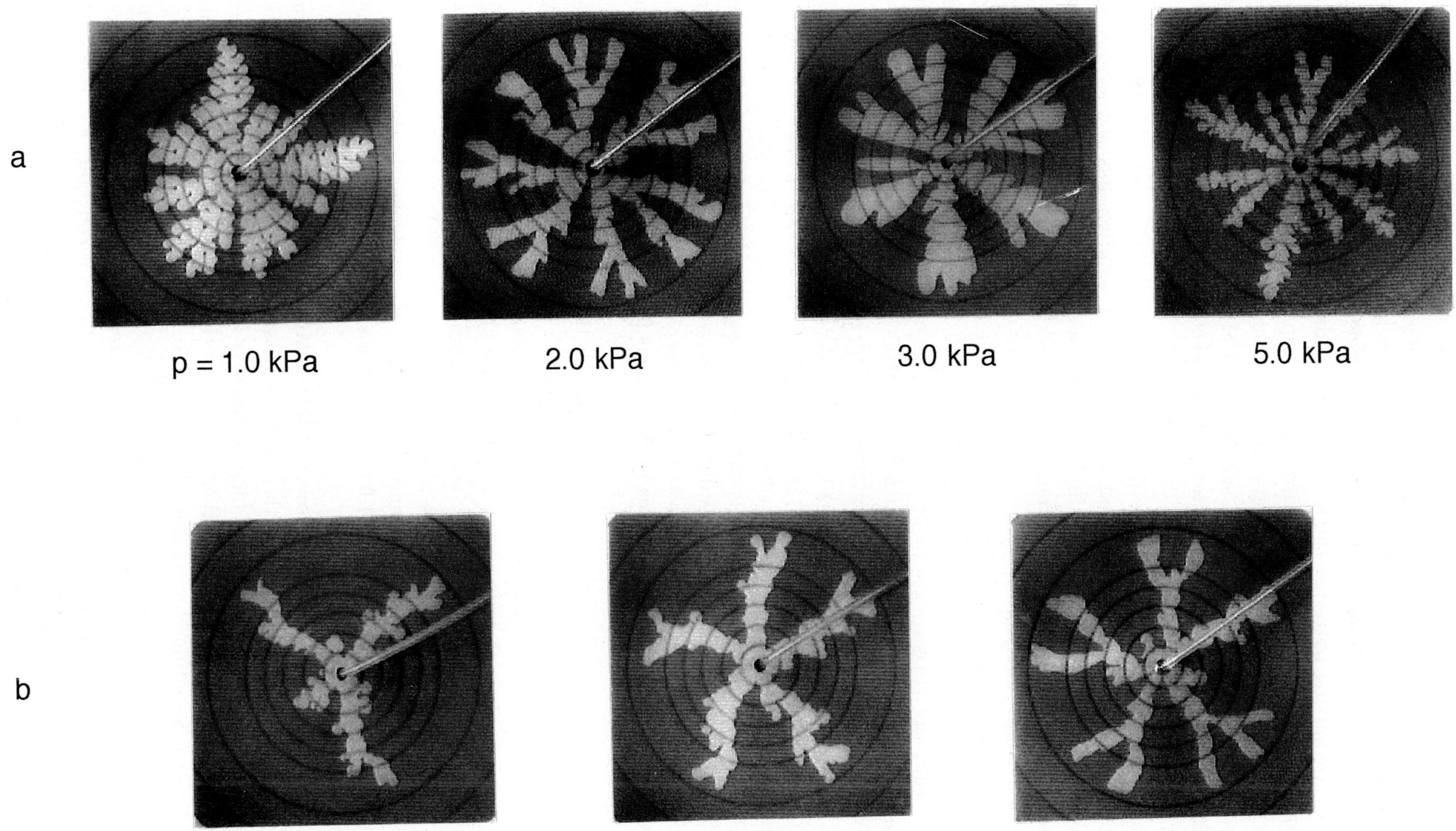


p = 1.0 kPa 2.0 kPa 3.0 kPa 5.0 kPa

p = 1.2 kPa 1.3 kPa 1.5 kPa

**Fig. 1** Morphology changes in various fingering patterns of the 0.5 wt. % HPMC-1 solution as a function of the injection pressure. The concentric circles in the photographs have the radii of 1, 2, 3, 5, 7, and 9 cm, respectively.

pressure. At 1.0 kPa, the pattern is typically highly branched and the finger tip advances with oscillation, which is similar to that observed previously.[16] At 2.0 and 3.0 kPa, the patterns are characteristically tip-splitting: the finger tip proceeds with the less rounded shape. At 5.0 kPa, the mode of the finger growth is quite different from both the highly branched and tip-splitting patterns: the growing tip broadens and flattens and around at the middle of flattening tip, a bump develops and grows up; we denote this as a skewering pattern. Thus, such changes in the fingering patterns are the first observation of homopolymer solutions. In contrast, a 1.5 g/100 mL solution of HPMC with the molecular weight of $251 \times 10^3$ shows only highly branched patterns at all injection pressures ranging from 1.0 to 5.0 kPa.[18]

The difference in the injection pressure dependence of the pattern may be related to the rheological properties of bulk solution. The HPMC-1 solution has a viscosity of 0.107 Pa·s at low shear rate and a shear-thinning onset at a shear rate above 4.0 $s^{-1}$ due to a chain entanglement. Thus, it will be expected that the chain entanglement, which contributes to the network junctions, may lead to an exponent of less than 2 for the frequency dependence of the dynamic storage modulus ($G'$). On the other hand, the low molecular weight HPMC solution has a zero-shear viscosity of 0.140 Pa·s at shear rates lower than 50 $s^{-1}$; above this, the viscosity gradually decreases with an increase in the shear rate.[16] Moreover, the $G'$ value is smaller than that for the HPMC-1 solution.[16]

In order to further check the crossover injection pressure for the pattern transition of the highly branched pattern to the tip-splitting one in HPMC-1 solutions, we have performed additional experiments at different injection pressures from 1.0 to 2.0 kPa. Figure 1(b) shows the typical fingering patterns at several injection pressures. When the injection pressure is larger than 1.3 kPa, the pattern observed is the tip-splitting one. At an injection pressure of 1.2 kPa, both highly branched and tip-splitting patterns are observed, but there is not observed only either a highly branched pattern or a tip-splitting pattern in several runs. To our knowledge, the coexisting two modes in the viscous fingering patterns has not been observed in homopolymer solutions.

It is interesting to note the crossover regimes for the transition of the patterns were also observed in clay suspensions,[11] in associating polymer solutions,[14] and in foams.[15] A crossover from a fingering to a fracturing pattern was observed for the clay and associating polymer experiments with injection water rate. Both clay suspensions and associated polymer solutions behaved as a viscoelastic body under shear deformation. In contrast, for the foam experiments a pattern transition from a jagged to a tip-splitting character was observed with an increase in injection pressure. The morphological changes in the HPMC-1 solution over the injection pressure range 1.0 to 3.0 kPa, seem to be similar to those in the foam experiments.

In order to understand the pattern crossover of the HPMC-1 solution, it is attempted to measure the tip velocity and the finger width. The data are summarized in Fig. 2 with phase boundaries among the three morphological phases. Changes in the tip velocity and the finger width correlate well with the morphological crossover. For the injection pressures where the tip-splitting pattern formation is dominant, the tip velocity as shown in Fig. 2(a) is nearly constant. When the skewering pattern is observed, the tip velocity increases rapidly. On the other hand, the finger width displayed in Fig. 2(b) increases with the injection pressure up to 3.0 kPa, above which is drastically decreases and reaches a constant, accompanied with the morphological changes from the tip-splitting pattern to the skewering pattern.

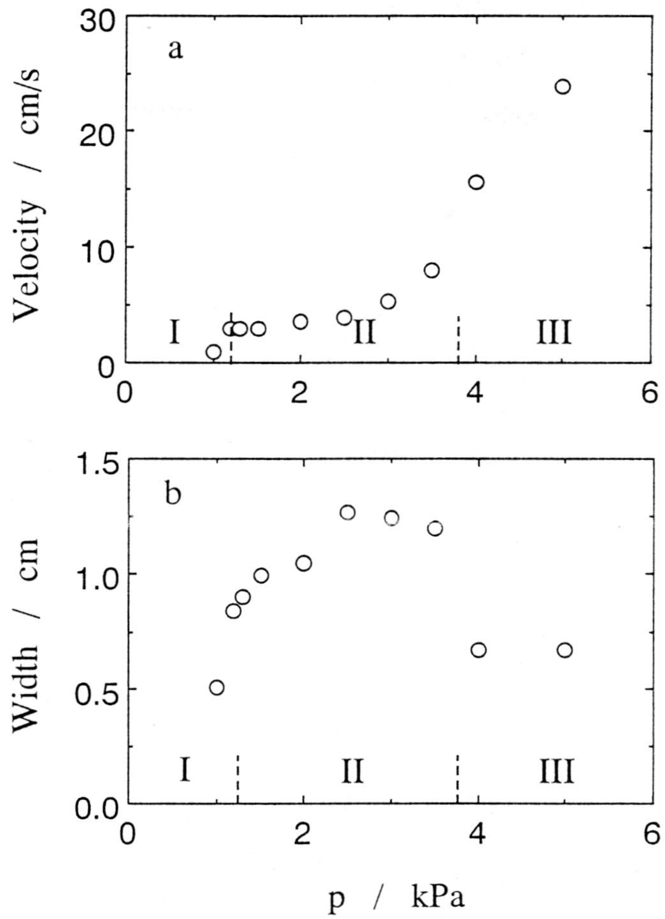


**Fig. 2** Tip velocity **(a)** and finger width **(b)** of the fingering pattern growth of 0.5 wt. % HPMC-1 solution as a function of the injection pressure. The dashed lines indicate the phase boundaries among three morphological phases, such as highly branched (I), tip-splitting (II), and skewering (III) patterns.

## REFERENCES


1. E. Ben-Jacob, R. Godbey, N. D. Goldenfeld, J. Koplik, H. Levine, T. Muller and L. M. Sander, *Phys. Rev. Lett.* **55**, 1315 (1985).
2. V. Horvath, T. Vicsek and J. Kertesz, *Phys. Rev.* **A35**, 2353 (1987).
3. J. Nittmann, G. Daccord and H. E. Stanley, *Nature* **314**, 141 (1985).
4. A. Buka, J. Kertesz and T. Vicsek, *Nature* **323**, 424 (1986).
5. Y. Couder, O. Cardoso, D. Dupuy, P. Tavernier and W. Thom, *Europhys. Lett.* **2**, 437 (1986).
6. S. Liang, *Phys. Rev.* **A33**, 2663 (1986).
7. J. D. Chen, *Exp. Fluids* **5**, 363 (1987).
8. G. Zocchi, B. E. Shaw, A. Libchaber and L. P. Kadanoff, *Phys. Rev.* **A36**, 1894 (1987).
9. S. K. Sarkar and D. Jasnow, *Phys. Rev.* **A39**, 5299 (1989).
10. E. Ben-Jacob and P. Garik, *Nature* **343**, 523 (1990).
11. E. Lemaire, P. Levitz, G. Daccord and H. Van Damme, *Phys. Rev. Lett.* **67**, 2009 (1991).
12. D. E. Smith, X. Z. Wu, A. Libchaber, E. Moses and T. Witten, *Phys. Rev.* **A45**, R2165 (1992).
13. H. Zhao and J. V. Maher, *Phys. Rev.* **A45**, R8328 (1992).
14. H. Zhao and J. V. Maher, *Phys. Rev.* **E47**, 4278 (1993).
15. S. S. Park and D. J. Durian, *Phys. Rev. Lett.* **72**, 3347 (1994).
16. K. Makino, M. Kawaguchi, K. Aoyama and T. Kato, *Phys. Fluids* **7**, 455 (1995).
17. T. Kato, T. Tokuya and A. Takahashi, *Kobunshi Ronbunshu* **39**, 293 (1982).
18. M. Kawaguchi and K. Makino, unpublished data.

Fractals, Vol. 4, No. 2 (1996) 187–192


# SURFACE EXPLOSION IN AUTOCATALYTIC MONOMER-MONOMER REACTIONS

M. KOLB
*Laboratoire Chimie Théorique, Ecole Normale Supérieure, 69364 Lyon, France*
*Institut de Recherches sur la Catalyse, 2, Ave. A. Einstein, 69626 Villeurbanne Cedex*

**Abstract**

A lattice model of the A-B type has been introduced recently to describe temporal oscillations and pattern formation for the NO–CO reaction catalyzed by a Pt surface. Under certain experimental conditions one observes the rapid reactional removal of an adsorbed layer upon nucleation — a direct consequence of the autocatalytic nature of the reaction. Here this surface explosion is described by the same model, but now only depending on a single parameter: the initial fraction of A sites. Simulations show that a reactive "cleaning" of the surface is possible for a wide range of A coverages. For low (high) values of the parameter, a transition towards a B (A) poisoned state is observed. Fractal reaction patterns appear in the vicinity of the transition points. Both transitions have large scale characteristics of standard percolation.

## 1. INTRODUCTION

Catalytic reactions in a well controlled environment have attracted much attention in recent years. For the best studied system, CO oxidation on a Pt catalyst, a rich variety of complex phenomena have been observed. Depending on the choice of the external parameters, spontaneous temporal oscillations appear both for monocrystalline and for polycrystalline samples.[1,2] In the first case, spatially resolved methods allow reaction fronts in the form of waves, spirals and chaotic patterns to be identified. Such patterns are related to specific nonlinear coupling between the reaction variables. They yield information on the underlying mechanism. Furthermore, these catalytic systems are excellent examples of the different spontaneous nonlinear spatio-temporal structures predicted by theory.

The modeling of catalytic processes and the associated patterns has been done either in terms of coupled nonlinear differential equations for the relevant surface variables,[3] with a diffusion term added for describing spatial features, or by microscopic lattice models which mimic the elementary reaction steps.[4] The former approach allows for analytic results in mean field approximation, the latter including microscopic fluctuations in a natural way.

More recently, the NO–CO reaction on Pt(100) was also found to exhibit spontaneous temporal oscillations and spatial pattern formation.[5–10] Close examination of the phase diagram shows that for some parameter values the state of the surface is instrumental in obtaining the oscillations, whereas for other parameter values (notably at lower temperature) the state of the surface is stable. A different mechanism was proposed for this case, based on the NO dissociation step.[7,8] In the following a lattice model, recently introduced to described this reaction,[11] will be used to characterize the autocatalytic surface explosion observed in experiments.

## 2. THE MODEL

The basic reaction steps for the Pt-catalyzed NO–CO reaction,[5–8] $NO + CO \rightarrow 1/2\ N_2 + CO_2$, are as follows:

$$CO \leftrightarrow CO|_{ads}$$

$$NO \leftrightarrow NO|_{ads}$$

$$NO|_{ads} + E \rightarrow N|_{ads} + O|_{ads}$$

$$N|_{ads} + N|ads \rightarrow N_2$$

$$CO|_{ads} + O|ads \rightarrow CO_2\,.$$

Both CO and NO adsorb on and desorb from a single surface site. The desorption rates of CO and NO differ little from each other. $NO|_{ads}$ dissociates provided there is room (E stands for an empty site) on the surface. This reaction is rapid (at sufficiently high temperature). $N|_{ads}$ diffuses across the surface, quickly reacts with another $N|_{ads'}$ and desorbs. Neighboring $CO|_{ads}$ and $O|_{ads}$ also react and desorb rapidly. Both $CO|_{ads}$ and $NO|_{ads}$ diffuse across the surface, whereas diffusion of $O|_{ads}$ can be neglected. We note that the stoichiometry of this process is of the A-B type rather than the A-$B_2$ type relevant for the CO oxidation reaction.

From these reaction steps, the following simplified A-B model incorporates all the essential steps[11]:

Each site of a square lattice can be in one of four states, E, A, B, B′. Four possible changes of the surface state are included:

$$E \rightarrow A$$

$$E \rightarrow B$$

$$B + E \rightarrow B' + E$$

$$A + B' \rightarrow 2\ E\,.$$

(1) A randomly chosen E site is changed into A with probability $p \equiv p_A = (1 - p_B)$ and into B with probability $p_B$ (adsorption).
(2) If a new A site has a nearest neighbor B′ site, both instantaneously transform into E sites (reaction). Among several neighboring B′ sites, one is selected at random.
(3) Similarly, if a new B site has one or more nearest neighbor E sites it changes into B′ (dissociation). The E site(s) remain unchanged.
(4) After step (2) the newly created E site may have nearest neighbor B sites. All of them immediately change into B′.
(5) Analogously, after step (3), the new B′ site may have one (or several) nearest neighbor A sites. This A-B′ pair (a randomly chosen A-B′ pair) is replaced by two E sites.
(6) The steps (4) and (5) are repeated sequentially until no neighboring B-E respectively. A-B′ pairs are left: first all B-E are transformed in B′-E, then all A-B′ into E-E, and so on.

Under certain conditions an avalanche of reactions may invade the whole surface. This is the case we investigate in detail. It is not necessary to include desorption or diffusion explicitly to model this surface explosion; nevertheless, it may be added for a more accurate comparison with the experiments. As for the CO-O model,[4] the reaction itself diffuses the reaction fronts.

The model studied has only two parameters, $p$ and the desorption rate $d$. One obtains both temporal oscillations and spatio-temporal patterns,[11] with the reaction cycle being: NO-poisoning from an active state; inactive NO to CO conversion; and nucleation of the active state.

Here we have considered as an initial state a dense mix of A and B (no E) with fraction $P_A$ and $P_B$, respectively and we neglect desorption. This corresponds to preparing the catalyst at low temperature, where desorption and dissociation are inactive. A single E site is then placed on the surface to nucleate the autocatalytic process. Monte Carlo simulations for different values of $P_A$ (and for different system sizes) allow us to draw a phase diagram and to characterize the transition points.

## 3. RESULTS

Before presenting the simulation results of the Monte Carlo simulations, the phase diagram of the dissociative A-B model is determined from mean field theory on a Bethe lattice, without loops. Each newly created E site on a $z$-coordinated lattice generates $(z-1)P_B$ $B'$ sites, each of which creates one new E site with probability $(z-1)P_A$. Percolation requires $(z-1)P_B > 1$ and $(z-1)^2 P_A P_B > 1$. Hence for a square lattice ($z = 4$), A poisoning occurs for $P_A > P_2 \cong 0.666$ and B poisoning for $P_A < P_1 \cong 0.127$. Between $P_1$ and $P_2$ the whole surface is invaded by the reaction.

In Fig. 1 the reaction rate $P_{A\text{-}B'}$, the number of A-B′ pairs reacted off per unit time, is shown as a function of time, for low $P_A$, at the transition point and in the domain where the reaction percolates. Time is defined in terms of the number of successive dissociation/reaction cycles. In the first case, the reaction dies out after a finite time because a layer of B′ next to a B layer prevents both dissociation and reaction. In the last case, the reaction propagates across the entire surface. In between, the reaction just marginally propagates and scaling holds. The number of reacted sites as a function of time (defined as

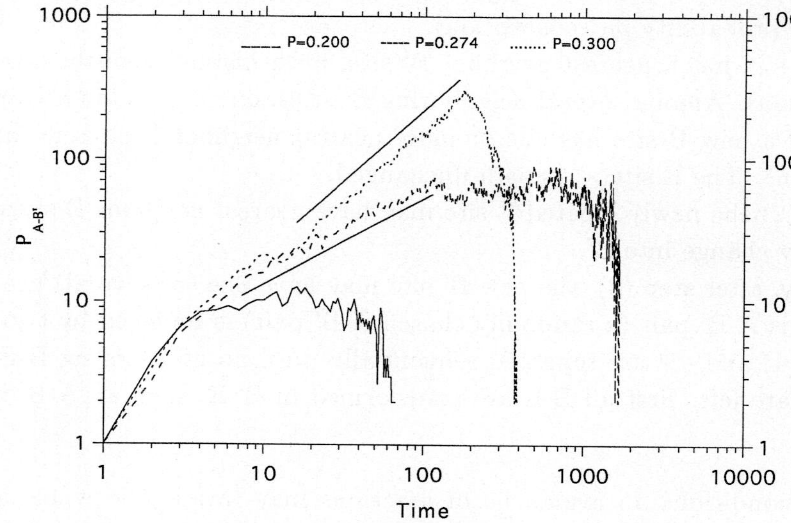


**Fig. 1** Reaction rate $P_{\text{A-B}'}$ (A-B′ formation) as a function of time (number of B-E → B′-E and A-B′ → E-E cycles) for $P_{\text{A}} = 0.200$, $P_{\text{A}} = P_{\text{A},1} \cong 0.274$ and $P_{\text{A}} = 0.300$.

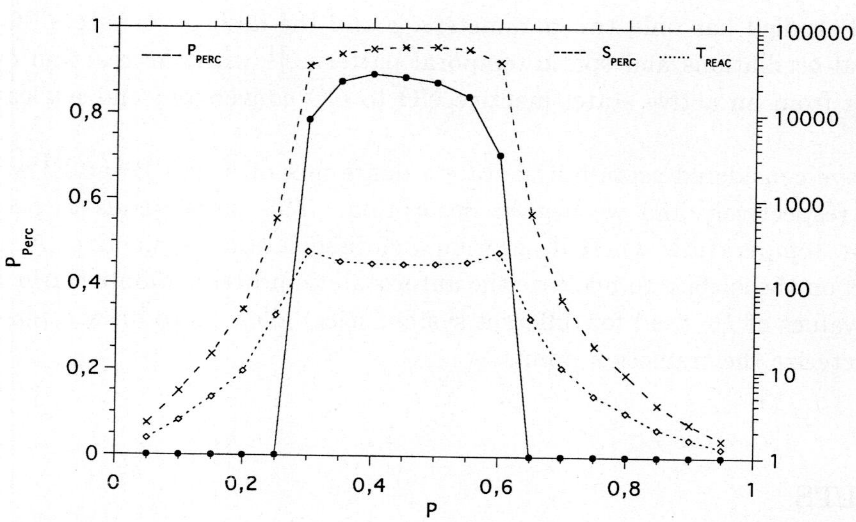


**Fig. 2** Phase diagram: Percolation probability $P_{PERC}$, "cleaned" surface area $S$ and total reaction time $T_{REAC}$, as a function of $P$.

the number of successive dissociation/reaction cycles), scales with an exponent consistent with predictions from percolation theory, $D/d_{\text{min}} - 1$. $D$ is the fractal dimension and $d_{\text{min}}$ the minimum path exponent (relating the chemical distance to the Euclidean distance), for percolation clusters. Above the critical point, the system is homogeneous and the reaction rate scales linearly with time (straight line). In Fig. 2 the percolation probability, the weight averaged "cleaned" surface, and the total reaction time are shown as a function of $P = P_{\text{A}}$. The two transitions are identified at $P_1 = 0.274(4)$ and $P_2 = 0.628(5)$. Note that the reaction time has a maximum at the critical points because of the divergence of the ratio of the minimum path over the linear system size. In Fig. 3 the scaling characteristics of the upper transition are shown. Beyond this point, an A-layer stops further reaction after a

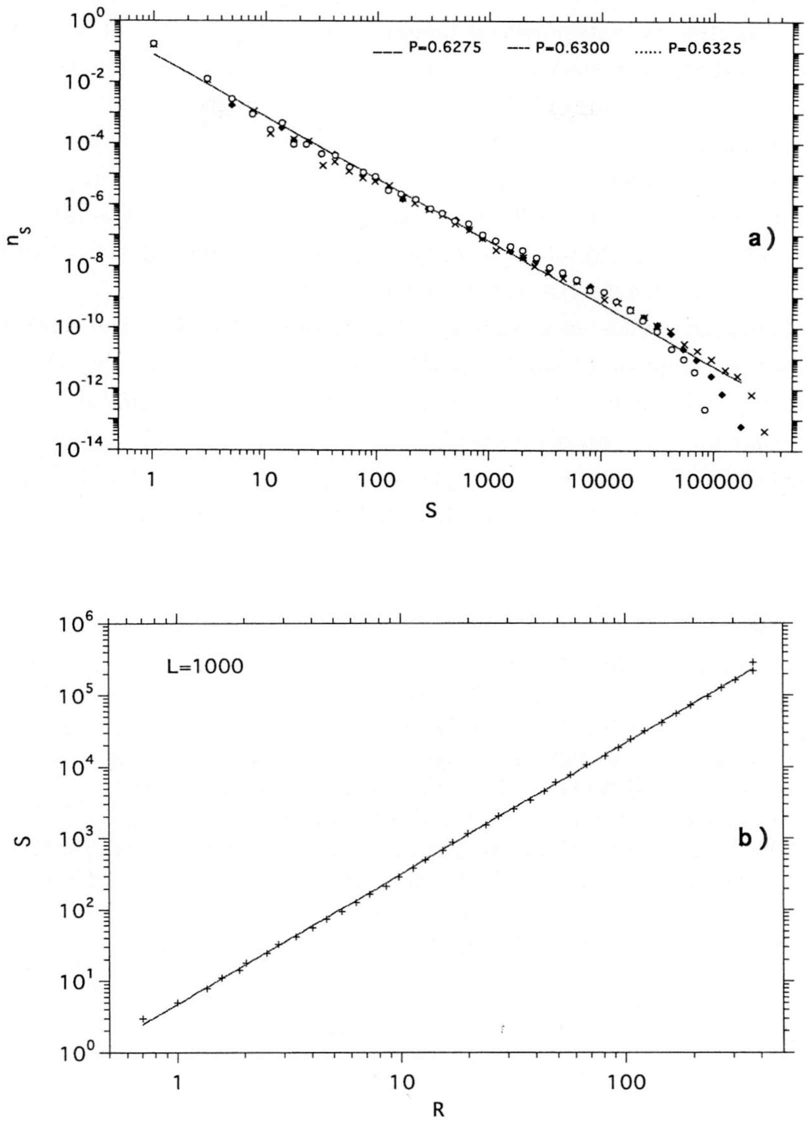


**Fig. 3** Scaling properties at the upper transition point. **(a)** Size distribution of the sites reacted off, $n_s$, for $P = 0.6275$, 0.300 and 0.6325 near $P_{A,2}$. **(b)** The mass-radius relation for the reacted off sites, $s \sim R^D$, for $P = 0.6275$. The slope $\tau = 2.04(4)$ in $n_s \sim s^{-\tau}$ in (a) and the measured value $D = 1.87(3)$ in (b) are consistent with the values predicted by percolation theory in two dimensions.

finite time. In Fig. 3(a) the domain size distribution $n_s$ of the "cleaned" catalyst surface is shown. At $P_2$ it apparently scales with an exponent characteristic of percolation. Similarly, in Fig. 3(b), the mass-radius relation of the empty surface area is consistent with standard percolation. Similar results hold for the lower transition point.

## 4. DISCUSSION

The data presented suggest that the poisoning that limits the surface explosion, both at low and at high $P$, can be modeled in terms of percolation. This despite the fact that the elementary reaction steps are entirely different from percolation: it can at best be described by a random, directed two-step (dissociation/reaction) process. The random choice of reaction bonds, furthermore, makes the cluster structure different for each realization; the lattice cannot be divided into mutually exclusive percolating domains in a unique way. The sequential dissociation/reaction process can be modeled by invasion percolation which, for the case relevant here (compressible fluid), has the same scaling properties as standard percolation. The directed nature of the "bonds" does not change the large scale nature of the problem, as there is no global preference as in directed percolation; the random choice of the directed bonds averages the effect to zero.

Finally, as compared with experiments, it is noteworthy that the surface explosion is possible for a large range of $P$ values, and — as is evident from Fig. 2 — there is little sensitivity to the precise value of the initial coverage. This explains one of the important experimental findings. The simulations show in addition, that in the explosive region the reaction is practically exhaustive: below the stoichiometric point practically no A sites are left after the reaction; above $P = 0.5$ no B and B′ sites are left.

## REFERENCES


1. G. Ertl, *Surf. Sci.* **299/300**, 742 (1994).
2. D. J. Kaul and E. E. Wolf, *J. Catal.* **91**, 216 (1985).
3. R. Dickman, *Phys. Rev.* **A34**, 4246 (1986).
4. R. M. Ziff, E. Gulari and Y. Barshad, *Phys. Rev. Lett.* **50**, 2553 (1986).
5. M. W. Lesley and L. D. Schmidt, *Surf. Sci.* **155**, 215 (1985).
6. S. B. Schwartz and L. D. Schmidt, *Surf. Sci.* **183**, L269 (1987).
7. Th. Fink, J.-P. Dath, R. Imbihl and G. Ertl, *J. Chem. Phys.* **95**, 2109 (1991).
8. Th. Fink, J.-P. Dath, M. R. Basset, R. Imbihl and G. Ertl, *Surf. Sci.* **245**, 96 (1991).
9. G. Veser and R. Imbihl, *J. Chem. Phys.* **96**, 7155 (1992).
10. G. Veser and R. Imbihl, *J. Chem. Phys.* **100**, 8483 (1994).
11. M. Kolb, preprint.

Fractals, Vol. 4, No. 2 (1996) 193–198


# PERCOLATIVE TREATMENT OF THE RESTRICTED TRIMOLECULAR REACTION $A + 2X \rightleftharpoons 3X$

A. Yu. TRETYAKOV
*School of Information Sciences, Tohoku University, Aramaki, Aoba-ku, Sendai, 980-77, Japan*

A. PROVATA*
*Faculté des Sciences and Center for Nonlinear Phenomena and Complex Systems, Univertité Libre de Bruxelles, Campus Plaine, CP 231, 1050 Brussels, Belgium*

### Abstract

We indicate the possibility of existence of a percolation-like transition in the behavior of the trimolecular reaction $A + 2X \rightleftharpoons 3X$ on a 2-d square lattice, with the initial concentration $p$ of $X$ particles serving as a control parameter. Using a renormalization group approach, we find that the transition value of $p$ falls into the interval $0 \leq p_c < 0.052$.

## 1. INTRODUCTION

The chemical reaction

$$A + 2X \rightleftharpoons 3X \tag{1}$$

and its behavior depending on the substrate geometry has received a lot of attention recently.[1–4] It was shown that on substrates with low enough dimensionality or

*Current address: Institute of Physical Chemistry, NRCPS, Demokritos 153 10 Ag. Paraskevi, Greece.

coordination number, the reaction (1) never reaches equilibrium and attains a non-equilibrium steady state characterized by an asymptotic $\langle A\rangle/\langle X\rangle$ ratio other than 1 (the value predicted by the mean field).

On the other hand, on a 2-dimensional square lattice the asymptotic $A/X$ ratio, starting from an all-$X$ covered configuration, has been numerically shown to be 1. One of us (A. P.) and J. W. Turner,[2] while noting that the configurations that cannot be reached from the all-$X$ covered configuration can be easily constructed, presented numerical evidence that, starting from a random mixture of $A$ and $X$ with high enough concentration of $X$, in a system of a big enough size, $A/X \to 1$ as $t \to \infty$. Reference 2 poses a problem, whether or not there exists a critical point at some density $p$ of $X$ particles, starting from below which the system never reaches equilibrium, similar to the situation in systems with restricted geometry.

## 2. THE MODEL

Let us consider the ordinary trimolecular system (1) on a square lattice. Every site of the lattice is occupied by either an $A$ or an $X$ particle (there are no empty sites). The dynamics proceeds as follows: at every time step we take a site at random; if the chosen site has 2 $X$-occupied neighbors, the state of the site is changed from $A$ to $X$ or from $X$ to $A$ depending on the current state of the site.

In the present work we impose the following restriction: reaction is possible only when 2 $X$-particles neighboring an $X$ or an $A$ particle are located at a straight angle. Thus reaction

$$\begin{matrix} & X & & & & X & \\ X & A & A & \rightleftharpoons & X & X & A \\ & A & & & & A & \end{matrix}$$

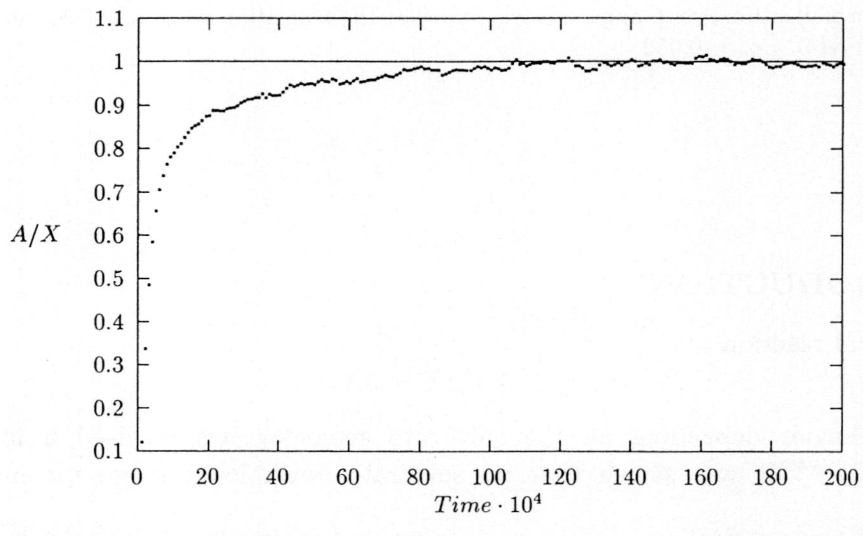


**Fig. 1** Time evolution of the trimolecular system on a square lattice, starting from an all-$X$ covered configuration.

is possible, but not reaction

$$\begin{array}{ccc} & A & \\ X & A & X \\ & A & \end{array} \rightleftharpoons \begin{array}{ccc} & A & \\ X & X & X \\ & A & \end{array} .$$

We call this new model the restricted trimolecular model. Although our definition is somewhat different from the one used in the previous work, where reaction is allowed for any configuration of $X$-occupied neighbors, we expect a qualitatively similar behavior.

Figure 1 shows the time development in the restricted trimolecular model starting from the all-$X$ covered configuration, obtained by a Monte Carlo simulation. As one can see, the asymptotic $A/X$ ratio has the mean field value of 1, as in the ordinary trimolecular model.[1,2]

## 3. TRIMOLECULAR REACTION AS A PERCOLATION PROBLEM

Let us note that it is possible to construct an infinite number of configurations on a square lattice, that are not statistically accessible from the all-$X$ covered configuration. Consider a rectangular block of $X$-particles completely surrounded by $A$ particles:

$$\begin{array}{ccccc} A & A & A & A & A \\ A & X & X & X & A \\ A & X & X & X & A \\ A & A & A & A & A \end{array} .$$

The reaction cannot spread from the block of $X$-particles outwards, because none of the $A$-particles has 2 $X$ neighbors. Any configuration formed by a combination of rectangular blocks of $X$-particles separated by a layer of $A$-particles, and all configurations connected by reaction to it, are not connected to the all-$X$ covered configuration. On the other hand, any configuration of $X$-particles is connected by reaction to some combination of square blocks of $X$ sites, which is attained if the maximal number of $A$ is allowed to react.

Let us define connectivity on the lattice as follows: two sites are connected if they belong to the same $X$-fillable rectangle. As the system size is increased, the finite clusters (blocks) of $X$-particles can connect (by means of trimolecular interactions) into an infinite cluster, spanning the system. Numerical evidence[2] suggests that it happens in the case of big enough concentrations of $X$.

## 4. RENORMALIZATION GROUP TREATMENT

Consider an $(nL) \times (nL)$ square lattice. It can be divided into $n \times n$ non-intersecting $L \times L$ blocks $S_{ij}$ as follows:

$$\begin{array}{cccc} S_{11} & S_{12} & \cdots & S_{1n} \\ S_{21} & S_{22} & \cdots & S_{2n} \\ \cdots & & & \\ S_{n1} & S_{n2} & \cdots & S_{nn} \end{array} .$$

Now, consider the trimolecular reaction in each of the blocks separately. If the configuration of block $S_{ij}$ is connected to an all-$X$ covered configuration, we replace it by symbol **X**, otherwise the block is replaced by symbol **A**. As a result, we obtain an $n \times n$ square lattice, filled by **A** and **X**, something like:

$$\begin{array}{cccc} A & X & \cdots & A \\ X & X & \cdots & A \\ \cdots & & & \\ A & A & \cdots & X \end{array} .$$

As one can easily see, if an **A**-block has two **X**-block neighbors, located at a straight angle, it can be reconstructed into an all-$X$ covered block even if it contains no $X$ sites. In other words, if in the renormalized system the dynamics is defined exactly in the same way as in the original system, then the configurations connected to the all-$X$ covered configuration in the renormalized system correspond to configurations connected to the all-$X$ covered configuration in the original system.

This observation allows us to make the following statement: if $f_n(p)$ is the probability that the trimolecular system on a $n \times n$ square lattice is connected to the all-$X$ covered configuration, taken as a function of the fraction of $X$-covered sites $p$, then an unstable (with respect to iterating $f_n$) solution of

$$f_n(p) = p \tag{2}$$

approximates the $p_c$ value in the infinite system from above.

For small $n$, $f_n(p)$ can be obtained exactly, using exact enumeration:

$$f_2(p) = 2p^2 - p^4 . \tag{3}$$

$$f_3(p) = 6p^3 + 9p^4 - 45p^5 + 48p^6 - 18p^7 + p^9 . \tag{4}$$

$$\begin{aligned} f_4(p) = 22p^4 &+ 160p^5 - 976p^6 + 1888p^7 - 1298p^8 - 992p^9 \\ &+ 2888p^{10} - 2816p^{11} + 1534p^{12} - 480p^{13} + 72p^{14} - p^{16} . \end{aligned} \tag{5}$$

The corresponding nontrivial solutions of Eq. (2) are obtained as $p_c^2 = 0.618034\ldots$, $p_c^3 = 0.482541\ldots$ and $p_c^4 = 0.40429\ldots$.

For big $n$ exact enumeration requires unrealistic computer resources, but good enough approximate solutions of Eq. (2) can still be obtained using the Monte Carlo approach. We used the following algorithm: for a fixed pseudo random sequence we find a $p$ value satisfying Eq. (2) with a certain precision ($f_n(p)$ is obtained as a fraction of the realizations that can be built up to the all-$X$ covered configuration, with the initial state prepared by occupying sites of a lattice with $X$-particles with probability $p$ and taking the rest of the sites as $A$-occupied). $p_c^n$ is taken as an average of approximate solutions obtained for a big enough number of different pseudo random sequences. This approach allowed us to obtain $p_c^n$ values for $n$ up to $n = 64$.

The results are shown in Fig. 2. One can see that as $n \to \infty$, $p_c^n$ tends to some value in the vicinity of zero. By construction of our renormalization procedure, each point in the plot constitutes an upper boundary for the true threshold value. While extrapolation of the data to $n \to \infty$ should provide an approximation of the $p_c$ value, our series of $p_c^n$ values is too short to judge what the error can be. On the other hand, since every member

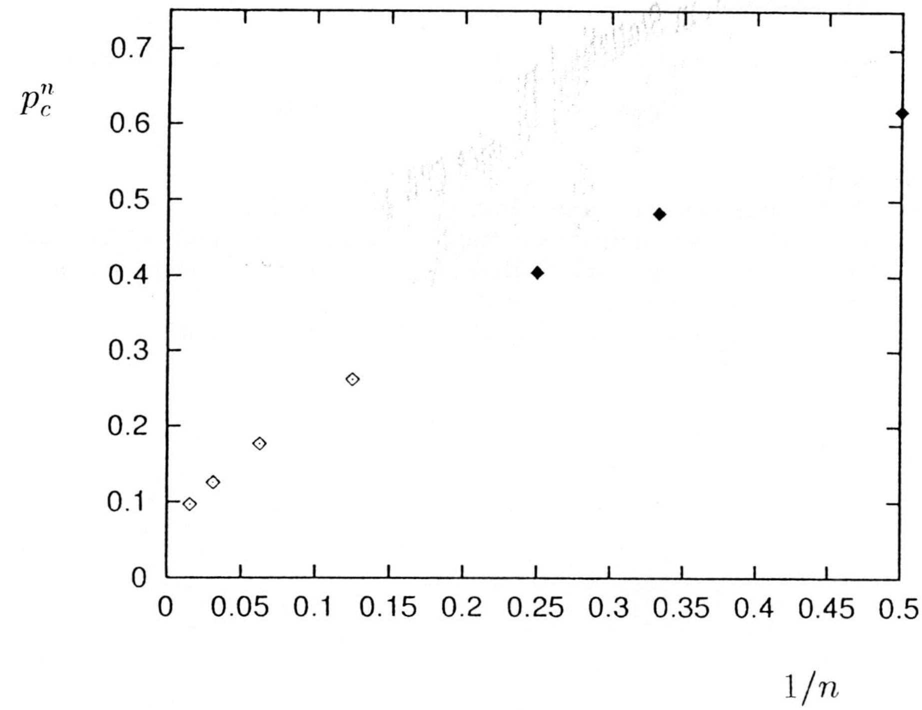


**Fig. 2** Renormalization group estimations of the $p_c$ value. Black points correspond to exact enumeration values; the rest were obtained using Monte Carlo approach.

in the sequence constitutes an upper bound, we find it reasonable to expect that the $p_c$ value obtained by interpolating the sequence should be shifted from the true value in the positive direction. Extrapolation (using power law as a fitting function) based on the four points corresponding to bigger $n$, allows us to place the $p_c$ value between 0 and 0.052, with 0 remaining a possibility.

## 5. CONCLUSION

We indicate the possibility of the existence of a percolation-like transition in the behavior of the trimolecular reaction on a 2-d square lattice, with the initial concentration $p$ of $X$ particles serving as a control parameter.

The existence of the percolation threshold would imply the following behavior for the reaction: for $p > p_c$, all sites of the systems are active. For $p < p_c$, the system breaks into an infinite number of active zones of rectangular shape, separated by inert areas covered by $A$-particles.

It should be expected that as the threshold is approached from below, the fraction of sites falling into inert areas tends to zero, and at $p_c$ the inert areas form a fractal structure with a zero density.

By using a renormalization group technique we show that the $p_c$ value falls into the interval $0 \leq p_c < 0.052$. At the current stage the possibility that $p_c$ is equal to zero cannot be excluded; providing a proof of the existence of the threshold remains an open problem.

## ACKNOWLEDGMENT

One of the authors (A. T.) acknowledges the support of the Inamori Foundation.

## REFERENCES

1. A. Provata, J. W. Turner and G. Nicolis, *Stat. Phys.* **70**, 1195 (1992).
2. A. Provata and J. W. Turner, in *Statistical Physics and Thermodynamics of Nonlinear Nonequilibrium Systems*, eds. W. Ebeling and W. Huschik (World Scientific, Singapore, 1993).
3. S. Prakash and G. Nicolis, submitted to the *J. Phys. Chem.*
4. A. Tretyakov, A. Provata and G. Nicolis, *J. Phys. Chem.* **99**, 2770 (1995).

Fractals, Vol. 4, No. 2 (1996) 199–204


# MOTT'S LIKE BEHAVIOR OF ELECTRON TRANSFER REACTIONS

J. MALINSKY and A. MOROZOVSKIY
*Department of Biomathematical Sciences,*
*Mount Sinai School of Medicine, NY*
*Physics Department of City University of NY (BCC), Bronx, NY*

**Abstract**

New model of Long Range Electron Transfer has been suggested to evaluate an electron transfer rate in some structures like proteins. Mott's like behavior has been obtained. It is possible that this description might yield better estimation for an electron rate than one which has been accepted in this field. This phenomenon may bridge proteins and doped semiconductors as far as their conductivity is concerned. Both Franck-Condon factor as well as an electron contribution have been optimized at the same footing. Fit with experimental data is considered to be reasonable. Fractal considerations allow one to account for intervening medium.

Recently, electron transfer studies have been developed which elucidate electron propagation through the protein backbone and sidechains. When the electron moves over large distances, as it does in case of protein, there are two dominant contributions in the electron transfer rate. The first one is defined by the Franck-Condon principle and depends on the reorganization energy of the nuclear part of the molecule. The second is the electron transfer term, which is basically the probability of the electron moving from donor to acceptor. Theories explaining electron transfer are divided into two main groups. Some theories consider the electron term to be known and unimportant, and describe experimental data as a function of the nuclear factor.[1] The others point out the importance of the electron term and assume the nuclear factor to be a known or optimized factor.[2] For any consistent theory, however, they must equally be taken into account.

In what follows, we consider the simplest model and show that even in this case the optimization of both of these factors is important and leads to new results. We can assume that the electron transfer rate will be larger when the Franck-Condon principle is fulfilled many times rather than just once, because the electron will arrive at an acceptor site after visiting several intermediate sites.

Let us consider two electron transfer events and try to estimate which one is optimal. First, we have to account for the usual electron movement when an electron jumps immediately from donor to acceptor. In this case the Franck-Condon principle is fulfilled once for an electron transfer between donor D and acceptor A. The theory for this situation has been fully developed and is well known.

The other process corresponds to a set of consecutive electron transfers when the Franck-Condon principle is fulfilled several times. Then we have several electron transfer reactions which are connected with each other.

Let us choose a relatively simple scheme, similar to scheme 1, from Ref. 3 and accept the possibility of several transfer reaction steps:

$$A + B \underset{k_{-12}}{\overset{k_{12}}{\rightleftarrows}} [A|B]_1 \underset{k_{-n_1}}{\overset{k_{n_1}}{\rightleftarrows}} [A|B]_2{}^* \underset{v_{eff_1}}{\overset{v_{eff_1}}{\rightleftarrows}} [A|B]_2 \underset{k_{-n2}}{\overset{k_{n_2}}{\rightleftarrows}} *** \underset{v_{eff_{i-1}}}{\overset{v_{eff_{i-1}}}{\rightleftarrows}}$$

$$\underset{v_{eff_{i-1}}}{\overset{v_{eff_{i-1}}}{\rightleftarrows}} [A|B]_i \underset{k_{-n_i}}{\overset{k_{n_i}}{\rightleftarrows}} [A|B]_i{}^* \underset{v_{eff_i}}{\overset{v_{eff_i}}{\rightleftarrows}} *** \underset{v_{eff m}}{\overset{v_{eff m}}{\rightleftarrows}} [A|B]_{m+1} \overset{k_{-n_{m+1}}}{\longrightarrow}$$

where A and B are the reactants, $A + B \underset{k_{-12}}{\overset{k_{12}}{\rightleftarrows}} [A|B]_1$ is the diffusion process for forming the precursor complex $[A|B]_1$, $[A|B]_i \underset{k_{-n_1}}{\overset{k_{n_1}}{\rightleftarrows}} [A|B]_i^*$ is the reorganization of the $i$th precursor complex into the configuration for the electron transfer, $[A|B]_i{}^* \underset{v_{eff_i}}{\overset{v_{eff_i}}{\rightleftarrows}} [A|B]_{i+1}$ is the electron transfer between $i$ and $i+1$ sites, and $v_{eff_i}$ is the effective electron hopping frequency for the $i$th jump.

We use steady-state approximation for the various states ($[A|B]_1$, $[A|B]_1^*$, $[A|B]_2$, $[A|B]_2^*$, $***$, $[A|B]_m^*$, $[A|B]_{m+1}$). For relatively low hopping frequency when:

$$v_{eff_i} \ll K_i,\ K_{-i}\,, \tag{1}$$

the expression for the electron transfer rate constant K is:

$$K = K_A K_1^* K_2^* *** K_m^* v_m\,, \tag{2}$$

where $K_A = K_{12}/K_{-12}$ describes the stability of precursor complex 1, $K_i^* = K_{n_i}/K_{-n_i}$ and $v_m$ is the hopping frequency for hopping between the $m$ and $m+1$ sites. We know that there are many another solutions for K in this system of linear equations under additional conditions like Eq. (1).

In order to simplify the result in Eq. (2), we use additional assumptions:

1. Assume that

$$K_1^* = K_2^* = *** = K_i^* = *** = K_m^* = K^*\,. \tag{3}$$

2. Use the standard Marcus' expression for $K_i^*$:

$$K_i^* = \exp\left[-\frac{\lambda}{4}\left(1+\frac{\Delta G_i^0}{\lambda}\right)^2 \frac{1}{kT}\right], \tag{4}$$

where $\lambda$ is the reorganization parameter and $\Delta G_i^0$ is the energy change through the $i$th reaction. From Eqs. (3) and (4) we have:

$$\Delta G_1^0 = *** = \Delta G_i^0 = *** = \Delta G_m^0 .$$

So if $\Delta G^0 = \Delta G_1^0 + \Delta G_2^0 + *** + \Delta G_m^0$, we can get

$$\Delta G_i^0 = \frac{\Delta G^0}{m} , \tag{5}$$

where $\Delta G^0$ is the net energy change for this sequence.

3. Assume for the sake of simplicity that all hopping frequencies are the same:

$$v_1 = v_1 = *** = v_m . \tag{6}$$

4. $v_m = v_0 \exp[-\beta R_m]$, where $R_m$ is the distance between the $m$ and $m+1$ sites. From this formula and Eq. (6) we have:

$$R_1 = R_2 = *** = R_m . \tag{7}$$

Up to this point, we have neglected the $v_0(T)$ dependence.

In order to get an analytical expression for K we should assume some additional dependence between $R$ and $R_m$, where $R$ is the distance between the 1st and $m+1$ sites. Here we consider here two simple connections between $R$ and $R_m$.

The simplest assumption is:

$$R = A_0 R_m m , \tag{8}$$

where $A_0 \to 1$.

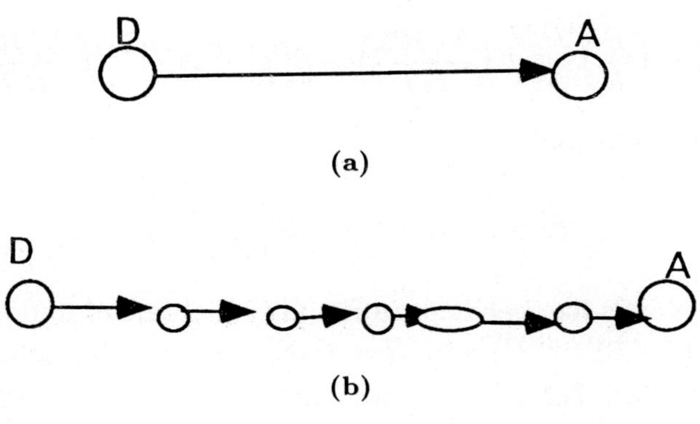


**Fig. 1** Possible electron's hopping between donor and acceptor: **(a)** direct-one step; **(b)** many steps.

**Fig. 2** Possible fractal electron path.

This assumption means that the electron's movement is quasi one-dimensional; however, proteins are complex objects, and the path of an electron can be much more complicated and irregular than one-dimensional movement. (Although this difference can be small.) The simplest assumption in this case is that the movement of the electron is fractal in its origin (Fig. 2).

5. The electron path between donor and acceptor is fractal in origin, and they are connected by the formula:

$$m = A\left(\frac{R}{R_m}\right)^D . \tag{9}$$

When $D = 1$, this expression in Eq. (9) coincides with Eq. (8), but Eq. (9) is more general. Using formula in Eq. (2) and assumptions 1–5 we obtain:

$$K = K_A v_0 \exp\left[-\frac{\lambda m}{4}\left(1+\frac{\Delta G^0}{m\lambda}\right)^2\frac{1}{kT} - \beta\left(\frac{A}{m}\right)^{1/D} R\right] . \tag{10}$$

To get the largest value for $K$ we should optimize $m$. A similar formula was obtained by Mott in his well-known investigations on doped semiconductors' conductivity. Optimization procedure has played an important role in electron hopping.[4] It is possible that such a coincidence is not accidental and comes closer to experimental observation than do currently accepted theoretical results.

By optimizing the value of $m$ at $D = 1$, we get the expression:

$$m = \sqrt{\left(\frac{\Delta G^0}{\lambda}\right)^2 + \frac{4kT\beta R}{\lambda}} . \tag{11}$$

In general, we can find $K$ only by assuming an additional small parameter. If, for instance:

$$\frac{\Delta G^0}{\lambda} \ll 1 , \tag{12}$$

we can write $K$ as:

$$K = K_0 v_0 \exp\left[-\left(\frac{\beta}{R}\right)^{1/1+D}\left(\frac{\lambda A R^D}{4kT}\right) + \beta^{D/1+D} D^{1/1+D}\left(\frac{\lambda A R^D}{4kT}\right)^{1/1+D}\right] . \tag{13}$$

For $D = 1$ we rewrite the formula:

$$K = K_A v_0 \exp\left[-B\left(\frac{B}{T}\right)^{1/2}\right] , \tag{14}$$

where $B$ is constant. Mott's-like behavior in Eq. (14) is obtained under the specific assumption that $K \sim \exp(-(T_0/T)^{1/(d+1)})$ with $d = 1$. Nonexponential behavior was observed in many experimental studies. For example, a formula like Eq. (14) can describe experimental data for the temperature dependence of the cytochrome oxidation rate[5] in a broader interval (200–290 K) than the usual experimental dependence (250–290 K). At low temperatures, when the length of the jump becomes the same as a D-A separation, it is expected that the rate will saturate. This fact is well known.[5] Also, specific quantum-mechanical effects such as interference of different pathways may change these results especially at low temperatures.[2] We would like to note that the coefficient $B$ from both the theoretical calculations and experimental data has the same order of magnitude value. In the case of doped semiconductors, the electrical conductivity possesses this typical Mott's-like behavior; now the other property, namely, the rate of electron transfer in macromolecules according to

Eq. (14), may be essentially the same. It would seem that both theoretical as well as experimental studies are required to understand the behavior of the electron rate on temperature, distance and structure.

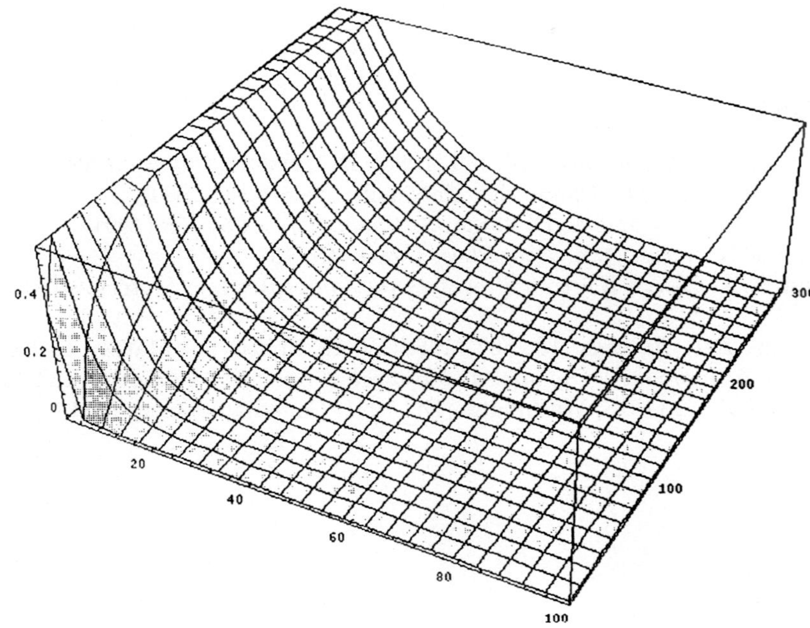


(a)

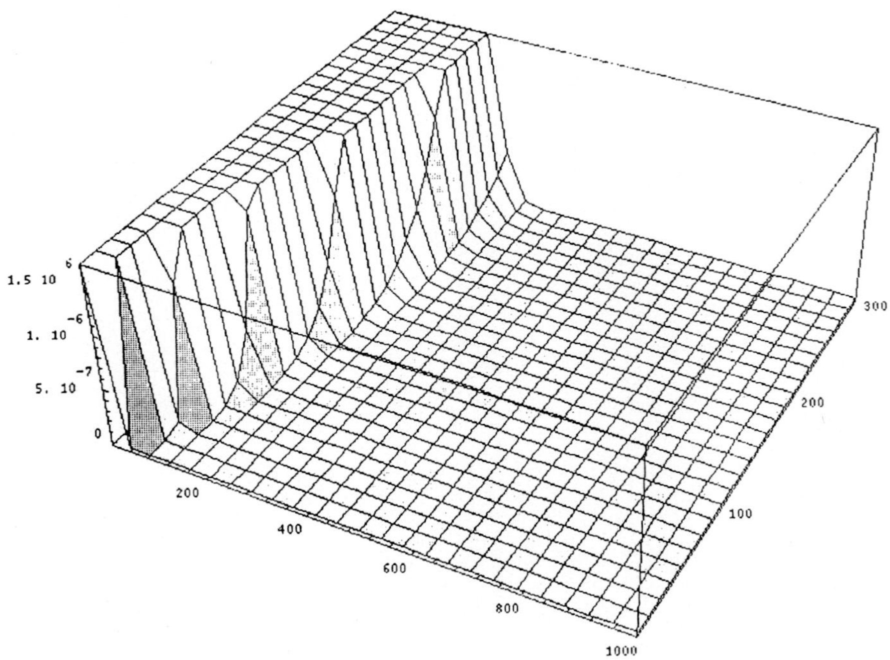


(b)

**Fig. 3** Dependence of electron transfer rate $K/k_A v_0$ on $BR^{1/2}$ and $T$ plot of $K/k_A v_0$ for different values of $BR^{1/2}$ and $T$.

## ACKNOWLEDGMENTS

The authors acknowledge with thanks the very helpful and stimulating discussions with Harry Gray, Jay Winkler of Beckman Institute at Caltech, Craig Benham of MSSM and Joseph L. Birman of CCNY. Our special thanks go to Toni Hill for putting our manuscript into readable form.

## REFERENCES

1. G. L. Closs and J. R. Miller, *Science* **240**, 440 (1988); D. DeVault, *Quantum Mechanical Tunneling in Biological Systems*, 2nd ed. (Cambridge University, New York, 1984).
2. Y. Magarshak, J. Malinsky and A. D. Joran, *J. Chem. Phys.* **95**, 418 (1991).
3. M. D. Newton and N. Sutin, *Ann. Rev. Phys. Chem.* **35**, 437 (1984).
4. N. F. Mott and E. A. Davis, *Electron Processes in Non-crystalline Materials* (Clarendon Press, Oxford, 1979).
5. D. DeVault, *Quar. Rev. Biophys.* **13**, 387 (1980). J. R. Bolton, N. Mataga and G. McLendon, *Electron Transfer in Inorganic, Organic, and Biological Systems*, Advances in Chemistry Series, CSC Symposium Series 2 (ACS, Washington DC 1991, Canadian Society for Chemistry, Ottawa, Canada).

Fractals, Vol. 4, No. 2 (1996) 205–211


# FRACTAL ROUGHNESS CHARACTERIZATION OF SUPER-GROUND Mn-Zn FERRITE SINGLE CRYSTALS

J. YU[*,†], Y. NAMBA[*] and M. SHIOKAWA[*]
[*] *Department of Mechanical Engineering, Chubu University, 1200 Matsumotocho, Kasugai, Aichi 487, Japan*
[†] *State Key Laboratory of Applied Optics, Changchun Institute of Optics and Fine Mechanics, Chinese Academy of Sciences, Changchun 130022, P. R. China*

### Abstract

The surface of superground Mn-Zn ferrite single crystal may be identified as a self-affine fractal in the stochastic sense. The rms roughness increased as a power of the scale from $10^2$ nm to $10^6$ nm with the roughness exponent $\alpha = 0.17 \pm 0.04$, and $0.11 \pm 0.06$, for grinding feed rate of 15 and 10 $\mu$m/rev, respectively. The scaling behavior coincided with the theory prediction well used for growing self-affine surfaces in the interested region for magnetic heads performance. The rms roughnesses increased with increase in the feed rate, implying that the feed rate is a crucial grinding parameter affecting the supersmooth surface roughness in the machining process.

## 1. INTRODUCTION

Manganese-zinc (Mn-Zn) ferrite single crystals have been widely used as a core material in magnetic heads for Video Cassette Recorders because of a high permeability in the frequency range of the video signals and a high wear resistance.[1,2] The magnetic heads have been expected to be superground by means of ductile-mode grinding[3,4] in order to meet the required dimensional tolerance. With the development of the magnetic heads machining technology, surface precise evaluation on the scales from nanometer to hundreds of micrometers, is becoming increasingly important since the influence of surface roughnesses may be crucial to performance.

In recent years the descriptions of scaling behaviour have become of crucial importance for many physical problems. A large number of natural surfaces are found to be self-affine, introduced by Mandelbort,[5] such as those encountered in surface growth models,[6] in landscape and erosion surfaces,[7] or in fracture surfaces.[8] The self-affine systems are invariant only under anisotropic magnifications, i.e., the roughness scales anisotropically in the vertical and transverse directions.[5,9–10] Therefore, a scaling analysis is important for the evaluation of surface roughness not only for grown surface[11] but also for super-machining surface, including X-ray optics[12] and the magnetic heads.

In this paper the self-affine roughness characterization of super-ground Mn-Zn ferrite single crystals is investigated with a scanning tunneling microscope and a optical interferometer, and the effect of feed rate $f$ on the roughness is also discussed.

## 2. EXPERIMENTAL DESCRIPTIONS

Mn-Zn ferrite single crystals of (001) face were cut into small dimension of 9.5 mm $\times$ 9.5 mm $\times$ 1 mm, a diamond saw wheel being used to grind single crystals, and lapped to be parallel with loose diamond powder. A ultra-precision surface grinder[4] having a glass-ceramic spindle of extremely low thermal expansion was used in this experiment. The machine has the vertical main spindle for a grinding wheel and a low-speed rotary table of $0.01 \sim 30$ rpm. The depth of cut can be controlled at a unit of 0.1 $\mu$m.

Mn-Zn ferrite samples have been ground under the ductile-mode conditions,[13] with the wheel speed $V_s = 1200$ m/min, depth of cut $a = 1$ $\mu$m (with a SD5000-75-B diamond wheel), and the feed rate $f = 10$, 15 $\mu$m/rev, respectively. The grinding direction was $\langle 001 \rangle$ on the (001) face.

The surface morphology of the Mn-Zn ferrite samples was studied in a commercial three-dimensional noncontact surface profiler TOPO-3D of WYKO Corp. and a Seiko SAM3500 scanning tunneling microscope (STM). The STM was designed for positioning big samples accurately under a Nomarski interference microscope and for observing details on an exact location by a sharp platinum probe. In order to investigate the scaling behavior: (1) STM measurements were carried out on various areas, i.e., 100 nm $\times$ 100 nm, 500 nm $\times$ 500 nm, 1 $\mu$m $\times$ 1 $\mu$m, 5 $\mu$m $\times$ 5 $\mu$m, 10 $\mu$m $\times$ 10 $\mu$m; (2) TOPO-3D measurements were carried out on the areas of 235 $\mu$m $\times$ 232 $\mu$m and 968 $\mu$m $\times$ 956 $\mu$m. STM measurements were performed using 0.1 bias voltage with the tip positive at 0.5 nA constant current. In each measurement, data were acquired and stored as digitized images with $256 \times 256$ pixels. The measuring direction is perpendicular to the grinding direction or marks.

## 3. RESULTS AND DISCUSSIONS

STM images of Mn-Zn ferrite single crystals with 15 $\mu$m/rev feed rate on the areas of 10 $\mu$m $\times$ 10 $\mu$m, 5 $\mu$m $\times$ 5 $\mu$m and 0.5 $\mu$m $\times$ 0.5 $\mu$m are shown in Figs. 1, 2 and 3, respectively. In these figures: (a) is a color gray scale topography of the surface; (b) is a vertical cross section profile of the surface; (c) is a transverse cross section profile of the surface; (d) is a combination of Abbott curve profile and height distribution profile; (e) is a surface top view image. The root-mean-square (rms) roughness values are shown on the bottom right-hand corner of these figures. The self-affine fractal roughness characterization involves the rms roughness, often called the surface width.[11] From the height deviations $z(x, y)$ of the actual surface from its perfect shape, one obtains the three-dimensional rms roughness $\sigma$ in a square section of size $L \times L$, i.e.,

$$\sigma(L) = \sqrt{\frac{1}{S} \int \int_s Z^2(x, y) dx dy} \tag{1}$$

where $S$ is the area of the scans, $S = L \times L$. As shown in Fig. 1, 2 and 3, with the decrease of scan length from 10 $\mu$m to 0.5 $\mu$m, i.e., shrinking along the transverse axis by a certain factor, the structures are invariant even under the anisotropic magnifications in a stochastic sense [see Fig. 1(c), Fig. 2(c) and Fig. 3(c)]. In other words, if we cut a small piece out of the original 10 $\mu$m $\times$ 10 $\mu$m picture, and rescale the transverse axis and vertical axis by a different factor, for example, from 5 $\mu$m $\times$ 5 $\mu$m or 0.5 $\mu$m $\times$ 0.5 $\mu$m picture, we will obtain the original picture in the stochastic view. This suggest that the surface of the superground Mn-Zn ferrite single crystals may be identified as a self-affine fractal.

Figure 4 shows the plot of $\sigma(L)$ vs. scanning length $L$ for $f = 15$ $\mu$m/rev of a Mn-Zn ferrite sample. On the scales of $10^2$ nm $\leqq L \leqq 10^6$ nm, the rms is dependent on the scale as a power law of $\sigma(L) \propto L^{\alpha}$, where the scaling exponent $\alpha = 0.17 \pm 0.04 < 1$. This implies an anisotropic scaling of the surface roughness, and therefore, the surface is identified as self-affine in this scanning region. The scaling behavior shown in Fig. 4 coincides with the theory prediction well used for growing self-affine surfaces,[11,14] i.e., the interface width

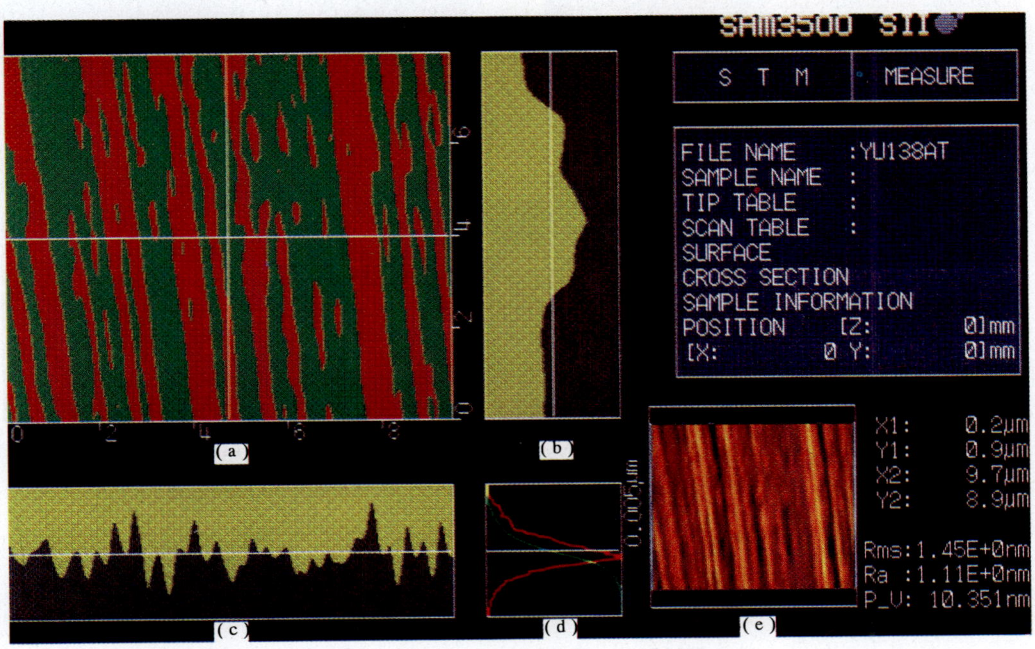


**Fig. 1** STM image of Mn-Zn ferrite single crystal with 15 $\mu$m/rev feed rate on the area of 10 $\mu$m $\times$ 10 $\mu$m. **(a)**, **(b)**, **(c)**, **(d)** and **(e)** were explained in the text.

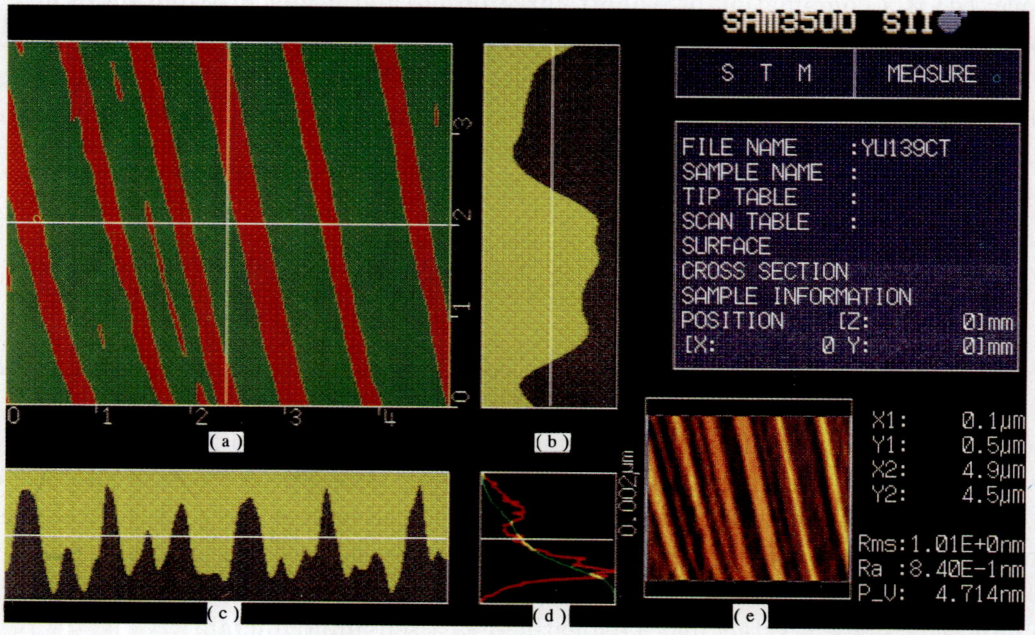


**Fig. 2** STM image of Mn-Zn ferrite single crystal with 15 $\mu$m/rev feed rate on the area of 5 $\mu$m $\times$ 5 $\mu$m.

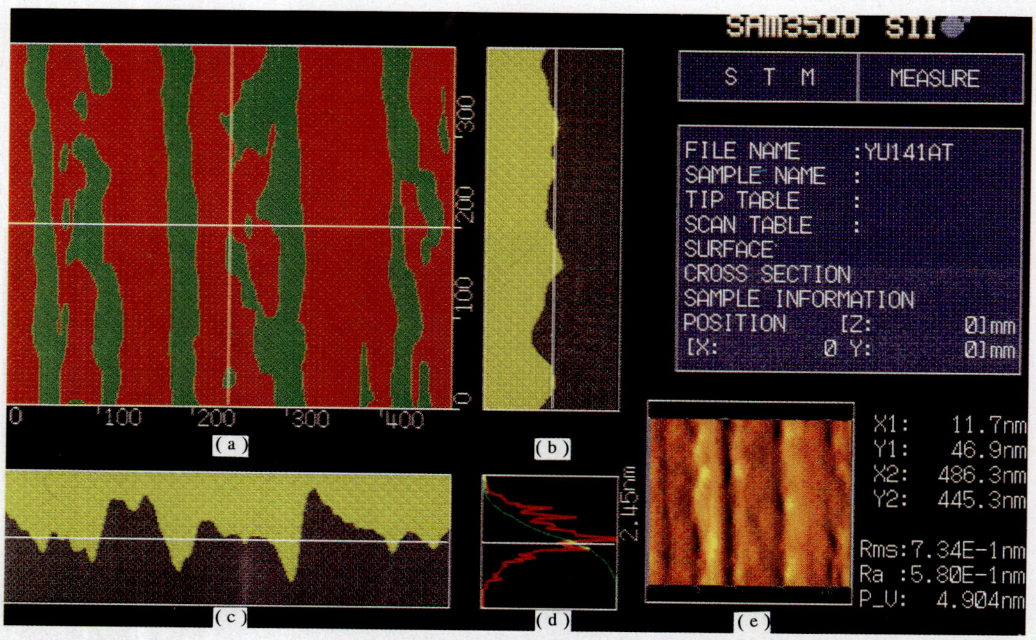


**Fig. 3** STM image of Mn-Zn ferrite single crystal with 15 $\mu$m/rev feed rate on the area of 0.5 $\mu$m$\times$ 0.5 $\mu$m.

(the rms roughness) $W(L, t)$ for length scale $L \ll L_c$ as

$$W(L, t) \propto L^{\alpha}, \quad \text{for} \quad L \ll L_c \tag{2}$$

where $L_c$ is the correlation length and $t$ is growth time. In Fig. 4 $L_c > 10^6$ nm.

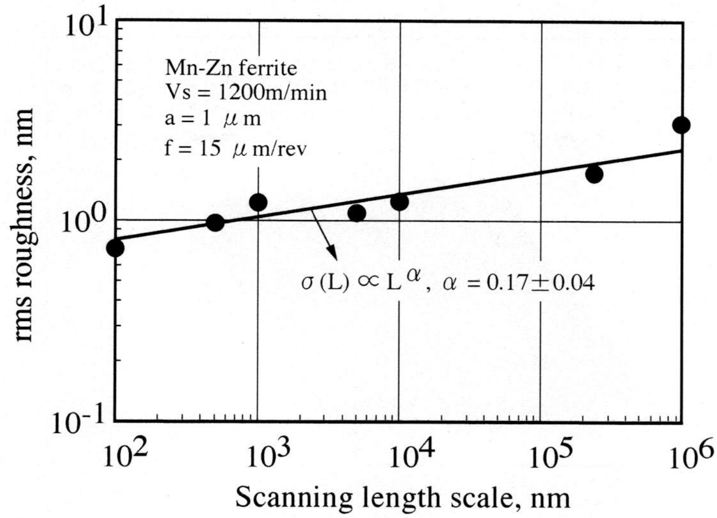


**Fig. 4** The rms roughness $\sigma(L)$ vs. scanning length $L$ for Mn-Zn ferrite single crystal of feed rate 15 $\mu$m/rev.

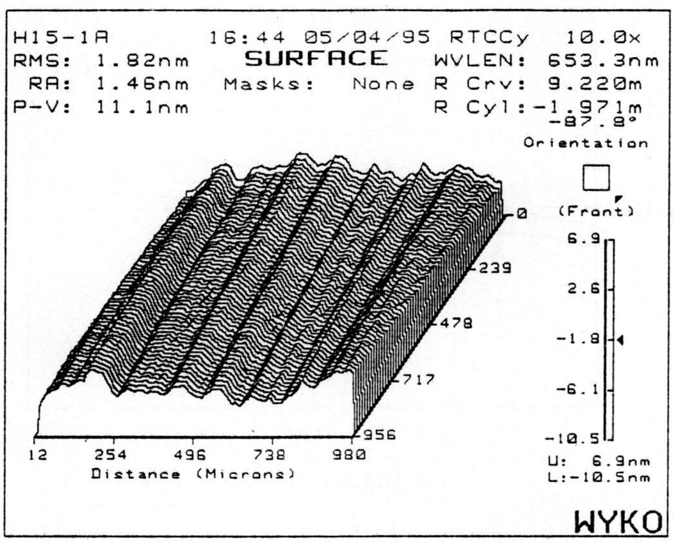


**Fig. 5** TOPO-3D topography of the Mn-Zn ferrite single crystal surface of feed rate 10 $\mu$m/rev. The area size is 968 $\mu$m × 956 $\mu$m.

In contrast to the growing films,[14–15] the surface $\sigma(L)$ of the super-ground Mn-Zn ferrite single crystals will "grow" with the increase in the feed rate $f$. Figures 5 and 6 show the TOPO-3D topography plots of the Mn-Zn ferrite surfaces of 968 $\mu$m × 956 $\mu$m, 10 and 15 $\mu$m/rev, respectively. In Figs. 5 and 6, we observe the maximum height difference between the highest peak and the deepest valley, i.e., P-V values in these figures, become larger from 11.1 nm to 17.4 nm with increasing feed rate, while the rms value is from 1.82 nm to 3.09 nm. This indicates that the feed rate is a crucial grinding parameter affecting the surface roughness for the ultra-precision grinding method.[13] $\sigma(L)$ becomes larger with increase in feed rate.

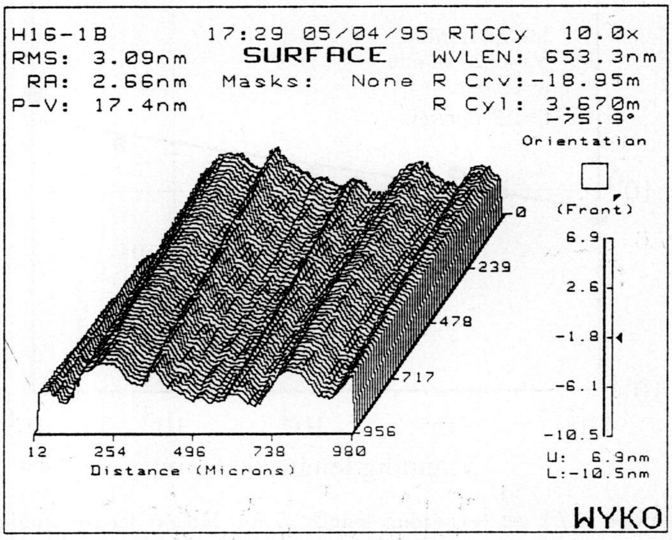


**Fig. 6** TOPO-3D topography of the Mn-Zn ferrite single crystal surface of feed rate 15 $\mu$m/rev. The area size is 968 $\mu$m × 956 $\mu$m.

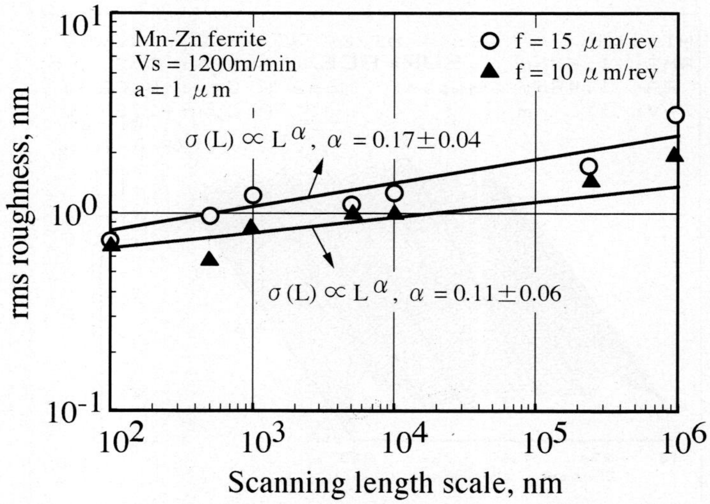


**Fig. 7** The rms roughness $\sigma(L)$ vs. scanning length $L$ for Mn-Zn single crystal surface of 10 $\mu$m/rev and 15 $\mu$m/rev feed rate under the same grinding conditions.

Figure 7 shows the plot of $\sigma(L)$ vs. $L$ for Mn-Zn samples of 10 $\mu$m/rev and 15 $\mu$m/rev under the same grinding conditions. For each feed rate, $\sigma(L)$ increases as a power of $L$, and the roughness exponent $\alpha$ of 10 $\mu$m/rev sample is $0.11 \pm 0.06$. Furthermore, Fig. 7 clearly shows on the interested scales, from a hundred nanometers to hundreds of micrometers, the rms roughness increasing with increase in the feed rate.

## 4. SUMMARY

In this study, the fractal roughness characterization of superground Mn-Zn ferrite single crystals was investigated. The experimental results show that the surface of the superground

Mn-Zn samples may be identified as a self-affine fractal in the stochastic sense. The rms roughness increased as a power of the scale from $10^2$ nm to $10^6$ nm with the roughness exponent $\alpha = 0.17 \pm 0.04$, and $0.11 \pm 0.06$, for feed rate of 15 and 10 $\mu$m/rev, respectively. The scaling behavior coincided with the theory prediction well used for growing self-affine surfaces in the interested region for magnetic heads performance. Such scaling behavior should be taken into account in the theoretical calculation of the surface performance, and also in metrology of roughness parameters. On the interested region, from $10^2$ nm to $10^6$ nm, the rms roughnesses increased with increase in the feed rate, implying that the feed rate is a crucial grinding parameter affecting the surface roughness in the machining process.

## ACKNOWLEDGMENTS

We would like to thank Prof. S. Miyazima for fruitful discussions. This work was supported in part by a Grant-in-Aid for Scientific Research for JSPS Fellows (Contract No. 95043) from the Ministry of Education, Science and Culture, Japan.

## REFERENCES

1. Y. Namba and H. Tsuwa, *Annals of the CIRP* **27/1**, 511–516 (1978).
2. Y. Namba and H. Tsuwa, *Proc. 4th Int. Conf. Prod. Engg.*, 1017–1022 (1980).
3. I. Stowers, R. Komanduri and E. Baird, *Proc. SPIE* **966**, 62–73 (1989).
4. Y. Namba, R. Wada, K. Unno and A. Tsuboi, *Annals of the CIRP* **38/1**, 331–334 (1989).
5. B. B. Mandelbort *Phys. Scr.* **32**, 257 (1985).
6. A. Bunde and S. Havlin (eds.), *Fractals in Science* (Springer-Verlag Berlin Heidelberg, 1994).
7. W. L. Newmann and D. L. Turcotte, *Geophys. J. Int.* **100**, 433 (1990).
8. B. B. Mandelbort, D. E. Passoja and A. J. Paullay, *Nature* **308**, 721 (1984).
9. B. B. Mandelbert, *The Fractal Geometry of Nature* (Freeman, New York, 1982).
10. J. Feder, *Fractals* (Plenum, New York, 1988).
11. T. Vicsek, *Fractal Growth Phenomena* (World Scientific, Singapore, 1989).
12. Y. Namba, J. Yu, K. Yamashita, H. Kunieda, Y. Tawara and K. Haga, OSA Annual Meeting and ILS-XI, MZ4, 1995.
13. Y. Namba, Y. Yamada, A. Tsuboi, K. Unno and H. Nakao, *Annals of the CIRP* **41/1**, 347–351 (1992).
14. A. Iwamoto, T. Yoshinobu and H. Iwasaki, *Phys. Rev. Lett.* **72**, 4025–4028 (1994).
15. L. Vazquez, R. C. Salvarezza, P. Ocon, P. Herrasti, J. M. Vara and A. J. Arvia, *Phys. Rev.* **49**, 1507–1511 (1994).

Fractals, Vol. 4, No. 2 (1996) 213–218


# FLUCTUATION OF SURFACE IN POLISHING PROCESS

Y. HASEGAWA, S. MIYAZIMA and Y. NAMBA*
*Department of Engineering Physics, Chubu University, Japan*
**Department of Mechanical Engineering, Chubu University, Japan*

**Abstract**

A change of surface pattern of materials during polishing is simulated and analyzed from the fractal point of view. We have proposed a simple two-dimensional model for microscopic processes of polishing. We have obtained a result that the time dependent fluctuation of surface scales with the momentum of the polishing particle.

## 1. INTRODUCTION

Polishing is one of the very old processing techniques which are used by mankind from ancient time. But it is still an exceedingly important technique because it is used for ultra-precise processing to get the smoothest surface today. Actually, a finished surface of silicon wafer which is a basic material of LSI is accomplished by applying this technique.[1,2]

It is an important problem to know how the surface changes during the polishing process. A schematic machine structure of float polishing is shown in Fig. 1.[1,2] A characteristic point of this machine is that there is no direct contact between a polished substance (sample) and the machine. There is a space between the diamond turn surface (see Fig. 1) and the polished substance. The space is filled with polishing fluid which is pure water containing many polishing particles. The polished substance is turned round the center axis of the machine, and also turned round the center axis of the sample holder. Therefore, the surface of polished substance is hit by particles from every direction in the polishing fluid. The polishing particles which have enough hardness to sharpen a material are contained in the polishing liquid. When the polishing particle hits the surface of a substance,

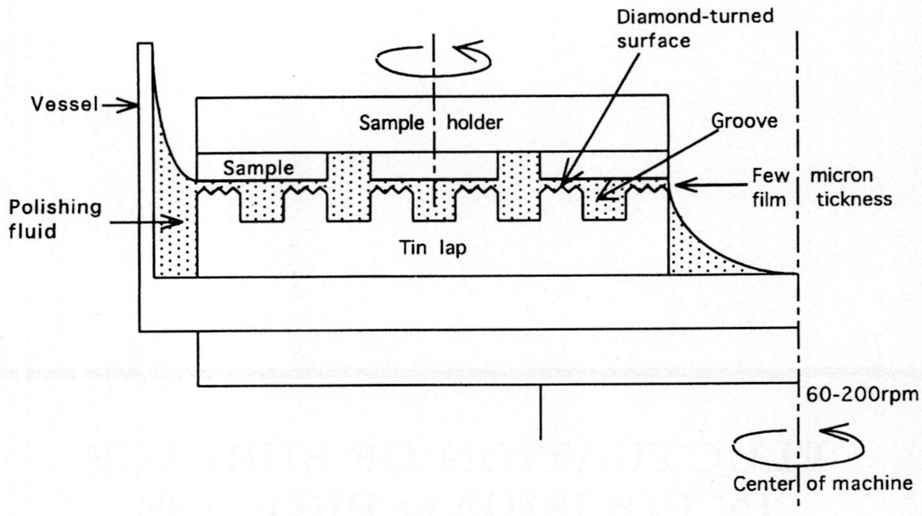


**Fig. 1** Schematic structure of float polishing machine.

some parts of the polished surface are removed. (The number of particles which are removed depends on the momentum of the polishing particles.) Finally, we get the smooth material surface, which has much less fluctuation.

## 2. MODEL FOR POLISHING

The basic mechanism of polishing is that the polishing particle removes some particles at the surface which it strikes.

We assume that the polished substance is constituted of many particles which occupy the lattice points of the two-dimensional square lattice. For the sake of simplicity, it is assumed that the polishing particles strike a particle at the lattice point from the left. A rough surface of substance is prepared which is made by using random numbers of Gaussian distribution in the beginning. A periodic boundary condition is used for the horizontal direction. When the polishing particle hits a lattice point of a substance, a part of the constituting particles in the substance is removed from the main body of the substance. The number of particles removed assumed to be proportional to the momentum of the polishing particle, and we represent the magnitude of the momentum of the polishing particle by the number of particles, $k$, which is removed from the surface. We suppose that the particle number is kept constant during a simulation. The time, $t$ is increased by one when the polishing particle strikes at a lattice point. We consider that the polishing particle can hardly run into the deep groove. The probability that the polishing particle strikes a point of a surface is inversely proportional to (height),[4] and the height is measured from the lowest groove.

Figure 3(a) shows an example of a polishing process when the particle number taken off is four. When the polishing particle strikes a part of a material from the left hand side, four particles are removed from the polished material. These particles are selected by priority of distance from the hit site and are removed from the polished material.

By the way, we have a case that some particles are separated from the polished part when some particles were removed, as shown in Fig. 3(b).

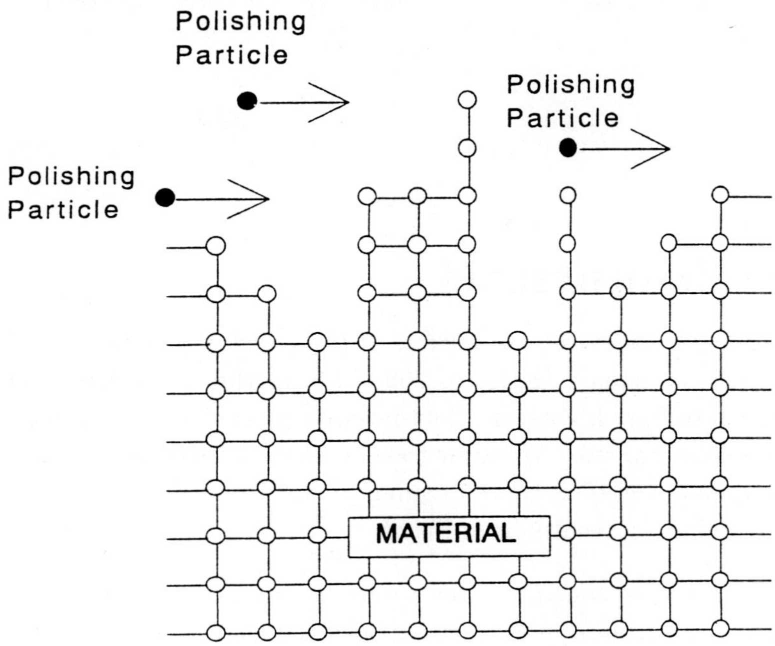


**Fig. 2** A polishing model on the square lattice. The roughness of surface of material is produced according to Gaussian distribution. The direction of polishing particles is fixed.

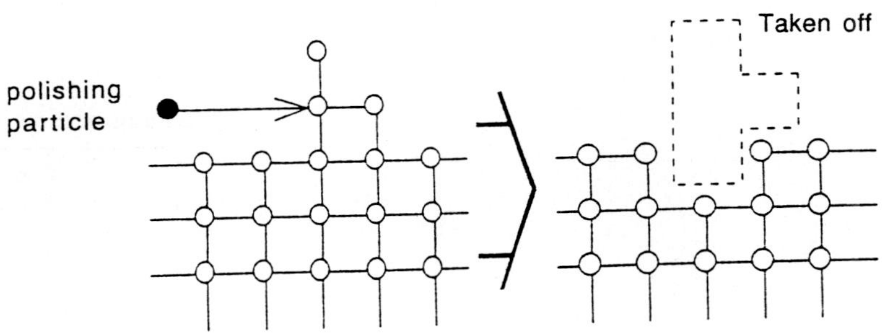


**Fig. 3** **(a)** Polishing process (1). The number of polished particles which are removed from the polished material is four. The polishing particle hits a site of material. Four particles are selected by priority of distance from hit site and are taken from material.

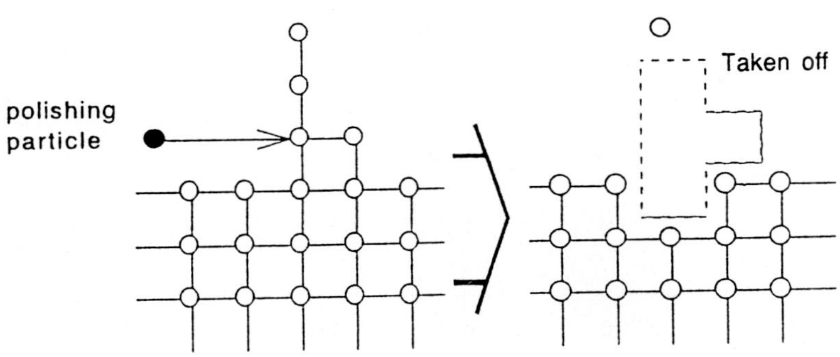


**Fig. 3** **(b)** Polishing process (2). The number of polished particles which are removed from material is four. After removal of four particles, a particle remains as part from the polished material. The particle is eliminated from the material.

The fluctuation $\sigma$ of the surface is calculated by the following expression:

$$\sigma = \sqrt{(h - \bar{h})^2}\,,$$

where $h$ is the height of surface from the bottom of material, $\bar{h}$ is an average of the height.

## 3. SIMULATIONAL RESULTS

The number of particles removed by a strike of the polishing particle is taken from two to ten on two-dimensional square lattice of $200 \times 140$ in this simulation. The result of this simulation is shown in Fig. 4, with the fluctuation $\sigma$ plotted along the vertical axis and the time $t$ along the horizontal axis. We draw the curves for $k$ from two to ten. The data shown in Fig. 4 are averaged over 1000 times of simulations.

We obtain the following results from Fig. 4.

- The fluctuation becomes constant after a cross-over time.
- The larger the momentum of polishing particle, the faster the surface is polished.
- The smaller the momentum of polishing particle, the smaller is the final fluctuation.

These characteristics can be found in real polishing processes.

We check the scaling collapse of the fluctuation of a rough surface.

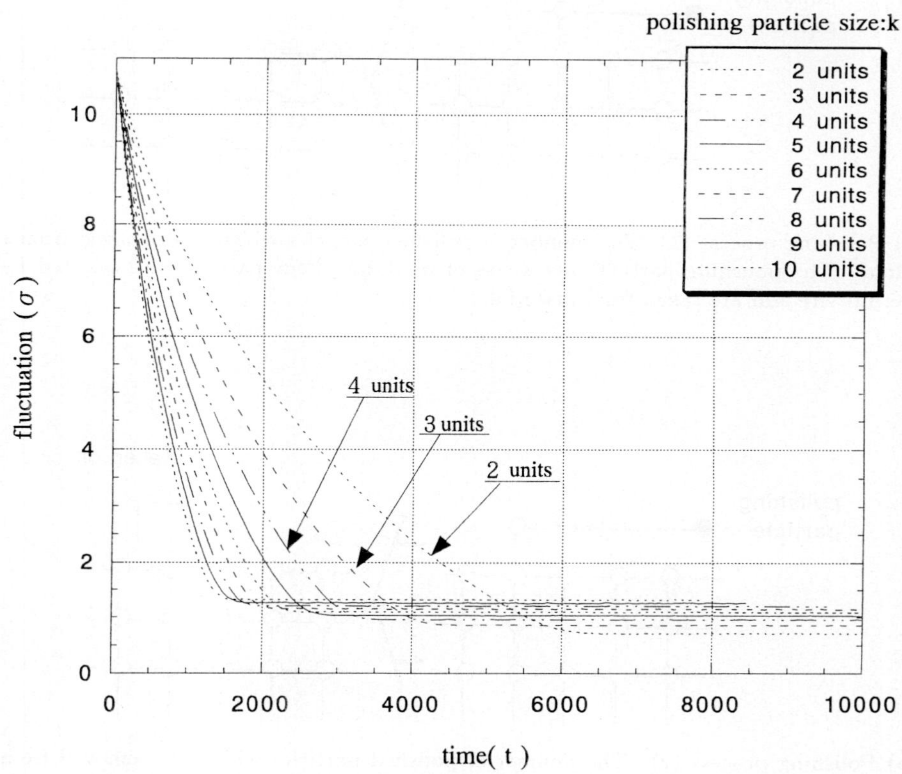


**Fig. 4** The fluctuation $\sigma$ vs. the time.

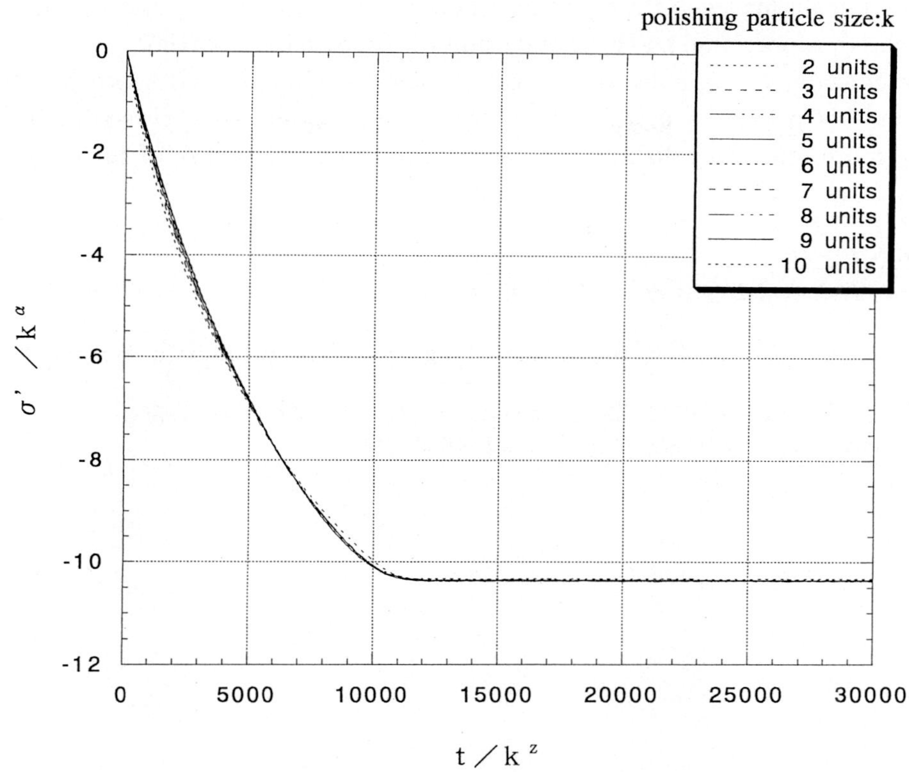


**Fig. 5** After scaling collapse. The vertical axis modifies that divide $\sigma'$ by $k^{\alpha}$, and the horizontal axis modifies that divide the time ($t$) by $k^z$, where $\alpha = -0.038$ and $z = -0.88$.

In Fig. 5, $\alpha = -0.038$ and $z = -0.88$. The data collapse for every line is satisfied.

## 4. DISCUSSION

We have proposed the simple simulation for a float polishing process. It has been found that the scaling function[3] for the fluctuation is expressed by:

$$\sigma(k,\, t) = -k^{-0.038} f\left(\frac{t}{k^{-0.88}}\right) + \overline{\sigma_0} \quad (\overline{\sigma_0} = 10.787)\,,$$

where

$$f(x) = \begin{cases} x & \text{for } x \ll 1000\,, \\ \text{const} & \text{for } x \gg 1000\,. \end{cases}$$

We get the relation:

$$\alpha + z = -0.038 - 0.88 \sim -0.92\,.$$

This result is different from the general relation:

$$\alpha + z = 2\,,$$

which is well known for the KPZ equation.[4] Our scaling parameter is not a sample width $l$, but $k$, which is proportional to the momentum of the polishing particle.

We have considered the polishing process in the two-dimensional space for the sake of simplicity in this paper. We will extend the present model to a three-dimensional one introducing more realistic polishing process such as potential energy of substance.

## REFERENCES


1. Y. Namba, *Technical Digest at the Topical Meeting on the Science of Polishing*, OSA, TuB-A2 (1984).
2. S. F. Soares, D. R. Baselt, J. P. Black, K. C. Jungling and W. K. Stowell, *Applied Optics* **33**, 89 (1994).
3. M. Kardar, G. Parisi and Y.-C. Zhang, *Phys. Rev. Lett.* **56**, 889–892 (1986).
4. T. Vicsek and F. Family, *Phys. Rev. Lett.* **52**, 1669 (1984).

Fractals, Vol. 4, No. 2 (1996) 219–225


# AC IMPEDANCE STUDY OF PATTERN FORMATION OF SILVER ELECTRODEPOSITS

IWAO MOGI and MASAO KAMIKO
*Institute for Materials Research, Tohoku University, Katahira, Sendai 980-77, Japan*

**Abstract**

Electrochemical impedance was examined in the two-dimensional growth of Ag electrodeposits to investigate the rate-determining step in the pattern formation of the aggregates. The frequency dispersions of the *in-situ* ac impedance during the electrodeposition demonstrated that the open-ramified growth is diffusion-limited, the dendritic growth is reaction-limited, the homogeneous growth is electric-migration-limited and the DBM (dense branching morphology) growth is both diffusion and electric-migration controlled.

## 1. INTRODUCTION

The diffusion-limited-aggregation (DLA) model proposed by Witten and Sander[1] has stimulated scientists to make the DLA pattern in laboratories. Two-dimensional growth of an electrodeposited metal is a good experimental candidate of the DLA model, and many experimental studies have been accomplished in various electrochemical conditions.[2–4] As a result, it was found that several representative patterns exist in the electrodeposits: DLA-like, dense branching morphology (DBM), dendritic, needle-like and homogeneous patterns. Many computer simulations have been done to reproduce such patterns.[5–9] However, the

E-mail: mogi@vostok.imr.tohoku.ac.jp

correspondence between the experimental results and the model simulations remains obscure because of the lack of detailed information about the electrochemical process for each pattern, especially concerning the rate-determining step.

The ac impedance measurement is useful for the investigation of the rate-determining step of the electrodeposition. In this paper we report the *in-situ* ac impedance behavior during the Ag electrodepositions, forming DLA-like, DBM, dendritic and homogeneous patterns.

## 2. EXPERIMENTAL

The Ag electrodeposition was performed at a constant electrode potential in a thin-layer (0.3 mm) solution cell between two glass plates, as shown in Fig. 1. The electrolytic solution was a $AgNO_3$ aqueous solution with the concentration of 0.01 ~ 0.1 M (M = mol/l), containing 0.05 ~ 0.5 M $NaClO_4$ as a supporting electrolyte. The cell consists of three electrodes: a carbon rod ($\phi$ 2 mm) as a working electrode, a $Ag/Ag^+$ (0.02 M) electrode as a reference electrode, and a Ag ring ($\phi$ 50 mm) as a counter electrode. The electrodeposition occurs on the working electrode whose electric potential was controlled against the reference electrode by a potentiostat in the Electrochemical Analyzer BAS100B/W. The growth pattern was observed by a CCD camera and recorded by a videotape recorder. The image data were transferred to a personal computer and digitized on it.

The *in-situ* ac impedance was measured during the electrodeposition at a constant potential by means of the noise-FFT method over a frequency range from 1 Hz to 1 kHz, with an ac voltage of 10 mV (peak-to-peak). Typical frequency dispersions of the complex impedance $Z = Z' + jZ''$ ($j = \sqrt{-1}$) for the electrode processes are shown in the Cole-Cole plot $-Z''$ vs. $Z'$ in Fig. 2 with the equivalent circuits,[10] which consist of the solution resistance $R_S$, the capacitance of the electric double layer at the electrode interface $C_{dl}$, the reaction resistance at the electrode $R$ and the Warburg impedance (diffusion impedance)

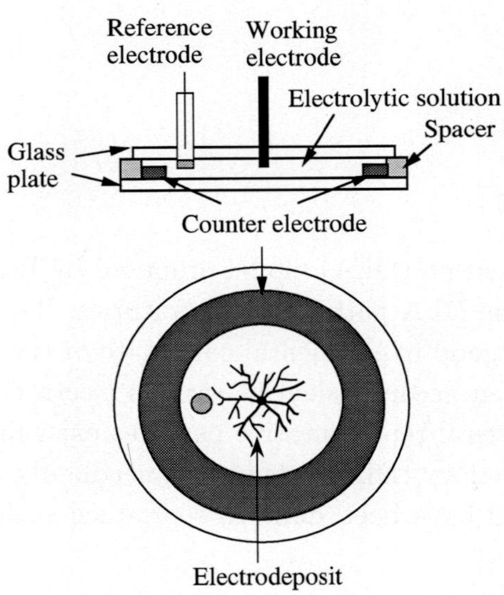


**Fig. 1** Schematic of the electrochemical thin-layer cell.

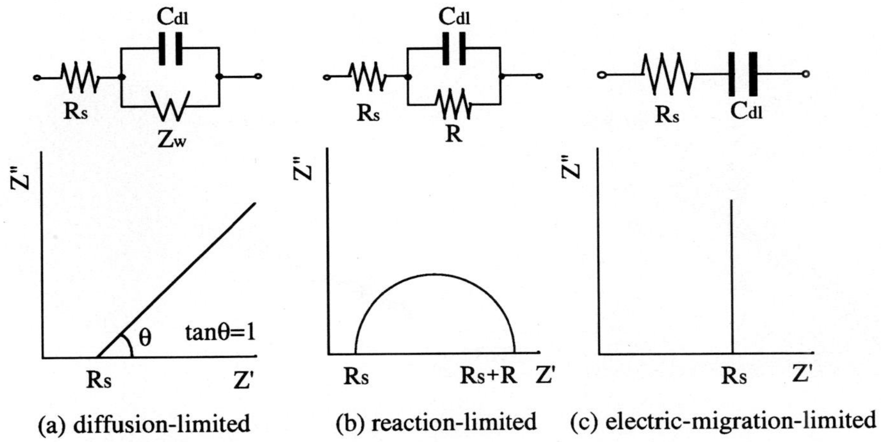


**Fig. 2** Representative equivalent circuits of the ac impedance $Z = Z' + jZ''$ for electrode processes and their frequency dispersion curve (the Cole-Cole plot).

$Z_W$. The Cole-Cole plot shows a straight line with the slope $\Delta = \tan\theta = 1$ in the diffusion-limited process [Fig. 2(a)] and a semicircle in the reaction-limited process [Fig. 2(b)]. If the electric migration is dominant in the mass-transfer-limited process, the Cole-Cole plot shows a vertical line [Fig. 2(c)].

## 3. RESULTS

Figure 3 shows four representative growth patterns of the Ag electrodeposits. The open-ramified DLA-like pattern appears at −500 mV in the 0.01 M $AgNO_3$ solution [Fig. 3(a)], and the DBM with a smooth circular periphery grows at −400 mV in 0.02 M $AgNO_3$ [Fig. 3(b)]. The dendritic growth is seen at −200 mV in 0.08 M $AgNO_3$ [Fig. 3(c)], and the homogeneous pattern without clear branches appears at −1000 mV in 0.1 M $AgNO_3$ [Fig. 3(d)]. The experimental results of the growth pattern is summarized in a phase diagram (the $Ag^+$ concentration vs. the electrode potential $E$) in Fig. 4, where dashed lines are not clear boundaries, and thus two patterns coexist near the boundaries.

Figure 5 shows the time dependence of the current ($i - t$ curve) during the electrodepositions corresponding to Fig. 3. In the DLA-like growth [Fig. 5(a)] the current decreases at the initial stages, and then makes a plateau, whose height increases with the growth of the pattern. In the DBM growth [Fig. 5(b)], the current increases gradually after the initial decrease. Characteristic current fluctuation is seen in the $i-t$ curve of the dendritic growth [Fig. 5(c)]. In contrast to above three cases, in the homogeneous growth the initial current decrease is not seen, and the current rises rapidly with the rapid growth of the deposit.

Figure 6 shows the Cole-Cole plots of the *in-situ* ac impedance during the electrodepositions corresponding to Fig. 3. The Cole-Cole plots of the DLA-like and the DBM growth are on nearly straight lines in low frequency ranges with their slopes $\Delta$ of $\Delta = 1$ and $\Delta > 1$, respectively. The plots are semielliptic in the dendritic growth, and they are on a vertical line in a low frequency range in the homogeneous growth.

**Fig. 3** Growth patterns of the Ag electrodeposits; **(a)** the DLA-like pattern at −500 mV in 0.01 M $AgNO_3$; **(b)** the DBM at −400 mV in 0.02 M $AgNO_3$; **(c)** the dendritic pattern at −200 mV in 0.08 M $AgNO_3$; and **(d)** the homogeneous pattern at −1000 mV in 0.1 M $AgNO_3$.

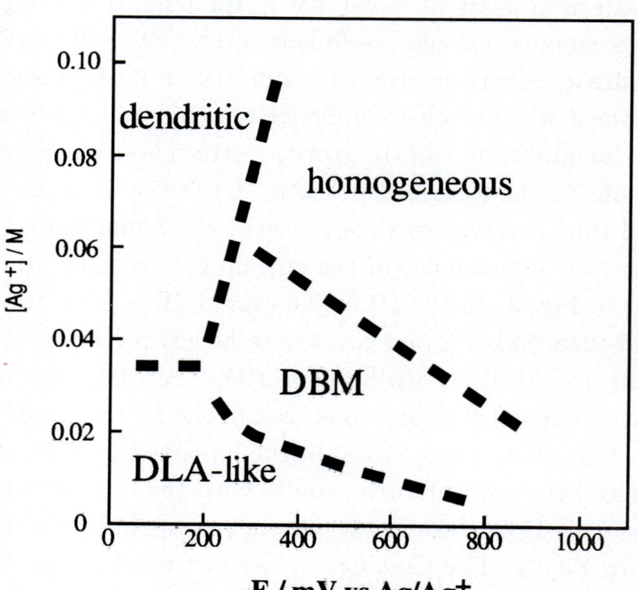


**Fig. 4** Phase diagram of the growth morphology of the Ag electrodeposits; the $Ag^+$ concentration $[Ag^+]$ vs. the electrode potential $E$ referred to the $Ag/Ag^+$ electrode.

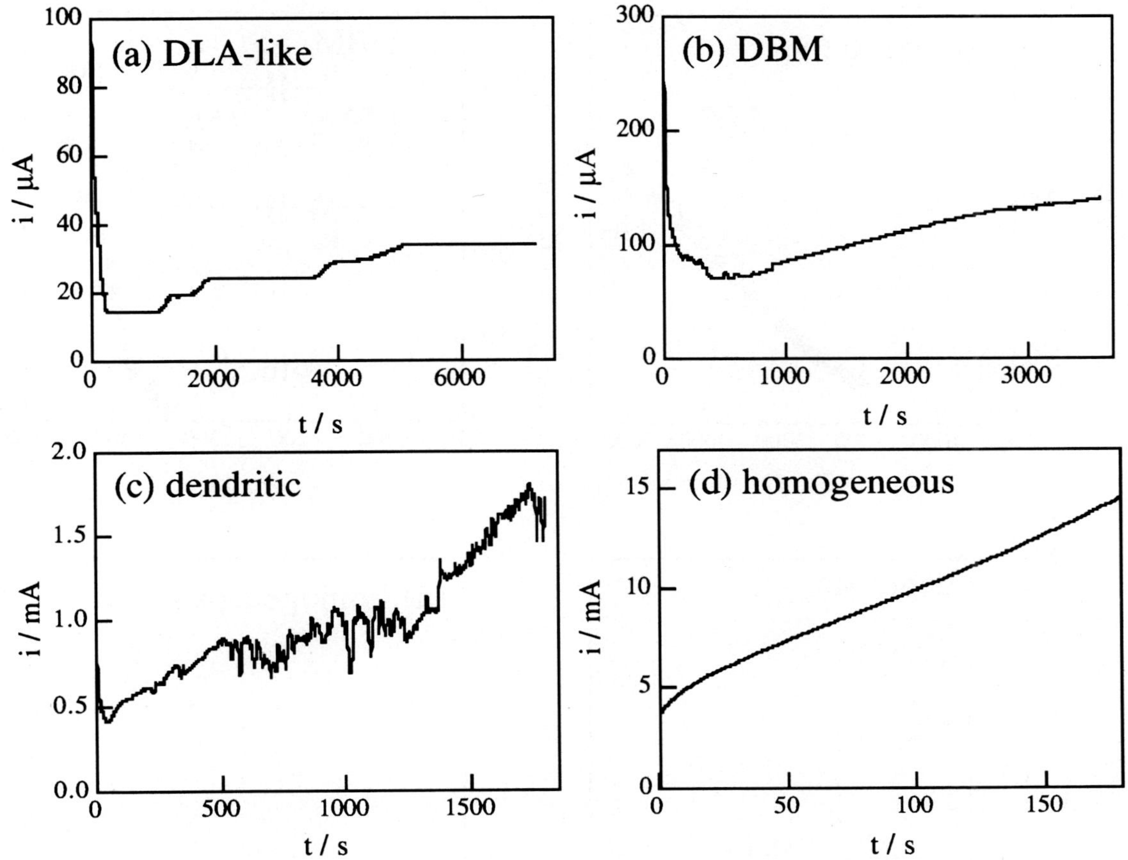


**Fig. 5** Time dependence of the current during the Ag electrodeposition for four growth patterns.

## 4. DISCUSSION

From the results of the ac impedance measurements, we can deduce the rate-determining step for each growth pattern. The Cole-Cole plots of the DLA-like growth [Fig. 6(a)] indicates that the electrodeposition is diffusion-limited [see Fig. 2(a)]. In the $i - t$ curve in Fig. 5(a), the initial current decrease is regarded as the formation of a diffusion layer and the plateaus indicate the diffusion current. Thus, the DLA-like growth has proved to be electrochemically diffusion-limited in the Ag electrodeposition.

The Cole-Cole plots of the dendritic growth [Fig. 6(c)] indicates that the electrodeposition is reaction-limited [see Fig. 2(b)], though the experimental plots are not on the perfect semicircle but on the depressed semicircle which comes from the geometrical inhomogeneity of the reaction surface.[11] In such a case, the equivalent circuit can be given by a ladder network as an inset in Fig. 6(c), where $C_n$ is the double-layer capacitance of the $n$th reaction site. On the other hand, the dendritic pattern was successfully reproduced by the model simulation taking account of the surface diffusion of particles,[8,9] in which the particles move along the growing surface and stick on kinks or steps. Thus, the surface diffusion process is considered to lead to the reaction-limited properties in the ac impedance measurement. The current fluctuation in the $i - t$ curve [Fig. 5(c)] is a characteristic of the dendritic growth and it arises from the fast growth of the needle parts and the slow branching processes.[12]

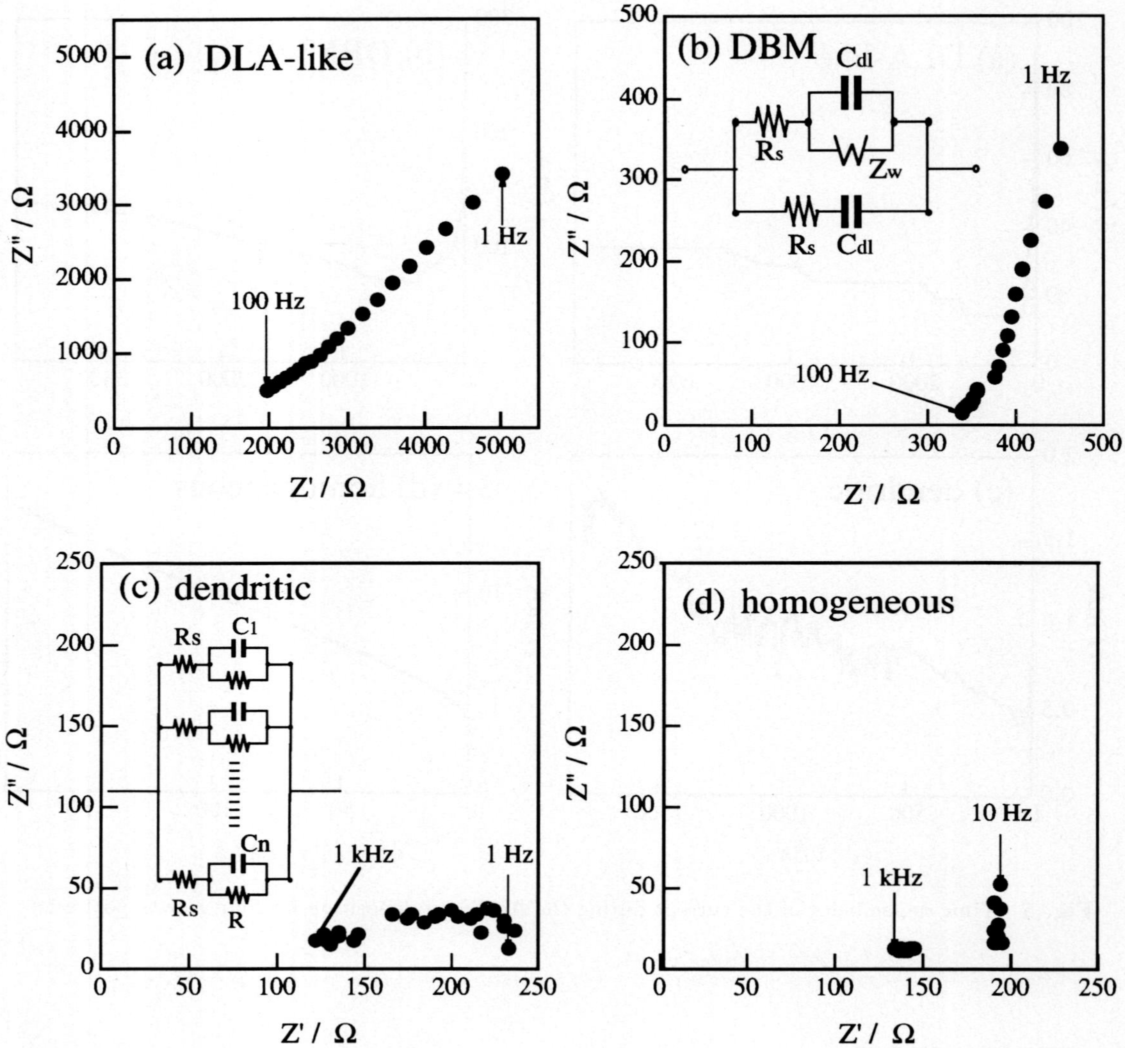


**Fig. 6** The Cole-Cole plots of the *in-situ* ac impedance during the Ag electrodepositions. The insets in (b) and (c) are proposed equivalent circuits for the electrodeposition processes.

The Cole-Cole plots of the homogeneous growth [Fig. 6(d)] indicates that the electrodeposition is electric-migration-limited (EML) [see Fig. 2(c)]. A characteristic of the EML process is reflected in the fact that the initial current decrease is unobservable in the $i - t$ curve in Fig. 5(d). The EML deposition corresponds to the DLA model simulations with radial drifts,[6,7] in which the DLA pattern changes into the homogeneous one with increasing probability of the radial drift.

The Cole-Cole plots of the DBM growth [Fig. 6(b)] show the line with the slope $\Delta > 1$. The equivalent circuit of such an electrode process can be written as a parallel circuit of Figs. 2(a) and 2(c) [see an inset in Fig. 6(c)]. Thus, the electrodeposition is controlled by both diffusion and electric migration. This is also suggested by the phase diagram of Fig. 4 in which the DBM phase is between the DLA-like and the homogeneous phases. The screening effect due to the diffusion leads to the deep valley between branches, and the radial drifts due to the electric migration suppress the geometrical fluctuation of the growth front. The balance of these two effects brings about the DBM growth.

## ACKNOWLEDGMENTS

We wish to thank Dr. Y. Fukunaka of Kyoto University for valuable discussion and Professor M. Motokawa for encouraging this work. We are indebted to Mr. S. Okubo for data processing on a computer.

## REFERENCES

1. T. A. Witten and L. M. Sander, *Phys. Rev. Lett.* **47**, 1400 (1981).
2. M. Matsushita, M. Sano, Y. Hayakawa, H. Honjo and Y. Sawada, *Phys. Rev. Lett.* **53**, 286 (1984).
3. Y. Sawada, A. Dougherty and J. P. Gollub, *Phys. Rev. Lett.* **56**, 1260 (1986).
4. D. Grier, E. Ben-Jacob, R. Clarke and L. M. Sander, *Phys. Rev. Lett.* **56**, 1264 (1986).
5. P. Meakin, J. Feder and T. Jøssang, *Phys. Rev.* **A43**, 1952 (1991).
6. P. Meakin, *Phys. Rev.* **B28**, 5221 (1983).
7. T. Nagatani, *Phys. Rev.* **A39**, 438 (1989).
8. S. Ohta, in *Pattern Formation in Complex Dissipative Systems*, ed. S. Kai (World Scientific, Singapore, 1992), p. 93.
9. J. M. García and F. Otáloa, *Physica* **A178**, 415 (1991).
10. M. Sluyters-Rehbach and J. H. Sluyters, *Comprehensive Treatise of Electrochemistry*, Vol. 9 (Plenum Press, 1984), p. 177.
11. Y. Zhou, R. Ichino, M. Okido and T. Oki, *Mat. Trans. JIM* **32**, 1159 (1991).
12. T. Horiuchi and R. Kammel, *Jpn. J. Appl. Phys.* **11**, 6 (1972).

Fractals, Vol. 4, No. 3 (1996) 227–235


# TRANSPORT PHENOMENA IN PERCOLATING STRUCTURES: A KEY FOR THE ANALYSIS AND REINTERPRETATION OF SOME PRACTICAL MODELING PROBLEMS

J. S. ANDRADE JR.[*,†,‡], D. A. STREET[*], Y. SHIBUSA[*] and N. ITO[§]

[*]*Department of Production Technology, Showa Denko K. K., 13-9 Shiba Daimon, 1-Chome, Minato-ku, Tokyo 105, Japan*

[†]*Cybernet Systems Co. Ltd., Nissay Otowa Bldg., 2-15-6 Otsuka, Bunkyo-ku, Tokyo 112, Japan*

[‡]*Departamento de Física, Universidade Federal do Ceará, 60451-970 Fortaleza, Ceará, Brazil*

[§]*Department of Applied Physics, Faculty of Engineering, The University of Tokyo, Hongo 7-3-1, Bunkyo-ku, Tokyo 113, Japan*

## Abstract

The percolating morphology is adopted as a reference conceptualization to evaluate the "fractality" effect on the transport phenomena in disordered systems. The relevance of this approach is demonstrated using two practical case studies: conducting polymers and porous media. In the first case, a conceptual model for conducting polymers is proposed in terms of a random resistor network subjected to percolation disorder. The effect of topological and morphological disorder on the conducting behavior of an idealized system is investigated and some insights, are given as to the way in which conducting polymers could be designed. In the second case, the examination of effective transport properties in percolation-like porous structures enables us to reinterpret the classical guidelines for the characterization of fluid flow through porous beds.

## 1. INTRODUCTION

According to Mandelbrot,[1] the percolation theory embodies an admirably complete fractal description in which physics and geometry can play almost indistinguishable roles. On its own, this statement fully justified the widespread application of the percolation paradigm in problems of practical interest in heterogeneous chemistry[2] and polymer science.[3] In most cases, the approach can rely entirely on the statistical and geometrical properties of percolation clusters as an idealized model framework.[4] In studying transport phenomena in disordered systems, however, one has to superimpose an additional process to the topological and morphological features of the percolative representation.[5] Typical examples include the well-known problems of flow through porous media and metal-to-insulator transition in polymer composites and alloys. *These are systems in which the interplay between structure and phenomenology has to be closely investigated, as it might be the ultimate factor determining optimum material properties.* For instance, when associated with the conductivity behavior of percolating random resistor networks (RRN), a multitude of fractal dimensions or, more specifically, the multifractal formalism[4] could also be viewed as a tool for the systematic categorization of this structural-phenomenological interaction. In the present paper, the aim is to corroborate these facts by demonstrating some potential perspectives on the utilization of the percolating morphology as a convenient geometrical archetype for understanding electrical transport in conducting polymers and fluid flow through porous media. This paper is organized as follows. In Sec. 2, a network model is introduced as a theoretical framework for studying the metal-to-insulator transition via doping in conducting polymers. In Sec. 3, an assessment is made of the influence of convection on the overall flow behavior of percolation-like porous structures. Finally in Sec. 4, we summarize the paper with some general concluding remarks.

## 2. ELECTRICAL TRANSPORT IN CONDUCTING POLYMER NETWORKS

The investigation of electrical transport mechanisms responsible for the wide range variations in conductivity with doping[6–8] has been the main focus of the most recent theoretical studies on the field of conducting polymers. However, there is a clear gap between micro-[9] and macroscopic[10] perspectives in representing the high conductivity state of such important materials. Few studies have been dedicated to providing a consistent description for the polymeric network morphology.[11] In a recent study,[12] the basic features of a random network model have been adopted as a hypothetical framework to evaluate the influence of spatial disorder on the conducting properties of computer generated polymer clusters. The results from this work clearly demonstrated the importance of including structural characteristics into the model to account for the intrinsic diversity of charge transfer mechanisms which dictate the conducting behavior of the polymer material. Here, the purpose is to study the effect of percolation disorder on the electrical behavior of conducting polymer networks (CPN). Basically, this representation is a modified version of the standard random resistor network model (RRN) where, instead of single bonds, the percolation elements are conducting polymer chains randomly allocated in the lattice.[12] The central idea is to show that the nature of the interchain connections can be a limiting factor influencing the size scaling properties of the overall maze conductivity at the critical point.

The polymer unit adopted in the simulations consists of a conjugated segment with a specified number $l$ of intrachain conductance elements ($\sigma_c$) surrounded by two possible types of interchain connections: hopping ($\sigma_h$) or doping ($\sigma_d$) conductances. In this idealization, an intrachain conductance element must be viewed as a directional unit comprising either one or several monomers of the macromolecule. A pictorial representation of our idealized physical system for a typical network realization right at the percolation threshold is displayed in Fig. 1. In all simulations, the proportion of doping interchain barriers replacing hopping conductances around each chain is kept constant and equal to 50%. By specifying the polymer chain length and values of the elementary conductances, the next step in the simulation refers to the generation of a polymeric network confined to a lattice of finite dimension and regular topology. Two-dimensional square lattices have been used and periodic boundary conditions have been applied to reduce the finite size effect in the transverse direction of the system ($y$ axis in Fig. 1). The polymeric units are sequentially allocated using a self-avoiding scheme so that interchain segments between different polymer units are restricted to hopping and/or doping conductances. For a given realization, the network generation is terminated when one of the polymer clusters percolates in the longitudinal direction ($x$ axis in Fig. 1). Subsequently, an arbitrary voltage difference between the two opposite sides in this direction is applied and Kirchhoff's current law is imposed at each nodal point of the resulting configuration. The effective conductivity of the network is then calculated by solving a large system of coupled linear algebraic equations.[4,12] We assume that equal values can be assigned for the intrachain and doping conductances ($\sigma_c = \sigma_d$), whereas the conductance magnitude due to hopping contacts can be systematically changed according to the parameter ratio $\alpha = \sigma_h/\sigma_d$. A change in this variable would be analogous to a modification in the temperature of the conducting polymer system.[10,12]

Simulations have been carried out using a number of realizations ranging from 4000 to 100 for several lattice dimensions going from $20 \times 20$ to $200 \times 200$ nodal points. Once the

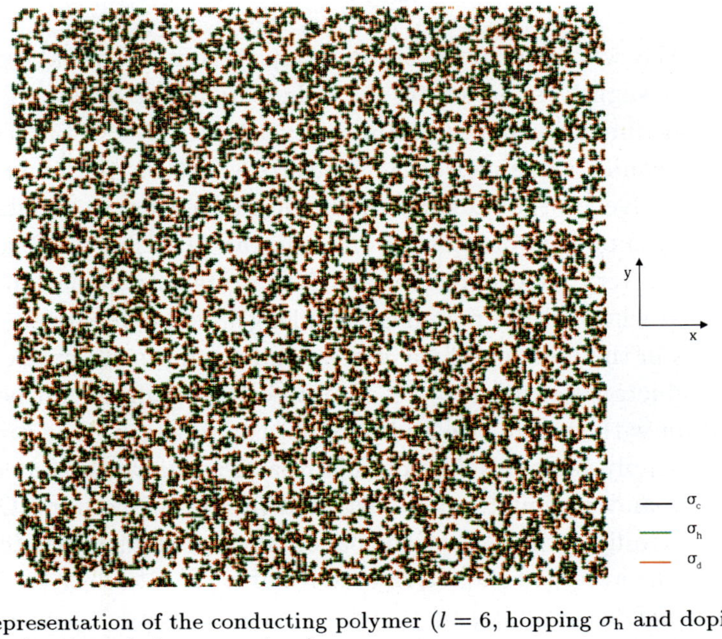


**Fig. 1** Pictorial representation of the conducting polymer ($l = 6$, hopping $\sigma_h$ and doping $\sigma_d$ conductances are equally distributed as interchain connections).

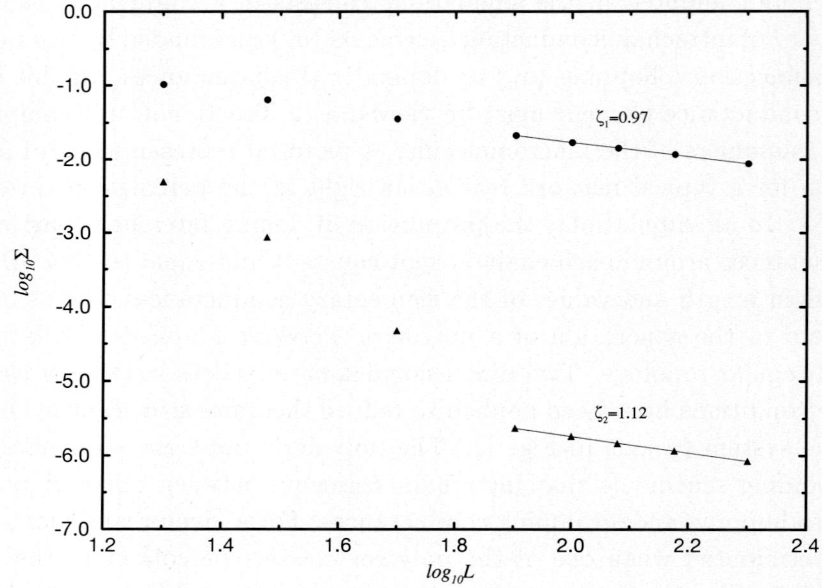


**Fig. 2** Logarithmic variation of the network conductivity $\Sigma$ with system size $L$ for two different values of the conductance ratio $\alpha$ ($l = 6$, hopping $\sigma_h$ and doping $\sigma_d$ conductances are equally distributed as interchain connections). Circles correspond to $\alpha = 1$ and triangles to $\alpha = 10^{-5}$.

spanning cluster is identified, the network conductivity is calculated for different values of the conductance ratio $\alpha = \sigma_h/\sigma_d$. In this paper, we show preliminary results with polymer chains of six units randomly distributed in the lattice. Further simulations using more realizations and different chain lengths are underway and will be reported in a subsequent publication.[13] Figure 2 shows the logarithmic dependence of the network conductivity $\Sigma$ on the system size for two different values of $\alpha$. At $\alpha = 1$, the scaling relationship,

$$\Sigma \propto L^{-\zeta} \tag{1}$$

seems to apply for the whole range of system sizes giving a critical exponent equal to $1.07 \pm 0.02$, which is slightly superior to the value predicted by standard Monte Carlo simulations with two dimensional percolating networks ($\zeta \cong 0.9745$, see Ref. 4). Such discrepancy can be readily explained as a finite-size effect on the conductivity of small lattices due to the relatively large polymer chain length used in the simulations. As shown in Fig. 2, the critical exponent obtained from a linear fit using only large system sizes ($\zeta_1 = 0.97 \pm 0.01$) is in very good agreement with the typical percolation value. Likewise, simulations performed with distinct polymer chain lengths showed entirely similar behavior, except for differences in the location of the corresponding finite-size crossovers.[13] At a very low value of the conductance ratio ($\alpha = 10^{-5}$), interesting changes can be observed in the conductivity variation with system size and polymer chain length. The presence of a sharp finite-size crossover simply indicates the minimum lattice size where a percolating path of high conductances $\sigma_c$ and $\sigma_d$ could not be detected in any realization. Obviously, at this value of $\alpha$ even an insignificant number of such anomalous configurations tends to produce a dramatic increase in the average conductivity values of small lattice sizes. Their occurrence is a result of the deficient sampling of "singly connected bonds" available as interchain bridges in the conducting backbones of small lattices.[13]

As in the previous case, the critical exponent has been calculated disregarding the conductivity data from small lattice sizes (see Fig. 2). Nevertheless, the discrepancy between the calculated value ($\zeta_2 = 1.12 \pm 0.02$) and the universal conductivity exponent can no longer be justified in terms of typical finite-size effects. The objective here is to demonstrate that at intermediate values of the ratio $l/L$, a decrease in the parameter $\alpha$ tends to increase the conductivity exponent towards the fractal dimension of the "minimal self-avoiding walk" through the conducting backbone.[4] At low values of the conductance ratio, the hopping mechanism is dominant due to the high resistance of these interchain barriers. Following the "link-nodes-blobs" model,[14] those paths through the "blobs" having a relatively large number of hopping connections will contribute very little to the electrical transport. At $\alpha = 10^{-5}$, this might be analogous to a "minimal self-avoiding walk" process since all paths through a given blob are virtually eliminated, excluding the one with the lowest resistance.

(a)

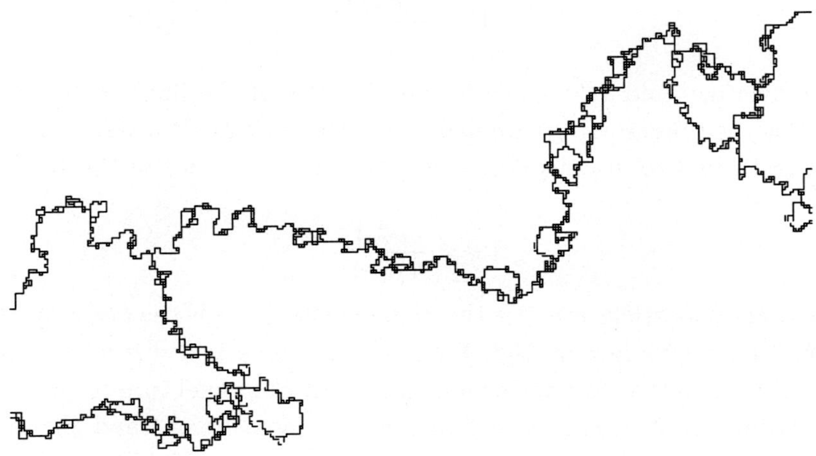

(b)

**Fig. 3** **(a)** Network carrying 99% of the total bond current through the original percolating lattice ($\alpha = 1$). **(b)** Network carrying 99% of the total bond current through the original percolating lattice ($\alpha = 10^{-5}$).

We would then expect a critical exponent close to $D_{\min} = 1.13$,[4] which is in very good agreement with the results of our simulations. By means of a suitable reconstruction technique,[15] a persuasive evidence for this fact can be visualized from Figs. 3(a) and 3(b), where the network of bonds carrying 99% of the total current in the original structure (see Fig. 1) is displayed for $\alpha$ values of 1 and $10^{-5}$ respectively. Compared with Fig. 3(a), the tenuous aspect of the conducting pathway in the 99% network shown in Fig. 3(b) is a clear indication that the relevant structures for transport are rather different for distinct $\alpha$ values. Clearly, at sufficiently low values of the ratio $l/L$, one should recover the universal exponent from the finite-size scaling of the network conductivity at the critical point ($\zeta \cong 0.9745$). Indeed, this behavior could also be verified from our simulation results utilizing short polymer chain lengths.[13]

We conclude this section by emphasizing the necessity for a detailed structural description of the electrical transport in real conducting polymers. From the simulation results with our idealized system, it seems that the size scaling properties of the material might be fairly susceptible not only to its morphological and topological features, but also to the operating physical conditions (e.g., temperature). Moreover, we also argue that the conducting backbone of the polymer network at the threshold can exhibit a multitude of fractal dimensions, which will intrinsically depend on the physical conditions of the system. Since the percolation threshold represents the average network realization with a minimum amount of polymer chains that still conducts electricity, this fact might be of great relevance for the design of improved conducting polymer materials.

## 3. FLUID FLOW THROUGH POROUS MEDIA

The modelling and simulation of transport phenomena through porous materials can be of foremost importance in several technological processes (e.g., filtration, catalysis, chromatography, spread of hazardous wastes, petroleum exploration and production, etc.). Concerning fluid flow through porous media, a fundamental index is the permeability coefficient $k$, usually defined in terms of Darcy's law[16]:

$$V = -\frac{k}{\mu}\frac{\Delta P}{L} \tag{2}$$

where $V$ is the average fluid velocity, $\mu$ is the viscosity of the fluid, $L$ is the length of the sample and $\Delta P$ is the pressure drop applied across the system. Precisely, the validity of this relationship is restricted to viscous flow conditions or low values of the Reynolds number, as defined by:

$$\mathrm{Re} = \frac{\rho d_p V}{\mu(1-\varepsilon)} \tag{3}$$

where $d_p$ is the average particle size, $\rho$ is the fluid density and $\varepsilon$ is the porosity of the medium. The breakdown of Darcy's law at high Reynolds numbers is a direct consequence of two nonlinear mechanisms influencing the overall transport of momentum in the interstitial pore space: the contribution of inertial forces to the flow (convection) and the development of turbulence. The evidence of fractal structures in natural and artificial porous media has been supported by several experimental studies.[17] Here, our objective is to provide a preliminary evaluation of the effect of convection on the fluid flow behavior through percolation-like porous structures. Although artificial, this geometrical model introduces well established

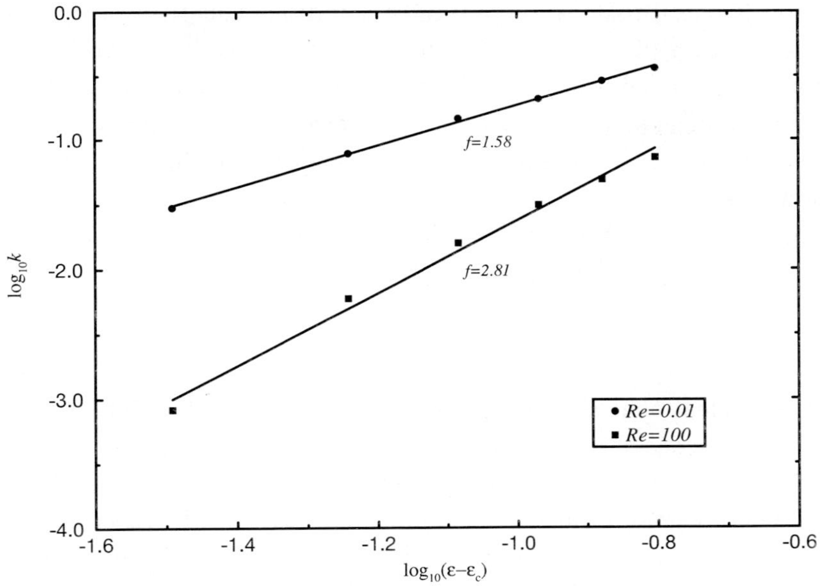


**Fig. 4** Logarithmic plot of $k$ vs $(\varepsilon - \varepsilon_c)$ for two different Reynolds values (from Ref. 20).

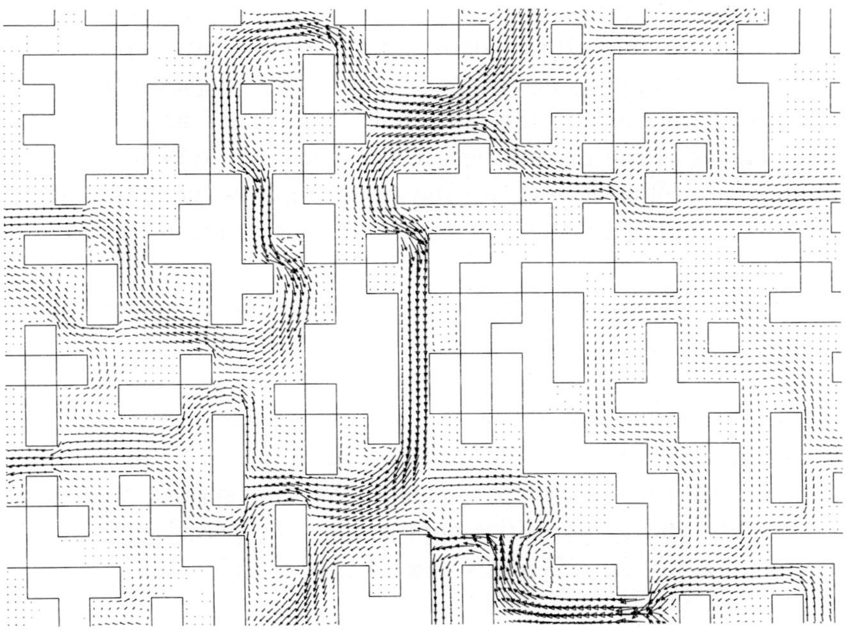

**Fig. 5** Close-up plot of the velocity flow field through a porous medium with low porosity at high Reynolds number ($\varepsilon = 0.65$, Re = 100, from Ref. 20).

features which can be qualitatively related to properties of real pore materials.[17,18] At the percolation threshold, the occurrence of multiple length scales qualifies the "infinite cluster" as a typical fractal object.[4] As a simplified conceptual framework, we adopt the picture of site percolation disorder in two-dimensional square lattices.

The celebrated Navier-Stokes and continuity equations constitute the phenomenological basis for the description of the local flow in the pore space.[19] A detailed account of the numerical scheme used to solve the model equations and calculate the network permeability is reported elsewhere.[20] The results presented in this work have been obtained from simulations performed with $50 \times 50$ square lattices randomly generated to produce different prescribed porosities.

Illustrated in Fig. 1 is the nonuniversal behavior observed by fitting low porosity data with the following power-law relationship:

$$k \propto (\varepsilon - \varepsilon_c)^f . \tag{4}$$

From this plot, it is clear that the two quantities disappear with different critical exponents as the porosity approaches the threshold value. The large discrepancy between the critical exponents at different Reynolds numbers clearly demonstrates that the interplay between structure and phenomenology must be carefully investigated in the analysis of fluid flow through porous materials. Consequently, it seems legitimate to conjecture that the critical exponent for permeability is a function of the Reynolds number, $f = f(\text{Re})$. This behavior can be partially justified in terms of the greater energy dissipation of the system in the presence of inertial forces, causing a higher resistance to flow at large Reynolds values.[20] Interesting features can also be detected from the visual inspection of the velocity field in non viscous conditions through a low porosity network. As shown in Fig. 2, the complicated topology of the porous structure at a low porosity value creates very tortuous pathways for fluid flow. At high Reynolds numbers, the intricate aspect of this channeling effect is enhanced by the appearance of several recirculation paths throughout the entire network. These zones attached to the main flow "backbone" are clearly coincident with the "dead-ends" regions of the percolating cluster. Since these dangling voids do not play any significant role in the fluid flow at low Reynolds numbers (viscous conditions), the permeability coefficient calculated at non viscous conditions must be strongly related to their occurrence.

Summarizing, the results presented in this section give cogent evidence of the nonuniversal behavior of the critical exponent relating flow permeability coefficients and porosities near the percolation threshold and above the range of Darcy's law validity. In analyzing flow through real percolating/fractal porous media, the strong nonlinear effect introduced by inertial terms contributing to the momentum transport can be responsible for large deviations from typical values of scaling exponents. We thus suggest that outside the range of Darcy's law validity and at porosities approaching the percolation threshold, different scaling relations shall apply for different values of the Reynolds number.

## 4. SUMMARY

This paper represents only one among many related works where the main objective is to develop a systematic methodology for modelling transport phenomena in disordered systems. As has been mentioned, this task requires a proper combination of structural and phenomenological descriptions to be consistently performed. The success of the percolation/fractal approach surely stems from the rich and practical geometrical insights it provides on the essential aspects of the problem. Hopefully, the modelling technique developed in this study will enable us to provide more useful information on the topological aspects of the electrical transport in conducting polymers and flow through porous materials. We believe that this information can be used to produce a more realistic picture about both physical phenomena and allow for a better understanding of their many connections with important applications in technology.

## REFERENCES


1. B. B. Mandelbrot, "An Overview of the Language of Fractals," in *The Fractal Approach to Heterogeneous Chemistry: Surfaces, Colloids, Polymers,* ed. D. Avnir (John Wiley & Sons, Chichester, 1989), Chap. 1.1, pp. 3–9.
2. D. Avnir (ed.), *The Fractal Approach to Heterogeneous Chemistry: Surfaces, Colloids, Polymers* (John Wiley & Sons, Chichester, 1989).
3. P. G. de Gennes, *Scaling Concepts in Polymer Physics* (Cornell University Press, Ithaca, 1979).
4. D. Stauffer and A. Aharony, *Introduction to Percolation Theory* (Taylor & Francis, London, 1992).
5. J. M. Ziman, *Models of Disorder* (Cambridge University, Cambridge, 1979).
6. H. Shirakawa, E. J. Louis, A. G. MacDiarmid, C. K. Chiang and A. J. Heeger, *J. Chem. Soc., Chem. Commun.* **578** (1977).
7. Y. W. Park, A. J. Heeger, M. A. Druy and A. G. MacDiarmid, *J. Chem. Phys.* **73**, 946 (1980).
8. C. O. Yoon, J. S. Yoo, Y. W. Park, K. Akagi and H. Shirakawa, *Synth. Met.* **55-57**, 4848 (1993).
9. Y. Lu, *Solitons and Polarons in Conducting Polymers* (World Scientific Publishing Co., Singapore, 1988).
10. A. B. Kaiser, *Phys. Rev.* **B40**, 2806 (1989).
11. R. H. Baughman and L. W. Shacklette, *J. Chem. Phys.* **90**, 7492 (1989).
12. J. S. Andrade Jr., Y. Shibusa, Y. Arai and A. F. Siqueira, *Synth. Met.* **68**, 167 (1995).
13. J. S. Andrade Jr., N. Ito and Y. Shibusa, in preparation.
14. H. E. Stanley, *J. Phys.* **A10**, L211 (1977).
15. R. F. Angulo and E. Medina, *J. Stat. Phys.* **75**, 135 (1994).
16. F. A. L. Dullien, *Porous Media — Fluid Transport and Pore Structure* (Academic Press, New York, 1979).
17. P. M. Adler, Flow in Porous Media, in *The Fractal Approach to Heterogeneous Chemistry: Surfaces, Colloids, Polymers* ed. D. Avnir (John Wiley & Sons, Chichester, 1989), Chap. 4.2.2, pp. 341–359.
18. M. Sahimi, G. R. Gavalas and T. Tsotsis, *Chem. Engng. Sci.* **45**, 1443 (1990).
19. R. B. Bird, W. E. Stewart and E. N. Lightfoot, *Transport Phenomena* (John Wiley & Sons, New York, 1960).
20. J. S. Andrade Jr., D. A. Street, T. Shinohara, Y. Shibusa and Y. Arai, *Phys. Rev.* **E51**, 5725 (1995).

Fractals, Vol. 4, No. 3 (1996) 237–239


# FRACTAL FORM ANALYSIS

MASUO SUZUKI
*Department of Physics, University of Tokyo, Bunkyo-ku, Tokyo 113*

**Abstract**

The concept of fractal form analysis is introduced, in which fractal metrication is defined for partial objects of fractals as $a\ \mathrm{cm}^D$ or $b\ \mathrm{m}^D$ with the fractal dimensionality $D$.

## 1. FRACTAL METRICATION

As is well known,[1] the fractal dimensionality $D$ is defined by the self-similar scaling exponent of an infinite object. Here, we study a finite fractal object, namely, fractals in a finite region. For example, we consider the Koch curve in a finite region, say, 1 cm. Then, we call[2,3] this partial fractal object the Koch curve of $1\ \mathrm{cm}.^D$ This naming seems quite natural, compared with ordinary geometrical objects. This fractal metrication $\mathrm{cm}^D$ (or $\mathrm{m}^D$, etc.) will be very convenient for studying finite fractal objects theoretically and experimentally. When the unit of metrication is changed, we have the following relation:

$$(3\ \mathrm{cm})^D = 3^D \mathrm{cm}^D = 4\ \mathrm{cm}^D \tag{1}$$

in the Koch curve, because $3^D = 4$ in this case.

In general we can define addition and subtraction of finite fractals with the same dimensionality $D$ as:

$$a\ \mathrm{cm}^D \pm b\ \mathrm{cm}^D = (a \pm b)\mathrm{cm}^D\,. \tag{2}$$

A geometrical interpretation of Eq. (2) is clear, when the fractal forms of the two relevant systems are completely the same. Note that a "negative fractal" appears when $a < b$. In this case,

$$(a - b)\mathrm{cm}^D = (-)(b - a)\mathrm{cm}^D\,. \tag{3}$$

Namely, when $\mathcal{F}$ is an ordinary "positive fractal", $(-)\mathcal{F}$ is called a negetive fractal and it may be interpreted physically as anti-matter, because

$$\mathcal{F} + (-)\mathcal{F} = 0\,. \tag{4}$$

Next we introduce multiplication of two partial fractal objects with the metrics $a\ \mathrm{cm}^{D_1}$ and $b\ \mathrm{cm}^{D_2}$ by:

$$a\ \mathrm{cm}^{D_1} \times b\ \mathrm{cm}^{D_2} = ab\ \mathrm{cm}^{D_1+D_2}\,. \tag{5}$$

For example, a Koch surface will be constructed by moving the Koch curve in another dimension, namely, we have:

$$a\ \mathrm{cm}^{D} \times b\ \mathrm{cm} = ab\ \mathrm{cm}^{D+1}\,. \tag{6}$$

These will be used in the following section.

## 2. FORM ANALYSIS

In order to introduce the concept of fractal form analysis, we define[2,3] the norm of a partial fractal $\mathcal{F}$ in terms of its metrication. For example, the norm $|\mathcal{F}|$ of a partial fractal $\mathcal{F}$ with the metrication a $\mathrm{cm}^D$ is defined as:

$$|\mathcal{F}| = a\,. \tag{7}$$

Thus, the magnitude of the norm $|\mathcal{F}|$ depends on the unit (say, $\mathrm{cm}^D$ or $\mathrm{m}^D$).

The continuity of the pattern which depends on the parameter $x$ is defined by the condition:

$$\lim_{x\to y} |\mathcal{F}(x) - \mathcal{F}(y)| = 0 \quad \text{or} \quad \lim_{y\to x} \mathcal{F}(y) = \mathcal{F}(x)\,. \tag{8}$$

The derivative of the fractal form $\mathcal{F}(x)$ with respect to the form $\mathcal{F}(x)$ and the parameter $x$, is defined[2,3] by:

$$\frac{d}{dx}\mathcal{F}(x) = \lim_{h\to 0} \frac{\mathcal{F}(x+h) - \mathcal{F}(x)}{h}\,. \tag{9}$$

Using this differential operation, we can study the change of a fractal object with respect to the relevant parameter $x$. A differential equation of the fractal object $\mathcal{F}(x)$ can be constructed and the solution of such an equation can be given using the integral defined by the inverse operation:

$$S(x) = \int_0^x \mathcal{F}(x)dt\,. \tag{10}$$

This kind of analysis is called[2,3] the fractal form analysis.

## 3. CONCEPTUAL RELATION BETWEEN THE FRACTAL FORM ANALYSIS AND THE CAM THEORY

It will be instructive to state the conceptual relation between the fractal form analysis and the CAM theory. In the CAM theory,[4–6] the mean-field critical coefficients obtained in a systematic series of generalized cluster mean-field approximations, show a coherent anomaly as the corresponding approximate critical points approach the true critical point.

These mean-field critical coefficients correspond conceptually to the norm of a fractal object. Critical exponents correspond to fractal dimensions.[7]

## 4. FRACTAL DECOMPOSITION OF EXPONENTIAL OPERATORS AND THE CAM SCHEME

Here we discuss the relation between the basic concept of the fractal decomposition of exponential operators[8–14] and the CAM scheme.[4–6] In general, the higher-order decomposition scheme proposed by the present author,[8–14] an $m$th order product formula, does not show any fractal property for finite m, but it becomes fractal in the infinite limit of $m$. Similarly, cluster mean-field approximations by themselves give only classical (non-fractal) critical exponents, but a systematic procedure of the CAM yields the asymptotic estimate of nonclassical true critical exponents. Thus, the two schemes have the above common feature.

## REFERENCES


1. B. B. Mandelbrot, *The Fractal Geometry of Nature* (Freeman, San Francisco, 1982).
2. M. Suzuki, in *Mathematical Approach to Fluctuations*, Vol. 2 ed. T. Hida (World Scientific, Singapore, 1995).
3. M. Suzuki, *Statistical Mechanics* (Iwanami, Tokyo, 1994 and English translation, Springer, 1997).
4. M. Suzuki, *J. Phys. Soc. Jpn.* **55**, 4205 (1986).
5. M. Suzuki, M. Katori and X. Hu, *J. Phys. Soc. Jpn.* **56**, 3092 (1987).
6. M. Suzuki, X. Hu, M. Katori, A. Lipowski, N. Hatano, K. Minami and Y. Nonomura, *Coherent Anomaly Method — Mean Field, Fluctuations and Systematics* (World Scientific, Singapore, 1995).
7. M. Suzuki, *Prog. Theor. Phys.* **69**, 65 (1983).
8. M. Suzuki, *Phys. Lett.* **A146**, 319 (1990); ibid **A165**, 387 (1992).
9. M. Suzuki, *J. Math. Phys.* **32**, 400 (1991).
10. M. Suzuki, *J. Phys. Soc. Jpn.* **61**, 3015 (1992).
11. M. Suzuki, in *Fractals and Disorder*, ed. A. Bunde (North-Holland, 1992) *Physica* **A191**, 501 (1992).
12. M. Suzuki, *Proc. Japan Acad.* **69**, Ser. B. 161 (1993).
13. M. Suzuki, *Physica* **A205**, 65 (1994) and references cited therein.
14. M. Suzuki, *Phys. Lett.* **A201**, 425 (1995).

Fractals, Vol. 4, No. 3 (1996) 241–250


# NUMERICAL SIMULATION OF MIXING BY RAYLEIGH-TAYLOR INSTABILITY AND ITS FRACTAL STRUCTURES

SUSUMU HASEGAWA and KATSUNOBU NISHIHARA
*Institute of Laser Engineering, Osaka University,*
*2-6 Yamada-oka, Suita, Osaka 565, Japan*

HITOSHI SAKAGAMI
*Computer Engineering, Himeji Institute of Technology, 2167, Shosha,*
*Himeji, Hyogo 671-22, Japan*

**Abstract**

Turbulent interface caused by the 2-dimensional Rayleigh-Taylor instability is investigated by direct numerical simulation. It is shown that the interface becomes fractal spontaneously in the case where there are initially multimode perturbations on the interface. The generalized dimensions and the singularity spectrum are obtained by applying the multifractal theory to the turbulent interface. The fractal dimension of the line interface is found to be 1.7–1.8, which is greater than that of turbulent/nonturbulent interface in a turbulent flow. Time evolution of the fractal dimensions of the interface is also investigated.

## 1. INTRODUCTION

The long-time behavior of chaotic, nonlinear dynamical systems can often be characterized by fractal or multifractal measures[1–5] which correspond, for example, to the invariant probability distribution on a strange attractor, the spatial distribution of dissipative regions in a turbulent flow,[6–7] and so on. Various "multifractal formalism" have recently been developed to describe the statistical properties of these measures in terms of their singularity spectrum $f(\alpha)$, or their generalized dimensions $D_q$.[8]

The Rayleigh-Taylor (RT) instability occurs when a perturbed interface between two fluids of different mass density is subjected to a normal pressure gradient.[9-15] A most spectacular occurrence of the RT instability appears in supernova SN1987A. It appears in other areas of astrophysics as well, such as models of neutron stars, the sun, and nebula. One area of current interest is the effect of the RT instability on the performance of inertial confinement fusion capsules. The aim is to obtain sufficiently high temperature and density in a DT mixture for thermonuclear reactions to take place. The RT instability arises in the acceleration phase of the implosion at an ablation surface[15] and in the stagnation phase of the implosion at a pusher-fuel interface.[10] In either case the instability growth reduces capsule performance.

We have investigated the stagnation phase of the implosion by direct numerical simulation with the use of a compressible hydrodynamic simulation code IMPACT.[10] It will be shown that a fully developed unstable interface becomes fractal due to mode coupling by the RT instability.[13,14]

Section 2 summarizes the multifractal formalism. In Sec. 3, numerical methods and simulation conditions are described. The nonlinear evolution of the RT instability is also shown. In Sec. 4, multifractal formalism is applied to the turbulent interface and thus we estimate the complexity of the turbulent interface. We conclude with a summary in Sec. 5.

## 2. MULTIFRACTAL FORMALISM

The multifractal formalism describes the statistical properties of these singular measures in terms of their singularity spectrum, or their generalized dimensions.[2,3,5-8] In particular, if we cover the support of the measure with boxes size $l$ and define $P_i(l)$ as the probability (integrated measure) in the $i$th box, then we can define an exponent (singularity strength) $\alpha_i$ by:

$$P_i(l) \sim l^{\alpha_i} . \tag{1}$$

If we count the number of boxes $N(\alpha)$, where the probability $P_i$ has singularity strength between $\alpha$ and $\alpha + d\alpha$, then $f(\alpha)$ can be loosely defined as the fractal dimension of the set of boxes with singularity strength $\alpha$ as:

$$N(\alpha) \sim l^{-f(\alpha)} . \tag{2}$$

This formalism leads to the description of a multifractal measure in terms of interwoven sets of Hausdorff dimension $f(\alpha)$ possessing singularity strength $\alpha$. On the other hand, the generalized dimensions $D_q$, which correspond to scaling exponents for the $q$th moments of the measure, provide an alternative description of the singular measure. Once again, if we cover the support of the measure with boxes of size $l$ and define $P_i(l)$ as the probability in the $i$th box, then one can define a series of exponents parameterized by $q$ according to:

$$\sum_i P_i^q(l) \sim l^{(q-1)} D_q . \tag{3}$$

For mathematical reasons, one is often interested in the limit of the box size approaching zero, leading to the conventional definition:

$$D_q = \frac{1}{q-1} \lim_{l \to 0} \frac{\ln \sum_j P_i^q(l)}{\ln(l)} . \tag{4}$$

In contrast to the complicated geometrical interpretation of the $D'_q$s, the $f(\alpha)$ singularity spectrum provides a precise and intuitive description of the multifractal measure in terms of interwoven sets, with differing singularity strengths $\alpha$, whose Hausdorff dimension is $f(\alpha)$.

The two sets of exponents $(\alpha, f(\alpha))$ and $(q, D_q)$ are related by:

$$\alpha = \frac{d}{dq}[(q-1)D_q], \quad f(\alpha) = q\alpha - (q-1)D_q . \tag{5}$$

These are precisely the Legendre transform relations between the generalized dimension and the original singularity spectrum $f(\alpha)$.

## 3. NUMERICAL METHODS AND SIMULATION CONDITIONS

We have used a two-dimensional Eulerian fluid code (IMPACT-2D[10]) using the Total Variation Diminishing (TVD) scheme, capable of capturing discontinuities without nonphysical oscillations and with second-order accuracy both in space and time.

In the classical RT instability,[9] in which two fluids are assumed to be incompressible, irrational, and have a sharp density discontinuity, the linear growth rate $\gamma^{cl}(\lambda)$ is given by:

$$\gamma^{cl}(\lambda) \equiv \sqrt{\frac{2\pi Ag}{\lambda}} , \tag{6}$$

where $\lambda$ is the wavelength of the perturbation, $g$ the acceleration, $A = (\rho_H - \rho_L)/(\rho_H + \rho_L)$ the Atwood number, and $\rho_H(\rho_L)$ the mass density of the heavy (light) fluid. If their initial amplitudes are much smaller than their wavelengths, the perturbations grow exponentially in time. As the amplitude of the RT instability becomes larger, nonlinear effects become important. In the multimode case, mode coupling among the modes excites new waves and breaks the original periodicity of the perturbations. The 2-D numerical simulation and the experiments of Read[11] and Youngs[12] at a wide range of density ratios, showed that the bubble penetration, i.e., the depth to which the turbulent mixing zone penetrates the denser fluid, is given by:

$$h_1 = \beta A g t^2 , \tag{7}$$

where $\beta$ is a constant.

We have investigated turbulent mixing by the RT instability and simulation conditions are as follows. The densities of the heavy and light fluids are $\rho_H = 5$ (g/cm$^3$) and $\rho_L = 0.5$ (g/cm$^3$), respectively, and so the Atwood number is approximate 0.818. A gravity is toward the minus $y$-direction (the absolute value of gravity is $g = 10^{15}$ (cm/sec$^2$)) and periodic boundary conditions are used in the $x$-direction. The number of the computational mesh is $L_x \times L_y = 516 \times 431$ and an initial position of the contact surface between the heavy and light fluids is $L_B = 291$, where the length of one mesh is 1 $\mu$m. An adiabatic variation is used initially throughout each fluid, i.e., $\partial p/\partial y = \rho g$, $p = \text{cons tan t} \times \rho^\gamma$, where $p$ and $\rho$ are the pressure and density and $\gamma$ is equal to 5/3. This gives neutral stability within each fluid. A compressibility parameter is given by $g\ L_x/c_s^2 \approx 0.154$, where $c_s^2 = \gamma p/\rho$ and this shows that the fluid is almost incompressible. The initial perturbations are of velocity alone (i.e., the initial interface perturbation is zero); the initial velocity perturbation is a superposition of modes of the form:

$$\mathbf{v}_m = (v_{x,m}, v_{y,m}) = \nabla\phi_m\,,$$

$$\phi_m = \frac{A_m \gamma^{cl}(\lambda_m)}{k_{x,m}} e^{-k_{x,m}|y-L_B|} \sin(k_{x,m}x + \delta_m)\,, \tag{8}$$

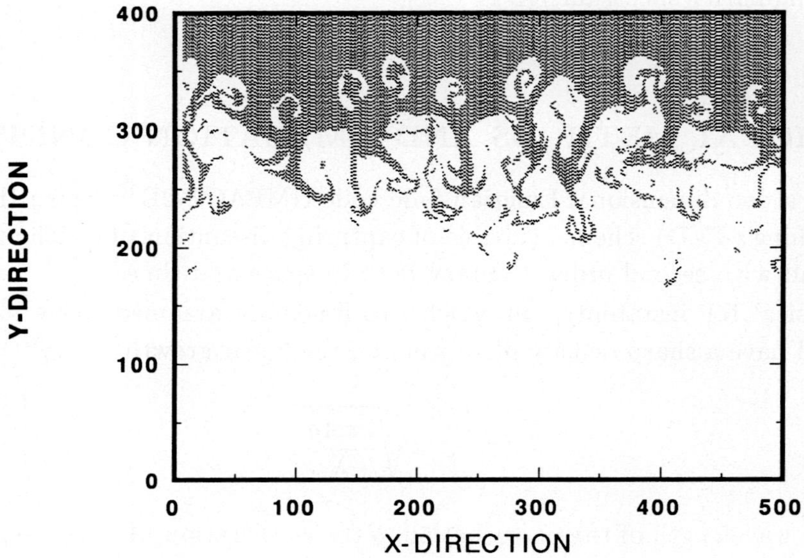


(a)

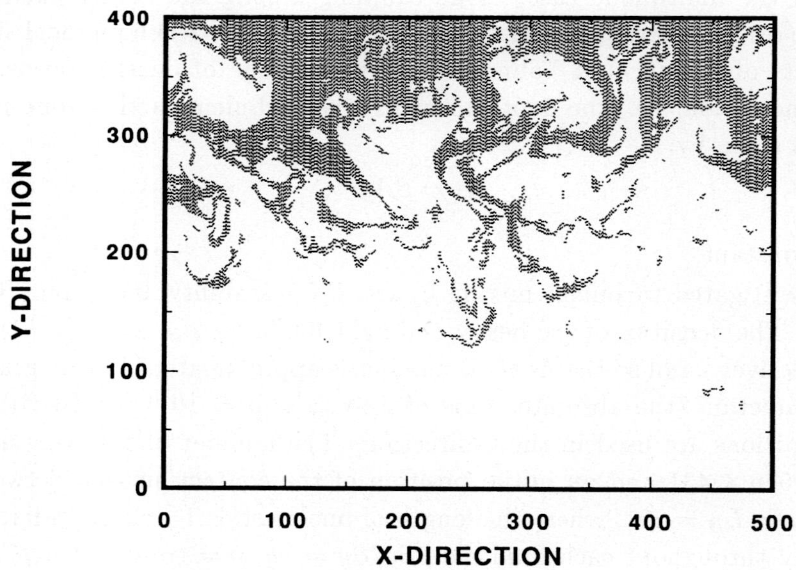


(b)

**Fig. 1** Points whose density is larger than the average density, $(\rho_H + \rho_L)/2$ are plotted. Time normalized to a linear growth rate $\gamma^{cl}(L_x)$ are **(a)** 4.735, **(b)** 6.313, **(c)** 7.891, and **(d)** 9.469. As time advances, the unstable surface becomes turbulent and has self-similar structures.

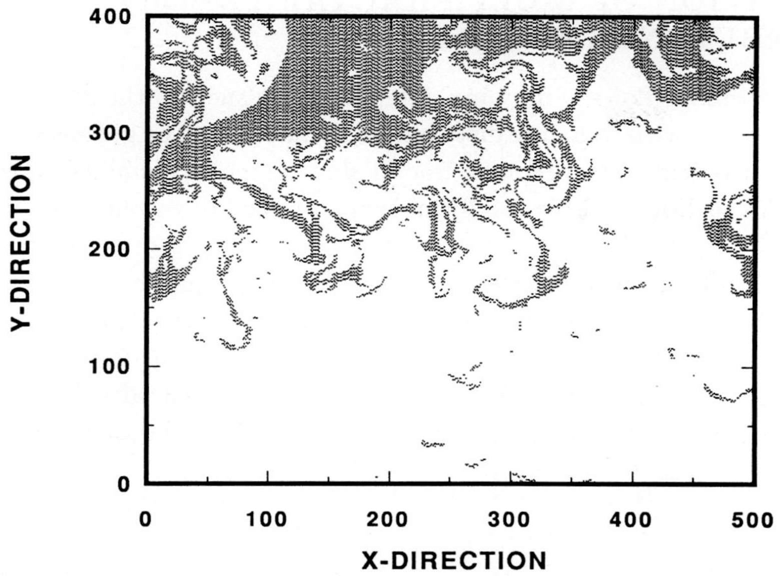


(c)

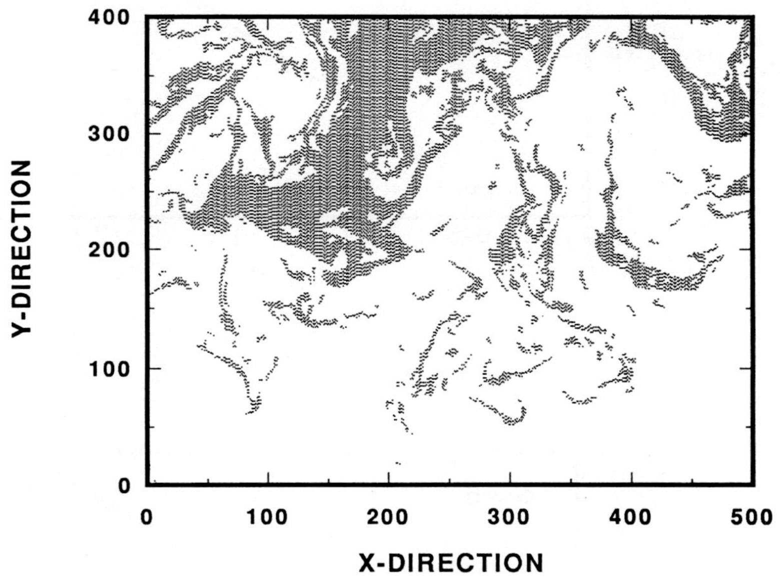


(d)

**Fig. 1** (*Continued*)

where $k_{x,m} = 2\pi m/L_x$, $\lambda_m = L_x/m$ and $m$ are a wavenumber, a wavelength and a mode number, respectively. And we give modes $m = 2 \sim 20$, $A_m = 10^{-5}\ \lambda_m$ and $\delta_m$ is random number.

Calculated results are plotted against normalized time to a linear growth rate $\gamma^{cl}(L_x)$. In Fig. 1, points whose densities are larger than the average density $(\rho_H + \rho_L)/2$ are plotted. As time advances, the unstable surface becomes turbulent and has self-similar structures. In the next section, multifractal theory is applied to the turbulent interface.

## 4. APPLICATION OF MULTIFRACTAL THEORY TO TURBULENT INTERFACE

Unstable hydrodynamic flows containing different fluids such as the RT instability rapidly evolve into turbulence with a chaotic distribution of material. The geometrical configuration of material in turbulent flow becomes fractal due to mode coupling caused by the RT instability. In this section, multifractal formalism is applied to turbulent interface calculated in Sec. 3.

Sampling points of turbulent interface are chosen as positions whose densities change from smaller (larger) to larger (smaller) in comparison with an average density (i.e., $(\rho_H + \rho_L)/2$). For example, the number of sampling points is approximately $10^4$ in the case of normalized time which is equal to 7.891. From these sampling points, generalized dimensions $D_q$ and $f-\alpha$ curve are obtained as follows. In Fig. 2, to get generalized dimensions $D_q$, we provide

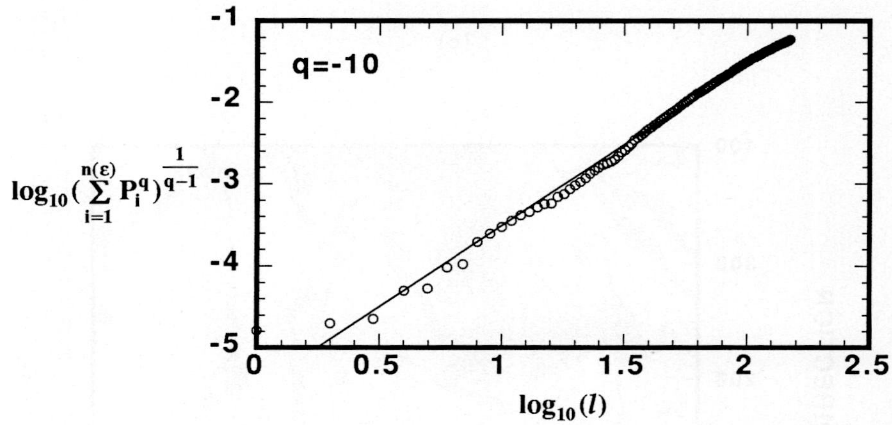


(a)

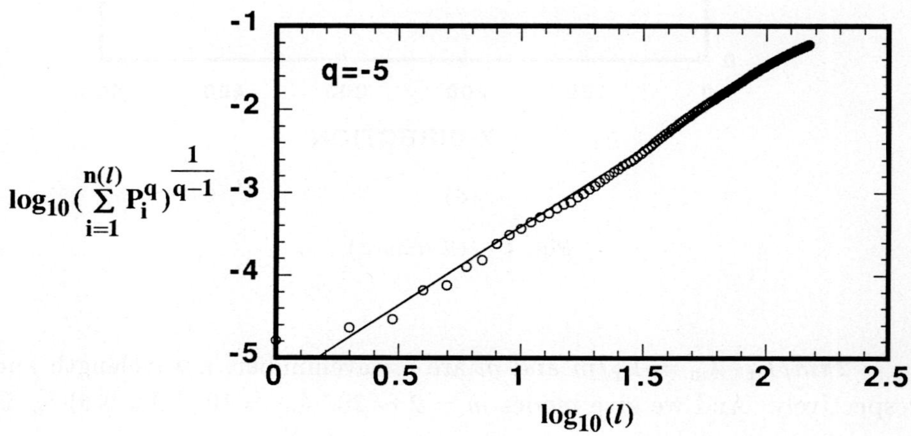


(b)

**Fig. 2** Plots of $\log_{10}(\sum_i^{n(l)} P_i^q(l))^{\frac{1}{q-1}}$ vs. $\log_{10}(l)$ for different values of $q$ (i.e., $q = -10, -5, +5, +10$) for the turbulent interface corresponding to Fig. 1 (c). Circles are points obtained from the simulation and the slopes of these graphs are $D_q$.

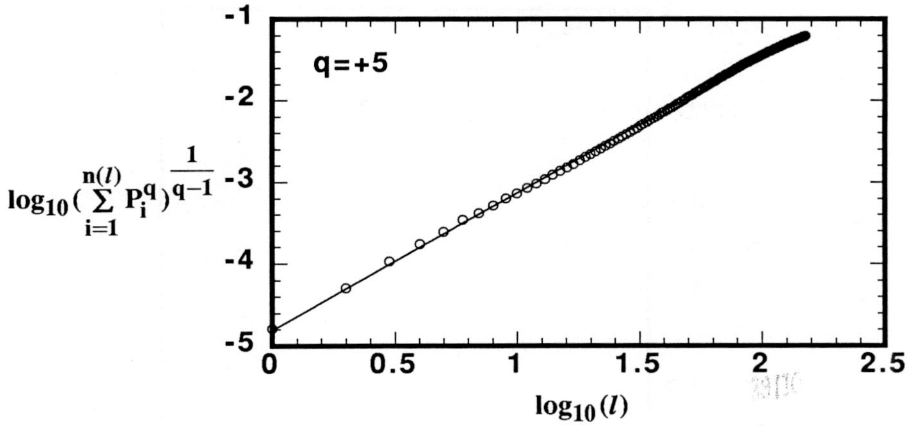


(c)

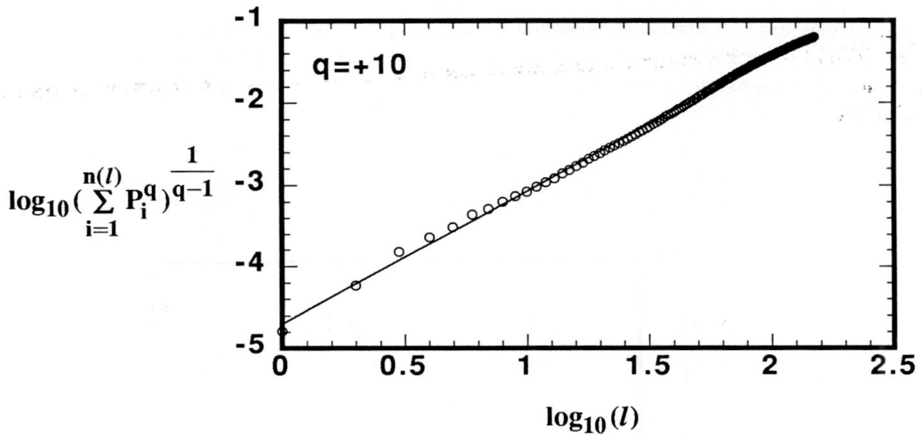


(d)

**Fig. 2** (*Continued*)

an example of the linear fit to $\log_{10}(\sum_i^{n(l)} P_i^q(l))^{\frac{1}{q-1}}$ vs. $\log_{10}(l)$ for different values of $q$ (i.e., $q = -10$, $-5$ $+5$, $+10$) to show that there is no ambiguity in the determination of the slope, where $n(l)$ is the number of circles of radius $l$. In Fig. 2, circles are points obtained from the simulation and the slopes of these graphs are $D_q$. In Fig. 3, open circles are calculated generalized dimensions and the solid line is the interpolation curve and as is well known, $D_0$, $D_1$ and $D_2$ are called the Hausdroff dimension (what we call fractal dimension), information dimension and correlation dimension, respectively. For homogeneous fractals, $D_q$ is a constant equal to the fractal dimension of the support, and does not depend on $q$. However, for inhomogeneous fractals, $D_q$ depends on $q$. From the obtained $D_q$ curves, the $f(\alpha)$ curves can be computed with the Legendre transformations (i.e., Eq. (5)). In Fig. 4, open circles are calculated $f(\alpha)$ singularity spectrum and the solid line is the interpolation curve.

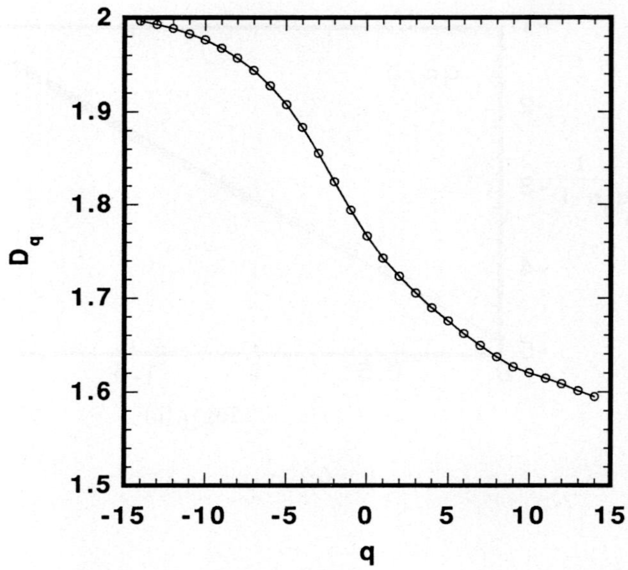


**Fig. 3** Open circles are calculated generalized dimensions and the solid line is the interpolation curve for the turbulent interface corresponding to Fig. 1(c). For homogeneous fractals, one has that $D_q$ is a constant equal to the fractal dimension of the support, and does not depend on $q$. However, for inhomogeneous fractals, $D_q$ depends on $q$.

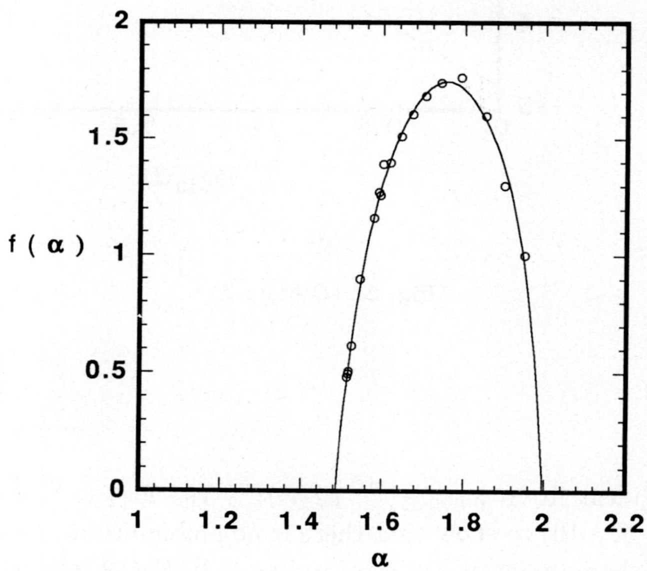


**Fig. 4** Open circles are calculated $f(\alpha)$ singularity spectrum and the solid line is the interpolation curve for the turbulent interface corresponding to Fig. 1(c). The two sets of exponents ($\alpha$, $f(\alpha)$) and ($q$, $D_q$) are related by Eq. (5).

Next, the time evolution of the fractal dimension ($D_0$) was obtained and plotted in Fig. 5. Open circles are calculated fractal dimensions and the dotted line is the interpolation curve. Initially, the fractal dimension ($D_0$) of the interface is obviously equal to one. As the interface becomes turbulent due to mode coupling caused by the RT instability, the fractal

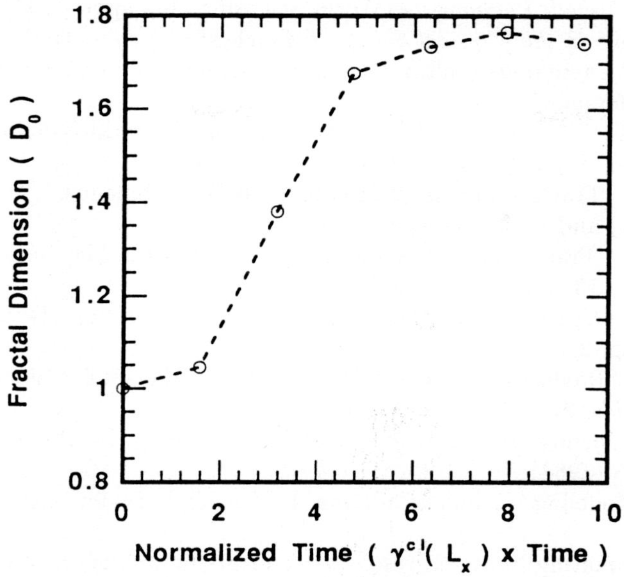


**Fig. 5** Open circles are calculated fractal dimensions and the dotted line is the interpolation curve. As the interface becomes turbulent due to mode coupling caused by the RT instability, the fractal dimension ($D_0$) begins to increase, but soon saturates to a constant (whose value is 1.7–1.8) in our simulation condition.

dimension ($D_0$) begins to increase, but soon saturates to a constant (whose value is 1.7–1.8) in our simulation condition, which is greater than that of turbulent/nonturbulent interface in a turbulent flow.[7] It will be interesting to investigate further nonlinear stage and the Atwood number dependence. We will investigate and clarify them in the next paper.

## 5. SUMMARY

Fully nonlinear stage of the turbulent interface caused by the Rayleigh-Taylor instability is investigated. We have shown that the fractal dimension and multifractal theory are useful concepts for estimating the area of very turbulent and chaotic surface in turbulent flow. The generalized dimensions and the $f(\alpha)$ curves of the interface are obtained by application of the multifractal theory. The fractal dimension of the line interface is found to be 1.7–1.8, which is greater than that of turbulent/nonturbulent interface in a turbulent flow. Time evolution of the fractal dimensions of the interface is also investigated.

## ACKNOWLEDGMENTS

We would like to express our thanks to Professor K. Mima and Professor S. Nakai for their interest and encouragement.

## REFERENCES

1. B. B. Mandelbrot, *The Fractal Geometry of Nature* (Freeman, San Francisco, 1982).
2. J. Feder, *Fractals* (Plenum, New York, 1988).

3. T. Vicsek, *Fractal Growth Phenomena* (World Scientific, Singapore, 1989).
4. H. Takayasu, *Fractals in the Physical Science* (Manchester University Press, Manchester, 1990).
5. G. Paladin, et al. "Anomalous scaling laws in multifractal objects," *Phys. Rep.* **156**, 147–225 (1987) and their references.
6. C. Meneveau, et al. "The multifractal spectrum to the dissipation field in turbulence flows," *Nucl. Phys. B (Proc. Suppl.)* **2**, 49–76 (1987).
7. K. R. Sreenivasan, "Fractals and multifractals in fluid turbulence," *Annu. Rev. Fluid Mech.* **156**, 539–600 (1991) and their references.
8. T. C. Halsey, et al. "Fractal measures and their singularities: The characterization of strange sets," *Phys. Rev. A.* **33**, 1141–1151 (1986).
9. S. Chandrasekhar, *Hydrodynamic and Hydromagnetic Stability* (Oxford University Press, London, 1968) Chap. 2.
10. H. Sakagami, et al. "Rayleigh-Taylor instability on the pusher-fuel contact surface of stagnating targets," *Phys. Fluids.* **26**, 2715–2730 (1990).
11. K. I. Read, "Experimental investigation of turbulent mixing by Rayleigh-Taylor instability," *Physica* **D12**, 45–58 (1984).
12. D. L. Youngs, "Modeling turbulent mixing to Rayleigh-Taylor instability," *Physica* **D37**, 270–287 (1989).
13. J. A. Viecelli, "Estimating limits on resolution and predictability in computations of hydrodynamically unstable flows," *J. Comput. Phys.* **74**, 456–476 (1988).
14. P. F. Linden, "Molecular mixing in Rayleigh-Taylor instability," *J. Fluid. Mech.* **265**, 97–124 (1989).
15. S. Hasegawa, et al. "Mode coupling theory in ablative Rayleigh-Taylor instability," to be published in *Phys. Plasmas* **2**, 4606–4616 (1995).

Fractals, Vol. 4, No. 3 (1996) 251–256


# POWER LAW RELAXATION OF PERIMETER LENGTH OF FRACTAL AGGREGATES

TOSHIHARU IRISAWA
*Computer Center, Gakushuin University, 1-5-1 Mejiro, Toshima-ku, Tokyo 171, Japan*

MAKIO UWAHA
*Department of Physics, Nagoya University, Furo-cho, Chikusa-ku, Nagoya 464-01, Japan*

YUKIO SAITO
*Department of Physics, Keio University, 3-14-1 Hiyoshi, Kohoku-ku, Yokohama 223, Japan*

**Abstract**

For a realistic aggregate grown under the diffusion control, the fractal scaling holds between two cutoff lengths. These cutoff lengths often control the dynamics of aggregation and relaxation. During thermal annealing, coarsening of the aggregate structure takes place, and the lower cutoff length increases. When the relaxation is limited by kinetics, we show by a simple dimensional argument that the perimeter length (or area) $A$ of the aggregate shrinks in a power law with time $t$ as $A(t) \sim t^{(d-1-D)/2}$ in a $d$-dimensional space, where $D$ is the fractal dimension of the aggregate. This prediction is tested by Monte Carlo simulation of the thermal relaxation of a two-dimensional diffusion-limited aggregation.

## 1. INTRODUCTION

Although fractal aggregation such as the Diffusion-Limited Aggregation (DLA) and the cluster-cluster aggregation have attracted much attention of physicists,[1] most studies are concentrated on their geometrical properties. As they are usually grown in a computer with rather arbitrary dynamics in which physical time is missing, the realistic growth mechanism cannot be discussed.

Usually DLA[2] is an idealization of the diffusional growth in the low-density and slow-growth limit. For the more realistic DLA growing in a finite-density gas,[3] we derived a growth law that relates the growth velocity $V$ of an aggregate to its fractal dimension $D$.[4–6] Due to the material conservation, the aggregate structure is fractal only up to an upper cutoff length $\xi_2$, which is related to the gas density $n_g$ as $\xi_2 \sim n_g^{-1/(d-D)}$, where $d$ is the spatial dimension. Since the low-density growth is realized within the range of the diffusion length $l_D \sim V^{-1}$, we identify $\xi_2$ with $l_D$, and then the growth velocity obeys the law $V \sim 1/\xi_2 \sim n_g^{1/(d-D)}$. There is an experimental evidence that this growth law holds for the real crystal growth system.[7] There is another example of fractal dendrite, where dynamical growth law is also controlled by the growth morphology of crystals.[8]

As well as the upper cutoff length $\xi_2$, there is a lower cutoff length $\xi_1$ for the fractal nature to hold for the realistic aggregate. This lower cutoff length can be the size of an atom, but it may change dynamically. If $\xi_1$ varies dynamically, it also controls the dynamics of a pattern formation.[9] Here we consider a shape relaxation process of a fractal aggregate. Experimentally, two-dimensional fractal crystals that are similar to the DLA can be made in thin film growth on a substrate under very large supersaturation such as the Molecular-Beam Epitaxy (MBE).[10] With thermal annealing, the fractal crystals change their shape into the compact equilibrium shape. The relaxation process can be regarded as coarsening of the fractal structure with a change of the lower cutoff $\xi_1$. A simple dimensional argument[9] predicts a power-law relaxation of the edge length of the two-dimensional fractal aggregate. In the present paper, we argue that for relaxation limited by kinetics the interface area $A(t)$ in $d$-dimensional space changes as the power of time $A(t) \sim t^{(d-1-D)/2}$. We demonstrate it with a Monte Carlo simulation in two dimension, which corresponds to the relaxation of a thin film crystal with fast diffusion on the substrate.

## 2. COARSENING OF A FRACTAL AGGREGATE AND DYNAMICAL SCALING LAW

When crystallization proceeds under very large supersaturation as in the MBE growth on a substrate, morphological instability in the surface diffusion field[11] occurs. The characteristic length of the instability can be very small compared to the diffusion length, and a fractal branching pattern appears.[12–15] The pattern is similar to that of DLA as observed in the experiment.[10] It is an unstable pattern formed far from equilibrium, and if the crystal is heated at a proper temperature, it slowly changes shape and relaxes to a compact equilibrium shape.

We regard thermal relaxation of a fractal aggregate to its equilibrium form as coarsening of a fractal, i.e., a simplifying process of a self-similar structure.[16,17] Realistic aggregation patterns exhibit self-similarity between the lower and the upper cutoff lengths, $\xi_1$ and $\xi_2$. The small length $\xi_1$ is typically the width of a fractal branch finger, and $\xi_2$ is the separation of the main branches determined by the density of the diffusing gas[5] or linear size of the whole aggregate when it grows in low-density gas. In the relaxation process, we may disregard the change of the upper cutoff (since the global configuration of the aggregate does not change except in the last stage) and consider the relaxation as coarsening accompanied by change of $\xi_1$. The interface area $A$ of a fractal pattern in a spatial dimension $d$ is given by:

$$A \sim \xi_1^{d-1-D} , \qquad (1)$$

which is a product of the area of a unit box, $\sim \xi_1^{d-1}$, and the number of boxes, $\sim \xi_1^{-D}$. Therefore the change of the cutoff length as a function of time $\xi_1(t)$ determines the change of the interface area $A(t)$.

In the relaxation process to the equilibrium there is no bulk driving force, and temporal evolution of the shape is necessarily driven by local deformation. When transport of material is fast enough so that the growth is limited by kinetics at the interface, the velocity of the interface is proportional to the local curvature $\kappa$ as in Ref. 1, $v \sim -\kappa$. Since the interface velocity determines the change of the characteristic length, $v \sim d\xi_1/dt$, and the curvature is determined by the characteristic length, $\kappa \sim 1/\xi_1$, the characteristic length increases with time as:

$$\xi_1 \sim t^{1/2} \,. \tag{2}$$

Combining Eqs. (1) and (2), we infer:

$$A(t) \sim t^{-\nu} \sim t^{-(1+D-d)/2} \,. \tag{3}$$

For the two-dimensional DLA fractal, $D \approx 1.72$[19] and the exponent is $\nu = 0.36$. When the transport process is limiting the relaxation rate, we expect different powers of time[9,18]: $\xi \sim t^{1/3}$ and $\nu = 0.24$ for surface diffusion on the substrate, and $\xi \sim t^{1/4}$ and $\nu = 0.18$ for edge diffusion along the perimeter.

## 3. MONTE CARLO SIMULATION

We performed Monte Carlo (MC) simulation to test the dynamical scaling in the relaxation limited by kinetics. For the starting fractal aggregate, we make a DLA pattern of the radius $\xi_2 \sim 10^3$ on a triangular lattice and thicken it by adding peripheral sites to the original pattern. We adopt the waiting-time method in the MC simulation. The rate of "evaporation" of the $i$th atom at the edge of the solid is:

$$R_i = \exp(-l_i E_b/T) \,, \tag{4}$$

where $l_i$ is the number of crystal neighbors and $E_b$ is the binding energy. The edge atoms are randomly chosen proportional to $R_i$ and evaporated, and the time increment for this step is $\Delta t = (\sum_i R_i)^{-1}$. The evaporated atom is added to one of the $N_p$ peripheral sites of the solid chosen randomly, with $\Delta t = \exp(3E_b/T)N_p^{-1}$. This value is chosen to balance the rate with Eq. (4) for a kink site. This algorithm approximately realizes the kinetics-limited relaxation with fast transport.

Typical patterns with $E_b/T = 3$ during relaxation are shown in Fig. 1. The change of the perimeter length as time is shown in Fig. 2(a) for ten different runs. We also show the result with other dynamics[9] in Fig. 2(b) for comparison. Coarsening of the fractal pattern is seen from (a) to (d) but the shape no longer has the fractal feature in (e) and (f). The fractal dimension of the initial DLA pattern is determined from the mass-radius relation $M(R)$[1,19] and $D = 1.7$. Thus the predicted power is $\nu = 0.35$. Figure 2(a) shows rather large scatter in the last stage, but for $13 < \ln t < 15.5$ (triangles) the scatter is small and the exponent in this period is $\nu = 0.38$, which is consistent with the theoretical value and in sharp contrast to the other cases shown in Fig. 2(b). For $\ln t > 16$, the exponent increases rapidly because the fractal feature is lost in this stage ($D = 2$ geves $\nu = 0.5$).

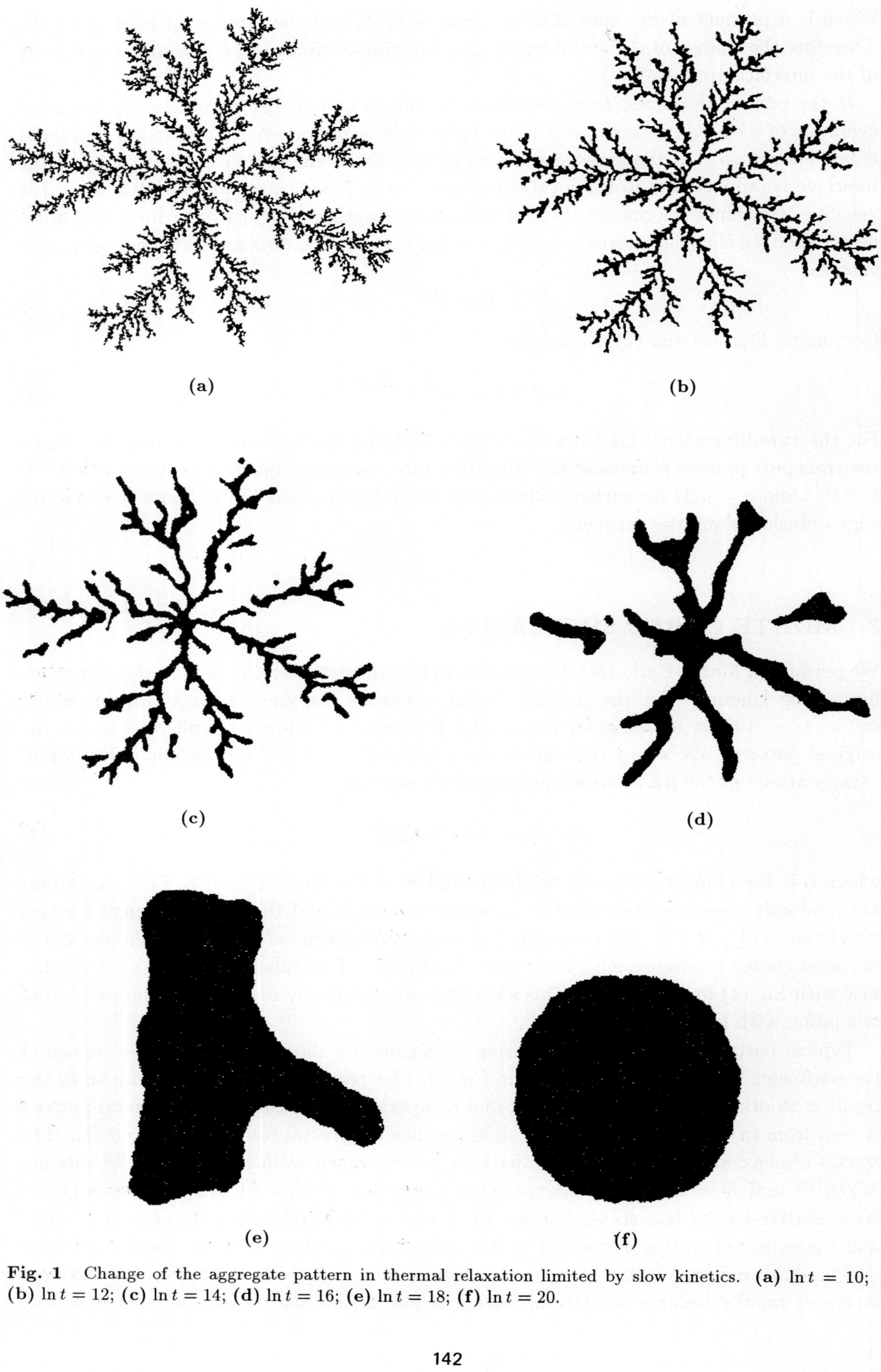

(a) (b) (c) (d) (e) (f)

**Fig. 1** Change of the aggregate pattern in thermal relaxation limited by slow kinetics. **(a)** $\ln t = 10$; **(b)** $\ln t = 12$; **(c)** $\ln t = 14$; **(d)** $\ln t = 16$; **(e)** $\ln t = 18$; **(f)** $\ln t = 20$.

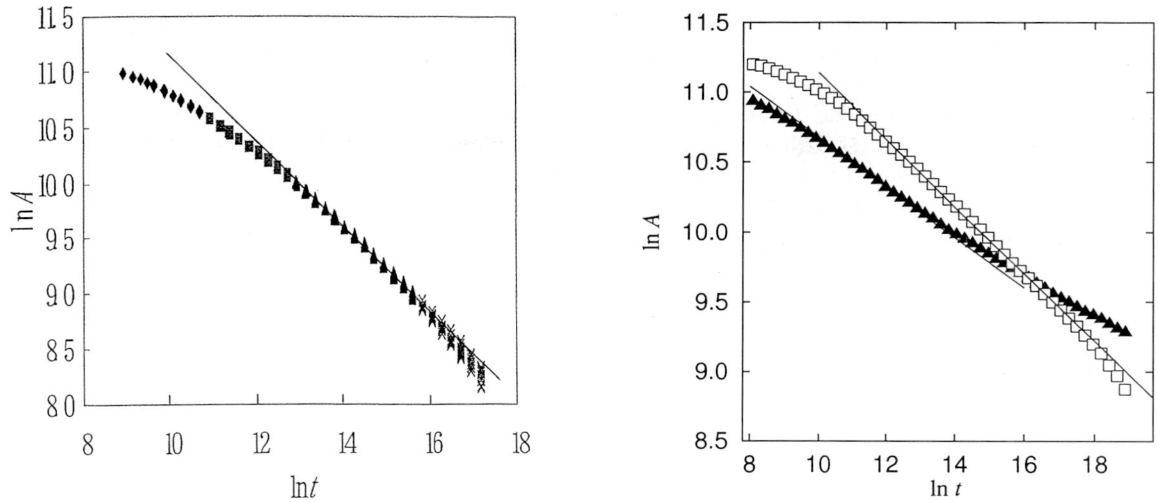


(a) (b)

**Fig. 2** Change of the perimeter length $A(t)$ as time. **(a)** Relaxation limited by kinetics (ten different runs). The straight line represents $\nu = 0.38$. **(b)** Relaxation limited by transport: surface diffusion (square) with the line $\nu = 0.24$ and edge diffusion (solid triangle) with the line $\nu = 0.18$.

The final round shape (f) corresponds to the equilibrium shape for $E_b/T = 3$ with thermal fluctuation.

In conclusion we demonstrated the dynamical scaling behavior for the relaxation of a DLA fractal in the limit of fast transport.

## ACKNOWLEDGMENTS

This work is supported by the Grant-in-Aid from the Ministry of Education, Science and Culture and by the Foundation for Promotion of Material Science and Technology of Japan. Y. S. and M. U. have benefited by the inter-university cooperative research program of IMR, Tohoku University.

## REFERENCES


1. For a review: T. Vicsek, *Fractal Growth Phenomena* (World Scientific, Singapore, 1989).
2. T. A. Witten and L. M. Sander, *Phys. Rev. Lett.* **47** (1981) 1400; *Phys. Rev.* **B27** (1983) 5686.
3. R. F. Voss, *J. Stat. Phys.* **36** (1984) 861.
4. M. Uwaha and Y. Saito, *J. Phys. Soc. Jpn.* **57** (1988) 3285.
5. M. Uwaha and Y. Saito, *Phys. Rev.* **A40** (1989) 4716.
6. M. Uwaha, Y. Saito and S. Seki, in *Dynamical Phenomena at Interfaces, Surfaces and Membranes*, ed. D. Beysens, N. Boccara and G. Forgacs (Nova Science, Commack, 1993) p. 177.
7. S. Miyashita, Y. Saito and M. Uwaha, unpublished.
8. E. Brener, H. Müller-Krumbhaar and D. Temkin, *Europhys. Lett.* **17** (1992) 535.
9. T. Irisawa, M. Uwaha and Y. Saito, *Europhys. Lett.* **30** (1995) 139.
10. R. Q. Hwang, J. Schröder, C. Günter and R. J. Behm, *Phys. Rev. Lett.* **67** (1991) 3297.
11. G. S. Bales and A. Zangwill, *Phys. Rev.* **B41** (1990) 5500.
12. Y. Saito and T. Ueta, *Phys. Rev.* **A40** (1989) 3408.

13. T. Irisawa, Y. Arima and T. Kuroda, *J. Cryst. Growth* **99** (1990) 491.
14. Y. Saito, T. Sakiyama and M. Uwaha, *J. Cryst. Growth* **128** (1993) 82.
15. A. Pimpinelli, J. Villain and D. Wolf, *J. Phys.* **I3** (1993) 447.
16. H. Toyoki and K. Honda, *Phys. Lett.* **111** (1985) 367.
17. For experiment, see: H. Orihara and Y. Ishibashi, *J. Phys. Soc. Jpn.* **56** (1987) 2340; H. Orihara, M. Nakamura and Y. Ishibashi, *J. Phys. Soc. Jpn.* **59** (1990) 2355.
18. C. Herring, *J. Appl. Phys.* **21** (1950) 301.
19. P. Meakin, in *Phase Transition and Critical Phenomena*, Vol. 12, eds. C. Domb and J. Lebovitz (Academic, London, 1988), p. 335.

Fractals, Vol. 4, No. 3 (1996) 257–264


# FRACTAL LIMIT DISTRIBUTIONS IN RANDOM TRANSPORTS

HIDEKI TAKAYASU and TAKUO KAWAKAMI
*Graduate School of Information Sciences, Tohoku University, Sendai 980-77, Japan*

Y.-H. TAGUCHI
*Department of Physics, Tokyo Institute of Technology, Oh-okayama, Meguro-ku, Tokyo 152, Japan*

TOMOO KATSUYAMA
*Department of Physics, Tokyo Metropolitan University, Minami-Ohsawa 1-1, Hachioji, Tokyo 192-03, Japan*

**Abstract**

We analyze a random transport model of a scalar quantity on a discrete space-time. By changing a parameter which is a portion of the quantity transported at a time, we observe a continuous change of steady-state distribution of fluctuations from Gaussian to a power-law when the mean value of the scalar quantity is not zero. In the symmetric case with zero mean, the steady-state converges either to a trivial no fluctuation state or to a Lorentzian fluctuation state with diverging variance independent of the parameter. We discuss a possible origin of the intermittent behaviors of fully-developed fluid turbulence as an application.

## 1. INTRODUCTION

It is well-known that fluctuations at thermal equilibrium often follow Gaussian distributions. From the view of mathematics, the reason is simply explained by applying the central limit theorem for the sum of many stochastic variables.[1] Linearity and reversibility of the system are playing essential roles in the decomposition of a fluctuation into summation of independent fluctuations.

Nonlinear effects become evident when the system is close to a critical point and the fluctuations deviate clearly from Gaussian.[2] A typical critical behavior is characterized by a power-law distribution with a diverging variance at the critical point.

Irreversibility can be found in every growth or dissipative systems. One of the most typical irreversible phenomena may be aggregation of sticky particles. Aggregation phenomenon has been analyzed intensively in relation to fractals. The geometry of random aggregates is well characterized by the fractal dimension[3] and the steady-state mass distribution of aggregates under continuous injection of fine particles has been proven to follow a power-law.[4]

In this paper we consider a general linear but irreversible random transport phenomenon and show that the Gaussian distribution appears only at an "infinitesimal transport limit". In all cases of finite transport, the effect of irreversibility enhances the tails of the distribution in steady-state fluctuations. Like the case of critical systems, the power-law distribution is also the key word in this system.

## 2. A RANDOM FINITE PORTION TRANSPORT MODEL

Let $u(j,\, t)$ be a scalar quantity on the $j$th site at time $t$. We assume a discrete transport dynamics as follows:

$$u(j,\, t+1) = u(j,\, t) \cdot \left(1 - \theta \cdot \sum_k W_{k,j}(t)\right) + \theta \cdot \sum_k W_{j,k}(t) \cdot u(k,\, t)\,, \tag{1}$$

where $\theta$ is a parameter that represents the portion of transport at a time and takes a value in the interval $(0, 1)$, $W_{k,j}(t)$ is a random number which takes either 1 or 0. $W_{k,j}(t) = 1$ means that there exists transport from the $j$th site to the $k$th site at time step $t$, and we require that $W_{k,j}(t)$ equals to 1 with only one of $k$ for each $j$, namely, in a time step a site can distribute a portion of its quantity to a chosen site. Here, we allow the possibility of $W_{j,j}(t) = 1$, which means no outflow at the $j$th site.

In the limit $\theta \longrightarrow 0$ with nearest neighbor transport, the model can be represented by the formalism of a motion of continuum in a random velocity field and the overall transport is described by the usual diffusion equation.[5] On the other hand in the case of $\theta = 1$, all quantity at a site is piled up on another site instantly, which is a typical irreversible aggregation process. Therefore, for $0 < \theta < 1$, we are considering a mixture of diffusion and aggregation effects.

The long term behaviors of this system are very different whether the mean value of $u(j,\, t)$ is 0 or not. We call the first case a symmetric case and the second an asymmetric case. The asymmetric case includes the case that $u(x,\, t)$ never takes a negative value, like mass or energy for example. When we consider momenta or charge transport, the symmetric case may be realizable.

In the asymmetric case, it is shown that the system converges to a non-trivial statistically steady-state independent of initial conditions. By applying a mean-field approximation and using a characteristic function technique,[5] we can show that the steady-state fluctuations follow a Gaussian around the mean value in the limit of $\theta \longrightarrow 0$. For $\theta \neq 0$, the distribution of fluctuation has a longer tail than Gaussian and the tail becomes larger for larger values of $\theta$. In the extreme limit of $\theta \longrightarrow 1$, we can confirm that the distribution converges to the power-law distribution, which appears in the case of $\theta = 1$ with uniform external injection.[4] In the present model, we do not have any injection terms but the small portion left behind at each transport $((1-\theta)u(j, t))$ may play the role of injection, while the dominant dynamics is approximated by the aggregation.

In the symmetric case the system always converges to the trivial steady-state, $u(j, t) = 0$ for all $j$, namely, no fluctuation remains in the infinite time limit. However, if we observe the fluctuations in a finite time with normalization by the standard deviation, we find that the normalized fluctuations follow a steady distribution for large $t$. The shape of the distribution is close to Gaussian for small $\theta$ and the deviation becomes evident as $\theta$ becomes larger. In the limit of $\theta \to 1$, we have a Lorentzian distribution.[5] For an intermediate $\theta$, the distribution shape can well be approximated by a stretched exponential function; however, the precise form cannot be represented by an explicit analytical function so far.

## 3. THE STOCHASTIC SHELL MODEL FOR TURBULENT CASCADE

Since the pioneering work of Kolmogorov,[6] the concept of energy cascade has been one of the central topics in the study of turbulence. The main idea of the energy cascade is very simple: in view of Fourier space the energy at a wave number is transported irreversibly to higher wave number components by nonlinear interactions of Navier-Stokes equation. Although the full numerical simulation of the Navier-Stokes equation has been tackled by many researchers, a complete confirmation of the energy cascade itself and detailed analyses of its statistical properties, are still far beyond the goal. There are two main difficulties for this numerical approach. One is the large degree of freedom due to 3-dimensionality and the other is the nonlinearity. Recently, a one-dimensional variant of the Navier-Stokes equation in Fourier space has been introduced and the process of energy cascade has been analyzed in detail.[7] It is called the shell model or the GOY model and is described by the following set of nonlinear differential equations,

$$\begin{aligned}\frac{d}{dt}u(j, t) = -\nu k_j^2 u(j, t) + f\delta_{j,4} + i\{ak_j u^*(j+1, t)u^*(j+2, t)\\ + bk_{j-1}u^*(j-1, t)u^*(j+1, t) + ck_{j-2}u^*(j-1, t)u^*(j-2, t)\}\,. \qquad (2)\end{aligned}$$

Here, $u(j, t)$ denotes the complex velocity Fourier component for wave number $k_j \equiv K^b$ with a positive number $K$. The first term on the right hand side shows a viscous dissipation effect and the second term corresponds to an external forcing. The remaining three terms are nonlinear terms which are designed to have similar nonlinearity with the Navier-Stokes equation. From these nonlinear terms, this deterministic system shows chaotic behaviors and the resulting statistics is known to give results consistent with real fluid such as the Kolmogorov energy spectrum and a higher order correction so called either the multifractal scalings[8] or the intermittency.[9]

The multifractal scalings and the intermittency are both introduced to describe the deviation of fluctuations from Gaussian distributions. Roughly speaking, these concepts describe the phenomenon that the fluctuations of $u(j, t)$ normalized by the standard deviation, are larger for larger wave number components. In order to show that such behavior can be a very general property when we consider transports of complex variables in Fourier space, we introduce a stochastic version of the shell model to be named as the stochastic shell model.[10]

The model is a generalization of Eq. (1) into complex variables with directed nearest neighbor transport. Instead of the chaotic nonlinear interactions, we introduce discrete stochastic process as follows;

$$u(j,\, t+1) = u(j,\, t) + \theta[R(j,\, j-1 :\, t)u(j-1,\, t) - |R(j+1,\, j;\, t)|u(j,\, t)]\,, \tag{3}$$

where $\theta$ is a non-negative constant less than 1 as before. $R(j,\, j-1;\, t)$ is a random number which expresses the momentum transfer from $k_{j-1}$ to $k_j$;

$$R(j,\, j-1;\, t) = \begin{cases} \exp\{i\phi(j,\, t)\}\,, & \text{Prob. } p \\ 0\,, & \text{Prob. } 1-p \end{cases}\,,$$

with $\phi$ being a random phase variable taking a value in the interval $[0,\, 2\pi)$, and $p$ is the probability of occurrence of transport which takes a value in the interval $(0, 1)$. Namely, we consider the random transport process of complex velocity component from smaller wave number components toward larger wave number components with phase randomization.

Now that we have a directed transport, we need to add an injection term at the smallest wave number component to realize a statistically steady state. As for this injection term, we make a complex random number at each time step and superpose it to the 0th component. We adopt the boundary condition that $u(N,\, t) = 0$ for $t > 0$, where $N$ is the total number of wave numbers. Compared with the ordinary exponential decay by viscous dissipation in large wave number regions, this artificial boundary condition of a sudden cut-off may look too crude. However, the boundary condition at the largest wave number does not affect the intermediate wave number components because our model introduces the directed one-way transport towards the larger wave numbers.

## 4. ANALYTICAL TREATMENT OF THE STOCHASTIC SHELL MODEL

We can analyze the basic properties of the stochastic shell model by introducing the characteristic function method. The characteristic function is a Fourier transform of the probability density and is defined in the present model as follows:

$$Y_j(\rho,\, t) = \langle \exp[i\rho u(j,\, t)] \rangle\,, \tag{4}$$

where $\langle \cdots \rangle$ denotes the average over realizations. Using this definition we have the following equation from Eq. (3).

$$Y_j(\rho,\, t+1) = \{(1-p)Y_j(\rho,\, t) + pY_j((1-\theta)\rho,\, t)\}\{1 - p + pY_{j-1}(\theta\rho,\, t)\}\,, \tag{5}$$

It is easy to show that the system quickly converges to a statistically steady state where all averaged quantities become independent of time steps. By expanding the characteristic function in terms of $\rho$ up to the second order, we have a rigorous steady state relation for energy, $E_j \equiv \langle |u(j,\, t)|^2/2 \rangle$, as

$$E_j = E_{j-1}\theta/(2-\theta)\,. \tag{6}$$

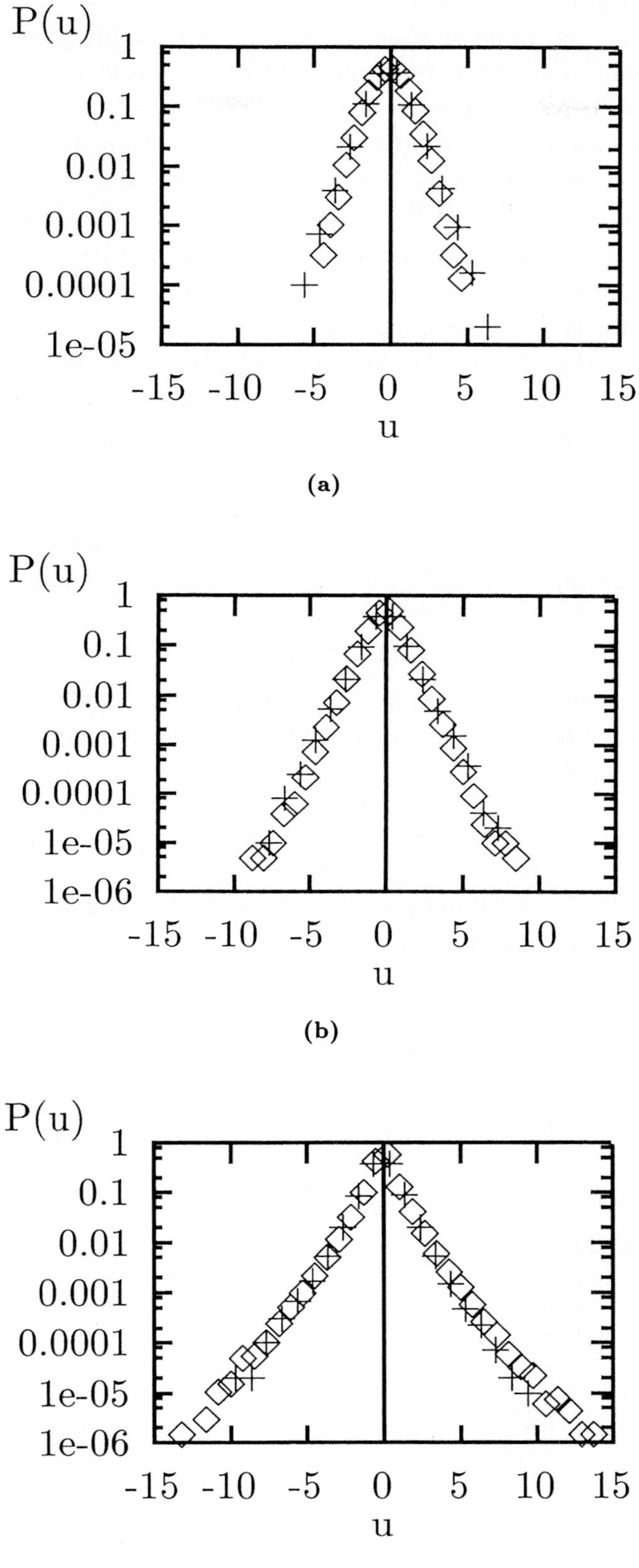


**Fig. 1** Comparison of probability density between the stochastic shell model (+) and experiments (◇). **(a)** Experiment observed with 0.15 kHz, the model with $j = 1$; **(b)** Experiment with 3 kHz, the model with $j = 2$; **(c)** Experiment with 10 kHz, the model with $j = 3$.

The energy spectrum follows the familiar power-law form $E_j = k_j^{-\zeta}$, with $\theta$ chosen to be $2/(1+K^{\zeta})$. We can fix the value of $K$ so as to satisfy the Kolmogorov spectrum.

Higher order moments can also be obtained from Eq. (5). It is difficult to solve the steady state values in explicit forms; however, we can evaluate them by a set of recursion relations. In the case that the distribution of $u(0,\,t)$ follows a Gaussian distribution, we can show that the normalized moments are larger for larger $j$, namely, the deviation from Gaussian is more evident for larger $j$ as expected.[10] In Fig. 1 numerically obtained probability densities in our model are plotted with real turbulent data obtained by frequency-band-pass-filtered velocity signals. In each wave number, the fits are very nice showing the tendency to deviate more from Gaussian for larger wave numbers.

The energy relation Eq. (6) is valid for all cases with finite variance, but we can also consider the cases with infinite variance theoretically. Assuming a stationary solution of Eq. (5), we expand the characteristic function by the following form:

$$Y_j(\rho) = 1 - c_j|\rho|^{\alpha} + \cdots . \tag{7}$$

Here, the exponent $\alpha$ can take a value in the interval $(0, 2]$ and the variance is finite only in the case of $\alpha = 2$. Introducing Eq. (7) into Eq. (5), we have the following relation for the $\alpha$th order of $\rho$.

$$c_j\{1-(1-\theta)^{\alpha}\} = c_{j-1}\theta^{\alpha} \tag{8}$$

It is easy to confirm that there exists a $j$-independent and $\theta$-independent solution in the case of $\alpha = 1$. The singular expansion of Eq. (7) with $\alpha = 1$ means that there exists long tails in the distribution of $u(j,\,t)$ as:

$$P(\geq |u|) = \int_{|u|}^{\infty} p(u')du' + \int_{-\infty}^{-|u|} p(u')du' \propto |u|^{-\alpha}\,, \quad \alpha = 1\,, \tag{9}$$

which is equivalent to the Lorentzian tails. Therefore, the Lorentzian tails can be a universal solution of our random discrete transports.

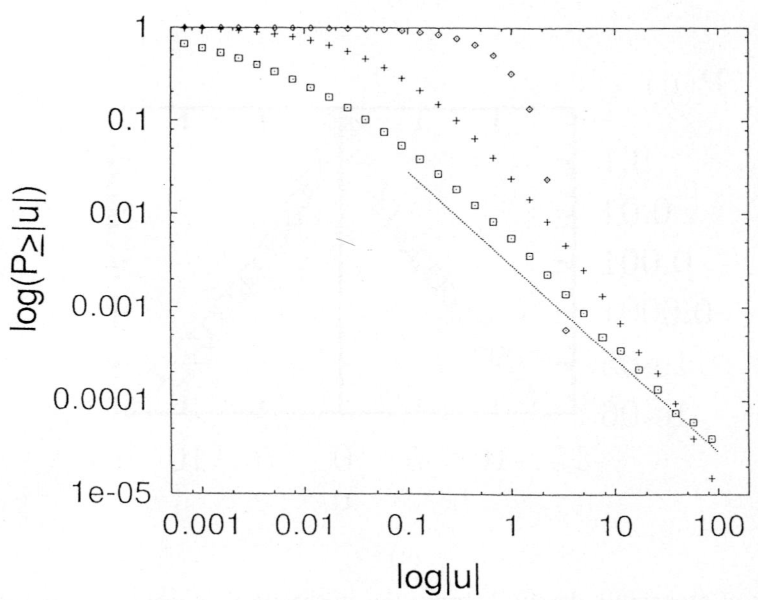


**Fig. 2** Convergence to the Lorentzian tails (dotted line) from Gaussian: $j = 0$ ($\Diamond$), $j = 50$ ($+$) and $j = 90$ ($\Box$).

This theoretical estimate can be checked numerically. When we inject a random external noise having Lorentzian distribution into 0th site, we observe the Lorentzian tails at every wave number component as expected. Moreover, when the distribution of the injection has a power-law tail decaying more slowly than the Lorentzian, we observe a convergence to the Lorentzian as $j$ gets larger (see Fig. 3). Therefore, the intermittency cannot be found in the case of Lorentzian injection.

## 5. DISCUSSION

The GOY model is a very rough simplification of the Navier-Stokes equation and we have learned from the model that the number of degree of freedom may not be an essential ingredient for the cascading energy transport in turbulence. Namely, discrete one-dimensional model can show the Kolmogorov spectrum and the intermittent behaviors. The stochastic shell model is still more simplified than the GOY model in the sense that it replaces the nonlinear deterministic continuous transport by a linear discrete stochastic transport in Fourier space. In spite of this bold simplification, the essential properties of cascading energy transport seem to be conserved. The Kolmogorov spectrum can be realized by choosing a parameter and the intermittent behaviors are observed clearly by setting the smallest wave number component to follow a nearly Gaussian distribution.

A natural question may arise, i.e., Is the stochastic shell model minimal or is there a more simplified model? Our answer for this question is yes. The variables $\{u(j,\, t)\}$ are not necessarily complex, namely, the cascading properties are conserved in a real number model. To demonstrate this, we replace the complex variables $\{u(j,\, t)\}$ by real variables and the values of $R(j,\, j-1;\, t)$ are modified as:

$$R(j,\, j-1;\, t) = \begin{cases} 1\,, & \text{Prob. } p/2 \\ -1\,, & \text{Prob. } p/2 \\ 0\,, & \text{Prob. } 1-p \end{cases} .$$

The energy at the $j$th site is given by the second order moment of $u(j,\, t)$ as before. We can easily confirm that the theoretical and numerical results for the complex version described above are all valid also in this real variable version.

In these stochastic shell models, the origin of intermittency is quite clear. The stochastic shell models are special cases of the random finite portion transport model which generally shows a property that the fluctuations tend to converge to the Lorentzian in the symmetric case. Whenever small wave number components have fluctuations with finite variance like the case of Gaussian, we observe a growth of the distribution tails for larger wave number components, which is recognized as the intermittency. However, the fluctuation tails do not always grow when the variance of the small wave number fluctuations is infinite. We can even observe an "inverse intermittency", i.e., the decay of tails becomes quicker when we inject power-law fluctuations whose power exponent is smaller than that of Lorentzian. In Fig. 3 we plot the case that the injection at $j=0$ follows a power-law fluctuation with $\alpha = 0.5$. We will find that the steeper slope of the distribution tails with $\alpha = 1$ appears as $j$ becomes larger. Therefore, we know that the essential point of the intermittency is not the growth of the distribution tails but the convergence to the Lorentzian.

The reason that the original GOY model shows intermittency may be recognized as follows. The nonlinear interaction in the GOY model produces chaotic transport towards the

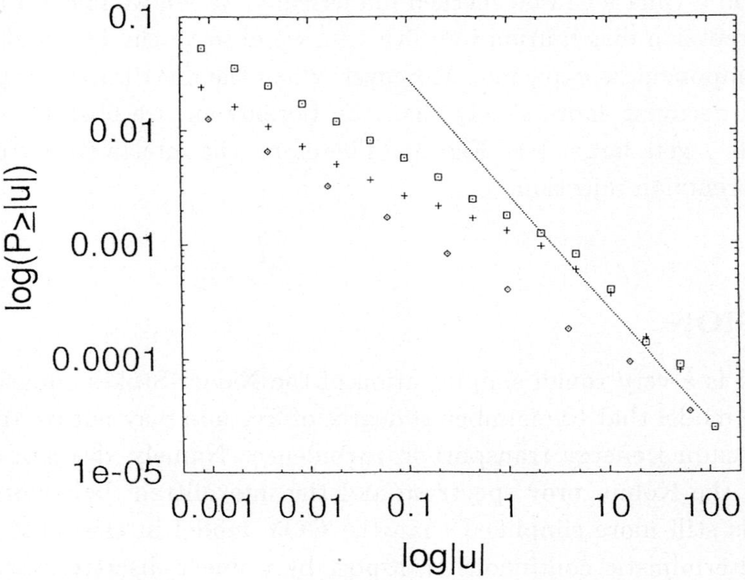


**Fig. 3** The case "inverse intermittency": $j = 0$ (◇), $j = 5$ (+) and $j = 7$ (□). The dotted line shows the Lorentzian tail.

higher wave numbers, which acts both as the phase randomization and as the random drift. The GOY model is continuous in time but the correlation time is finite so that each transport continues for a finite time. Contrary to the ordinary thermal equilibrium fluctuations, the amount of transported quantity cannot be infinitesimal. Also, the injection term gives an ordinary random noise having a finite variance; therefore the fluctuation distribution follows a Gaussian by the central limit theorem. Thus, the GOY model fulfills all the necessary properties to realize the intermittency from the standpoint of the stochastic shell model. We believe that these conditions will be applicable also in Navier-Stokes turbulence and in the real turbulence.

## REFERENCES

1. W. Feller, *An Introduction to Probability Theory and Its Applications* (Willey, New York, 1966).
2. H. E. Stanley, *Introduction to Phase Transitions and Critical Phenomena* (Clarendon Press, 1971).
3. R. Jullien and R. Botet, *Aggregation and Fractal Aggregates* (World Scientific Publishing, Singapore, 1987).
4. H. Takayasu, M. Takayasu, A. Provata and G. Huber, *J. Stat. Phys.* **65**, 725 (1991).
5. H. Takayasu and Y.-H. Taguchi, *Phys. Rev. Lett.* **70**, 782 (1993): M. Takayasu, H. Takayasu and Y.-H. Taguchi, *Int. J. Mod. Phys.* **B8** 3887 (1994).
6. A. Monin and A. Yaglom, *Statistical Fluid Mechanics* (MIT Press, Cambridge, 1971).
7. M. Yamada and K. Ohkitani, *Phys. Rev. Lett.* **60**, 983 (1988).
8. Z.-S. She and E. Leveque, *Phys. Rev. Lett.* **72**, 336 (1994).
9. G. K. Batchelor, *Homogeneous Turbulence* (Cambridge University Press, Cambridge, 1953).
10. H. Takayasu, Y.-H. Taguchi and Tomoo Katsuyama, preprint.

Fractals, Vol. 4, No. 3 (1996) 265–271


# CALCULATION OF THE POWER LAW INDEX FOR A TIME SERIES BY MEANS OF ITS FRACTAL DIMENSION

SHINICHI YASUE, KAZUOKI MUNAKATA,
MASASHIGE KATO and SATORU MORI
*Department of Physics, Faculty of Science, Shinshu University,
Matsumoto 390, Japan*

**Abstract**

The relationship between the power-law index for a time series and the fractal dimensions for both the original and the integrated time series, is investigated by using a numerical experiment. This relationship is extended and applied to the cosmic ray time series recorded at the deep underground site at Matsushiro.

It is shown that we can use the present method in order to obtain the stable and reliable value of the power-law index of a time series, even if the time series has relatively large statistical fluctuations.

## 1. INTRODUCTION

The power spectrum of a cosmic ray time series observed on the earth can be expressed by a single power law: $P(f) \propto f^{-\alpha}$, in a wide frequency range, and the power-law index $\alpha$ can be used as an index to describe the irregularity of the time series.[1–4] The time series also shows the property of a fractal curve. When the fractal length is expressed as $L(\tau) \propto \tau^{-D}$ ($\tau$ is a time interval), the fractal dimension $D$ is also used to represent the irregularity of the time series.[5] The power-law index $\alpha$ shows one-to-one correspondence to the fractal dimension $D$, and the relationship between $\alpha$ and $D$ has been extensively investigated by Higuchi.[6]

In the present report, we deal with the cosmic ray time series obtained at the deep underground site at Matsushiro, Japan.[7] The irregularity of these time series gives important information on the magnetic field fluctuations in interplanetary and/or interstellar space through which cosmic rays propagate.[8–10] However, it is not easy to evaluate a stable and precise value of $\alpha$ by the following reasons. First, because of the large statistical fluctuations due to small cosmic ray counts, the power spectrum for the underground stations contains relatively large white noise level over a wide frequency range. Thus, when we apply the power spectral analysis to those data, we are obliged to estimate the index $\alpha$ using the power spectrum in a limited frequency range. Secondly, when we apply the method of fractal analysis to the data, the fractal dimension is also contaminated by the random fluctuations due to the counting rate statistics. The purpose of the present paper is to develop a method to overcome these difficulties, by using a numerical experiment.

## 2. RELATIONSHIP BETWEEN THE FRACTAL DIMENSION AND THE POWER LAW INDEX

### 2.1 Methods for Calculating the Fractal Dimension

Burlaga and Klein[5] have presented a method to calculate the length of a curve and obtained stable values of the fractal dimension of large-scale fluctuations of the interplanetary magnetic field. Following them, we define the length $L_{BK}$ of a curve $B(t)$ as,

$$L_{BK}(\tau) = \sum_{k=1}^{n} |\bar{B}(t_k + \tau) - \bar{B}(t_k)|/\tau \tag{1}$$

where $\bar{B}(t_k)$ denotes the average value of $B(t)$ between $t = t_k$ and $t_k + \tau$. This length is a function of $\tau$, and for statistically self-affine curves, the length is expressed as, $L_{BK}(\tau) \propto \tau^{-D}$. Using this relation, we can estimate the value of $D$. The fractal dimension thus derived and the power-law index $\alpha$ of the power spectrum are related as, $D = (5 - \alpha)/2$ which is called the Berry's expression.[5]

The other method which also gives a stable and reasonable value of the fractal dimension has been presented by Higuchi.[11] He modifies Burlaga and Klein's method, and defines the length as follows. He first constructs a new time series $X_\tau^m$ from a finite set of time series $X(i)$, $(i = 1, 2, \ldots, N)$,

$$X_\tau^m;\ X(m),\ X(m+\tau),\ X(m+2\tau), \ldots,\ X\left(m + \left[\frac{N-m}{\tau}\right] \cdot \tau\right) \quad (m = 1, 2, \ldots, \tau) \tag{2}$$

where [ ] denotes the Gauss' notation, and defines the length of the curve $X_\tau^m$ as,

$$L_m(\tau) = \left\{ \left( \sum_{i=1}^{[\frac{N-m}{\tau}]} |X(m+i\tau) - X(m+(i-1)\cdot\tau)| \right) \frac{N-1}{\left[\frac{N-m}{\tau}\right] \cdot \tau} \right\} \Big/ \tau. \tag{3}$$

The average value $\langle L(\tau) \rangle$ over $\tau$ sets of $L_m(\tau)$ is defined as the length of the curve for the time interval $\tau$. If $\langle L(\tau) \rangle \propto \tau^{-D}$, then the curve is fractal with dimension $D$. He has examined the relationship between the fractal dimension $D$ and the power-law index

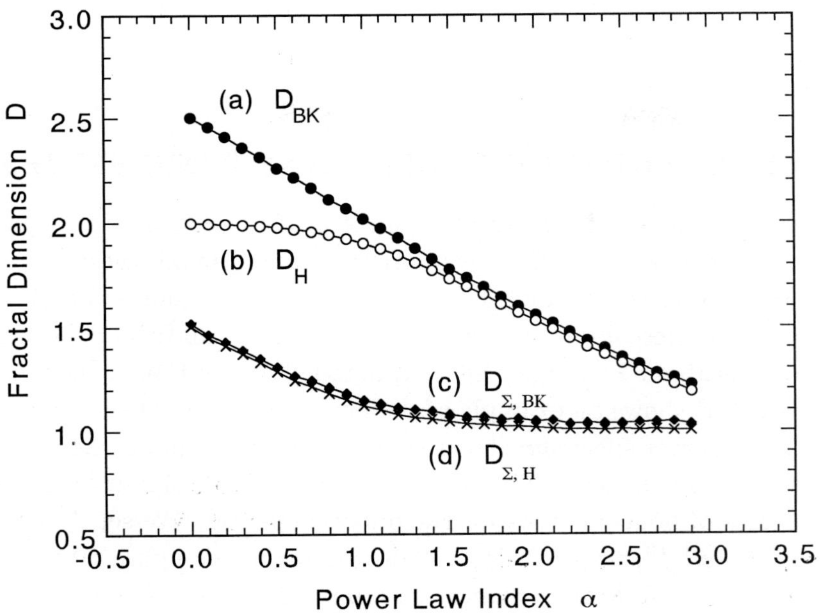


**Fig. 1** The calculated fractal dimension $D$ as a function of the power-law index $\alpha$. We assume a single power law, and we set $N = 2^{13}$.

$\alpha$, by calculating the fractal dimension of the simulated time series which follows a single power-law spectrum.

The fractal dimension takes a different value depending on the definition of the length of a time series. We have examined the relationship between the power-law index and the fractal dimensions calculated using the above two definitions of the length. Figure 1 shows the relationship obtained between $\alpha$ and $D_{BK}$ (a), and between $\alpha$ and $D_H$ (b), respectively, for the range $0 \leqslant \alpha \leqslant 3$. We assume here a single power law, and we set $N = 2^{13}$. We can see in the figure that, for the range, $0 \leqslant \alpha \lesssim 1$, the curve (b) of $D_H$ tends to flatten and approaches a constant value $D_H = 2$ with decreasing $\alpha$, whereas the curve (a) of $D_{BK}$ increases with decreasing $\alpha$ along a straight line which is close to the Berry's expression.

## 2.2 Fractal Dimension of the Integrated Time Series

Higuchi[6] has proposed an integrated series of the original time series which is given by:

$$X_\Sigma(j) = \sum_{i=1}^{j} X(i) \quad (j = 1, 2, \ldots, N).$$

He designates the fractal dimension of the integrated time series, $X_\Sigma$ by $D_\Sigma$, and the power-law index for $X_\Sigma$ is denoted by $\alpha_\Sigma$. As $\alpha_\Sigma$ is related to $\alpha$ by $\alpha_\Sigma = \alpha + 2$, $D_\Sigma$ is expressed as, $D_\Sigma = (3 - \alpha)/2$. We can estimate the fractal dimensions $D_{\Sigma,BK}$ and $D_{\Sigma,H}$ by applying Eqs. (1) and (3) respectively, for the length of the integrated time series, $X_\Sigma$.

In Fig. 1, the relationship between $\alpha$ and $D_{\Sigma,BK}$ (c), and between $\alpha$ and $D_{\Sigma,H}$ (d) are also shown for $0 \leqslant \alpha \leqslant 3$. It is noted that two curves (c) and (d) are in good accordance with each other in the range, $0 \leqslant \alpha \lesssim 1$, in spite of a large difference between the two curves (a) and (b) in that range.

## 3. POWER SPECTRUM OF COSMIC RAY TIME SERIES

The cosmic ray telescope at the Matsushiro underground (220 m.w.e. depth, geographic latitude 36.5° N, longitude 138.0° E) is a multi-directional muon telescope which consists of two (upper and lower) layers, each having 25 scintillation counters of 1 $m^2$ area. The muon counts for each telescope is recorded every one hour, and the counting rate for the vertical component is about $2 \times 10^4$ per hour. A detailed description of the telescope is given elsewhere.[7] Figure 2(a) shows an example of the power spectrum of the time series which is simulated so that its power spectrum shows similar characteristics as those obtained by the muon telescope at Matsushiro.[6,12] The power spectrum is calculated by the FFT method using the time series of hourly counts for the one-year period. We set $N = 2^{13}$. Two lines with a slope of $f^{-1}$ and $f^0$, respectively, are also shown in the figure. We can see that the slope of the power spectral density changes at around $10^{-5}$ Hz from $f^{-1}$ to $f^0$, where the flat level is caused by the white noise due to Poisson error of cosmic ray counts.

Figure 2(b) shows an example of the fractal length $L_{BK}(\tau)$ defined by Eq. (1) of the same time series as Fig. 2(a), plotted against $\tau$. We can see that the slope also changes at around $\tau = 20$ h from $\tau^{-2.5}$ to $\tau^{-2.0}$, as shown in the figure with the straight lines. The slope $D$ of the length changes depending on the corresponding slope $\alpha$ of the power spectrum. We can expect that the value of $D$ can be obtained much more accurately than the value of $\alpha$, because the correlation between $L(\tau)$ and $\tau$ is much better than that between $P(f)$ and $f$, as is shown in Fig. 1. However, the slope $D$ is contaminated not only by the random

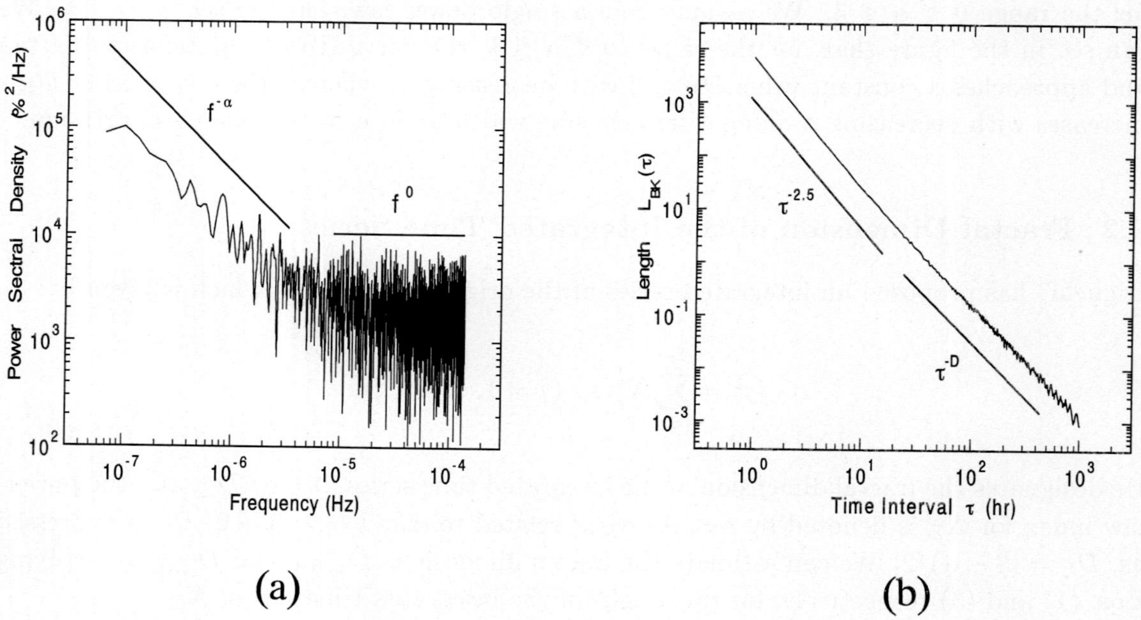


**Fig. 2** **(a)** The power spectrum of the time series similar to the Matsushiro data. We set $N = 2^{13}$. **(b)** The fractal length $L_{BK}(\tau)$ vs. $\tau$ for the same time series as in Fig. 1(a).

fluctuation caused by the counting rate error, but also on the data period or the number of the data point. In order to apply the present method to the data such as shown in Fig. 2, we need an additional numerical experiment, i.e., we need to examine the relationship between $D$ and $\alpha$ for the case in which the power spectrum has a break at a certain frequency due to the random fluctuations.

## 4. RELATION BETWEEN $\alpha$ AND $D$ FOR THE TIME SERIES WITH RANDOM FLUCTUATIONS

We first simulate a time series of which power spectral density changes from $f^{-\alpha}$ to $f^0$ at a fixed frequency $f_0 = 10^{-5}$ Hz. The range of $\alpha$-values is selected for $0 \leqslant \alpha \leqslant 3$. We calculate four types of fractal dimensions described above for each value of $\alpha$. Figure 3 shows the relationship between $\alpha$ and $D_{BK}$ (a), between $\alpha$ and $D_H$ (b), between $\alpha$ and $D_{\Sigma,BK}$ (c), and between $\alpha$ and $D_{\Sigma,H}$ (d), respectively, for the range $0 \leqslant \alpha \leqslant 3$. We set $N = 2^{13}$, $\tau_{\min} = 20$ and $\tau_{\max} = 200$.

We calculate the four types of fractal dimensions using data from the vertical telescope at Matsushiro. The year-to-year variations of the four fractal dimensions obtained are shown in Fig. 4. We can see that the yearly variation of the fractal dimensions are similar to each other. In particular, the changing manner of two fractal dimensions (c) and (d) for the integrated time series is quite similar to each other. Using four values of the fractal dimensions and the curves shown in Fig. 3, we can obtain four values of $\alpha$ for each year. The values of $\alpha$ thus obtained are similar to each other. The mean value and its standard error are shown in Fig. 5 for the years from 1985 to 1993. We emphasize here that much more

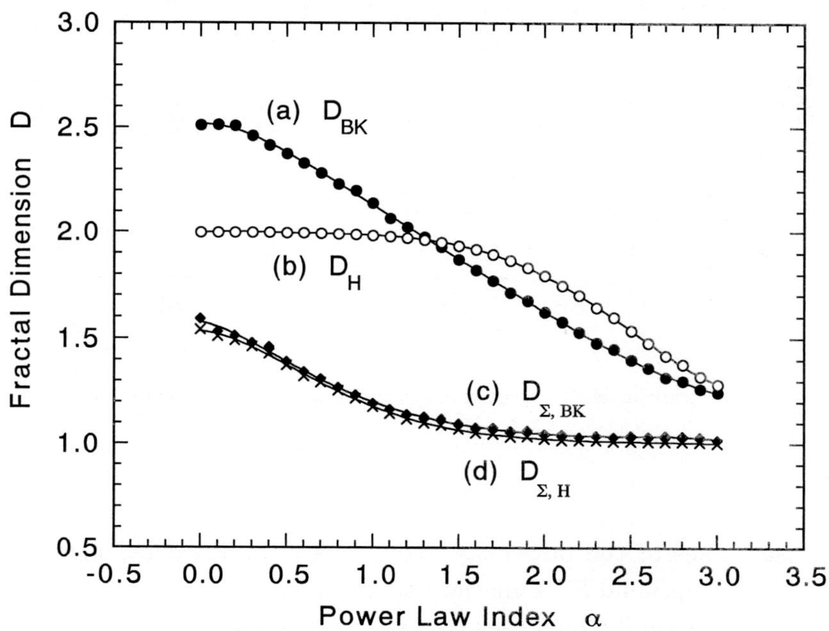


**Fig. 3** The calculated fractal dimension $D$ as a function of the power-law index $\alpha$. We assume a power spectrum as shown in Fig. 2(a).

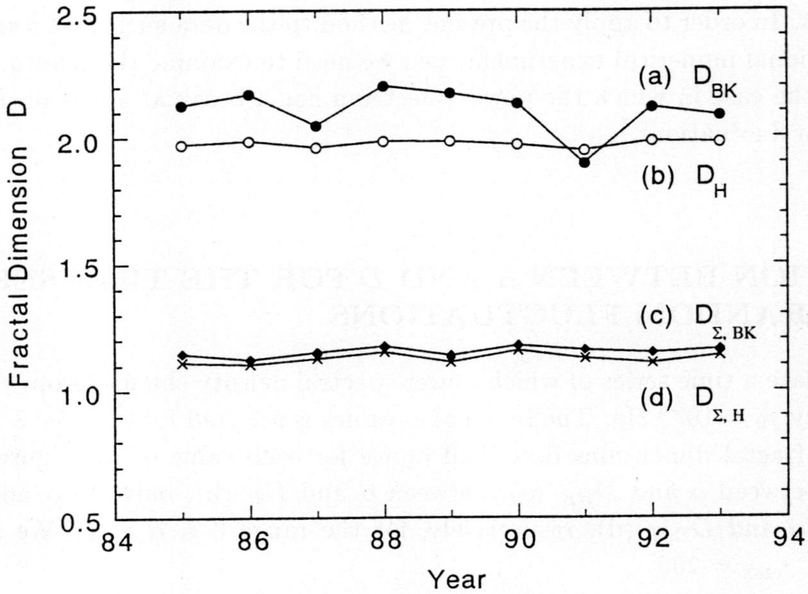


**Fig. 4** The year-to-year variation of the four types of fractal dimensions calculated for the cosmic ray time series from Matsushiro.

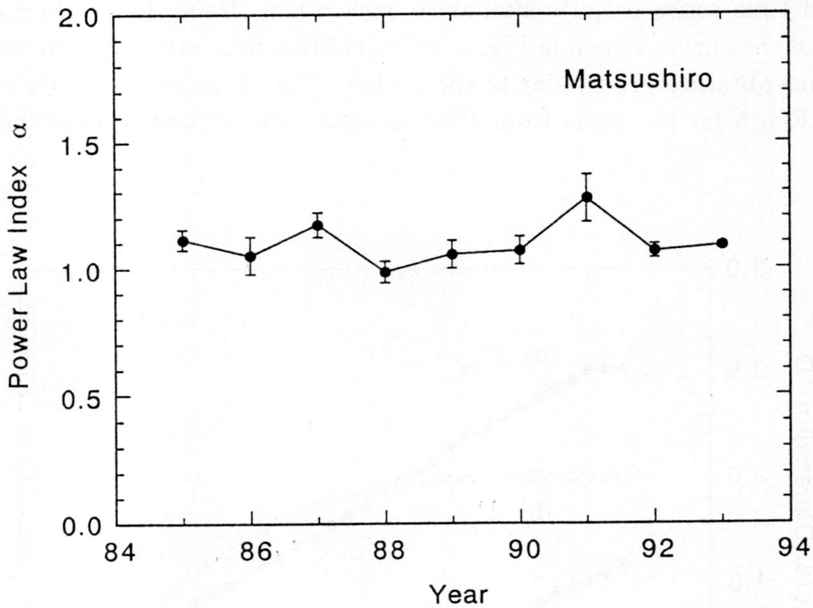


**Fig. 5** The year-to-year variation of the mean value of the power-law indices $\alpha$'s for Matsushiro, calculated based on Figs. 3 and 4.

stable and precise values of the power-law index $\alpha$ are obtained by using fractal analysis than by using the conventional FFT method separately.[6,12]

In summary, a new method which uses fractal analysis is developed in order to obtain the stable and reliable value of the power-law index of the cosmic ray time series which usually accompanies large statistical fluctuations.

## REFERENCES


1. L. Bergamasco et al., *J. Geophys. Res.* **95**, 2419 (1990).
2. S. Yasue and S. Mori, *21st Int. Conf. Cosmic Rays, Adelaide* **6**, 276 (1990).
3. K. Kudela et al., *J. Geophys. Res.* **96**, 15871 (1991).
4. M. Aglietta et al., *J. Geophys. Res.* **98**, 15241 (1993).
5. L. F. Burlaga and L. W. Klein, *J. Geophys. Res.* **91**, 347 (1986).
6. T. Higuchi, *Physica* **D46**, 254 (1990).
7. S. Mori et al., *J. Fac. Sci.*, Shinshu University **24**, 1 (1989).
8. A. J. Owens, *J. Geophys. Res.* **79**, 895 (1974).
9. A. J. Owens and J. R. Jokipii, *J. Geophys. Res.* **79**, 907 (1974).
10. S. Yasue et al., *24th Int. Conf. Cosmic Rays, Rome* **4**, 968 (1995).
11. T. Higuchi, *Physica* **D31**, 277 (1988).
12. S. Yasue et al., *23rd Int. Conf. Cosmic Rays, Calgary* **3**, 617 (1993).

Fractals, Vol. 4, No. 3 (1996) 273–278


# EFFECT OF BOUNDARY TETHERING ON VIBRATIONAL MODES OF FRACTALS

SONALI MUKHERJEE and HISAO NAKANISHI
*Department of Physics, Purdue University,*
*West Lafayette, Indiana 47907, USA*

**Abstract**

We introduce a mapping between a *tethered* scalar elastic network and a diffusion problem with permanent traps. Various vibrational properties of progressively tethered *fractals* are discussed using this analogy both in terms of scaling ansatz and numerically by approximately diagonalizing the corresponding large random matrices, with the critical percolation cluster as an example of a fractal.

## 1. ANALOGY TO RANDOM WALK WITH TRAPS

In this work we consider the effects on the *scalar* elasticity of a disordered fractal when its boundary is progressively tethered. We will discuss the behaviors of the lowest frequency mode and the vibrational density of states as well as that of the localization properties of the normal modes under these conditions.

By scalar elasticity, we mean the vibrational problem where the local displacement variable $u_i(t)$ at site $i$ is a scalar and satisfies the equation of motion:

$$\frac{\partial^2}{\partial t^2} u_i(t) = \sum_j \frac{k_{ij}}{m_i}(u_j - u_i)\,, \tag{1}$$

where the summation is over the neighbors $j$ connected elastically to the site $i$ (on a discrete lattice), $m_i$ is the mass attached to site $i$, and $k_{ij}$ is the spring constant of the bond $ij$, assumed harmonic. Although such a *scalar* model would not be appropriate for macroscopic network vibration problem in general, it would be a good start for a problem particularly if the vibration is nearly isotropic *and* harmonic. Disorder can be introduced by making $k_{ij}$ a random variable, extreme examples being a missing bond ($k_{ij} = 0$) or a rigid link ($k_{ij} \to \infty$).

The problem of vibrations in a disordered system is in itself of much physical interest. Granular systems,[1] such as sand or snow piles or crops stored in silos, are among the systems of potential application. For example, sound propagation in a granular system was recently modeled by Leibig[2] using a scalar elastic network with a random distribution of spring constants and analyzed in terms of the normal modes. Another potential interest might be due to the recently discovered anomalous behavior of the superfluid transition when aerogels[3] are embedded in liquid $^4$He or $^3$He, where the aerogels may impose a boundary condition over a fractal boundary.

We study this problem making use of a well-known mathematical mapping to the diffusion problem.[4] It is simplest to introduce a fictitious self-coupling $k_{ii} > 0$ for each site in such a way that $\sum_j' k_{ij}$ is a constant independent of $i$, where the summation now includes the diagonal term $j = i$. Normalizing and Fourier transforming in time, we get:

$$-c_i\omega^2\tilde{u}_i = \sum_j K_{ij}(\tilde{u}_j - \tilde{u}_i) \tag{2}$$

where $\sum_j' K_{ij} = 1$ when the summation includes the diagonal term and $c_i \propto m_i$ is a constant. If $m_i$ is uniform, this equation is of the same form as the Laplace transform of a spatially discrete diffusion equation, where $K_{ij}$ can be interpreted as the (symmetric) hopping rate from site $j$ to $i$ and $-\omega^2$ plays the role of the Laplace variable conjugate to the time $t$. The mapping to diffusion leads naturally to replace the problem with that of the time evolution of the corresponding diffusion field $P_i(t)$,

$$P_i(t+1) = \sum_j{}' [p_{ij}P_j(t) + (\delta_{ij} - p_{ji})P_i(t)] = \sum_j{}' W_{ij}P_j(t)\,, \tag{3}$$

where the vibration amplitudes are replaced by $P_i(t)$ and the normalized spring constants $K_{ij}$ by the (discrete) hopping probability $p_{ij}$. Note that, if $p_{ij} = p_{ji} = 1/z$, where $z$ is the lattice coordination number, this equation describes what is known as the *blind ant*. The eigenvalues $\lambda$ of the matrix $\mathbf{W} \equiv [W_{ij}]$ are related to those of the original linear equation by:

$$\lambda = 1 - \omega^2 \approx \exp(-\omega^2) \tag{4}$$

where the last approximation corresponds to the long time limit. This formulation has certain numerical advantages since the spectral region of interest $\mathbf{W}$ is the region of the maximum eigenvalues.[5]

A boundary site $i$ is tethered when there exist one or more bonds (*tethers*) connecting it to external (*perimeter*) sites which do not vibrate. To model this situation, we consider for each boundary site $i$ all *potential* tether bonds $ij'$ (where $j'$ denotes a perimeter site). A normalized bond strength of $K_{j'i}$ is then temporarily assigned to each potential tether, replacing the self-coupling $K_{ii}$, so that $\sum_j K_{ji} + \sum_{j'} K_{j'i} = 1$ for each $i$. However, we add the corresponding terms to Eq. (3) only for those sites $i$ actually chosen to be tethered. This can be shown to amount to the (partial) leakage of probability for the corresponding diffusion problem where, for a tether $ij'$, $p_{j'i} = K_{j'i} > 0$ but $p_{ij'} = 0$. Although the corresponding $\mathbf{W}$ could be enlarged to incorporate the tether anchors $j'$, only the submatrix corresponding to the network sites (without the additional tether anchors) determines the normal modes for the network. This submatrix is, however, altered so that the diagonal terms for tethered boundary sites are zero, since each tether bond reduces the corresponding diagonal term. In particular, if all boundary sites are fully tethered and if all hopping probabilities are $1/z$, then we recover the matrix $\mathbf{W}$ for the so-called ideal chain problem.[6] This discussion implies the equivalence of the tethered vibration problem with the diffusion with permanent *traps*, where the traps are the external sites where the tethers are anchored.

## 2. SLOWEST MODE AND THE DENSITY OF STATES OF THE TETHERED NETWORK

We first discuss the slowest mode, i.e., the eigenstate with the smallest $\varepsilon \equiv |\ln \lambda|$ of the matrix $\mathbf{W}$. In the limit of no tethering (or the blind ant), the true slowest mode has $\omega = 0$, while the slowest *relaxation mode* scales with the size $S$ of the network as a power law, $S^{-d_w/d_f}$,[7] where $d_w$ and $d_f$ are the walk and fractal dimensions, respectively. For full boundary tethering, on the other hand, we have a logarithmic dependence[8] $\varepsilon_1 \sim (\ln S)^{-2/d_0}$, where $d_0$ is the exponent which describes the stretched exponential behavior of the density of states for the ideal chain problem.[6] As the fraction of tethered sites (randomly chosen from the boundary) increases, we expect a crossover from the ant to ideal chain regimes.

Clearly, one important length scale is that of the average separation between the tethered sites, which scales as:

$$l \sim \left(\frac{S}{N}\right)^{1/d_f}, \tag{5}$$

where $N$ is the number of tethered sites. On the other hand, there is another important length scale $L$ for a fractal network like the critical percolation cluster[9] which consists of *blobs* of essentially compact regions and the singly connected links stringing the blobs. This is the linear size of the maximal blob (or those in the scaling regime), say,

$$L \sim S^y, \tag{6}$$

where $y$ is an appropriate exponent, left undetermined for the moment. We propose a physical picture as follows: if $l \gg L$, there is a vanishing effect of tethering and $\varepsilon_1 \approx 0$, and for $l \ll L$, we have essentially an ideal chain where $\varepsilon_1$ must depend on $S$ logarithmically. For $l \sim L$, the stationary mode of the *ant* problem is replaced by its slowest relaxation mode whose behavior is dominated by the largest blob. Thus, our proposed scaling form of $\varepsilon_1$ is:

$$\varepsilon_1 \sim S^{-d_w/d_f} f\left(\left(\frac{L}{l}\right)^{d_f}\right) = S^{-d_w/d_f} f\left(\frac{N}{S^x}\right), \tag{7}$$

where $x = 1 - yd_f$ and the scaling function $f(z) \to 0$ for $z \to 0$ and $f(z) \sim z^{d_w/[d_f(1-x)]} \cdot (\ln z)^{-2/d_0}$ for $z \to \infty$.

This type of scaling has been numerically investigated for the critical percolation cluster[9] on the square lattice. The normal modes of $\mathbf{W}$ have been obtained by an approximate diagonalization method known as Arnoldi-Saad algorithm,[5,10] where all $p_{ij}$ is taken to be $1/z = 1/4$. Our data clearly show data collapsing of the form of Eq. (7) (see Fig. 1) (only) for a range of $x$ slightly larger than 0.5. However, they lack sufficient precision to make an accurate estimate of the value of $x$ at this time. We observe that a relationship between the

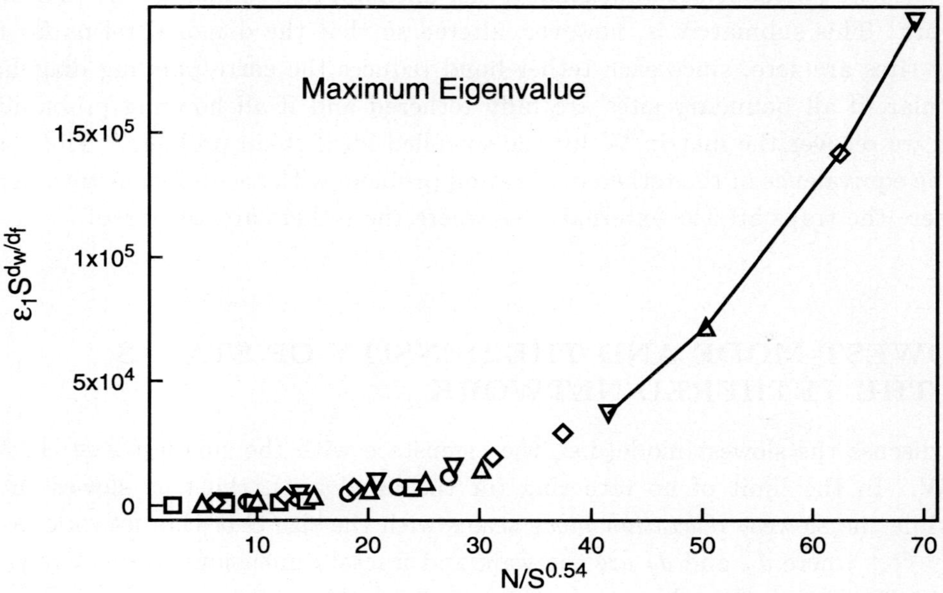


**Fig. 1** Scaling of the slowest mode is shown by plotting $\varepsilon_1 S^{d_w/d_f}$ vs. $N/S^x$ for the critical percolation cluster on the square lattice. The value of $x$ used here is 0.54. The symbols □, ○, △, ◇, and ▽, correspond to $S = 1000$, 4000, 5000, 8000, and 10 000, respectively. The solid line shows a power-law with the exponent equal to $1.54/0.46 \approx 3.35$.

corresponding exponent $y \approx 0.25$ and the standard percolation exponents, may be obtained from a scaling picture of the *blobs* based on certain assumptions.[11]

We next consider the vibrational density of states, i.e., the normal mode density $\rho(\varepsilon)$ of the matrix **W**. For fractal network with *no* tethers (blind ant), it is known to generally have a power-law increase toward the maximum eigenvalue of 1.[7] On the other hand, if all boundary sites are tethered, we recover the *ideal chain* result of a stretched exponential *decrease* toward the maximum eigenvalue (which is less than 1 for any finite network).[6] Thus, when progressively more of the boundary sites are tethered, a crossover between the two limits is expected. For an intermediate amount of tethering, we expect an initial increase of $\rho$ down to some value of $\varepsilon$ because the modes are too localized to be affected by the tethers, while for smaller $\varepsilon$ the tethers act to drag down the density of states. Thus a peak in $\rho(\varepsilon)$ is expected, say, at $\varepsilon_p$, which gradually increases from 0 in the ant limit toward the ideal chain limit. The existence of this feature has implications on the possible resonances in response to an external source of vibrational energy. The crossover in $\varepsilon_p$ is then expected to obey the same type of scaling as for the lowest mode $\varepsilon_1$, and

$$\varepsilon_p \sim S^{-d_w/d_f} g\left(\frac{N}{S^x}\right), \tag{8}$$

with the same value of $x$ as in Eq. (7).

This form of scaling has also been investigated for the critical percolation cluster on the square lattice. However, the density of states is rather flat causing difficulty in locating the peak with any precision. We do, however, find some evidence of data collapsing also with the value of $x$ in the vicinity of 0.5 as shown in Fig. 2.

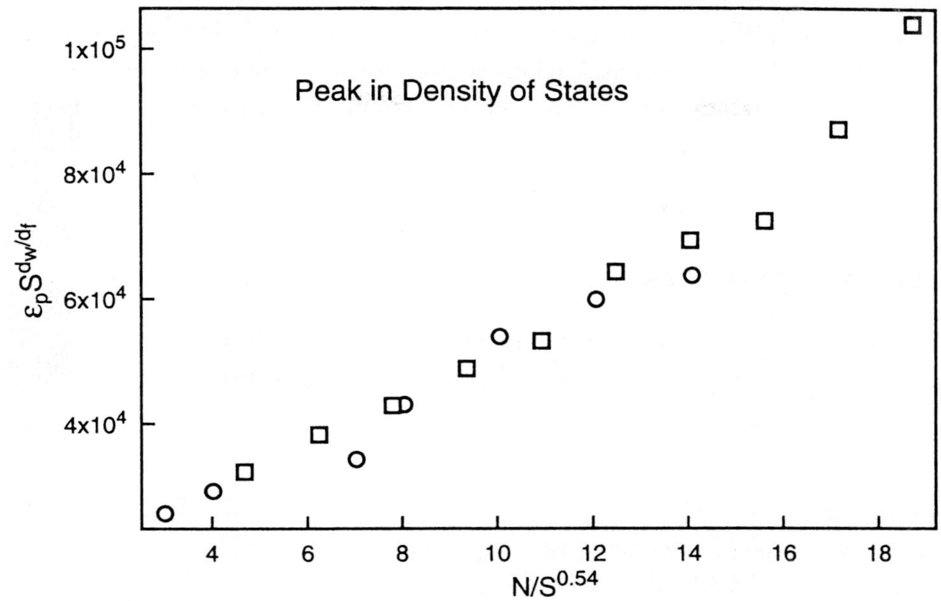


**Fig. 2** Scaling of the peak in the density of states is shown by plotting $\varepsilon_p S^{d_w/d_f}$ vs. $N/S^x$ for the critical percolation cluster on the square lattice. The value of $x$ used here is 0.54. The symbols ○ and □ correspond to $S = 5000$ and 8000, respectively.

## 3. OTHER PROPERTIES OF THE NORMAL MODES

There are many other interesting properties of the normal modes. For example, their geometrical localization property as a function of $\varepsilon$ is instructive. One simple measure of localization is given by:

$$S_\varepsilon \equiv \frac{\left(\sum_{i=1}^{S} u_i(\varepsilon)^2\right)^2}{\sum_{i=1}^{S} u_i(\varepsilon)^4} \tag{9}$$

where $u_i(\varepsilon)$ is the $i$th component of the eigenvector of $\mathbf{W}$ with the eigenvalue $\lambda = e^{-\varepsilon}$. $S_\varepsilon$ is small for well localized states and is equal to $S$ for a uniformly extended state. Thus it gives a measure of the geometrical size (number of sites) contributing significantly to the particular normal mode.

For the limit of no tethering (*ant*), $S_\varepsilon$ for the critical percolation cluster in two and three dimensions shows a power-law increasing behavior as $\varepsilon$ is lowered from a relatively large value (about 0.01) but then becomes flat well before the length scale reaches values corresponding to that of the cluster. Since we know independently[7] that even the modes of small $\varepsilon$ are localized just enough to reflect the fractal dimension of the cluster, we could interpret the behavior of the ant's $S_\varepsilon$ as indicating that the modes become less dense towards small $\varepsilon$ but keeping a similar fractal structure. For the fully tethered limit, on the other hand, $S_\varepsilon$ is very small and nearly constant, indicating extreme localization, except for modes of $\lambda = 0$.[12]

Another interesting aspect is the effect of *where* on the boundary to attach tethers. We have found drastic differences when the tethers are only attached to the external boundary sites (called *hull*) which have unique properties in two dimensions. For the hull tethering, the *blob* length scale appears to be no longer the dominant length, but rather the full cluster length $S^{1/d_f}$ replaces it. As a result, various scaling forms change, and also the localization properties are very much affected.[11] The detailed results on these points will be published elsewhere.

## ACKNOWLEDGMENTS

We thank N. H. Fuchs, A. Giacometti and D. Rintoul for helpful discussions. This work was supported in part by a David Ross Fellowship from the Purdue Research Foundation.

## REFERENCES

1. See, e.g., H. J. Herrmann, *Physica* **A191**, 263 (1992).
2. M. Leibig, *Phys. Rev.* **E49**, 1647 (1994).
3. N. Mulders, R. Mehrotra, L. S. Goldner and G. Ahlers, *Phys. Rev. Lett.* **67**, 695 (1991); S. B. Kim, J. Ma and M. H. W. Chan, *Phys. Rev. Lett.* **71**, 2268 (1993).
4. E. W. Montroll, *Ann. Math. Stat.* **18**, 1 (1947); S. Alexander and R. Orbach, *J. Phys. Lett. (Paris)* **43**, 625 (1982).
5. See, e.g., H. Nakanishi in *Annual Reviews of Computational Physics*, Vol. 1, ed. D. Stauffer (World Scientific, Singapore, 1994).
6. A. Giacometti and H. Nakanishi, *Phys. Rev.*, **E50**, 1093 (1994) and references therein; see also its erratum [*Phys. Rev.* **E51**, 6328 (1995)] where two figures are replaced due to technical errors in the originals and also an error in the numerical estimation of some exponents is reported.
7. S. Mukherjee, H. Nakanishi and N. H. Fuchs, *Phys. Rev.*, **E49**, 5032 (1994) and references therein.
8. S. Mukherjee and H. Nakanishi *Phys. Rev.* **E53**, 1470 (1996).
9. D. Stauffer and A. Aharony, *Introduction to Percolation Theory*, 2nd ed. (Taylor and Francis, London, 1992).
10. Y. Saad, *Linear Algebra Appl.* **34**, 269 (1980) and W. E. Arnoldi, *Quart. Appl. Math.* **9**, 17 (1951).
11. S. Mukherjee and H. Nakanishi, to be published.
12. A. Giacometti and H. Nakanishi, to be published.

Fractals, Vol. 4, No. 3 (1996) 279–283


# SELF-ORGANIZED CRITICALITY IN 1D TRAFFIC FLOW

TAKASHI NAGATANI
*College of Engineering, Shizuoka University, Hamamatsu 432, Japan*

Received July 16, 1995

**Abstract**

Annihilation process of traffic jams is investigated in a one-dimensional traffic flow on a highway. The one-dimensional fully asymmetric exclusion model with open boundaries for parallel update is extended to take into account stochastic transition of cars, where a car moves ahead with transition probability $p_t$. Near $p_t = 1$, the system is driven asymptotically into a steady state exhibiting a self-organized criticality. Traffic jams with various lifetimes (or sizes) appear and disappear by colliding with an empty wave. The typical lifetime $\langle m \rangle$ of traffic jams scales as $\langle m \rangle \approx \Delta p_t^{-0.54\pm0.04}$, where $\Delta p_t = 1 - p_t$. It is shown that the cumulative lifetime distribution $N_m(\Delta p_t)$ satisfies the scaling form $N_m(\Delta p_t) \approx \Delta p_t^{1.1} f(m\Delta p_t^{0.54})$.

## 1. INTRODUCTION

Recently, phenomena exhibiting self-organized criticality have attracted considerable attention.[1,2] Bak et al.[1] showed that the sandpile naturally evolves into a critical state through a self-organization process. This critical state is not characterized by intrinsic length or time scales.

Very recently, Nagel and Herrmann[3] showed that the 1D traffic flow model with open boundaries exhibits a self-organized criticality, providing enough input and output of cars at the boundaries. Nagel[4] studied the lifetimes of simulated traffic jams for freeway traffic and found emergent traffic jams with a self-similar appearance. However, the traffic flow model is not simple since it is described by the cellular automaton rule of seven states. The dependence of scaling behavior on the system size is unclear.

Nagatani[5] showed that introducing injection or extraction into the 1D fully asymmetric exclusion model with periodic boundary, drives the system asymptotically into a steady state exhibiting a self-organized criticality. In the 1D model with periodic boundary, the injection or extraction of particles maintains the self-organized critical state. Also, it was found that adding a temporary stopping of particles to the 1D fully asymmetric exclusion model with open boundaries drives the system asymptotically into a steady state exhibiting a self-organized criticality.[6] In the 1D model with open boundaries, the self-organized critical state is maintained by introducing cars at the inlet and extracting cars at the exit. It was shown that the typical lifetime $\langle m \rangle$ of traffic jams scales as:

$$\langle m \rangle \approx L^{0.52\pm0.05}, \tag{1}$$

where $L$ is the system size. The cummulative lifetime distribution $N_m(L)$ satisfies the scaling form:

$$N_m(L) \approx L^{-1} f\left(\frac{m}{L^{0.52}}\right). \tag{2}$$

In this paper, we present the 1D stochastic fully asymmetric exclusion model to take into account stochastic transition of particles. A particle moves ahead with transition probability $p_t$ if the foreward nearest neighbor is not occupied. The model describes the 1D traffic flow on a highway. Cars move into the inlet and move out at the exit on a highway where cars move or sometimes stop. When $p_t = 1$, the model is consistent with the 1D deterministic fully asymmetric exclusion model. At $p_t = 1$, the traffic flow is driven into the threshold of the maximal current. Near $p_t = 1$, the traffic jams appear and are annihilated by colliding with empty waves. We show that the self-organized criticality occurs near $p_t = 1$. We study the scaling behavior of traffic jams.

## 2. MODEL AND SIMULATION

We extend the 1D deterministic fully asymmetric simple-exclusion model with open boundaries for parallel update to take into account fluctuating velocity. Its fluctuating velocity is taken into account as the stochastic transition of particles in the 1D fully asymmetric exclusion model. Cars move ahead with probability $p_t$ if the forward nearest neighbor is empty; otherwise, cars stop with probability $1 - p_t$. Cars move or stop stochastically. Near $p_t = 1$, cars move nearly with the maximal current phase in which they are distributed in the configuration with alternate spacing. Some cars stop temporarily. Then, a traffic jam and an empty wave are created at the same time. The traffic jam consists of two connected cars and propagates backward. The empty wave consists of two connected empty sites and propagates forward. When the traffic jams collide with the empty waves, they disappear. The traffic jams and empty waves appear one after the other. Thus, introducing stochastic transition into the deterministic asymmetric exclusion model drives the system into a steady state. We study the scaling behavior of the creation and annihilation of the traffic jams in the steady state.

Our traffic flow model is defined on a one-dimensional lattice of $L$ sites with open boundaries. Each site is either empty or occupied by one car. For an arbitrary configuration, one update of the system is performed in parallel for all cars. A car is added at the inlet site 1 if the inlet site is empty. If the exit site $L$ is occupied by one car, the car is removed from the exit site.

## 3. SIMULATION RESULT

We performed simulations of the traffic flow model starting with an ensemble of random initial conditions, where the system size is $L = 10^3 - 10^5$ and the initial density of cars is $p_0 = 0.0 - 1.0$. Each run is calculated up to $10^4$–$10^5$ time steps. Figure 1 shows the typical pattern for the initial density $p_0 = 0.2$ and the transition probability $p_t = 0.99$ up to 500 time steps where the system size is $L = 300$. The vertical direction indicates that in which cars move ahead. The horizontal direction is that of time. A car is indicated by a dot. The trajectory of a car is represented by a curve. In the pattern, the grey region indicates that in which cars move with an interval of two sites. The local density of cars is $p = 0.5$. Cars within the region move with the maximum velocity 1. The black region represents the traffic jam in which cars are stopped because their progress is blocked by another car. The traffic jam propagates backward and consists of two connected cars. The white region indicates the empty wave which consists of two connected empty sites. Its empty wave propagates forward. The traffic jam disappears by colliding with the empty wave. The coalescence process between the traffic jam and empty wave is similar to a ballistic two-species annihilation reaction, $A + B \to \phi$. The traffic jam disappears with various lifetimes. For $\Delta p_t (= 1 - p_t) \ll 1$, the traffic jam is described by the ballistic two-species annihilation process. However, with increasing $\Delta p_t$, the density fluctuates greatly and the coalescence process between the traffic jam and empty wave becomes obscure. The coalescence process cannot be described by the ballistic annihilation process.

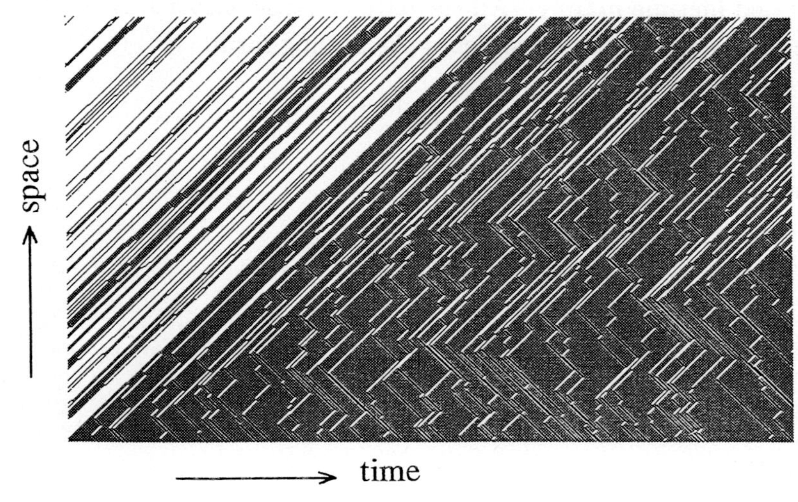


**Fig. 1** The typical pattern of cars up to 500 time steps for an initial density $p_0 = 0.2$ and a transition probability $p_t = 0.99$, where the system size is $L = 300$. The vertical and horizontal directions indicate, respectively, space and time. A car is represented by a dot. The trajectory of a car is indicated by a curve. The black regions represent the traffic jams in which cars are stopped because their progress is blocked by another car. The white regions represent the empty waves. The traffic jam and empty wave are created at the same time. The traffic jam disappears by colliding with the empty wave.

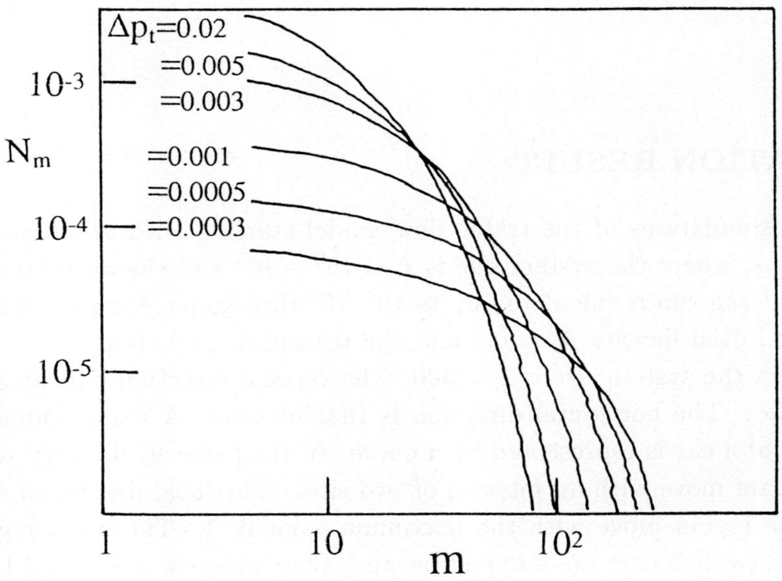


**Fig. 2** The log-log plot of the cumulative lifetime distribution $N_m$ against lifetime $m$ for $\Delta p_t(=1-p_t) = 0.02$, 0.005, 0.003, 0.001, 0.0005 and 0.0003. All data collapse onto a single curve when the rescaled cumulative lifetime distribution $N_m \Delta p_t^{-1.1}$ is plotted aginst the rescaled lifetime $m\Delta p_t^{0.54}$.

We study the scaling behavior of the lifetimes in the steady state near $p_t = 1$ after a sufficient number of time steps. The cumulative lifetime distribution $N_m$ is defined as $N_m \equiv \sum_{m'=m}^{\infty} n_{m'}$, where $m$ is the lifetime of traffic jams and $n_m$ is the lifetime distribution (the density of traffic jams with lifetime $m$ per one time step). Figure 2 shows the log–log plot of the cumulative lifetime distribution $N_m$ against lifetime $m$ for $\Delta p_t = 0.02$, 0.005, 0.003, 0.001, 0.0005, 0.0003. We plot the rescaled cumulative lifetime distribution $N_m \Delta p_t^{-1.1}$ against the rescaled lifetime $m\Delta p_t^{0.54}$ All the data collapse onto a single curve. We find that the cumulative lifetime distribution satisfies the scaling form:

$$N_m(\Delta p_t) \approx \Delta p_t^{1.1} f(m\Delta p_t^{0.54}) \,. \tag{3}$$

The typical lifetime $\langle m \rangle$ of traffic jams scales as:

$$\langle m \rangle \approx \Delta p_t^{-0.54 \pm 0.04} \,, \tag{4}$$

where $\langle m \rangle \equiv \sum_{m=1}^{\infty} m^2 n_m / \sum_{m=1}^{\infty} m n_m$.

## 4. SUMMARY

We presented the 1D traffic flow model with fluctuating velocity for freeway traffic. We showed that the system is driven asymptotically into a steady state exhibiting a self-organized criticality near $p_t = 1$. We found that the typical lifetime $\langle m \rangle$ scales as $\langle m \rangle \approx \Delta p_t^{-0.54}$ and the cumulative lifetime distribution satisfies the scaling form $N_m(\Delta p_t) \approx \Delta p_t^{1.1} f(m\Delta p_t^{0.54})$.

## REFERENCES

1. P. Bak, C. Tang and K. Wiesenfeld, *Phys. Rev.* **A38**, 364 (1988).
2. L. P. Kadanoff, S. R. Nagel, L. Wu and S. Zhou, *Phys. Rev.* **A39**, 6524 (1989).
3. K. Nagel and H. J. Herrmann, *Physica* **A199**, 254 (1993).
4. K. Nagel, *Int. J. Mod. Phys.* **C5**, 567 (1994).
5. T. Nagatani, *J. Phys.* **A28**, L119 (1995).
6. T. Nagatani, *Physica* **A218**, 145 (1995).

Fractals, Vol. 4, No. 3 (1996) 285–292


# TIME-REVERSAL DUALITY AND PLANAR LATTICE DUALITY IN NON-EQUILIBRIUM LATTICE MODELS

MAKOTO KATORI
*Department of Physics, Faculty of Science and Engineering, Chuo University, Kasuga, Bunkyo-ku, Tokyo 112, Japan*

### Abstract

The contact process (CP) is a simple mathematical model for the spread of infection of a contagious disease. Though it has only nearest-neighbor interactions, phase transitions occur even in the one-dimensional system and non-equilibrium stationary states appear in supercritical phase. This implies violation of detailed balance. The appearance of such non-equilibrium states is related to directed percolation problems on the spatio-temporal plane. In the present paper, we study discrete-time versions of the CP, the two-neighbor stochastic cellular automata (SCA), and clarify this viewpoint. We use two kinds of duality relations, the time-reversal duality and the planar lattice duality on the spatio-temporal plane, and give a good lower bound for the critical line of non-equilibrium phase transitions in the two-neighbor SCA.

## 1. INTRODUCTION

In 1986 Ziff, Gulari and Barshad (ZGB) introduced a simple model for the oxidation of carbon monoxide CO on a catalytic surface, which has been widely studied in non-equilibrium statistical physics.[1] This model is defined as an irreversible process on an infinite lattice with two absorbing states, the *poisoned states* completely covered by oxygen O or by CO. ZGB showed, by performing the Monte Carlo simulation, that this model exhibits phase transitions from an active steady state into poisoned states. From the theoretical point of view, however, rigorously proved things about the ZGB model are limited.[2] The difficulty may be due to the fact that we have to treat two species, O and CO, and that the ZGB model should be considered on surfaces, since there is no active phase in the one-dimensional version of the ZGB model. Is there any other model which is simpler but still shows non-equilibrium phase transitions similar to those in the ZGB model?

The one-dimensional contact process (CP) is the simplest but non-trivial model showing non-equilibrium phase transition. The basic CP was proposed by Harris[3,4] as a simple mathematical model for the spread of infection of a contagious disease. The same kind of models has been studied in a wide variety of fields.[5–8] Here we consider the generalized CP.[9–14] It is a continuous-time Markov process on an integer lattice $\mathbf{Z}$. The state at time $t$ is given by a set of the lattice sites $\eta_t \subset \mathbf{Z}$ which are occupied by particles. The state evolves as follows:

(i) If $x \in \eta_{t-}$, then $x$ becomes vacant at rate 1.
(ii) If $x \notin \eta_{t-}$, then $x$ becomes occupied at the rate $f(N_x(\eta_{t-}))$ which depends on the number of the nearest-neighbor particles $N_x(\eta_{t-}) = |\eta_{t-} \cap \{x-1, x+1\}|$, where $|A|$ denotes the number of sites in a set $A$, as $f(N) = 0$ if $N = 0$, $= \lambda$ if $N = 1$, and $= \theta\lambda$ if $N = 2$.

Here $\lambda$ and $\theta$ are non-negative parameters. For an epidemic model, an individual at $x \in \mathbf{Z}$ is regarded as infected if $x \in \eta_t$ and healthy if $x \notin \eta_t$. The parameter $\lambda$ is the infection rate in the case that only one of the neighbors is infected ($N_x = 1$). The parameter $\theta$ is the ratio of the infection rate in the case $N_x = 2$ to that in the case $N_x = 1$. It is proved for each $\theta \geq 1$, that there is a unique critical value $\lambda_c(\theta)$ so that any process becomes extinct with probability 1 for $\lambda < \lambda_c(\theta)$, but all processes starting from non-empty initial states have positive probabilities of survival for $\lambda > \lambda_c(\theta)$. Katori and Konno[10] gave rigorous upper and lower bounds for the critical line $\lambda = \lambda_c(\theta)$ with $1 \leq \theta \leq 2$, and showed explicitly that there exist two distinct phases. The *extinction phase* ($\lambda < \lambda_c(\theta)$) corresponds to extermination of the disease, and the *survival phase* ($\lambda > \lambda_c(\theta)$) does the state where the infection is spreading and healthy individuals and infected ones coexist. In the context of chemical reactions, the former corresponds to the poisoned state and the latter the active steady state of catalytic surfaces.

As its name suggests, the CP has only nearest-neighbor interactions. However, we observe phase transitions even in the one-dimensional system. This implies that the survival state of the CP cannot be described by the ordinary Gibbs state with any finite-ranged potentials. In other words, detailed balance is violated in the stationary states which appear in the survival phase. The appearance of such non-equilibrium stationary states is related to directed percolation (DP) problems[15] on the spatio-temporal plane. In order to clarify this viewpoint, we will study the discrete-time version of the generalized CP in this paper.

## 2. TWO-NEIGHBOR SCA

We introduce a set of discrete-time version of CP, $\xi_t$, on a spatio-temporal plane $V = \{(x, t) \in \mathbf{Z}^2 : x + t = \text{even}, t = 0, 1, 2, \ldots\}$, in which the sites evolve in parallel. The discrete-time stochastic process, in which all sites are updated simultaneously, are usually called *stochastic cellular automata* (SCA).[16] We consider the two-valued SCA; each site $x$ takes one of the two states 0 (vacant) and 1 (occupied by a particle). The state at time $t$ is thus given as a set of sites occupied by particles. Let $\mathbf{Z}_e = \{\ldots, -4, -2, 0, 2, 4, \ldots\}$ and

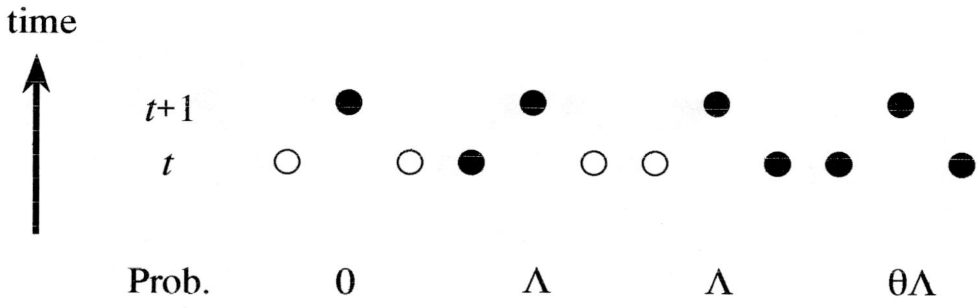


**Fig. 1** The elementary processes of the two-neighbor SCA. The full (resp., open) circles denote particles (resp., vacancies).

$\mathbf{Z}_o = \{\ldots, -3, -1, 1, 3, \ldots\}$. Then $\xi_t \subset \mathbf{Z}_e$ for $t =$ even and $\xi_t \subset \mathbf{Z}_o$ for $t =$ odd. The time evolution is given by:

$$P(x \in \xi_{t+1}|\xi_t) = g(N_x(\xi_t)), \tag{1}$$

where $P(\omega_1|\omega_2)$ denotes the conditional probability of $\omega_1$ given $\omega_2$. We assume that:

$$g(N) = \begin{cases} 0 & \text{if } N = 0 \\ \Lambda & \text{if } N = 1 \\ \theta\Lambda & \text{if } N = 2, \end{cases} \tag{2}$$

where $0 \leq \Lambda \leq 1$ and $0 \leq \theta\Lambda \leq 1$. Figure 1 shows the elementary processes. We call this SCA the two-neighbor SAC.[16,17]

Let $\Omega_L = \{1, 2, \ldots, L\}$ be a finite integer lattice with size $L$ and periodic boundary condition. The two-neighbor SCA on this finite set $\Omega_L$ are completely determined by the $2^L \times 2^L$ transition matrices $M_L$; $M_L(A, B) = P(\xi_{t+1} = B|\xi_t = A)$, where $A$ and $B \in Y_L \equiv$ a collection of subsets of $\Omega_L$.

## 3. TIME-REVERSAL DUALITY

We introduce another SCA, *coalescing branching random walk* (CBRW), $\tilde{\xi}_t$. Each particle branches with probability $p$, hops to the left or to the right neighbor site with probability $q$ for each, or annihilates with probability $1 - p - 2q$. If two particles land on the same site, they coalesce into one (see Fig. 2). Let $\tilde{M}_L$ be the transition matrix of $\tilde{\xi}_t$ defined on a finite system $\Omega_L$.

We can prove the following statement. If $\Lambda = p + q$ and $\theta\Lambda = p + 2q$, then the equalities:

$$\sum_{C \cap B \neq \emptyset} M_L(A, C) = \sum_{C \cap A \neq \emptyset} \tilde{M}_L(B, C), \tag{3}$$

hold for all $A$, $B \in Y_L$ with any $L \geq 2$.[18] We noticed that these are identified with the relation which is called the *time-reversal duality* or specifically the *coalescing duality* relation in probability theory.[4,15,17] If this time-reversal duality holds, $M_L$ and the transpose of

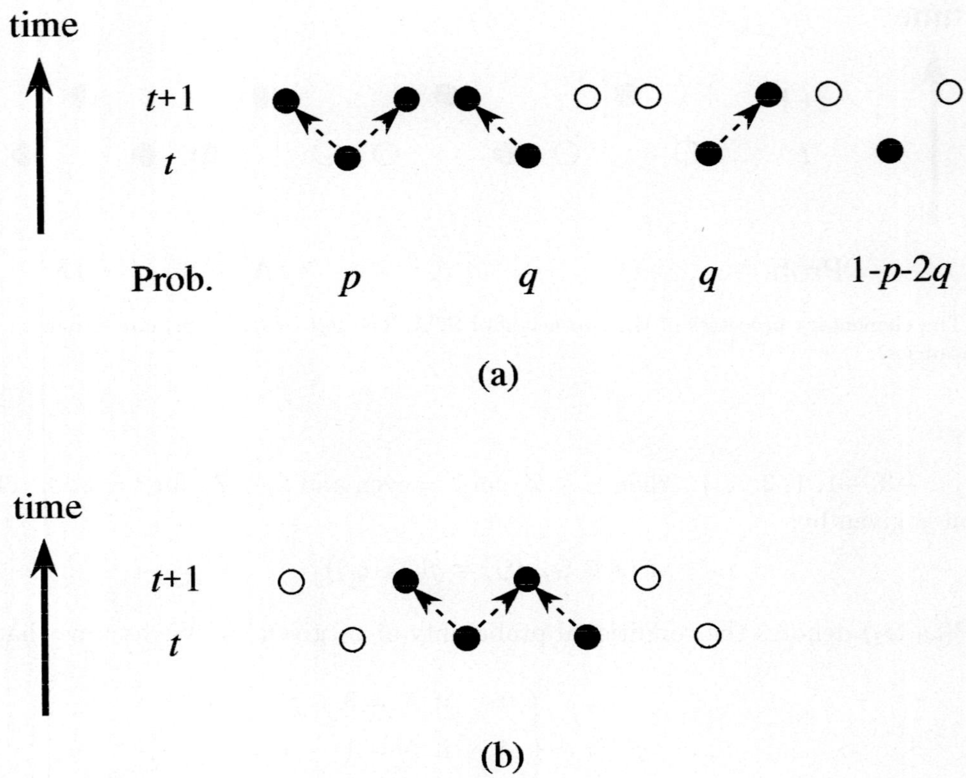


**Fig. 2** (a) The elementary processes of the CBRW. (b) If two particles land on the same site, they coalesce into one.

$\tilde{M}_L$ are conjugate and thus the characteristic polynomials are equal to each other,

$$\det(M_L - zI) = \det(\tilde{M}_L - zI)\,, \tag{4}$$

for arbitrary $L \geq 2$, where $I$ is the identity matrix.[18]

Critical values and critical exponents are determined by asymptote of eigenvalues of the transition matrix in the infinite-size limit $L \to \infty$.[16,19] Therefore the following statement should be true. If we observe a phase transition at $(p_c,\, q_c)$ in the CBRW $\tilde{\xi}_t$, then we have a phase transition in the two-neighbor SCA $\xi_t$ at $(\Lambda_c,\, \theta_c)$ with $\Lambda_c = p_c + q_c$ and $\theta_c \Lambda_c = p_c + 2q_c$. Moreover, if the transitions are continuous and critical phenomena are observed, then these two critical phenomena belong to the same universality class.[18] It should be remarked that since $0 \leq p,\, q \leq 1$, the above statement has meaning if and only if $1 \leq \theta \leq 2$, which will be assumed from now on.

## 4. PLANAR LATTICE DUALITY

The CBRW $\tilde{\xi}_t$ can be defined as a two-dimensional statistical model on the spatio-temporal plane $V$ as follows.[17] At each triangle of three points $(x,\, t)$ $(x-1,\, t+1)$ and $(x+1,\, t+1)$, we place the following "gadgets" independently. Two arrows with probability $p$, one arrow with probability $q$ (the left and right ones, respectively) and nothing with probability $1 - p - 2q$

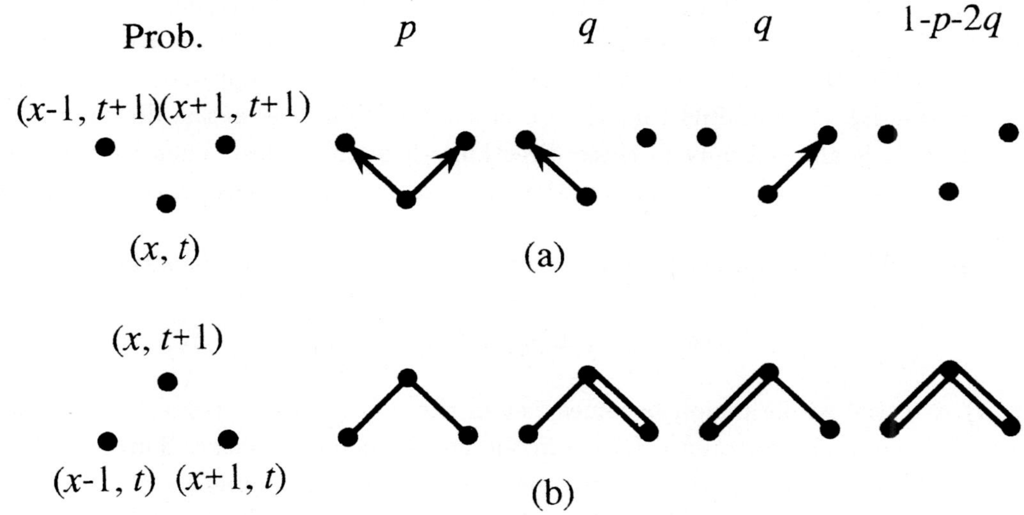


**Fig. 3** **(a)** The gadgets for constructing the CBRW on the spatio-temporal plane $V$. **(b)** For each gadget in (a), we assign a new gadget which consists of resistors (double lines) and/or diodes (single lines). These gadgets are placed on triangles of three points on the lattice $V^*$ which is the planar lattice dual of $V$.

[see Fig. 3(a)]. The arrow from $(x,\, t)$ to $(x-1,\, t+1)$ or that from $(x,\, t)$ to $(x+1,\, t+1)$ denotes that the directed bond is open. We say "there is an open path from $(x,\, t)$ to $(y,\, t{+}n)$ $(n \geq 1)$", if there is a sequence $(x_0,\, t) = (x,\, t),\, (x_1,\, t+1), \ldots,\, (x_n,\, t+n) = (y,\, t+n)$ of points in $V$, such that for each $0 \leq i \leq n-1$, the bond from $(x_i,\, t+i)$ to $(x_{i+1},\, t+i+1)$ is open, and write $(x,\, t) \to (y,\, t+n)$ for short. If we define a process by setting:

$$\tilde{\xi}_t^x = \{y : (x,\, 0) \to (y,\, t)\} \quad \text{and} \quad \tilde{\xi}_t^A = \bigcup_{x \in A} \tilde{\xi}_t^x\,, \tag{5}$$

then the resulting system is nothing but the CBRW starting from the state $A$.

It is easy to confirm that the above set of processes includes the site and bond DP models as special cases. Therefore the CBRW $\tilde{\xi}_t$ is regarded as the generalized DP on the square lattice.

Here the relation between the SCA and the DP problems is clarified. The extinction (resp. survival) phase of the SCA corresponds to the non-percolation (resp. percolation) phase of the DP. Non-equilibrium phase transitions from a trivial absorbing state to the active stationary states are the percolation transitions of directed systems on the spatio-temporal plane. Now we can use the technique, which has been developed for percolation problems, to analyze non-equilibrium phase transitions.

Dhar[20] invented a series of strategies which improve successively lower bounds for the bond DP. His method is based on the observation by Dhar, Barma and Phani[21] that bond DP is the *planar lattice dual* of a diode-resistor percolation model. Dhar et al.[21] introduced an angle $\phi$ which characterizes the asymptotic behavior of the spread of diode-resistor percolation region. They argued that if $\phi > \frac{\pi}{2}$, then whole region becomes "wet" in the diode-resistor percolation and it follows that the corresponding bond DP is in the non-percolation phase, while if $\phi < \frac{\pi}{2}$, then the percolation probability is positive in the DP model. Dhar's method is for obtaining lower bounds for $\phi$.

We find that Dhar's method can be extended for the above mentioned generalized DP.[22] The diode-resistor percolation model, which is planar lattice dual to the generalized DP, is defined on the lattice $V^* = \{(x, t) \in \mathbf{Z}^2 : x + t = \text{odd}, t = 0, 1, 2, \ldots\}$, as follows. For each gadget of $\tilde{\xi}_t$ in Fig. 3(a), we assign a new gadget as shown in Fig. 3(b). Here double lines represent resistors (two-way conducting bonds) and single lines represent diodes (one-way conducting bonds). We assume that the orientation of diodes is in the direction (1, 1) or (1, −1); current is allowed only in these directions through diodes. Then we put a source of fluid at the point $(1, 0) \in V^*$ and consider the percolation problem on this diode-resistor system.

The result which is obtained by generalizing Dhar's first strategy is stated as follows.[22] Let

$$F(p, q) = (p + 2q) - \frac{1 - 2(p+q)^3}{1 - (p+q)^2} \,. \tag{6}$$

If $F(p, q) < 0$, the whole region becomes wet in the diode-resistor percolation model and its planar lattice dual, generalized DP, is in the non-percolation phase. Through the time-reversal duality relation, this result is mapped to the following bound.

Assume that $1 \le \theta \le 2$ and let $\Lambda = \Lambda_c(\theta)$ be the critical line which separates the extinction phase ($\Lambda < \Lambda_c(\theta)$) and the survival phase ($\Lambda > \Lambda_c(\theta)$) of the two-neighbor SCA. Define $\Lambda_L(\theta)$ as the smallest root of the cubic equation:

$$(2 - \theta)\Lambda^3 + \theta\Lambda - 1 = 0 \,. \tag{7}$$

Then $\Lambda_L(\theta) \le \Lambda_c(\theta)$.

Recently Liggett[23] applied the so-called Holley-Liggett argument[4,10,14,24] to the present systems and proved the following upper bound.

$$\Lambda_c(\theta) \le \Lambda_U(\theta) = 1 - \frac{\theta}{4} \,. \tag{8}$$

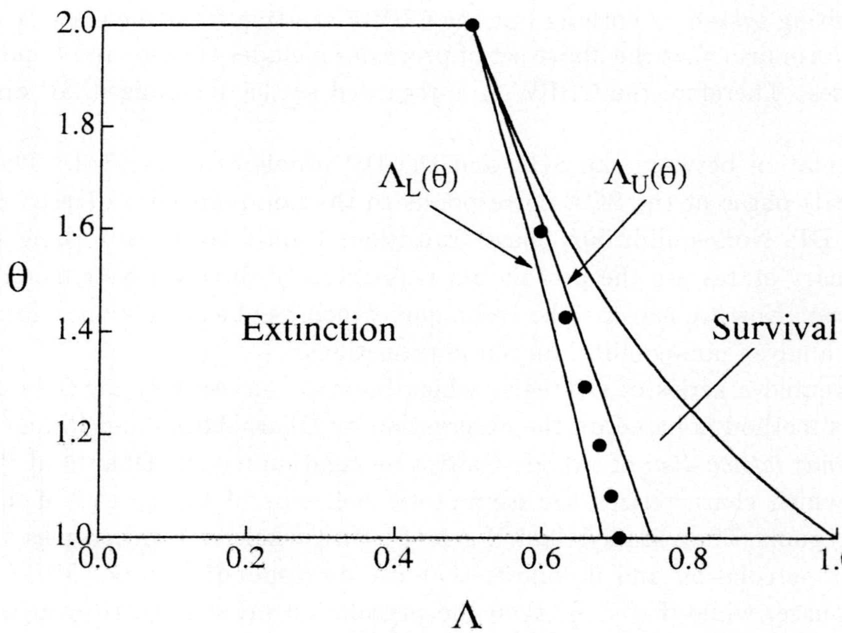


**Fig. 4** The lower bound $\Lambda = \Lambda_L(\theta)$ given in this paper and the upper bound $\Lambda = \Lambda_U(\theta)$ by Liggett are shown for the critical line $\Lambda = \Lambda_c(\theta)$. The critical values evaluated by the Monte Carlo simulations are denoted by full circles, which fall between the bounds.

Figure 4 shows the above results. It should be remarked that the region which is bounded by $\Lambda_L(\theta)$ and $\Lambda_U(\theta)$ is very narrow. We performed Monte Carlo simulation.[22] The evaluated critical values are denoted by full circles in Fig. 4. These results fall between the bounds and the difference from our lower bound is within a few percent of each value.

## 5. SUMMARY AND FUTURE PROBLEMS

Figure 5 is a summary of the procedure which I explained in this paper. First the time-reversal duality was discussed. The CBRW, which is the time-reversal dual of the two-neighbor SCA, can be defined as a generalized DP on the spatio-temporal plane. This DP has a planar lattice dual, which is generalized version of diode-resistor percolation model. We succeeded in extending Dhar's method and obtained a new result for this diode-resistor percolation. Then the result is backpropagated to the two-neighbor SCA and we can give a good lower bound for the critical line of non-equilibrium phase transitions in the discrete-time version of CP.

Can we apply the same procedure to the continuous-time CP and the ZGB model? The generalized CP has a time-reversible dual, which is a coalescing branching process with single annihilation.[9] We can give its graphical representation on the spatio-temporal plane $\mathbf{Z} \times [0, \infty)$, which is a kind of DP model.[11] The planar lattice dual can be considered,[25] but it has not yet been reported that Dhar's method is applied to such a diode-resistor percolation on $\mathbf{Z} \times [0, \infty)$. Since the ZGB model is the three-valued model, we have to

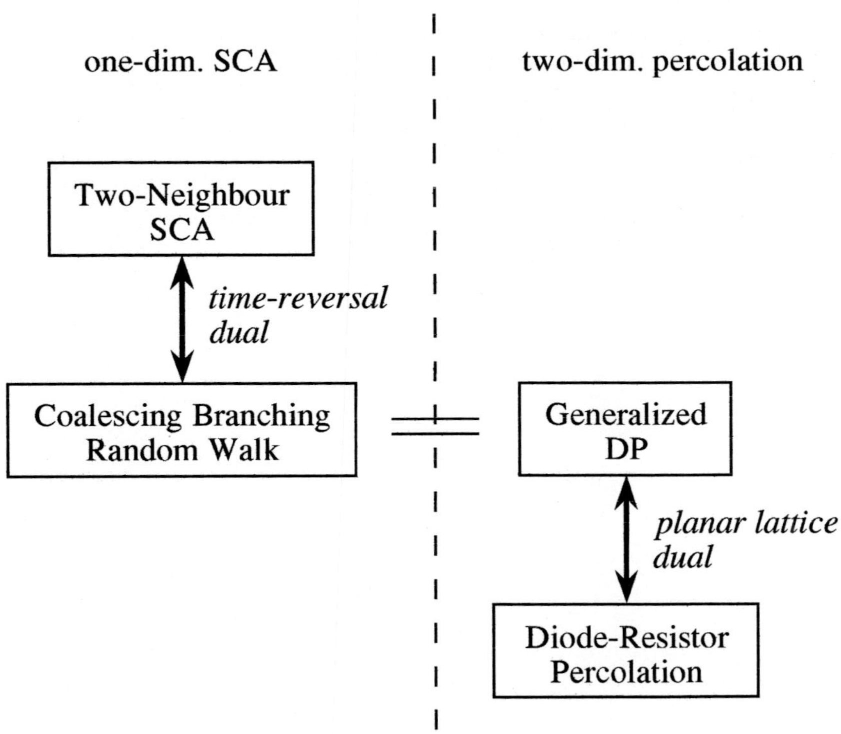


**Fig. 5** Summary of two kinds of duality: the time-reversal duality and the planar lattice duality.

generalize the theory of time-reversal duality. The time-reversal dual process of the ZGB model shall not be a simple branching process.[2] It is a branching process of sets of sets of sites.[13,26] We do not know useful graphical representation of such a branching process, which represents a bundle of possible paths on the spatio-temporal hyper-plane. I hope that the theory of duality relations will be developed and applied to many models which show interesting non-equilibrium phase transitions in the near future.

## REFERENCES


1. R. M. Ziff, E. Gulari and Y. Barshad, *Phys. Rev. Lett.* **56**, 2553 (1986).
2. E. R. Grannan and G. Swindle, *J. Stat. Phys.* **61**, 1085 (1990).
3. T. E. Harris, *Ann. Prob.* **2**, 969 (1974).
4. T. M. Liggett, *Interacting Particle Systems* (Springer, New York, 1985).
5. R. C. Brower, M. A. Furman and M. Moshe, *Phys. Lett.* **B76**, 213 (1978).
6. P. Grassberger and A. de la Torre, *Ann. Phys.* **122**, 373 (1979).
7. F. Schlögl, *Z. Physik* **253**, 147 (1972).
8. P. Grassberger and M. Scheunert, *Fortschr. Phys.* **28**, 547 (1980).
9. R. Durrett and D. Griffeath, *Ann. Probab.* **11**, 1 (1983).
10. M. Katori and N. Konno, *J. Phys. A: Math. Gen.* **26**, 6597 (1993).
11. M. Katori and N. Konno, in *Formation, Dynamics and Statistics of Patterns*, Vol. 2, eds. K. Kawasaki and M. Suzuki (World Scientific, Singapore, 1993), pp. 23–72.
12. I. Jensen and R. Dickman, *Physica* **A203**, 175 (1994).
13. M. Katori, *J. Phys. A: Math. Gen.* **27**, 3191 (1994).
14. N. Konno, *Phase Transitions of Interacting Particle Systems* (World Scientific, Singapore, 1994).
15. R. Durrett, *Ann. Probab.* **12**, 999 (1984).
16. W. Kinzel, *Z. Phys.* **B58**, 229 (1985).
17. R. Durrett, *Lecture Notes on Particle Systems and Percolation* (Wadsworth and Brooks/Cole, Pacific Grove, CA, 1988).
18. N. Inui, M. Katori and T. Uzawa, *J. Phys. A: Math. Gen.* **28**, 1817 (1995).
19. D. Ben-Avraham, R. Bidaux and L. S. Schulman, *Phys. Rev.* **A43**, 7093 (1991).
20. D. Dhar, *J. Phys. A: Math. Gen.* **15**, 1849 (1982).
21. D. Dhar, M. Barma and M. K. Phani, *Phys. Rev. Lett.* **47**, 1238 (1981).
22. M. Katori and H. Tsukahara, *J. Phys. A: Math. Gen.* **28**, 3935 (1995).
23. T. M. Liggett, *Ann. Appl. Probab.* **5**, 613 (1995).
24. R. Holley and T. M. Liggett, *Ann. Probab.* **6**, 198 (1978).
25. R. Durrett and R. H. Schonmann, *Ann. Probab.* **16**, 1570 (1988).
26. L. Gray, *Ann. Probab.* **14**, 371 (1986).

Fractals, Vol. 4, No. 3 (1996) 293–303


# AN INTERFACE OBTAINED BY COLLISION OF TWO INDEPENDENT ROUGH GROWING SURFACES — IMBIBITION OF INK INTO THE PAPER

T. NAGAMINE and S. MIYAZIMA
*Department of Engineering Physics, Chubu University, Kasugai, Aichi 487, Japan*

## 1. INTRODUCTION

In our daily life, we see many complicated patterns such as branches of trees, forms of clouds, shapes of lightning, smoke of cigarettes, ridges of mountains, etc. But it is not yet clear how their patterns and structures are formed and how the roughness of their boundaries can be quantitatively expressed. B. B. Mandelbrot paid attention to such complicated structures, becoming the originator of the concept of fractal geometry at the end of 1960[1,2]; this concept has since been expanded upon during the past ten years by the researchers all over the world. Many growth models such as the Eden model, the ballistic deposition model, etc., were studied. At the present time, fractal analysis is widely applied to biology (e.g., the formation of bacterium[3] and fungus[4]); medical science (e.g., the complexity of bronchus,[5] human retinal vessels,[6] and the size distribution of a patch of mycosis fungoides on the skin surface[7]); chemistry (e.g., the formation of islands made by Ag atoms deposited on Pt[8]) etc.[1,2,8,9] We think that the analysis of rough growth surfaces is one of the very important research problems on the fractal phenomena. The morphological properties and dynamics of such rough growth surfaces are investigated not only by the KPZ theories[10] and simulations, but also by experiments such as DLA,[11] viscous fingering,[12] burning problem[13] and imbibition experiment of liquid[14,15] etc.[1,2,8,9]

In this paper, we report the fractality of the rough interface formed when ink propagates into the fabric of paper. Furthermore, we also investigate an interface formed by collision of two rough growing fronts. The interface is called the collision interface hereafter. We have obtained the roughness exponent $\alpha$ of the growth surface made by ink invading the paper, and the roughness exponent $\alpha'$ of the collision interface.[15] The roughness exponent characterizes the morphological property of the stationary growth surface. In this experiment, we will obtain the growth exponent $\beta$ of the growth surface made by ink and the growth exponent $\beta'$ of the collision interface made by ink. The growth exponent characterizes the growth property of dynamical growth surface. From the results of our experiments and simulation, we discuss not only the scaling relation $\alpha + z = 2$ ($z = \alpha/\beta$), but also the scaling relation $\alpha' + z' = 2$ ($z' = \alpha'/\beta'$) obtained by our experiments. $z$ and $z'$ are the dynamical exponents of the growth surface and the collision interface, respectively.

## 2. SCALING RELATION OF GROWTH SURFACE

The Eden model is one of the simplest growth models and the Eden growth surface made from a line seed is very well-known for the rough growth surface with self-affine property. Therefore we have thought that the scaling property of the Eden growth surface corresponds to the scaling property of the rough growth surface made by nonequilibrium phenomena. It is well-known that the Eden growth surface has two scaling properties with respect to the roughness of growth surface. In the first place, the roughness of the Eden growth surface increases such that it obeys the power law of the average height of the Eden growth surface; in the second place, the roughness of the Eden growth surface saturates independently of the increase of the averaged growth height. These scaling properties of the Eden growth surface are supported by the simulations[8,9] and are also observed in some experiments.[8,9,14,15]

In general the roughness $\sigma$ of a rough surface is defined by the root mean square (rms) of the height of the growth surface, which is measured from the starting line within a width

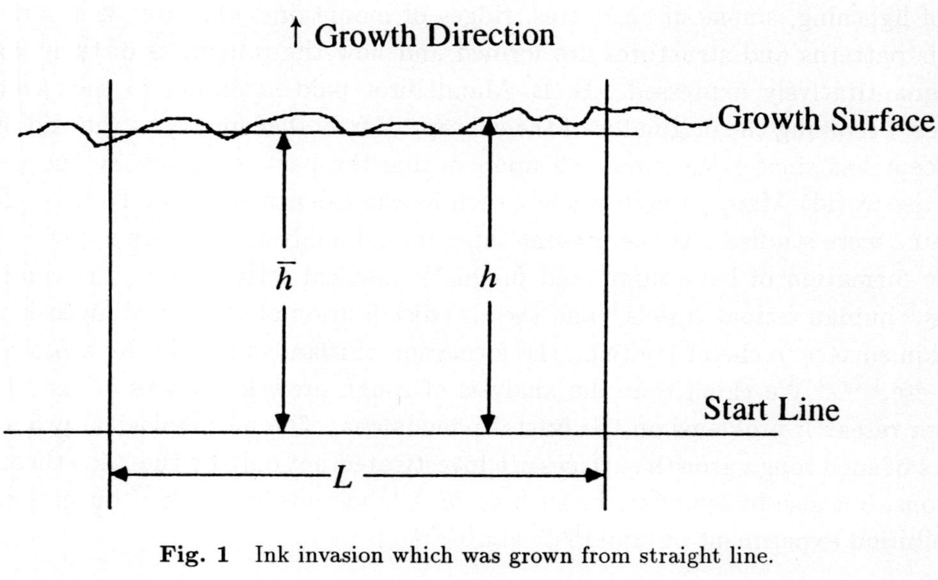


**Fig. 1** Ink invasion which was grown from straight line.

of $l$. $l$ is an arbitrary width of equal sections into which the entire system size is divided (see Fig. 1).

The equation of the roughness $\sigma$ of a rough frontal broundary is expressed as follows:

$$\sigma(\bar{h}, l) = \sqrt{\frac{1}{N} \sum (h - \bar{h})^2} .$$

Here $h$ and $\bar{h}$ are the height and the average height of the growth surface, respectively. $N$ is the number of cells into which the growth surface is divided. The self-affine property of the Eden growth surface which starts from the straight line seed is expressed by the following:

$$\sigma(\bar{h}, l) \propto l^{\alpha} \cdot f \frac{\bar{h}}{l^z} ,$$

where $z(= \alpha/\beta)$ is called the dynamical exponent. Here, if the average height $\bar{h}$ of growth surface is much smaller than the width $l$, the scaling function $f(x)$ is proportional to $x^z$, and if $\bar{h}$ is much larger than $l$, $f(x)$ is constant. Therefore the above equation is simplified as follows:

$$\sigma(\bar{h}, l) \propto \bar{h}^{\beta} \quad \text{when} \quad \bar{h} \ll l ,$$

$$\sigma(\bar{h}, l) \propto l^{\alpha} \quad \text{when} \quad \bar{h} \gg l .$$

Further the two exponents $\alpha$ and $z$ satisfy the following scaling relation:

$$\alpha + z = 2 \quad \left(z = \frac{\alpha}{\beta}\right) .$$

**Table 1 Results of theory and simulation.**

| | Theory[9] | Simulation[9] |
|---|---|---|
| $\alpha$ | $\frac{1}{2}$ | $0.51 \pm 0.025$ |
| $\beta$ | $\frac{1}{3}$ | $0.33 \pm 0.015$ |
| $\alpha + z \left(z = \frac{\alpha}{\beta}\right)$ | 2 | $2.05 \pm 0.031$ |

In this experiment, we will obtain the growth exponents $\beta$ and $\beta'$ of the growth surface and the collision interface, respectively. Experimental roughness exponents $\alpha$ and $\alpha'$ were obtained by the previous experiment.[15]

## 3. EXPERIMENT

We have made an experimental apparatus which can display the growing front of the ink invading from the edge of the paper. The ink used in our experiments is called the India ink or Chinese ink.

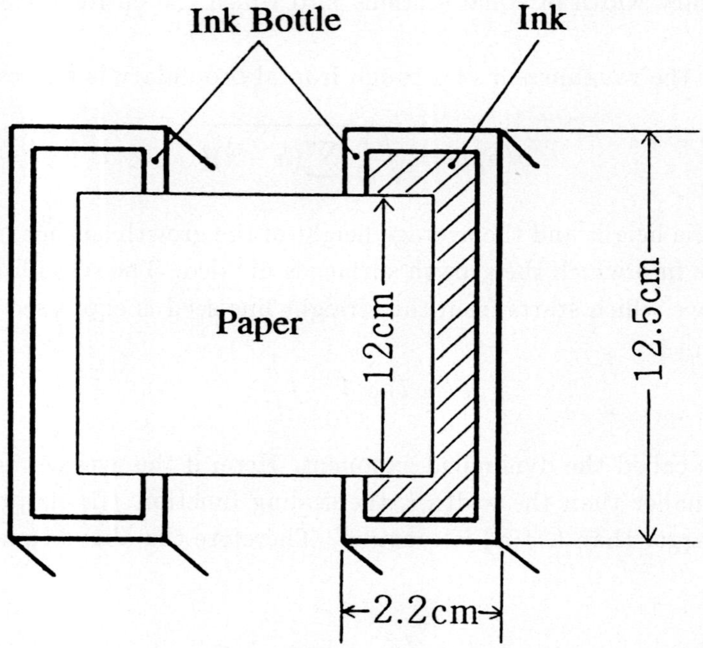


**Fig. 2** Illustration of apparatus of the first experiment.

**3.1** Our first experiment is to investigate the growth property of the surface formed when ink invaded from an edge of the paper. Consequently, we have to obtain the growth exponent of the surface formed by the frontal boundary of the invading ink. Figure 2 shows the outline of our apparatus of the first experiment. In Fig. 2, a yellow line indicates a starting line from which ink starts to invade into the horizontally laid paper. The ink bottle on the right hand side is filled up with ink. The paper is put on the lifting device and is kept in a horizontal position. Ink starts invading from one side of the paper.

**3.2** When the paper goes down horizontally with the lift, ink must be fed from the right hand side of the paper, as shown in Fig. 3. A black region of the paper in Fig. 3 is the part invaded by ink and becomes wet. The boundary between the black and white (wet and dry) regions forms a very interesting pattern. We video tape the ink invasion process using 8 mm video movie (SONY; CCD-TR850).

**3.3** We reproduce the ink invasion process on a TV monitor, and trace the boundary between the wet and dry regions (the growth surface of the ink) on the transparent paper, as shown in Fig. 4. The yellow line indicates the start line of ink invasion. Each curve represents the growth surface traced at an intermittent time and the growing direction of the ink is expressed by the big arrow. Both edges of the paper must be cut to eliminate any edge effects.

**3.4** Each curve of Fig. 4 is digitized by a scanner, and we can calculate the roughness $\sigma$ and the average height $\bar{h}$ of each curve. After that, by plotting $\sigma$ against the average height $\bar{h}$, we obtained the following graph (Fig. 5).

**Fig. 3** Ink invasion from right side of paper.

By fitting the curve by a straight line, we obtained its slope, which gives us the growth exponent $\beta$.

**3.5** Next, we performed the second experiment. The purpose of the second experiment is to study the growth property of the collision interface formed when two

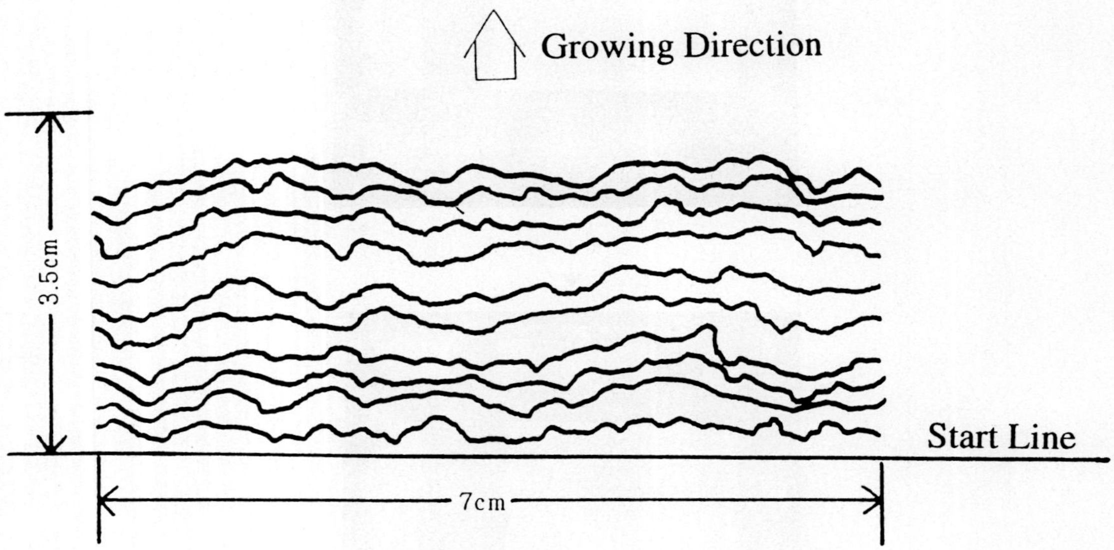


**Fig. 4** Growth surfaces of ink.

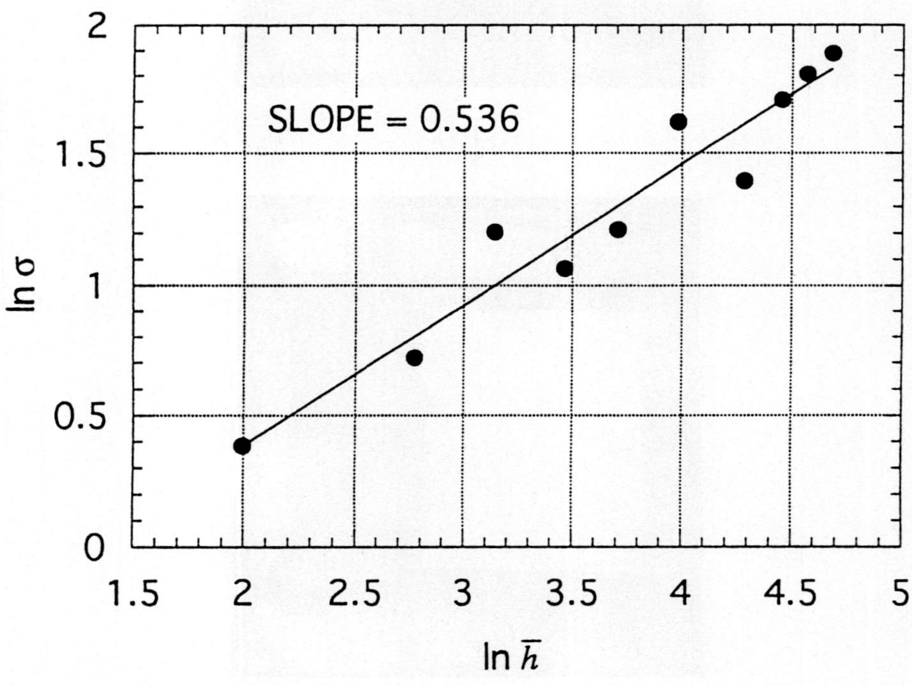


**Fig, 5** Log-log plot of $\sigma$ vs. $\bar{h}$ of growth surface.

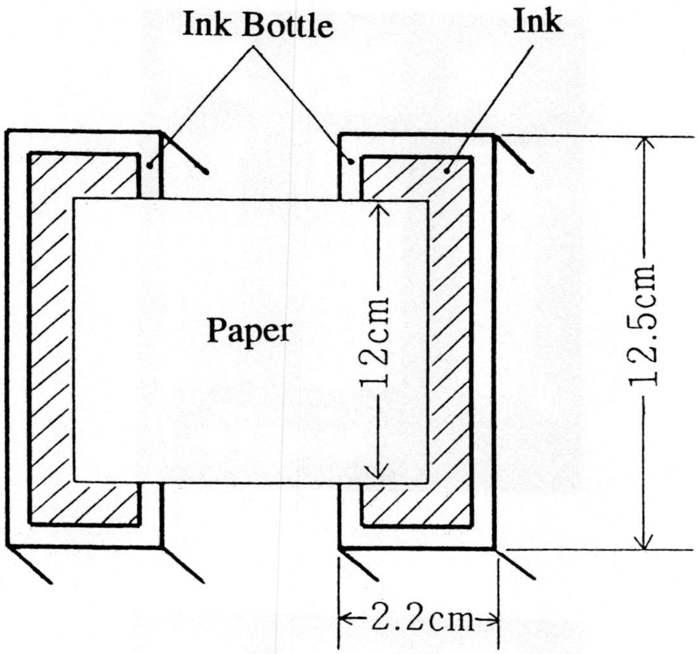


**Fig. 6** Illustration of apparatus of collision experiment.

independent growth surfaces made by ink imbibition collide. For that purpose, we calculated the growth exponent $\beta'$ of the collision interface. In Fig. 6, two ink bottles are filled with ink.

When ink is fed onto the paper, ink forms a very complex surface, as shown in Fig. 7. In Fig. 7, when two black regions (wet regions) grow, two black regions eventually collide somewhere near the center of the paper and the collision interface is formed by tracing the collision points, which are determined by the collision of the two black regions.

In order to draw a log-log plot of the roughness against the average height of the collision interface, we must get the collision interfaces with different average heights. Therefore, by changing the distance between the two ink bottles, we made the collision experiment of ink (see Fig. 8). We could then draw the graph as shown in Fig. 9. We calculate the slope of the straight line in Fig. 9, which corresponds to the growth exponent $\beta'$ of the collision interface of ink.

## 4. RESULT

In the first experiment, we obtained the growth exponent $\beta$ of the rough growth surface produced by ink and in the second experiment, we obtained the growth exponent $\beta'$ of the collision interface of ink, shown in Table 2. $\beta$ and $\beta'$ are $0.55 \pm 0.08$ and $0.47 \pm 0.10$, respectively. $\alpha$ and $\alpha'$ shown in Table 2 are recalculated from the previous experimental results. $\alpha$ and $\alpha'$ are $0.72 \pm 0.02$ and $0.67 \pm 0.02$, respectively. Furthermore, we could confirm the scaling quantity $\alpha + z(z = \alpha/\beta)$, too.

**Fig. 7** Ink invasion from both sides of paper.

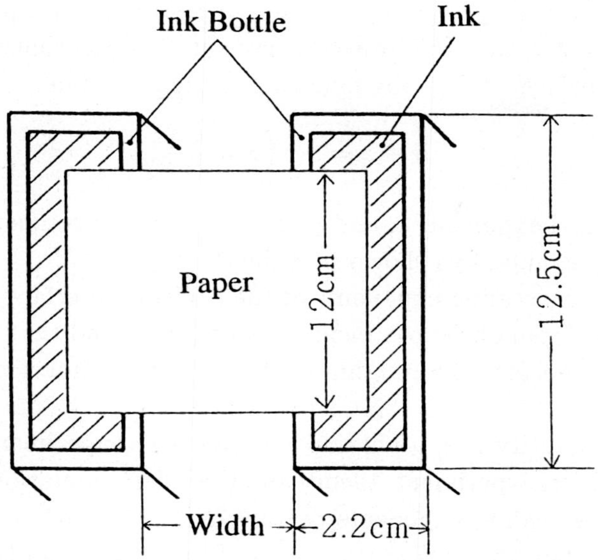


**Fig. 8** Illustration of apparatus of collision experiment with the changeable width between two ink bottles.

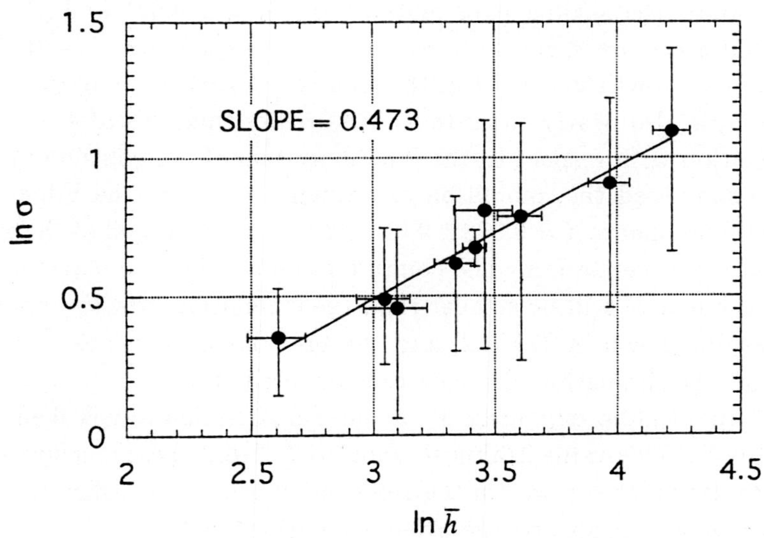


**Fig. 9** Log-log plot of $\sigma$ vs. $\bar{h}$ of collision surface.

**Table 2 Results of our ink imbibition experiment.**

| | Growth Surface | Collision Interface |
|---|---|---|
| Roughness exponent | $\alpha = 0.72 \pm 0.02$ | $\alpha' = 0.67 \pm 0.02$ |
| Growth exponent | $\beta = 0.55 \pm 0.08$ | $\beta' = 0.47 \pm 0.10$ |
| $\alpha + z \left( z = \dfrac{\alpha}{\beta} \right)$ | $2.03 \pm 0.13$ | $2.09 \pm 0.24$ |

## 5. DISCUSSION

The roughness exponent $\alpha$ and the growth exponent $\beta$ which characterize the fluctuation of the Eden growth surface, satisfy the following scaling relation:

$$\alpha + z = 2 \quad \left(z = \frac{\alpha}{\beta}\right),$$

where $z$ is the dynamical exponent. $\alpha$ and $\beta$ are 1/2 and 1/3, respectively, according to the KPZ equation and by simulations shown in Table 1.

In Table 2, both of the scaling exponents of the collision interface are smaller than those of the growth surface. From our experimental results, we found that when two independent rough growth surfaces collide, the roughness of the collision interface is reduced by about 0.707 times.

Next, the scaling quantity $\alpha + z$ obtained by our first experiment is $2.03 \pm 0.13$. This means that ink imbibition experiment, theory and the Eden simulation have the same scaling relation. But both the scaling exponents $\alpha$ $(= 0.72 \pm 0.02)$ and $\beta$ $(= 0.55 \pm 0.08)$ of our experiment are larger than the scaling exponents of the theory and the simulation shown in Table 1. We guess that this difference between the experimental and the simulation results is explained by introducing some noise effects in the computer simulation. In fact, Zhang simulated the growth model with a noise effect obeying the power-law distribution[16] and Zhang's simulation result and our experimental results yielded the same value.

Furthermore, both the scaling exponents $\alpha'$ $(= 0.67 \pm 0.02)$ and $\beta'$ $(= 0.47 \pm 0.10)$ of the collision interface are different from the scaling exponents $\alpha$ $(= 0.72 \pm 0.02)$ and $\beta$ $(= 0.55 \pm 0.08)$ of the growth surface. But the scaling quantity $\alpha' + z'$ of the collision interface is $2.09 \pm 0.24$, and is also nearly equal to 2. We predict that this $\alpha' + z' = 2$ indicates the super scaling relation with respect to the growth system of the collision interface.

We have also simulated the imbibition experiment of ink by the Eden process and obtained the growth exponents $\beta = 0.343 \pm 0.01$ and $\beta' = 0.336 \pm 0.03$ of the growth exponent and the collision interface made by the computer simulation. We expected that the simulation growth exponent $\beta'$ will be smaller than the simulation growth exponent $\beta$ from the experimental results shown in Table 2, and the previous conclusions.[15,17] As a result, we can conclude that the simulation also confirms our prediction.

Furthermore, the scaling exponent $\alpha'$ of our Eden simulation is $0.48 \pm 0.10$, thereby $\alpha' + z' = 1.87 \pm 0.27$. This value is almost equal to 2. From these conclusions we can guess that the scaling relation $\alpha' + z' = 2$ of the collision interface is satisfied not only by the ink imbibition experiment, but also by the Eden simulation.

## REFERENCES


1. B. B. Mandelbrot, *Nature of Fractal Geometry* (Freeman, San Francisco, 1982).
2. J. Feder, *Fractals* (Plenum Press, New York, 1988).
3. H. Fujikawa and T. Matsushita, *J. Phys. Soc. Japan* **58**, L3875 (1989).
4. S. Matsuura and S. Miyazima, *Fractals* **1**, 11 (World Scientific Publishing, Singapore, 1993).
5. H. Kitaoka and H. Itoh, *Forma* **6**, 181 (1991).
6. F. Family, B. R. Masters and D. E. Platt, *Physica* **D38**, 98 (1989).
7. P. Mcginley, R. G. Smith and J. C. Landry, *Fractals* **2**(4), 493 (World Scientific Publishing, Singapore, 1994).
8. A.-L. Barabási and H. E. Stanley, *Fractal Concepts in Surface Growth* (Cambridge University Press, Cambridge, 1995).

9. T. Vicsek, *Fractal Growth Phenomena* (World Scientific Publishing, Singapore, 1989).
10. M. Kardar, G. Parisi and Y.-C. Zhang, *Phys. Rev. Lett.* **56**, 889 (1986).
11. M. Matsushita, M. Sano, Y. Hayakawa, H. Honjo and Y. Sawada, *Phys. Rev. Lett.* **53**, 286 (1984).
12. J. P. Stokes, D. A. Weitz, J. P. Gollub, A. Dougherty, M. O. Robbins, P. M. Chaikin and H. M. Lindsay, *Phys. Rev. Lett.* **57**, 1718 (1986).
13. J. Zhang, Y.-C. Zhang, P. Alstrøm and M. T. Levinsen, in *Surface Disordering: Growth, Roughening and Phase Transitions* (Nova Science, New York, 1992).
14. S. V. Buldyrev, A.-L. Barabási, F. Caserta, S. Havlin, H. E. Stanley and T. Vicsek, *Phys. Rev.* **A45**, R8313 (1992).
15. T. Nagamine, K. Maruyama and S. Miyazima, *Fractals* **1**(4), 998 (World Scientific Publishing, Singapore, 1993).
16. Y.-C. Zhang, *J. Phys. France* **51**, 2129 (1990).
17. Y. Oota, S. Miyazima and H. Nishihara, *Journal of Research Institute for Science and Technology* **5**, 21 (Research Institute for Science and Technology, Chubu University, 1993).

Fractals, Vol. 4, No. 3 (1996) 305


# NUCLEAR RELAXATION IN IRREGULAR OR FRACTAL PORES

B. SAPOVAL, S. RUSS, J. P. KORB and D. PETIT
*Laboratoire de Physique de la Matière Condensée, Ecole Polytechnique, C.N.R.S., 91128 Palaiseau, France*

## Abstract

We apply a fractal description of pore surface irregularity to study the nuclear relaxation of a liquid confined in an irregular pore. Our calculation on a pre-fractal volume with infinite surface relaxation rate shows that the long time relaxation behavior is dominated by an exponential mode characterizing a free diffusive volume. At short time, the calculated magnetization follows closely the power-law behavior previously proposed by de Gennes. For finite surface relaxation rate, we extend our studies of the efficiency of exchange on fractal membranes to nuclear relaxation in the same geometry. We predict a decay involving two characteristic times. For very slow surface relaxation, there will exist exponential relaxation governed by the surface relaxation. These laws show that the pore surface irregularities may play an important role in nuclear relaxation.

Fractals, Vol. 4, No. 3 (1996) 307–319


# AVALANCHES IN THE DIRECTED PERCOLATION DEPINNING AND SELF-ORGANIZED DEPINNING MODELS OF INTERFACE ROUGHENING

S. V. BULDYREV*, L. A. N. AMARAL*, A.-L. BARABÁSI*,
S. T. HARRINGTON*, S. HAVLIN*,†, J. KERTÉSZ‡,
R. SADR-LAHIJANY* and H. E. STANLEY*
*Center for Polymer Studies and Department of Physics, Boston University, Boston, MA 02215 USA
†Minerva Center and Department of Physics, Bar-Ilan University, Ramat Gan, Israel
‡Institute of Physics, Technical University, Budapest, Budafoki ut 8, H-1111, Hungary

**Abstract**

We review the recently introduced Directed Percolation Depinning (DPD) and Self-Organized Depinning (SOD) models for interface roughening with quenched disorder. The differences in the dynamics of the invasion process in these two models are discussed and different avalanche definitions are presented. The scaling properties of the avalanche size distribution and the properties of active cells are discussed.

## 1. INTRODUCTION

Recently the growth of rough interfaces has witnessed an explosion of theoretical, numerical, and experimental studies, fueled by the broad interdisciplinary aspects of the subject.[1–6] Applications can be as diverse as imbibition in porous media, fluid–fluid displacement, bacterial colony growth, fire front motion, and the motion of flux lines in superconductors.[7–18]

In general, a $d$-dimensional self-affine interface, described by a single-valued function $h(x, t)$, evolves in a $(d+1)$-dimensional medium. Usually some form of disorder $\eta$ affects the motion of the interface leading to its roughening. Two main classes of disorder have been discussed in the literature. The first, called thermal or "annealed", depends only on time. The second, referred to as "quenched", is frozen in the medium. Early studies focused on time-dependent uncorrelated disorder as being responsible for the roughening. Here, we focus on the effect of quenched disorder on the growth.

The roughening process can be quantified by studying the *global* interface width $W$:

$$W(L, t) \equiv \langle(\overline{h^2(x, t)} - \overline{h(x, t)}^2)^{1/2}\rangle \,, \tag{1.1}$$

where $L$ is the system size, the bar denotes a spatial average, and the brackets denote an average over realizations of the disorder. The study of discrete models[19–22] and continuum growth equations[23,24] leads to the observation that during the initial period of the growth, i.e., for $t \ll t_\times(L)$, the width grows with time as:

$$W(t) \sim t^\beta \quad [t \ll t_\times] \,, \tag{1.2}$$

where $\beta$ is the *growth* exponent. For times much larger than $t_\times$, the width saturates to a constant value. It was observed that the saturation width of the interface $W_{sat}$ scales with $L$ as:

$$W_{sat} \sim L^\alpha \quad [t \gg t_\times] \,, \tag{1.3}$$

where $\alpha$ is the *roughness* exponent. The dependence of $t_\times$ on $L$ allows the combination of Eqs. (1.2) and (1.3) into a single scaling law[19]:

$$W(L, t) \sim L^\alpha f_1(t/t_\times) \,, \tag{1.4a}$$

where

$$t_\times \sim L^z \,. \tag{1.4b}$$

Here $z = \alpha/\beta$ is the *dynamical* exponent, and $f_1(u)$ is a universal scaling function that grows as $u^\beta$ when $u \ll 1$, and approaches a constant when $u \gg 1$.

An alternative way of determining the scaling exponents is to study the *local* width $w$ in an observation window of length $\ell < L$. The scaling law [Eq. (1.4)], and the fact that the interface is self-affine, allow us to conclude:

$$w(\ell, t) \sim \ell^\alpha f_2(\ell/\ell_\times) \,, \tag{1.5a}$$

where

$$\ell_\times \sim t^{1/z} \quad [t \ll t_\times] \,, \tag{1.5b}$$

or

$$\ell_\times \sim L \quad [t \gg t_\times] \,. \tag{1.5c}$$

Here $f_2(u)$ is a universal scaling function that decreases as $u^{-\alpha}$ when $u \gg 1$ and approaches a constant when $u \ll 1$.

The simulation of discrete models[19–22] gives exponents that agree with the predictions of phenomenological continuum approaches, such as the Edwards-Wilkinson (EW) equation[23] and the Kardar-Parisi-Zhang (KPZ) equation.[24] However, experimental studies find exponents significantly larger than the predictions of theory, e.g., for $(1+1)$ dimensions, Refs. 23 and 24 predict $\alpha = 1/2$ but experiments show $\alpha \sim 0.6 - 1.0$.[7–18] Although various explanations have been proposed — long-range correlations,[25] power-law distribution[26] for the disorder, or coupling of the interface to impurities[27] — it is currently accepted that *quenched* disorder plays an essential role in those experiments.[13–18,28,30–38,40–44]

The presence of quenched disorder allows an interesting analogy with critical phenomena. The continual motion of the interface requires the application of a driving force $F$. There exists a critical value $F_c$ such that for $F < F_c$, the interface will become pinned by the disorder after some finite time. For $F > F_c$, the interface moves indefinitely with a constant velocity $v$. This means that the motion of driven rough interfaces in disordered media can be studied as a phase transition — called the depinning transition. The velocity of the interface $v$ plays the role of the *order parameter*, since as $F \to F_c^+$, $v$ vanishes as:

$$v \sim f^{\theta}, \tag{1.6}$$

where $\theta$ is the *velocity* exponent, and $f \equiv (F - F_c)/F_c$ is the *reduced force.*

For $F \to F_c^+$, large but finite regions of the interface are pinned by the disorder. Near the transition, the characteristic length $\xi$ of these pinned regions diverges,

$$\xi \sim f^{-\nu}, \tag{1.7}$$

where $\nu$ is the *correlation length* exponent.

Several models in which quenched disorder plays an essential role have been proposed recently.[13–18,28,30–44] For one class of models,[13,14,28,40] in $(1+1)$ dimensions, $\alpha$ can be obtained by mapping the interface, at the depinning transition onto *Directed Percolation* (DP). In higher dimensions the interface can be mapped to *Directed Surfaces* (DS).[16] In $(1+1)$ dimensions, DP and DS are equivalent. We refer to this class of models as the *Directed Percolation Depinning* (DPD) universality class.

Recent numerical studies,[40] supported by analytical arguments,[41] showed that this class of models can be described by a stochastic differential equation of the KPZ type[38]:

$$\frac{\partial h}{\partial t} = F + \nabla^2 h + \lambda(\nabla h)^2 + \eta(x, h), \tag{1.8}$$

where $\eta(x, h)$ represents the quenched disorder, and the coefficient $\lambda$ of the nonlinear term diverges at the depinning transition.[40] This equation was originally proposed in the context of interface roughening in the presence of quenched disorder in Ref. 38. The numerical integration of Eq. (1.8) yielded exponents in agreement with the calculations for the models in the DPD universality class.[38] A recent attempt[39] at using a renormalization group approach for Eq. (1.8) suggests that there is a strong coupling fixed point at which the coupling constant associated with the KPZ nonlinearity diverges at the threshold, which is in agreement with Ref. 40. Hence, the high-dimensional behavior of Eq. (1.8) remains an unsolved problem. For a number of different models[30–35,42] belonging to a second universality class — referred to as *isotropic growth* models — we have either $\lambda = 0$ or $\lambda \to 0$ at the depinning transition.[40] Therefore, near the depinning transition, they can be described by an equation of the EW type with quenched disorder[45]:

$$\frac{\partial h}{\partial t} = F + \nabla^2 h + \eta(x, h). \tag{1.9}$$

This equation has been studied by means of the functional renormalization group,[43,44] yielding $\alpha = \epsilon/3$, $\nu = 1/(2-\alpha)$, and $z = 2 - 2\epsilon/9$, where $\epsilon = 4 - d$. Thus the upper critical dimension for Eq. (1.9) is 4. When $F \gg F_c$, the size of the pinned regions in the interface

$\xi$ decreases to values much smaller than the system size $L$. For length scales $\ell$ larger than $\xi$, the quenched disorder becomes irrelevant and time-dependent noise dominates the roughening process. This means that for $\ell \ll \xi$, we should recover the results of either the EW or the KPZ equation with annealed noise (depending on the absence or presence of nonlinear terms). This behavior has been observed in some of the experiments[7–12] and in simulations of discrete models.[13–16,28,30–35,42]

Theoretical interest in the behavior of Eq. (1.8) in high dimensions motivated the investigation of the DPD problem on the Cayley Tree,[46] which is usually believed to describe the infinite dimensional limit of the system. The exact solution on the Cayley Tree yields $\alpha = 0$ and suggests that the upper critical dimension is infinity, the conclusion being supported by numerical simulations.[47] The roughness exponent $\alpha$ remains positive for every finite dimension similar to the original KPZ equation, but unlike Eq. (1.9), in which $\alpha = 0$ at $d \geq 4$. However, recent numerical results[48] support the latter conclusion. The relation between Eqs. (18) and (19) and the DPD model is also discussed in Ref. 49. The numerical and theoretical results of Refs. 48 and 49 reveal small but significant discrepancies compared with Ref. 47 which await further investigation.

The DPD model, discussed in this paper, was introduced in Refs. 13 and 14 to explain a set of simple imbibition experiments — a somewhat different model, belonging to the same universality class, was independently introduced by Tang and Leschhorn.[28] In these experiments a colored suspension (coffee or ink) imbibes a sheet of paper, in the $(1+1)$-dimensional case — or a porous, spongy-like brick, in the $(2+1)$-dimensional case.[13–18] The experimentally measured roughness exponents are in good agreement with the predictions of the DPD models.[13–16,28] However, a number of experimental features cannot be explained by this model[29] (See also Ref. 11). For example, in the experiments, the saturation width and the average height of the pinned interface depend on the rate of evaporation, which is not taken into account in the DPD model. A variant of the DPD model that explains the experimental results in terms of the effect of evaporation is discussed in Ref. 18.

A self-organized variant of the DPD model has been introduced by Havlin et al.[13] who used the analogy with invasion percolation. Similar model was studied by Sneppen and others.[50] We will call these models the models of Self-Organized Depinning (SOD). In these models growth proceeds by avalanches, whose properties are of interest not only for the study of interface roughening,[50–53] but also for other fields, including biological evolution in ecological systems.[54–57] In this paper we will address some open questions concerning static and dynamic properties of the DPD and SOD models.

## 2. AVALANCHES IN DPD AND SOD MODELS

In both the DPD and SOD models the fluid interface propagates in a lattice with quenched disorder. Each cell on the lattice has a pinning threshold $\eta_i$ that is uniformly distributed between 0 and 1. However, the rules of update in these models are different. The question is: What are the similarities and the differences in the dynamics of these models?

In the DPD model, at each time step, each dry cell that is a nearest-neighbor to a wet cell becomes wet if its threshold $\eta_i$ is less than the value of a driving force $p$. In addition, there is a rule of erosion of overhangs: all the cells that are below a wet cell become wet instantaneously.

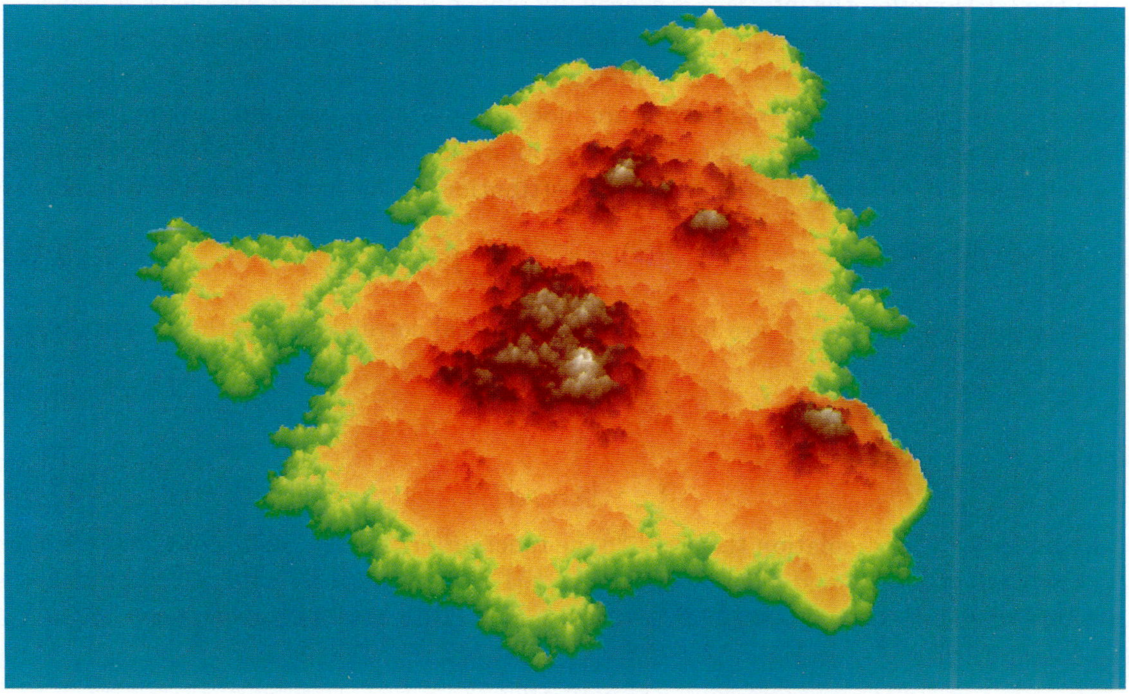

**Fig. 1** **(a)** A three dimensional "airplane" view of a directed surface that completely pinned an avalanche in $(2+1)$ dimensions for $p = 0.801$, presented as an island surrounded by a flat "sea". We used definition B, i.e., the avalanche was started at the center of the figure from the initial flat horizontal surface shown in blue. The Sneppen update rule was used. The diameter of the avalanche is approximately $2^{10}$. The uniform blue area shows the region left unchanged since the beginning of the process. The color code for heights corresponds to a conventional color code often used in geographical maps. The white "snow caps" corresponds to the largest heights of the interface. Note that the interfaces of a completely pinned avalanche are the same in the DPD and the SOD models.

Other rules of erosion may be implemented: the bounded slope rule implemented by Tang and Leschhorn,[28] or the rule of instantaneously-adjusting slopes introduced by Sneppen.[50] The DPD model can be introduced on other types of lattices, e.g., on a tilted square or cubical lattice. In this case the DPD model becomes equivalent to the diode-resistor percolation problem introduced more than a decade ago[58,59]; see also Refs. 17 and 46. The initial condition of the DPD model may be different. One way to start the process is from a single wet cell on a flat interface[17] (see Fig. 1). Another way is to start from a flat interface with periodic boundaries in which each cell is wet.[16]

In the SOD model, at each time step all the cells which are in contact with wet cells are examined and the one with the minimal value of $\eta_i$ is selected and becomes wet. In addition there is a rule of eroding overhangs[13] or adjusting slopes introduced by Sneppen[50]: all the neighboring cells become wet until all neighboring cells on the interface satisfy a condition that the difference in heights of neighboring points is less than or equal to one. In other words, cells instantaneously become wet if their heights $h_i$ and horizontal coordinates $x_{ij}$, $(1 \le j \le d)$ satisfy the inequality:

$$h_i \le h_0 - \sum_{j=1}^{d} |x_{ij} - x_{0j}| \,, \qquad (2.1)$$

where $h_0$ and $x_{0j}$ are coordinates of the cell that becomes wet due to minimal value of $\eta_i$.

In the DPD model, the interface is eventually pinned if the value of $p$ is below $p_c$, where $p_c$ is the critically directed percolation threshold. Otherwise the interface in the infinite system propagates indefinitely. In the DPD model, one can, however, modify a rule of growth[16] and allow the cells whose $\eta_i > p$ (those that are blocked) to erode at an infinitesimal rate. Thus the removal of a random blocked cell produces an avalanche of growth until, once again, all the cells on the interface are blocked. The distribution of the total avalanche volume $s$ is:

$$P(s) \sim s^{-\tau} f\left(\frac{s}{s_c}\right), \tag{2.2}$$

where $s_c \sim \xi_{\|}^{d} \xi_{\perp}$, and $\xi_{\|}$, $\xi_{\perp}$ are horizontal and vertical correlation lengths that scale as:

$$\xi_{\|} \sim |p - p_c|^{-\nu_{\|}}, \tag{2.3}$$

and

$$\xi_{\perp} \sim |p - p_c|^{-\nu_{\perp}}. \tag{2.4}$$

In analogy with percolation theory,[60–62] the function $f(x)$ in Eq. (2.2) is a cutoff function: $f(x) = 1$ when $x \ll 1$ and $f(x) = 0$ when $x \gg 1$. The typical volume of an avalanche $s_c$ diverges when $p \to p_c$ as $|p - p_c|^{-1/\sigma}$, where $1/\sigma = d\nu_{\|} + \nu_{\perp}$. The average avalanche size $\langle s \rangle$ diverges when $p \to p_c$ as:

$$\langle s \rangle \sim |p - p_c|^{-\gamma}, \tag{2.5}$$

where

$$\gamma = \frac{\tau - 2}{\sigma} = \frac{1}{\sigma} - \beta_p, \tag{2.6}$$

and $\beta_p$ is an exponent of the order parameter. The physical meaning of the order parameter is the probability for an avalanche to percolate through the characteristic volume $s_c$. In analogy with percolation $\beta_p = (d - d_c)\nu_{\|}$, where $d_c$ is the fractal dimension of points on the interface which can start a percolating avalanche. Using Eq. (2.6) one gets[63]:

$$\tau = 1 + \frac{d - d_c}{d + \alpha}. \tag{2.7}$$

For length scales smaller than $\xi_{\|}$, the interface is self-affine with roughness exponent of a completely blocked interface:

$$\alpha = \frac{\nu_{\perp}}{\nu_{\|}}. \tag{2.8}$$

Another way of generating avalanches in the DPD model is to start every avalanche with a single unblocked cell on a flat surface and wait until the interface is entirely blocked by a directed surface.[17] We will call this definition of the avalanche definition "B" to contrast it with the original avalanche definition[16] described above, which we will refer to as definition A. One may expect that the exponent $\sigma$ and $\tau$, characterizing the distribution of the avalanches, should be the same for both definitions of the avalanches. However numerical studies of Gat and Olami[48] showed that $\tau_A < \tau_B$, leading to the inequality $\gamma_A > \gamma_B$. Note that for the Cayley tree model, the exact results of the exponents for definition B are $\nu_{\|} = 1/4$ (for $p > p_c$), $\nu_{\|} = 3/4$ (for $p < p_c$), $\nu_{\perp} = 0$, $\alpha = 0$, $\tau_B = 2$, $\theta = 0$, $\beta_p = \infty$ and $\gamma_B = 0$.

Definition A appears to be also exactly solvable on the Cayley Tree,[64] and it yields exponents identical to those in definition B except for $\gamma$; $\gamma_A = 1$. Exponent $\tau_A$ is equal to 2 in both cases, but Eq. (2.2) has different logarithmic corrections, thus yielding an average cluster size that diverges in model A ($\gamma_A = 1$) and converges in model B ($\gamma_B = 0$).

The numerical results in high dimensions in avalanche definition A demonstrate that $\gamma_A$ monotonically approaches 1 from above when dimensionality increases, and for avalanche definition B that $\gamma_B \to 0$ for $d \to \infty$. For both definitions, exponents $\nu_{\|}$ and $\nu_{\perp}$ are the same, and in both cases the hyperscaling relation $(d-1)\nu_{\|} + \nu_{\perp} = 1/\sigma = \infty$ suggests the same upper critical dimension, $d = \infty$.

In the SOD model, the inteface organizes itself to a critical state in which it becomes self-affine on arbitrarily large length scales with the same exponent $\alpha$. The maximum value of $\eta_i$ of a cell that has been selected for erosion never exceeds $p_c$ for an infinite system.

It should be pointed out that at any moment of time when the next record $\eta_{\max}$ of $\eta_i$ is reached, the interface is the same as it would be in the DPD model in which the value of the driving force equals $\eta_{\max}$. In this definition of SOD, no avalanches occur. However, one can artificially define an avalanche as a time span during which the minimum of $\eta_i$ on the interface is below a certain value $p$. See also the definition of an associated process in Ref. 52.

The scaling formulae for the avalanche size distribution are the same as in DPD, since the avalanches are bounded by the surfaces of the same geometry. When the avalanche starts or stops, the interface is the same as one of the interfaces that completely stops the growth in the DPD model. One can expect that the exponents $\tau$, $\nu_{\|}$, $\nu_{\perp}$, and $\alpha$ are exactly the same in the DPD for definition A and SOD models, since they can be derived from the properties of the same subset of interfaces. Numerical simulations support this point of view.

If one starts the SOD model from a single point on a flat interface, every other cell having the strongest possible value of blocking force $\eta_i = 1$, then the distribution of the first avalanches, i.e., the distribution of the amount of material that becomes wet at the moment when the cell with $\eta_i \geq p$ was removed for the first time, would be the same as in definition B for DPD.

For the SOD model it is possible to derive an additional scaling relation[52,53]:

$$\gamma_A = 1 + \nu_{\perp} \,. \tag{2.9}$$

Using Eq. (2.6), one gets:

$$\tau_A = 1 + \frac{d - 1/\nu_{\|}}{d+\alpha} \,. \tag{2.10}$$

Analogous relations are well known for invasion percolation[65]:

$$\tau_A = 1 + \frac{d_h - d_r}{D} \,, \tag{2.11}$$

where $d_h$, $d_r = 1/\nu$, and $D$ denotes the fractal dimensions of the hull, the red bonds, and the percolation cluster, respectively. Equations (2.9) and (2.10) seem to be in agreement with numerical simulations in high dimension for definition A of DPD, and with the infinite-dimension limit on the Cayley tree. Comparing relations (2.7) and (2.10), one concludes that for definition A, $d_c = 1/\nu_{\|}$. Thus $d_c$ for definition A is analogous to the fractal dimension of the red bonds in percolation.[63]

For definition B, neither Eq. (2.9) nor Eq. (2.10) is valid. However, for both definitions, $\tau$ and $\gamma$ are still related by Eq. (2.6). Moreover, for both definitions, relation (2.7) is satisfied, but for definition B, $d_c$ is not equal to $1/\nu_{\|}$. The same phenomenon is observed for invasion percolation,[65] where the exponents characterizing the invaded regions analogous to avalanches of definition A are not equal to the exponents of the cluster size distribution of classical percolation which can be grown from a single wet cell by the Leath algorithm in analogy to definition B. In invasion percolation, this simply means that $d_h - d_r$ is not equal to $d - D$, and thus $\tau = d/D \neq \tau_A$. It is also known[66] that the exponents $\tau$ and $\gamma$ of a percolation cluster started on the half plane boundary differ from the classical values for clusters started from the bulk. The reason for this discrepancy is the complex self-similar or self-affine structure of the interface from which the avalanches of definition A start to grow. Note that for the Bak-Sneppen model of biological evolution,[55,56] both definitions of avalanches yield the same exponents, since the properties of the interface in this model do not affect the growth.

## 3. ACTIVE CELLS AND DYNAMIC EXPONENTS IN THE DPD AND SOD MODELS

In the DPD model, during an avalanche that starts at time $t = 0$, at each time step there is a certain number $n_a(t)$ of "active" columns on the interface that contain unblocked cells on the interface. In the SOD model the active cells are those whose $\eta_i < p$, where $p$ is the value of the first cell removed during a particular avalanche. In order to give time in the SOD model a physical meaning, one must redefine the time interval needed to erode one cell as:

$$dt_p = \frac{1}{n_a} . \tag{3.1}$$

This definition of time is called parallel time $t_p$.[55]

The number of active columns scales with time $t$ and volume $s$ as:

$$n_a(t) \sim t^{\delta} \sim s^{d_s} . \tag{3.2}$$

The active columns form a fractal dust with a fractal dimension $d_F$.

The durations of the avalanches are distributed according to the scaling law:

$$P(t) \sim t^{-\tau_{surv}} f\left(\frac{t}{t_c}\right) , \tag{3.3}$$

where $t_c = \xi_{\|}^{z}$.

The dynamical exponents of the avalanches and active cells appear to differ in SOD and DPD models. However, in both classes models, $z$, $\delta$, $d_s$, $\tau$, and $\tau_{surv}$ are related in the same way. Indeed, the size of avalanches and the time are related as:

$$s \sim \xi_{\|}^{d} \xi_{\perp} \sim \xi_{\|}^{d+\alpha} \sim t^{d+\alpha/z} . \tag{3.4}$$

Because $p(s)ds = P(t)dt$, the avalanche distributions are related as:

$$(\tau - 1)\frac{d+\alpha}{z} = (\tau_{surv} - 1) . \tag{3.5}$$

In the DPD and SOD models some of the avalanche sites are invaded by oncoming fluid when these sites are active sites on the interface; other sites are occupied due to an erosion rule. It is interesting to note that the number of eroded sites for each invaded site is a fixed, model-dependent constant which does not grow with the total size of an avalanche.[49] Thus the total volume of an avalanche is proportional to the sum of all active cells $n_a$ invaded during the avalanche:

$$s \sim \int_0^t n_a(t)dt = \int_0^t t^\delta dt = t^{(d+\alpha)/z} = t^{\delta+1}, \tag{3.6}$$

then:

$$\delta = \frac{d+\alpha}{z} - 1. \tag{3.7}$$

Finally, we can also relate the velocity of the interface to the number of active cells, as defined in Eq. (3.2). The velocity at each instant can be obtained as the number of active cells divided by the size of the parallel projection of the invaded region:

$$v \sim n_a(t)/\xi_\|^d \sim t^\delta/\xi_\|^d \sim \xi_\|^{z\delta-d} \sim f^{-\nu_\|(z\delta-d)}. \tag{3.8}$$

Comparing with Eq. (1.6), we obtain:

$$\theta = \nu_\|(d - z\delta). \tag{3.9}$$

The fractal dimension of the live cells is related to the distribution of voids between them, i.e., the distribution of linear sizes of the avalanches. If the probability density used to find the void of length $\ell$ is equal to $\ell^{-\tau_\|}$, then:

$$d_F = \tau_\| - 1. \tag{3.10}$$

If we assume that the number of active points in an avalanche of linear size $\ell$ scales as $\ell^{d_F}$, we conclude that:

$$n_a(t) \sim \ell^{d_F} \sim t^{d_F/z}. \tag{3.11}$$

Hence from Eq. (3.7) we conclude that:

$$d_F = d + \alpha - z. \tag{3.12}$$

Also, since $s \sim \ell^{\alpha+d}$, we have [see Eq. (3.2)]:

$$d_s = d_F/(\alpha + d), \tag{3.13}$$

and

$$(\tau_\| - 1) = (\tau - 1)(d + \alpha). \tag{3.14}$$

After some algebra, we get[55]:

$$z = (2 - \tau)(d + \alpha). \tag{3.15}$$

However, the assumption $n(t) \sim \ell^{d_F}$ is not necessarily true.

We know that in DPD, $z$ is very close to the chemical dimension of the regular isotropic percolation in $d$-dimensions.[67] Thus, in $d = 1$, $z = 1$. However, plugging into Eq. (3.15) the numerical values[16,52] for $\tau = 1.29$ and $\alpha = 0.63$ in $d = 1$, we get $z = 0.71 \cdot 1.63 = 1.16$. This apparent contradiction happens because in DPD the live cells clump together in blocks within which the density of active cells is constant while the blocks are forming a fractal set. The correlation function $G(r)$ of active cells in DPD undergoes crossover from small slope for small $r$ to large slopes for large $(r)$: $G(r) \sim r^{d_F-d}$. In $d = 1$, using Eqs. (3.10) and (3.14) we get $G(r) \sim r^{2-\tau_\|} \sim r^{-0.53}$. This is consistent with our numerical calculations of $G(r)$. However, for small $r$, $G(r)$ is almost constant. Note that the shape of the moving front

(a)

(b)

**Fig. 2** Illustration of the dynamics of the DPD model at $p = p_c = 0.539$ **(a)**, and the SOD model **(b)**, in $(1+1)$ dimensions. In both cases the Sneppen update rule is used. In both cases the interface is shown after 1000 time steps starting from a single cell near the center on a flat interface (definition B). For the SOD model, the parallel definition of time is used. The "Rainbow" color code indicates time of invasion. Regions invaded at the beginning of the process are displayed in red. The color code changes with time from red to orange, yellow, green, cyan, blue and finally to violet. Note that in DPD model the invasion is caused by lateral propagation of steep slopes that corresponds to constantly changing color in lateral direction. In contrast, in the SOD model the invasion jumps constantly from one part of the interface to the next, that can be seen in part (b) as small alternating "patches" of different colors.

within an avalanche in the SOD model differs from that of the DPD model (see Fig. 2). In the SOD model, the interface is consistently self-affine with the exponent $\alpha$, and the activity constantly jumps from place to place according to a Lévy flight rule.[55,56] In the DPD model, the moving interface is not self-affine; it consists of blocked parts interrupted by steep slopes moving as large quasi-particles. Both the horizontal and vertical dimensions of such slopes are of typical size $\xi_\perp$. In $d = 1$ they move with a constant speed $z = 1$. In higher dimensions $z$ is probably equal to the chemical dimension of $d$-dimensional regular percolation.[67] In the SOD model, the set of active cells is fractal on all length scales and Eq. (3.11) is correct. Thus it is the self-organization rule — by which we select a site with the smallest value of $\eta_i$ present in the interface — that causes the value of $z$ in SOD to differ from that of DPD.

To this end, it seems that $z$ cannot be completely determined with the help of only static properties of completely pinned avalanches. In contrast, $z$ depends on the dynamics of active cells that involve the unknown exponent $\delta$ [see Eq. (3.7)]. Hence, for any value of $\delta$, one can always satisfy this relation by taking:

$$z = \frac{d+\alpha}{\delta+1}\,. \tag{3.16}$$

That is why it is quite possible that in DPD the upper critical dimension is equal to 6 for dynamic exponents[67] and infinity for static exponents.[46] (See also Ref. 48 for different interpretation.) The dynamics of the SOD model in which the update rule assumes an infinitely fast propagation of the information along the interface or an infinitesimally slow "creeping"[68] growth of the interface, differs from the dynamics of KPZ with quenched disorder and fixed driving force $F$ [see Eq. (1.8)], which is probably in the same universality class as DPD.

## ACKNOWLEDGMENTS

We thank O. Gat, Z. Olami, I. Procaccia and R. Zeitak for communicating with us their results prior to publication and helpful contributions and T. Vicsek, G. Huber, B. Sapoval, J. F. Gouyet, M. Paczuski, and H. Leschhorn for valuable discussions. The Center for Polymer Studies is supported by the National Science Foundation. SH acknowledges support from the Israel Science Foundation and JK from OTKA T016568 and the US-Hungarian Joint Fund (contract No. 352).

## REFERENCES

1. T. Vicsek, *Fractal Growth Phenomena*, 2nd ed., Part IV (World Scientific, Singapore, 1992); *Dynamics of Fractal Surfaces*, eds. F. Family and T. Vicsek (World Scientific, Singapore, 1991); J. Kertész and T. Vicsek, in *Fractals in Science*, eds. by A. Bunde and S. Havlin (Springer-Verlag, Heidelberg, 1994).
2. J. Krug and H. Spohn, in *Solids Far From Equilibrium: Growth, Morphology and Defects*, edited by C. Godréche (Cambridge University Press, Cambridge, 1991).
3. P. Meakin, *Phys. Rep.* **235**, 189 (1993).
4. T. Halpin-Healey and Y.-C. Zhang, *Phys. Rep.* **254**, 215 (1995).
5. A.-L. Barabási and H. E. Stanley, *Fractal Concepts in Surface Growth* (Cambridge University Press, Cambridge, 1995).
6. J.-F. Gouyet, M. Rosso and B. Sapoval, in *Fractals and Disordered Systems*, eds. A. Bunde and S. Havlin (Springer-Verlag, Heidelberg, 1991).
7. J. P. Stokes, A. P. Kushnick and M. O. Robbins, *Phys. Rev. Lett.* **60**, 1386 (1988).
8. M. A. Rubio, C. A. Edwards, A. Dougherty and J. P. Gollub, *Phys. Rev. Lett.* **63**, 1685 (1989); V. K. Horváth, F. Family and T. Vicsek, *Ibid.* **65**, 1388 (1990), M. A. Rubio, C. A. Edwards, A. Dougherty and J. P. Gollub, *Ibid.*, 1389 (1990).
9. T. Vicsek, M. Cserzö and V. K. Horváth, *Physica* **A167**, 315 (1990).
10. V. K. Horváth, F. Family and T. Vicsek, *J. Phys.* **A24**, L25 (1991); *Phys. Rev. Lett.* **67**, 3207 (1991).
11. S. He, G. L. M. K. S. Kahanda and P.-Z. Wong, *Phys. Rev. Lett.* **69**, 3731 (1992).
12. J. Zhang, Y.-C. Zhang, P. Alstrøm and M. T. Levinsen, *Physica* **A189**, 383 (1992).
13. S. Havlin, A.-L. Barabási, S. V. Buldyrev, C. K. Peng, M. Schwartz, H. E. Stanley and T. Vicsek, in *Growth Patterns in Physical Sciences and Biology* [Proc. 1991 NATO Advanced Research Workshop, Granada], eds. J. M. Garcia-Ruiz, E. Louis, P. Meakin and L. M. Sander (Plenum Press, New York, 1993).
14. S. V. Buldyrev, A.-L. Barabási, S. Havlin, F. Caserta, H. E. Stanley and T. Vicsek, *Phys. Rev.* **A45**, R8313 (1992).
15. S. V. Buldyrev, A.-L. Barabási, S. Havlin, J. Kertész, H. E. Stanley and H. S. Xenias, *Physica* **A191**, 220 (1992).
16. A.-L. Barabási, S. V. Buldyrev, S. Havlin, G. Huber, H. E. Stanley and T. Vicsek, in *Surface Disordering: Growth, Roughening and Phase Transitions*, eds. R. Jullien, J. Kertész, P. Meakin and D. E. Wolf (Nova Science, New York, 1992).
17. S. V. Buldyrev, S. Havlin and H. E. Stanley, *Physica* **A200**, 200 (1993); S. V. Buldyrev, S. Havlin, J. Kertész, A. Shehter and H. E. Stanley, *Fractals* **1**, 827 (1993).
18. L. A. N. Amaral, A.-L. Barabási, S. V. Buldyrev, S. Havlin and H. E. Stanley, *Phys. Rev. Lett.* **72**, 641 (1994); *Fractals* **1**, 818 (1993).
19. F. Family and T. Vicsek, *J. Phys.* **A18**, L75 (1985).
20. F. Family, *J. Phys.* **A19**, L441 (1986).
21. P. Meakin, P. Ramanlal, L. M. Sander and R. C. Ball, *Phys. Rev.* **A34**, 5091 (1986).

22. J. M. Kim and J. M. Kosterlitz, *Phys. Rev. Lett.* **62**, 2289 (1989).
23. S. F. Edwards and D. R. Wilkinson, *Proc. R. Soc. Lond.* **A381**, 17 (1982).
24. M. Kardar, G. Parisi and Y.-C. Zhang, *Phys. Rev. Lett.* **56**, 889 (1986).
25. E. Medina, T. Hwa, M. Kardar and Y.-C. Zhang, *Phys. Rev.* **A39**, 3053 (1989); C. K. Peng, S. Havlin, M. Schwartz and H. E. Stanley, *Phys. Rev.* **A44**, 2239 (1991); J. G. Amar, P.-M. Lam and F. Family, *Phys. Rev.* **A43**, 4548 (1991).
26. Y.-C. Zhang, *J. Physique* **51**, 2129 (1990); S. V. Buldyrev, S. Havlin, J. Kertész, H. E. Stanley and T. Vicsek, *Phys. Rev.* **A43**, 7113 (1991); S. Havlin, S. V. Buldyrev, H. E. Stanley and G. H. Weiss, *J. Phys.* **A24**, L925 (1991).
27. A.-L. Barabási, *Phys. Rev.* **A46**, R2977 (1992).
28. L.-H. Tang and H. Leschhorn, *Phys. Rev.* **A45**, R8309 (1992).
29. V. K. Horváth and H. E. Stanley, "Temporal Scaling of Interfaces Propagating in Porous Media," *Phys. Rev.* **E52**, 5166 (1995).
30. M. Cieplak and M. O. Robbins, *Phys. Rev. Lett.* **60**, 2042 (1988); N. Martys, M. Cieplak and M. O. Robbins, *Phys. Rev. Lett.* **66**, 1058 (1991).
31. C. S. Nolle, B. Koiller, N. Martys and M. O. Robbins, *Phys. Rev. Lett.* **71**, 2074 (1993); B. Koiller, M. O. Robbins, H. Ji and C. S. Nolle, in *New Trends in Magnetic Materials and their Applications*, eds. J. L. Moran-Lopez and J. M. Sanchez (Plenum, New York, 1993).
32. D. A. Kessler, H. Levine and Y. Tu, *Phys. Rev.* **A43**, 4551 (1991).
33. G. Parisi, *Europhys. Lett.* **17**, 673 (1992); L. A. N. Amaral (unpublished).
34. H. Leschhorn, *Physica* **A195**, 324 (1993).
35. M. Dong, M. C. Marchetti, A. A. Middleton and V. Vinokur, *Phys. Rev. Lett.* **70**, 662 (1993).
36. M. Benoit and R. Jullien, *Physica* **A207**, 500 (1994).
37. D. Spasojevic and P. Alstrøm, *Physica* **A201**, 482 (1993).
38. Z. Csahók, K. Honda and T. Vicsek, *J. Phys.* **A26**, L171 (1993); Z. Csahók, K. Honda, E. Somfai, M. Vicsek and T. Vicsek, *Physica* **A200**, 136 (1993).
39. S. Stepanow, *J. Phys. II* (France) **5**, 11 (1995).
40. L. A. N. Amaral, A.-L. Barabási and H. E. Stanley, *Phys. Rev. Lett.* **73**, 62 (1994); L. A. N. Amaral, A.-L. Barabási, H. A. Makse and H. E. Stanley, *Phys. Rev.* **E52**, 4087 (1995).
41. L.-H. Tang, M. Kardar and D. Dhar, *Phys. Rev. Lett.* **74**, 920 (1995).
42. H. A. Makse and L. A. N. Amaral, *Europhys. Lett.* **31**, 379 (1995).
43. T. Nattermann, S. Stepanov, L.-H. Tang and H. Leschhorn, *J. Phys. II* (France) **2**, 1483 (1992); H. Leschhorn, T. Nattermann, S. Stepanov and L.-H. Tang, *Phys. Rev.* **E** (submitted).
44. O. Narayan and D. S. Fisher, *Phys. Rev.* **B48**, 7030 (1993).
45. M. V. Feigel'man, *Sov. Phys. JETP* **58**, 1076 (1983); R. Bruinsma and G. Aeppli, *Phys. Rev. Lett.* **52**, 1547 (1984); J. Koplik and H. Levine, *Phys. Rev.* **B46**, 280 (1985).
46. S. V. Buldyrev, S. Havlin, J. Kertész, R. Sadr-Lahijany, A. Shehter and H. E. Stanley, *Phys. Rev.* **E52**, 373 (1995); see also the application of the Cayley tree to avalanches in lung inflation [A.-L. Barabási, S. V. Buldyrev, H. E. Stanley and B. Suki (preprint)].
47. L. A. N. Amaral, A.-L. Barabási, S. V. Buldyrev, S. Havlin, S. T. Harrington, R. Sadr-Lahijany and H. E. Stanley, *Phys. Rev.* **E51**, 4655 (1995).
48. O. Gat and Z. Olami (unpublished).
49. Z. Olami, I. Procaccia and R. Zeitak, *Phys. Rev.* **E52**, 3402 (1995).
50. K. Sneppen, *Phys. Rev. Lett.* **69**, 3539 (1992); L.-H. Tang and H. Leschhorn, *Ibid.* **70**, 3832 (1993); K. Sneppen and M. H. Jensen, *Ibid.* **70**, 3833 (1993); K. Sneppen and M. H. Jensen, *Phys. Rev. Lett.* **71**, 101 (1993).
51. H. Leschhorn and L.-H. Tang, *Phys. Rev.* **E49**, 1238 (1994).
52. Z. Olami, I. Procaccia and R. Zeitak, *Phys. Rev.* **E49**, 1232 (1994).
53. S. Maslov and M. Paczuski, *Phys. Rev.* **E50**, R643 (1994).
54. P. Bak, C. Tang and K. Wiesenfeld, *Phys. Rev. Lett.* **59**, 381 (1987).
55. P. Bak and K. Sneppan, *Phys. Rev. Lett.* **71**, 4083 (1993); M. Paczuski, S. Maslov and P. Bak, *Europhys. Lett.* **27**, 96 (1994); S. Maslov, M. Paczuski and P. Bak, *Phys. Rev. Lett.* **73**, 2162 (1994); M. Paczuski, P. Bak and S. Maslov, *Phys. Rev. Lett.* **74**, 4253 (1995); M. Paczuski, S. Maslov and P. Bak, *Phys. Rev.* **53**, 414 (1996).
56. T. Ray and N. Jan, *Phys. Rev. Lett.* **72**, 4045 (1994); B. Jovanović, S. V. Buldyrev, S. Havlin and H. E. Stanley, *Phys. Rev.* **E50**, R2403 (1994).
57. B. Suki, A.-L. Barabási, Z. Hantos, F. Peták and H. E. Stanley, *Nature* **368**, 615 (1994).

58. D. Dhar, M. Barma and M. K. Phani, *Phys. Rev. Lett.* **47**, 1238 (1981).
59. S. Redner, *Phys. Rev.* **B25**, 3242 (1982).
60. D. Stauffer and A. Aharony, *Introduction to Percolation Theory*, 2nd ed. (Taylor & Francis, London, 1992).
61. *Fractals and Disordered Systems*, eds. A. Bunde and S. Havlin (Springer-Verlag, Heidelberg, 1991); S. Havlin and D. Ben-Avraham, *Adv. Phys.* **36**, 695 (1987).
62. J. W. Essam, K. De'Bell, J. Adler and F. M. Bhatti, *Phys. Rev.* **B33**, 1982 (1986); J. W. Essam, A. J. Guttmann and K. De'Bell, *J. Phys.* **A21**, 3815 (1988).
63. G. Huber, M. H. Jensen and K. Sneppen, *Phys. Rev.* **E52**, R2133 (1995).
64. S. V. Buldyrev et al. (unpublished).
65. J. F. Gouyet, *Physica* **A168**, 581 (1990).
66. J. L. Cardy, *Nuclear Phys.* **B240**, 514 (1984).
67. S. Havlin, L. A. N. Amaral, S. V. Buldyrev, S. T. Harrington and H. E. Stanley, *Phys. Rev. Lett.* **74**, 4205 (1995).
68. S. I. Zaitsev, *Physica* **A189**, 411 (1992).

Fractals, Vol. 4, No. 3 (1996) 321–329


# A FRACTAL MODEL FOR THE FIRST STAGES OF THIN FILM GROWTH

PABLO JENSEN[*,||], ALBERT-LÁSZLÓ BARABÁSI[†], HERNÁN LARRALDE[‡], SHLOMO HAVLIN[§] and H. EUGENE STANLEY[¶]

[*] *Départment de Physique des Matériaux, Université Claude Bernard Lyon-1, 69622 Villeurbanne Cedex, France*

[†] *University of Notre Dame, Department of Physics, Notre Dame, IN 46556, USA*

[‡] *Physics Department, Cavendish Laboratory, Cambridge University, Madingley Road, Cambridge CB3 0HE, England*

[§] *Physics Department, Bar Ilan University, Ramat Gan, Israel*

[¶] *Center for Polymer Studies and Department of Physics, Boston University, Boston, MA 02215, USA*

## Abstract

In this paper, we briefly review a model that describes the diffusion-controlled aggregation exhibited by particles as they are deposited on a surface. This model allows us to understand many experiments of thin film deposition. In the Sec. 1, we describe the model, which incorporates deposition, particle and cluster diffusion, and aggregation. In Sec. 2, we study the dynamical evolution of the model. Finally, we analyze the effects of small cluster mobility and show that the introduction of cluster diffusion dramatically affects the dynamics of film growth. Some of these effects can be tested experimentally.

## 1. INTRODUCTION

Understanding the processes underlying the growth of thin films has led to widespread interest,[1] both from the theoretical[1–8] and the experimental points of view.[9–12] Recently,

[||] E-mail: jensen@dpm.univ-lyon1.fr

improvements in experimental techniques — such as scanning tunneling microscopy — permit the investigation of atomic details of the embryonic "sub-monolayer" stages of film growth. Recent experimental works[10–12] have shown that *fractal* structures can be formed during the first stages of the growth (Fig. 1). Then, it is interesting to try to analyze the growth in the framework of fractal models. One might consider the use of the percolation

**Fig. 1** Experimental images of the first stages of growth for films prepared by **(a)** Atomic deposition of Ag on Pd[11]; and **(b)** Deposition of Sb *compact* aggregates containing 2300 atoms on graphite.[12]

model[13,14] to describe certain experiments of surface deposition.[15] However, percolation assumes that particles do not diffuse after being deposited, when in fact not only diffusion but also aggregation of the diffusing particles takes place. There exist models of diffusing particles that aggregate, but such "Cluster-Cluster Aggregation" (CCA) models[16] do not allow the continual injection of new particles via deposition. Here, we develop a model that incorporates the three physical ingredients of thin film growth: Deposition, Diffusion and Aggregation (DDA). Similar models that neglect the shape of the islands or the possibility of cluster diffusion were studied independently.[4,6] We have shown in detail elsewhere[5] how the DDA model generates a wide variety of fractal structures characteristic of different models such as percolation, Diffusion Limited Aggregation (DLA)[17,18] or CCA. Here, we focus on the dynamics of film growth and the importance of (small) island diffusion.

## 2. MODEL DESCRIPTION AND JUSTIFICATION

The DDA model is defined as follows (Fig. 2):

(1) *Deposition.* Particles are deposited at randomly-chosen positions of the surface at a flux $F$ per lattice site per unit time.
(2) *Diffusion.* All particles and clusters (sets of connected particles) are chosen at random and attempted to move North, East, South or West by one lattice constant per unit time. The probability that they actually move is proportional to their mobility, which we assume to be given by $D_s = D_1 s^{-\gamma}$. Here $s$ is the number of particles in the cluster, $D_1$ is the diffusion coefficient for a monomer ($s = 1$), and the parameter $\gamma$ characterizes the dependence of $D_s$ on cluster size.
(3) *Aggregation.* If two particles come to occupy neighboring sites, they (and therefore the clusters to which they belong) stick irreversibly.

We call *particles* the isolated atoms (or monomers) that are deposited on the surface, *clusters* any set of connected particles (including the monomers) and *islands* the clusters containing more than one particle. Physically, two competing mechanisms are introduced in the model, each one with its own time scale: deposition and diffusion. It is useful to introduce the normalized flux defined as the number of particles deposited per unit site per *diffusion time* $\tau$, where $\tau$ is the mean time needed by a monomer to jump by a lattice site. The monomer diffusion coefficient is then given by $D_1 = 1/(4\tau)$, and the normalized flux by $\phi = F\tau$. Then, from experimental values of $F$ and $D_1$, it is possible to calculate $\phi$ and the morphologies predicted by our model. The program actually calculates a probability for dropping a particle: $p_{drop} = \phi L^2/(\phi L^2 + N_{cl})$, where $L$ is the system size and $N_{cl}$ is the total number of clusters present in the system. A random number $p$ is chosen and compared with $p_{drop}$. If $p < p_{drop}$, a particle is added at a random position on the lattice. If $p > p_{drop}$, a cluster or a particle is chosen at random and attempted to move. In both cases, the time is increased by $\tau/(\phi L^2 + N_{cl})$.

It should be stressed that this is only a "zeroth-order" model which has the ambition to give a feeling on the relative influence of deposition and diffusion on the growth properties of films. Details specific to certain experimental systems, such as the existence of the Schwoebel barrier, the precise dependence of cluster diffusion on size, etc., are not carefully taken into account since we want to keep the DDA model as general as possible (see Ref. 5).

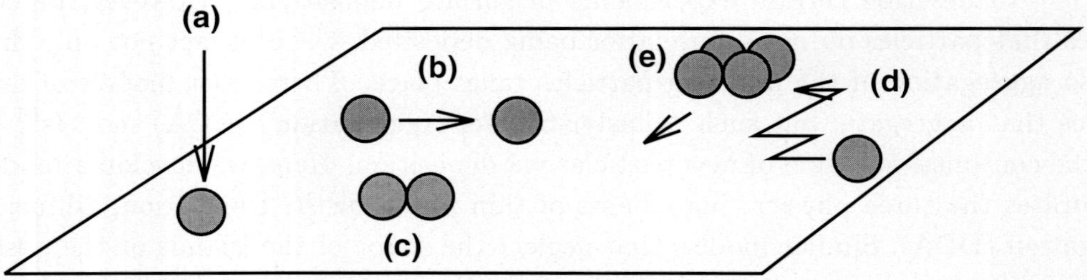


**Fig. 2** Schematic representation of the basic processes considered in this model: **(a)** deposition; **(b)** and **(d)** particle diffusion; **(e)** island diffusion; and **(c)** aggregation. (b) corresponds to *nucleation* (i.e., a new island (c) is created), while (d) corresponds to *growth* of an already existing island (see the text for details).

## 3. DYNAMICAL EVOLUTION

We present here snapshots of the system at different times to show the formation of the islands. These images were obtained for $\phi = 1.2\ 10^{-8}$ and $\gamma = \infty$ (i.e., only monomers are allowed to move). A detailed presentation of the dynamical evolution of the model has been given elsewhere.[5] Actually, it is more interesting to use the surface coverage $\theta$ instead of the time as the evolution parameter. $\theta$ is defined as the ratio of the number of occupied sites to the total number of sites on the surface, and for the times studied here we have $\theta \sim Ft$.

At very short times ($\theta \leq 0.001$), mainly monomers (isolated particles) are found on the substrate, since they did not yet meet another one to form a cluster. Later, small clusters are homogeneously grown on the surface [Fig. 3(a)], and the island density (i.e., the number of islands per lattice site) starts to grow. These small clusters can be considered as the "nucleation centers" for the growth. As time increases, large clusters grow on these nucleation centers, by addition of single particles [Fig. 3(b)]. These clusters are very similar to those obtained experimentally [Figs. 1(a) and 1(b)]. We have found that their fractal dimension is 1.65, very close to the fractal dimension of the DLA clusters. This (DLA-like) growth mechanism goes on until the linear dimension of the clusters becomes comparable to the separation between them [Fig. 3(c)]. Then, many particles start to fall *inside* the clusters and their fractal dimension rapidly increases. Eventually, a cluster of a size comparable to the system size is built and the system spans [Fig. 3(d)]. At that time, the effective fractal dimension of the spanning cluster reaches a value close to 1.9.

It is interesting to note that the growth of the film for $\gamma = \infty$ is in some loose sense "self-regulated" since islands tend to avoid each other, by growing preferentially in the directions where no island is present. The reason is that monomers are more abundant in those regions since no island captures them. Then, the probability for a given point of an island to grow is higher if it is far from other islands. This effect can be quantified in the following way. We first grow a system with a given coverage and we freeze the growth. Then, we deposit a monomer and we move it exactly as during the growth, but each time the monomer reaches an island, we record the point where it touched the island and *remove* it. By counting the number of monomers that hit each point, we can measure the growth probability within each island. Figure 4(a) shows a global view of these probabilities for a large system at a coverage of 0.04. The white points correspond to those point which grow slower and the bright blue points show the regions of high growth probability. Figure 4(b)

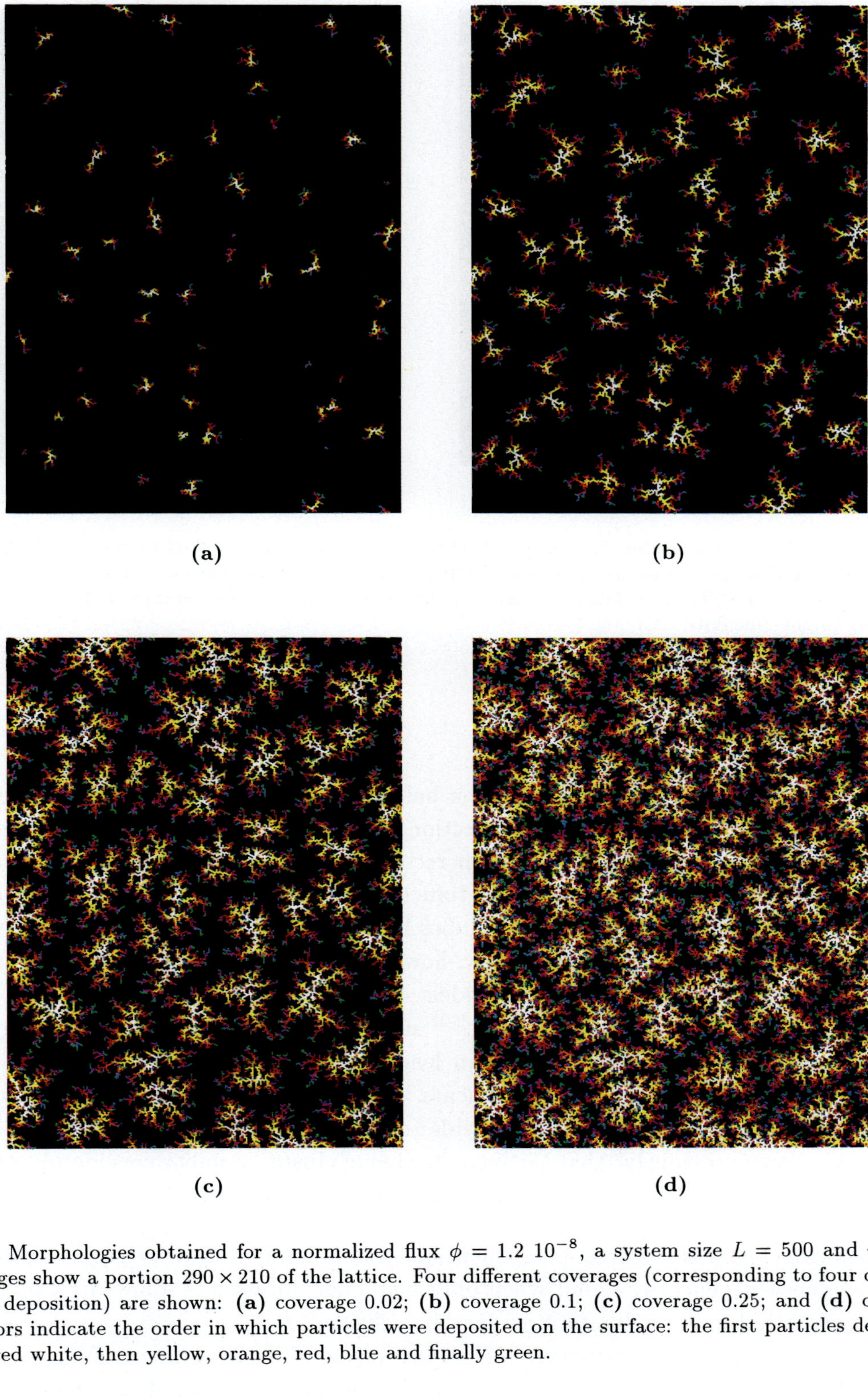

**Fig. 3** Morphologies obtained for a normalized flux $\phi = 1.2\ 10^{-8}$, a system size $L = 500$ and $\gamma = \infty$. The images show a portion $290 \times 210$ of the lattice. Four different coverages (corresponding to four different times of deposition) are shown: **(a)** coverage 0.02; **(b)** coverage 0.1; **(c)** coverage 0.25; and **(d)** coverage 0.4. Colors indicate the order in which particles were deposited on the surface: the first particles deposited are colored white, then yellow, orange, red, blue and finally green.

shows in detail the growth probabilities for one of the islands. The fact that each island has its own "capture zone" and grows inside it, has some interesting consequences on the size distribution of the islands. This will be discussed below.

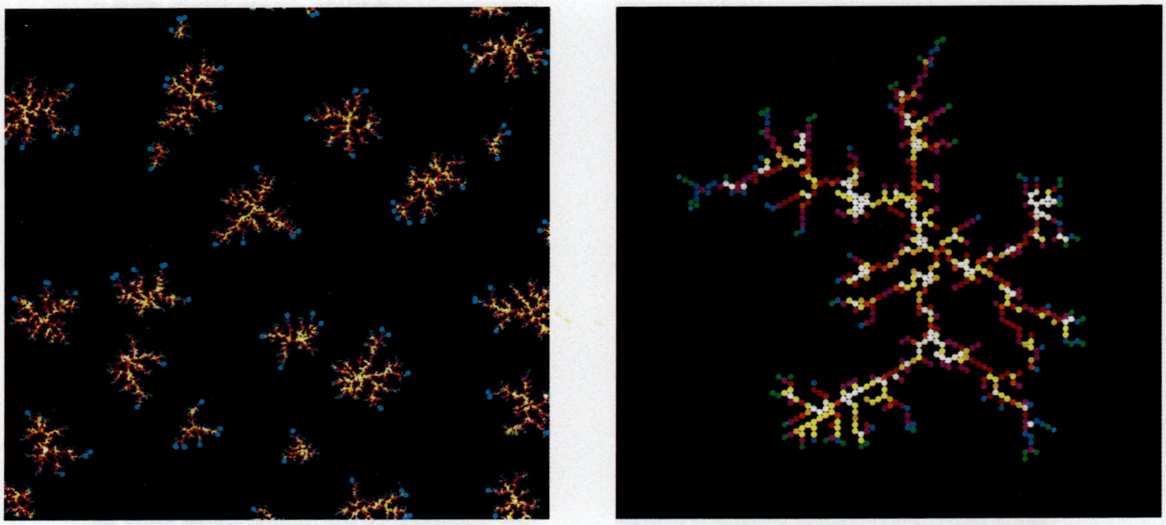

(a) (b)

**Fig. 4** Growth probabilities for different points of the islands at a coverage 0.04 ($\gamma = \infty$). **(a)** global view of the surface; **(b)** detailed view of an island. Colors give the probability that a given site of the island has to catch a monomer. The probability increases in the order: white, yellow, orange, red, blue, green and finally big bright blue dots.

## 4. ISLAND DIFFUSION

In the last paragraph, we have studied the behavior of the model in the case where only monomers are allowed to move. In this section, we want to address some consequences of (small) cluster diffusion. It has long been recognized that cluster diffusion can influence the growth of the films,[2,19] even if at that time experimental proofs were lacking. Recently, experiments[20] and molecular dynamics studies have shown that small clusters can move on the surfaces without breaking.[21] Generally, however, the experimental results are analyzed within the framework of the different models containing only monomer diffusion. These models can include *reversible* aggregation,[8,22] meaning that two monomers that aggregate can detach after a certain time. This can lead to wrong interpretations if small cluster mobility is indeed present in the experiments.[23] Then, it is important to understand the consequences of this mobility so as to be able to identify them in the experiments.

We have shown previously[5] that the introduction of cluster mobility considerably changes the growth dynamics of the film. Specifically, we have showed that:

(1) The mean cluster size increases *exponentially* as a function of the coverage. A power-law dependence is generally found when only monomer diffusion is allowed. This effect is intriguing and may be due to large cluster diffusion.[5]

(2) The maximum island density depends on the incident flux according to $N_{\text{max}} \sim F^{\zeta}$ with $\zeta = 0.42$, as opposed to $\zeta = 0.33$ found with monomer diffusion. It is also known that when only *dimer* diffusion is introduced, $\zeta = 0.4$.[7] The problem is that generally the accuracy of the experimental determinations of $N_{\text{max}}$ is not high enough to distinguish between the different exponents.[12]

(3) More interesting experimentally: the evolution of the island density as a function of coverage for low normalized fluxes (typically less than $10^{-5}$). It is known[4] that the

Fractals, Vol. 4, No. 3 (1996) 331–337


# SCALING THEORY ON GROWING ROUGH SURFACES

K. HONDA
*Department of Mathematical Sciences, Shinshu University, Matsumoto 390, Japan*

**Abstract**

We study the dynamics of growth processes of rough surfaces based on mathematical models such as the Kardar-Parisi-Zhang equation. The white-noise assumption in the KPZ equation is, however, noted to fail for higher dimensional cases. A careful continuum limit leads to a smooth surface solution for the cases. We develop the scaling theory to give an insight into the problem, by means of the intermediate asymptotics of the second kind, which is a very useful notion for the purpose. We find the roughness exponent and the dynamic exponent as functions of the substrate dimensionality for a model equation applicable to quenched disorder systems.

## 1. INTRODUCTION

The roughening phenomena of growing surfaces (interfaces) with self-affine symmetry have recently brought about much interest. A lot of vigorous studies by computer simulation, experiment, and theory have accumulated.[1,2] The reasons are considered as follows:

(1) These are typical nonequilibrium phenomena seen universally in very wide fields of science and technology and in nature.
(2) Self-affine patterns themselves are interesting as a natural extension of isotropic (self-similar) fractals to be studied.
(3) Last but not least, this kind of problem is expected to be understood through an idea of scaling hypothesis, which is a familiar concept to statistical physicists.

The scaling law in this problem, proposed first by Family and Vicsek,[3] is concerned with the averaged height fluctuation of the surface, $w(L, t) \equiv \langle \{h(\mathbf{r}, t) - \langle h(\mathbf{r}, t)\rangle\}^2\rangle^{1/2}$, which is expressed as:

$$w(L, t) = t^{\beta}\Psi(L/t^{1/z}), \tag{1}$$

where the height $h(\mathbf{r}, t)$ is measured at time $t$ from a position $\mathbf{r}$ ($\in [0, L]^d$) of the $d$-dimensional substrate of size $L$. The scaling function $\Psi(x)$ increases as $x^{\alpha}$ for sufficiently small $x$, and approaches a saturation value as $x$ is increased beyond a crossover $x_c \simeq 1$. The roughness exponent $\alpha$ (or Hurst exponent for self-affine fractals) is given as a product of the temporal scaling exponent $\beta$ and the dynamic exponent $z$, i.e., $\alpha = \beta z$.

The theoretical breakthrough brought by Kardar, Parisi and Zhang for the problem of growing rough surfaces is a proposal of a simple nonlinear Langevin equation[4]:

$$\frac{\partial h(\mathbf{r}, t)}{\partial t} = \nu \nabla^2 h(\mathbf{r}, t) + \frac{\lambda}{2}(\nabla h(\mathbf{r}, t))^2 + \eta(\mathbf{r}, t). \tag{2}$$

The first term on the right-hand side describes relaxation of a surface by the surface tension $\nu$. The second, nonlinear term arises from the tendency of the surface to grow locally normal to itself. The noise $\eta(\mathbf{r}, t)$ has zero mean and its correlation is given as:

$$\langle \eta(\mathbf{r} + \mathbf{r}', t + t')\eta(\mathbf{r}', t')\rangle = 2D\delta^d(\mathbf{r})\delta(t), \tag{3}$$

i.e., we usually assume the Gaussian white noise for $\eta(\mathbf{r}, t)$.

The KPZ equation allows us to attack the problem from a theoretical aspect. The tilting transformation of the surface by a small angle leads to a scaling relation,

$$\alpha + z = 2. \tag{4}$$

It has been believed that $\alpha = 1/2$ and $\beta = 1/3$ are exact for $d = 1$. The KPZ equation itself is interesting to study, because it is equivalent to some equations studied in different fields.

The KPZ equation gives the exponents in good agreement with computer simulation results of the Eden model and the ballistic deposition model. However, we have no experimental evidence which guarantees the KPZ equation as an acceptable model for real systems. The most promising continuum model for dynamics of rough surfaces in random media is a quenched version of the KPZ equation.[5] Namely, we replace the noise term $\eta(\mathbf{r}, t)$ in Eq. (2) with $\eta(\mathbf{r}, h(\mathbf{r}, t))$, which reflects the fact that impurities or defects have already been distributed in space before the surface arrives at the position. Therefore we will study

$$\frac{\partial h(\mathbf{r}, t)}{\partial t} = \nu \nabla^2 h(\mathbf{r}, t) + \frac{\lambda}{2}(\nabla h(\mathbf{r}, t))^2 + f + \eta(\mathbf{r}, h(\mathbf{r}, t)). \tag{5}$$

The driving force term $f$ is indispensable because it cannot be transferred into $h(\mathbf{r}, t)$ in this case. The noise $\eta(\mathbf{r}, h)$ cannot be assumed to have a $\delta$-function type correlation, as explained below. In short, we call Eq. (5) the KPZQ equation.

This paper is organized as follows: Sec. 2 is devoted to point out that the white-noise assumption for the KPZ equation fails for $d > 2$. In Sec. 3, the scaling theory based on the intermediate asymptotics of the second kind is applied to the KPZ equation. Although a final goal cannot be arrived, we throw a new light on the problem. In Sec. 4, we study the KPZQ problem. In this case, we have a definite solution for the exponents.

## 2. EXAMINATION OF THE WHITE-NOISE ASSUMPTION

To the end we study the linearized version ($\lambda = 0$), the so called Edwards-Wilkinson (EW) equation.[6] The conclusion is also valid for the KPZ model. For the EW equation we have, using $d$-component wave vector $\mathbf{k} = 2\pi\mathbf{n}/L$ with $d$-dimensional integer $\mathbf{n} = (n_1, \ldots, n_d)$,

$$w^2(L, t) = \frac{D}{\nu} \int_{2\pi/L \le |k_j| (j=1,2,\ldots,d)} \frac{d^d\mathbf{k}}{(2\pi)^d} \{1 - \exp(-2\nu k^2 t)\}/k^2 \,, \tag{6}$$

for large $L$, which, however, brings about a divergence for $d > 2$. In order to avoid the difficulty, the continuum model with Eq. (3) should be abandoned because the divergence arises from modes with very short wavelength (ultraviolet divergence). We have to return to a discrete version and take the continuum limit more carefully. Equation (3) should be replaced with

$$\langle \eta(\mathbf{r} + \mathbf{r}', t + t')\eta(\mathbf{r}', t)\rangle = 2D[R(\ell)\Delta_\ell(\mathbf{r})][T(\tau)\Delta_\tau(t)] \,, \tag{7}$$

where $\Delta_\ell(\mathbf{r})(\Delta_\tau(t))$ is a sharply localized function of $\mathbf{r}(t)$ around the origin with narrow width $\ell(\tau)$, which corresponds to some microscopic size existing in phenomena under consideration such as a lattice constant or the size of impurity. For example, $\Delta_\ell(\mathbf{r}) = \exp[-r^2/(2\ell^2)]$ and $\Delta_\tau = \exp[-t^2/(2\tau^2)]$. As the microscopic sizes, $\ell$ and $\tau$, are decreased, or the continuum limit is approached, the respective magnitude is assumed to grow as:

$$R(\ell) \sim \ell^{-\phi} \,, \tag{8}$$

$$T(\tau) \sim \tau^{-\psi} \,. \tag{9}$$

It should be noted that Eq. (3) indicates $\phi = d$ and $\psi = 1$.

We observe the width $w(L, t)$ of the rough surface from a macroscopic point of view after taking limits of $\ell \to 0$ and $\tau \to 0$. The use of Eq. (7) to the EW equation leads to:

$$\begin{aligned} w^2(L, t) &= 2D \lim_{\ell \to 0} \lim_{\tau \to 0} R(\ell) T(\tau) \\ &\times \int_{2\pi/L \le |k_j| \le 2\pi/\ell (j=1,2,\ldots,d)} \frac{d^d\mathbf{k}}{(2\pi)^d} \int_0^t dt_1 \int_0^t dt_2 e^{-\nu k^2(t-t_1)} e^{-\nu k^2(t-t_2)} \\ &\times \prod_{j=1}^{d} \left\{ \int_0^L dr_j (1 - r_j/L) 2\cos(r_j k_j) \right\} \Delta_\ell(\mathbf{r}) \Delta_\tau(t_1 - t_2) \,. \end{aligned} \tag{10}$$

For $d < 2$, if we put

$$\phi = d \quad \text{and} \quad \psi = 1 \,, \tag{11}$$

Equation (10) gives the well-known result, Eq. (6), with $\alpha = (2 - d)/2$ and $\beta = (2 - d)/4$. The integration in Eq. (6) is finite when $d < 2$. In the case of $d > 2$, however, we have to choose another set of $\phi$ and $\psi$ to obtain the finite $w(L, t)$ through vanishing limits of $\ell$ and $\tau$. Let us show that for $d > 2$, if the condition:

$$4 - \phi - 2\psi = 0 \,, \tag{12}$$

is satisfied, we can fulfill the purpose. To the end it is useful to change the variables into $\tilde{\mathbf{r}} = \mathbf{r}/\ell$, $\tilde{\mathbf{k}} = \mathbf{k}\ell$ and $\tilde{t} = t/\tau$. Each component of $\tilde{\mathbf{k}}$ is bounded between $2\pi\ell/L \le |\tilde{k}_j| \le 2\pi$

$(j = 1, \ldots, d)$. Then we have from Eq. (10) with Eq. (12):

$$w^2(L, t) = \sigma^{1-\psi}(D/\nu) \int_{0 \le |\tilde{k}_j| \le 2\pi (j=1,\ldots,d)} \frac{d\tilde{\mathbf{k}}}{(2\pi)^d} \tilde{k}^{-2}$$
$$\times \int_0^\infty d\tilde{t}\, \exp(\tilde{\nu}\tilde{k}^2\tilde{t}) \Delta_\tau(\tilde{t}) \prod_{j=1}^{d} \left\{ \int_0^\infty d\tilde{r}_j 2\cos(\tilde{k}_j \tilde{r}_j) \right\} \Delta_\ell(\tilde{\mathbf{r}}) \,. \quad (13)$$

Here, we have performed two limits of $\ell$ and $\tau$ under fixed $\sigma = \tau/\ell^2$ and used $\tilde{\nu} = \sigma\nu$. The resulting $w(L, t)$ is noted to be finite for $d > 2$, while an infrared divergence occurs for $d < 2$. Equation (13) shows that $w(L, t)$ is independent of both $L$ and $t$, corresponding to $\alpha = \beta = 0$. The statement here, which is closely related to a theorem known as the Sobolev's lemma among mathematicians,[8] is also valid for the KPZ equation. The detailed discussion will be reported elsewhere.

## 3. SCALING THEORY ON THE KPZ EQUATION

Let us first review briefly the intermediate asymptotics of the second kind,[7] which is a very useful idea to explore a scaling solution for a certain class of nonlinear problems. For some problems, we can find exact scaling exponents without solving the equation directly. Suppose a problem expressed in dimensionless variables, $\Pi$, $\Pi_0$, $\Pi_1, \ldots, \Pi_n$, with a solution:

$$\Pi = F(\Pi_0, \Pi_1, \ldots, \Pi_n) \,. \quad (14)$$

The asymptotics is said to be of the second kind if, with an appropriate choice of exponents $x$, $x_1, \ldots, x_n$, the asymptotic behavior is of the form:

$$\Pi \sim \Pi_0^x G\left(\frac{\Pi_1}{\Pi_0^{x_1}}, \ldots, \frac{\Pi_n}{\Pi_n^{x_n}}\right) , \quad \text{as} \quad \Pi_0 \to 0 \,, \quad (15)$$

where $G$ is a scaling function. It should be noted that we can automatically obtain a scaling form.

In order to look for the dimensionless variables in the KPZ problem, let us first calculate *dimensions* of quantities appearing in Eqs. (2) and (7). We have from them the following relations:

$$[h]/[t] = [\nu][h]/[r]^2 = [\lambda][h]^2/[r]^2 = [\eta] \,, \quad (16)$$
$$[\eta]^2 = [D][R][T] \,, \quad (17)$$

where the dimension of $q$ is denoted by $[q]$. It is necessary to represent $\ell$, $L$, $\tau$ and $w(L, t)$ in dimensionless forms. The above three equations lead to:

$$[w] = [h] = [D]^{1/2}[R]^{1/2}[T]^{1/2}[t] \,,$$
$$[\ell] = [L] = [r] = [\nu]^{1/2}[t]^{1/2} \,, \quad (18)$$
$$[\lambda] = [\nu][D]^{-1/2}[R]^{-1/2}[T]^{-1/2}[t]^{-1} \,.$$

From Eq. (18), a dimensionless form of $w(L, t)$ is represented from a microscopic point of view by:

$$w_{\rm micro}(L,\ t;\ \nu,\ \lambda,\ D;\ \ell,\ \tau) = D^{1/2}tR^{1/2}(\ell)T^{1/2}(\tau) \times \Phi_{\rm micro}\left(\frac{L}{\sqrt{(\nu t)}},\ \frac{\lambda D^{1/2}tR^{1/2}(\ell)T^{1/2}(\tau)}{\nu},\ \frac{\ell}{\sqrt{(\nu t)}},\ \frac{\tau}{t}\right). \tag{19}$$

The intermediate asymptotics of the second kind allows one to obtain a finite limit, that is,

$$w(L,\ t) = \lim_{\ell\to 0}\lim_{\tau\to 0} D^{1/2}tR^{1/2}(\ell)T^{1/2}(\tau)(\ell/\sqrt{(\nu t)})^{x}(\tau/t)^{y} \times \Phi\left(\frac{L}{\sqrt{(\nu t)}}\left(\frac{\ell}{\sqrt{(\nu t)}}\right)^{x_1}\left(\frac{\tau}{t}\right)^{y_1},\ \frac{\lambda D^{1/2}tR^{1/2}(\ell)T^{1/2}(\tau)}{\nu}\left(\frac{\ell}{\sqrt{(\nu t)}}\right)^{x_2}\left(\frac{\tau}{t}\right)^{y_2}\right), \tag{20}$$

takes a finite value if:

$$x = x_2 = \phi/2\,,\quad y = y_2 = \psi/2\,,\quad x_1 = y_1 = 0\,, \tag{21}$$

are satisfied. Summarizing the remaining terms gives:

$$w(L,\ t) = D^{1/2}\nu^{-\phi/4}t^{(4-\phi-2\psi)/4}\tilde{\Phi}_\lambda\left(\frac{L^2}{\nu t},\ \frac{\lambda^2 D t^{(4-\phi-2\psi)/2}}{\nu^{(4+\phi)/2}}\right). \tag{22}$$

By, instead of $\lambda$, the alternative choice $\nu$ as an *independent* quantity, we have:

$$w(L,\ t) = D^{2/(4+\phi)}\lambda^{-\phi/(4+\phi)}t^{(4-\phi-2\psi)/(4+\phi)}\tilde{\Phi}_\nu\left(\frac{L^{4+\phi}}{\lambda^2 D t^{4-\psi}},\ \frac{\nu^{4+\phi}}{\lambda^4 D^2 t^{4-\phi-2\psi}}\right). \tag{23}$$

From the above, we can derive some important conclusions. One of them is concerned with the fact that the relation Eq. (12) between $\phi$ and $\psi$ leads two vanishing exponents on $t$, which finds again $\alpha = \beta = 0$ for $d > 2$. This conclusion implies clearly the absence of any roughening transition from a smooth surface for $d > 2$ in this model, although there exist the predictions that it should occur as $\lambda$ is increased.[1] Secondly, the solution for the linear version or the EW equation can be simply obtained:

$$w_{\rm EW}(L,\ t) = D^{1/2}\nu^{-\phi/4}t^{(4-\phi-2\psi)/4}\tilde{\Phi}_\lambda\left(\frac{L^2}{\nu t},\ 0\right), \tag{24}$$

which indicates that $\alpha = (2-d)/2$, $z = 2$ for $d < 2$ and $\alpha = 0$, $z = 2$ for $d > 2$. Finally, by applying again the intermediate asymptotics of the second kind, the steady state solution is given in a scaling form as follows: Using $a = -1/2$ and $b = 2/(4-\phi-2\psi)$,

$$w(L,\ \infty) = \lim_{t\to\infty} D^{1/2}\nu^{-\phi/4}t^{(4-\phi-2\psi)/4}\left(\frac{\lambda^2 D t^{(4-\phi-2\psi)/2}}{\nu^{(4+\phi)/2}}\right)^{a}\Phi_{\rm st}\left(\frac{L^2}{\nu t}\left(\frac{\lambda^2 D t^{(4-\phi-2\psi)/2}}{\nu^{(4+\phi)/2}}\right)^{b}\right)$$

$$= (\nu/\lambda)\tilde{\Phi}_{\rm st}(gL^{(4-\phi-2\psi)})\,, \tag{25}$$

where $g$ is the coupling constant defined by $g = \lambda^2 D/\nu^{4-\psi}$. In the alternative representation we also have:

$$w(L,\ \infty) = \left(\frac{D}{\nu^{2-\psi}}\right)^{1/2} L^{(4-\phi-2\psi)/2}\hat{\Phi}_{\rm st}(gL^{(4-\phi-2\psi)})\,. \tag{26}$$

In the weak coupling limit, the EW solution is recovered, which implies that $\tilde{\Phi}_{st}(x) \sim \sqrt{x}$ and $\hat{\Phi}_{st}(x) \sim$ constant for very small $x$. However, it is a very difficult problem to predict the large $x$ behavior. If possible, we can obtain the final solution for the KPZ model. We can easily have the temporal scaling law $w(\infty, t)$ as well from Eqs. (22) and (23) through letting $L$ be infinity.

## 4. SCALING THEORY ON THE KPZQ EQUATION

Before developing the scaling theory on the KPZQ equation, Eq. (5), we summarize the facts which have already been clarified.[5] The driving force $f$ plays a role of the control parameter. Below a critical $f_c$, the surface after some characteristic time becomes pinned, i.e., due to the randomly distributed negative value of $\eta$, it stops completely. On the other hand, if $f > f_c$, the surface keeps moving for arbitrarily large times. However, the growth behavior changes after a crossover time $t^*$ is passed, i.e., a slope in the $\log w$ vs. $\log t$ plane decreases abruptly around $t^*$. The crossover time $t^*$ becomes larger as $f$ is approached $f_c$ from above. We denote the exponents as $\alpha_Q$ and $\beta_Q$, respectively, for $t \ll t^*$. Numerical integration of the KPZQ equation in the case $d = 1$ yielded $\alpha_Q \simeq 0.71$ and $\beta_Q \simeq 0.61$ when $f \simeq f_c$, which agree well with experimental results.[5]

From a microscopic point of view, the correlation for $\eta(\mathbf{r}, h)$ is denoted in the same way with Eq. (7) as:

$$\langle \eta(\mathbf{r}+\mathbf{r}', h+h')\eta(\mathbf{r}', h')\rangle = 2D[R(\ell)\Delta_\ell(\mathbf{r})][H(\rho)\Delta_\rho(h)]\,, \tag{27}$$

where

$$H(\rho) \sim \rho^{-\varphi}\,, \tag{28}$$

is assumed. However, we put in this case that:

$$\phi = d \quad \text{and} \quad \varphi = 0\,. \tag{29}$$

It should be noted that the KPZQ equation does not describe a stochastic process although it includes random variables $\{\eta(\mathbf{r}, h)\}$. Remember that a numerical integration of the KPZQ equation is calculated irrespective of the lattice size or the mesh size along the $h$ direction, from which Eq. (29) is deduced.

For the KPZQ equation, we have following dimension equations:

$$[h]/[t] = [\nu][h]/[r]^2 = [\lambda][h]^2/[r]^2 = [f] = [\eta]\,, \tag{30}$$

$$[\eta]^2 = [D][R][H]\,. \tag{31}$$

From these we have:

$$\begin{aligned}
[\rho] = [w] = [h] &= [D]^{1/2}[R]^{1/2}[H]^{1/2}[t] \\
[\ell] = [L] = [r] &= [\lambda]^{1/2}[D]^{1/4}[R]^{1/4}[H]^{1/4}[t] \\
[\nu] &= [\lambda][D]^{1/2}[R]^{1/2}[H]^{1/2}[t] \\
[f] &= [D]^{1/2}[R]^{1/2}[H]^{1/2}\,.
\end{aligned} \tag{32}$$

In the similar way to Eq. (20), after taking limits of $\ell \to 0$ and $\rho \to 0$, the scaling form for $w(L, t)$ can be obtained as:

$$w(L,\, t) = D^{2/(4+\phi+2\varphi)}\lambda^{-\phi/(4+\phi+2\varphi)}t^{(4-\phi)/(4+\phi+2\varphi)} \times \Phi_Q\left(\frac{L^{4+\phi+2\varphi}}{D\lambda^{2+\varphi}t^{4+\varphi}},\, \frac{(f-f_c)^{4+\phi+2\varphi}t^{2(\phi+\varphi)}}{D^2\lambda^{-\varphi}},\, \frac{\nu^{2(\phi+\varphi)}(f-f_c)^{4-\phi}}{D^2\lambda^{\phi+2\varphi}}\right), \tag{33}$$

where $f_c$ has been introduced. The third argument has been recombined to eliminate $t$, which is then regarded as a system parameter. Moreover, we consider such limited cases that $f \simeq f_c$ or $t \ll t^*$, where the crossover time is defined by:

$$t^* = \{(f-f_c)^{4+\phi+2\varphi}\lambda^{\phi}/D^2\}^{-1/2(\phi+\varphi)}\,. \tag{34}$$

Substituting Eq. (29) into Eq. (33) gives:

$$w(L,\, t) = D^{2/(4+d)}\lambda^{-d/(4+d)}t^{(4-d)/(4+d)}\tilde{\Phi}_Q(L^{4+d}/D\lambda^2 t^4)\,, \tag{35}$$

with

$$\alpha_Q = (4-d)/4 \quad \text{and} \quad \beta_Q = (4-d)/(4+d)\,, \tag{36}$$

giving already estimated values for $d = 1$. It is interesting to note that they satisfy the scaling relation Eq. (4)

## ACKNOWLEDGMENTS

I would like to express my gratitude to M. Matsushita and T. Vicsek for their continual collaboration.

## REFERENCES


1. T. Vicsek, *Fractal Growth Phenomena* (2nd ed., World Scientific, Singapore, 1992).
2. *Dynamics of Fractal Surfaces*, eds. F. Family and T. Vicsek (World Scientific, Singapore, 1991).
3. F. Family and T. Vicsek, *J. Phys.* **A18**, L75 (1985).
4. M. Kardar, G. Parisi and Y.-C. Zhang, *Phys. Rev. Lett.* **56**, 889 (1986).
5. Z. Csahók, K. Honda and T. Vicsek, *J. Phys.* **A26**, L171 (1993) and references therein.
6. S. F. Edwards and D. R. Wilkinson, *Proc. Roy. Soc. Lond.* **A381**, 17 (1982).
7. N. D. Goldenfeld, O. Martin, Y. Oono and F. Liu, *Phys. Rev. Lett.* **64**, 1361 (1990).
8. See for example, M. Reed and B. Simon, *Fourier Analysis, Self-Adjointness*, Methods of Modern Mathematical Physics II (Academic Press, New York, 1975).

Fractals, Vol. 4, No. 3 (1996) 339–348


# SELF-ORGANIZED CRITICAL BEHAVIOR OF TWO-DIMENSIONAL FOAMS

KYOZI KAWASAKI
*Department of Natural Science and Mathematics, Chubu University, Kasugai, Aichi 487, Japan*

TOHRU OKUZONO
*Department of Physics, Ochanomizu University, Tokyo 112, Japan*

**Abstract**

A simple dynamical model for two-dimensional dry foam rheology is constructed for which surface tension effects and viscous dissipation at Plateau borders (intersections of three cell boundaries) are taken into account and is studied by computer simulation. Under externally applied shear strain increasing at small rates, the system exhibits avalanche-like release of stress that has been accumulating under increasing strain. There is a close similarity with earthquake models that show self-organized criticality (SOC). We discuss related simulation of two-dimensional wet foams under statically applied strain by Hutzler et al. [*Phil. Mag.* **B71**, 277 (1995)] showing critical behavior.

## 1. INTRODUCTION

The concept of self-organized criticality (SOC) made popular by Bak and coworkers[1] is considered to be one of those key concepts which underlie a large class of natural phenomena which seem to lack characteristic length and time scales and are governed by some sorts of power laws. The concept was first illustrated for a simple sand pile model and was applied to various other problems including earthquakes.[2] On the other hand, the existing theoretical models of sand piles, earthquakes, etc. appear to be inadequate in view of great complexity of the systems under consideration, and it is not obvious whether and how these simplified models can be improved so as to better simulate reality. It is certainly true that in all kinds of critical behavior one is interested in the universal aspects and not in irrelevant details. This point of view has been vindicated in the conventional critical phenomena from which we now have a good understanding of what constitute relevant and irrelevant features of a model.[3] As long as such an understanding is lacking, one can only guess and hope that the chosen model contains all the features relevant for universality, which often turns out not to be the case afterward. Under such circumstances, it is highly valuable to have a system whose microscopic physics is fairly well understood. We think that a high quality soap froth is a good example of such a model, as has also been emphasized by Gopal and Durian recently.[4] A typical foam consists of finite domains of liquid or gas surrounded by another liquid which forms a continuous region. When the volume fraction of enclosed fluid regions is almost unity (dry foam), the system exhibits a high degree of elasticity under small strain. However, if the strain is increased indefinitely, the steadily accumulating stress is released, triggered by a local topology change of cellular pattern such as neighbor switching (T1 process). This process can take place in an avalanche-like event with large numbers of topology changes.[5,6] In the case of the Princen–Prud'homme model of two-dimensional foam rheology with a periodic array of identical hexagonal bubbles, neighbor switchings occur in unison at a particular value of strain, releasing all the stress accumulated previously.[7] This event can be regarded as one giant avalanche spreading over the entire system. Real foams are generally irregular and topology changes occur as random local events. Thus the interesting question is: How will this giant avalanche be decomposed into smaller ones for irregular foams? Will the decomposition be complete in the sense that in the end each event consists of a finite numbers of elementary topology changes such that the system exhibits only the usual plastic behavior? Or will the decomposition be incomplete so that avalanches of indefinitely large sizes are still possible though with small frequencies?

In this paper we report our recent computer simulation works for a particular dynamical model of two-dimensional foam to answer the abovementioned question at least for this particular case.[5] Our model is a variant of the vertex models that we developed originally to study coarsening of two-dimensional dry foams where cell boundaries are approximated by straight lines (edges).[8] The main differences for the case of rheology are:

(1) The area of each cell was kept fixed assuming no diffusion of liquid or gas through cellular boundaries.

(2) Dissipation occurs only at the Plateau borders, i.e., at intersections of three edges.

For the Plateau border dissipation, we take the result of Schwartz and Princen[9] who considered the viscous dissipation associated with the liquid flow in the continuous region near a Plateau border. We found it particularly important to make the model invariant under Galilei transformation which permits large scale foam flow associated with an avalanche.

## 2. DYNAMICAL MODEL

Throughout this work, we only consider purely dissipative dynamics. The equation of motion is generally expressed in the form of a force balance equation as:

$$\mathbf{f}_F = \mathbf{f}_S\,, \tag{1}$$

where $\mathbf{f}_F$ and $\mathbf{f}_S$ are the frictional and static forces, respectively. If these forces are derivable from a Rayleigh dissipation function $R(\{v\},\,\{q\})$ and a potential function $H(\{q\})$, where

$\{q\}$ is a set of appropriate dynamical variables $q_i$, $i = 1, 2, \ldots$ and $\{v\}$ is the set $v_i \equiv \mathrm{d}q_i/\mathrm{d}t$, $i = 1, 2, \ldots$, Eq. (1) takes the form:

$$\frac{\partial R}{\partial v_i} = -\frac{\partial H}{\partial q_i}\,. \tag{2}$$

Here $\{v\}$ and $\{q\}$ may not be discrete variables but can be functions, and then $R$ and $H$ become functionals.

The vertex approach to two-dimensional foam proposes to take locations of the vertices (Plateau borders) $\{\mathbf{r}_i\}$ and their velocities $\{\mathbf{v}_i\} = \{\mathrm{d}\mathbf{r}_i/\mathrm{d}t\}$ as the dynamical variables $\{q\}$ and $\{v\}$, respectively, where cell boundaries are approximated by straight lines. The nature of such approximations and corrections has been discussed elsewhere[10] and will not be repeated here. We only point out that the vertex model is the simplest of the series of approximation schemes in which a number of supplementary points are introduced in the originally curved cell boundaries and treated like additional vertices (virtual vertices). This refined model approaches to the original curved interface model in the limit of dense virtual vertices.

Now, the energetics and dissipation occurring in real foams are rather complex even for incompressible cases as, for example, can be seen in the recent analyses by Buzza et al.[11] We consider here the most simple case (as explained in Sec. 1) in which the static forces come from surface tensions of cell boundaries. The vertex equation of motion now becomes (see Refs. 8 and 12 for details of derivation):

$$\frac{\partial}{\partial \mathbf{v}_i}R + \frac{\partial}{\partial \mathbf{r}_i}\left[\sum_{\langle ij\rangle} 2\sigma|\mathbf{r}_{ij}| - \sum_{\alpha}\lambda_\alpha A_\alpha\right] = 0\,, \tag{3}$$

$$\frac{\partial R}{\partial \mathbf{u}_{ij}} = \frac{\mathrm{d}A_\alpha}{\mathrm{d}t} = 0\,, \tag{4}$$

with

$$R = \frac{6}{5}\sigma^{1/3}(3\mu)^{2/3}\sum_i \sum_j^{(i)} U_{ij}^{5/3}[1.2215\theta(U_{ij}) + 1.1866\theta(-U_{ij})]\,, \tag{5}$$

$$U_{ij} \equiv (\mathbf{v}_i - \mathbf{u}_{ij})\cdot(\mathbf{r}_i - \mathbf{r}_j)/|\mathbf{r}_i - \mathbf{r}_j|\,, \tag{6}$$

where $\sigma$ and $\mu$ are the surface tension and the shear viscosity of liquid in the continuous region, respectively. Here $A_\alpha$ is the area of cell $\alpha$ and $\lambda_\alpha$ the Lagrange multiplier. The sum with respect to $\langle ij\rangle$ is over all the edges $\langle ij\rangle$ and the second summation in Eq. (5) is over the vertices $j$ connected to the vertex $i$ by a single edge. $\theta(x)$ is the step function equal to 1 and 0 for $x > 0$ and $x < 0$, respectively. $\mathbf{u}_{ij}$ turns out to be the velocity of the midpoint of the edge $\langle ij\rangle$. This guarantees the Galilei invariance of Eq. (3).

## 3. SIMULATION AND RESULTS

We numerically solve Eqs. (3)–(6) under the sheared periodic boundary condition. Under this boundary condition, a unit system relatively moves to the adjacent systems above and below by a constant speed $\dot{\gamma}L_y$, where $\dot{\gamma}$ is a constant shear rate and $L_y$ the dimension of the unit system in $y$ direction, i.e., the position $\mathbf{r}^{(m)}$ in a unit system specified by a pair of integers $\mathbf{m} = (m_x, m_y)$ is related to the position $\mathbf{r}^{(0)}$ in the centered unit system in the following manner, at time $t$,

$$\mathbf{r}^{(m)} = \mathbf{r}^{(0)} + \begin{pmatrix} 1 & \dot{\gamma}t \\ 0 & 1 \end{pmatrix} \begin{pmatrix} m_x L_x \\ m_y L_y \end{pmatrix}, \tag{7}$$

where $L_x$ is the dimension of the unit system in direction $x$. Throughout this simulation, the time and length are scaled by $r_{av}\mu/\sigma$ and $r_{av}$, respectively, where $r_{av}$ is the average linear size of cells. For detailed numerical method, see Refs. 5, 8 and 13.

In a previous study,[5] we reported that our model system under a weak shear exhibits an avalanche-like phenomenon and the distribution of avalanche size and the power spectra of response functions obey power laws. We examine here how the shear rate and the finite system size affect the avalanche-like behavior. We carried out 4 runs for an initial state which contains 1071 cells with different shear rates: $\dot{\gamma} = 1.0 \times 10^{-5}$, $5.0 \times 10^{-5}$, $1.0 \times 10^{-4}$, $5.0 \times 10^{-4}$, and a run for another initial state which contains 4271 cells with a shear rate $\dot{\gamma} = 1.0 \times 10^{-4}$. Hereafter we refer to the former 4 runs as run (a), run (b),..., run (d), respectively, and the last run as run (L).

In Fig. 1 we plot the probability distribution $P(s; \dot{\gamma})$ of avalanche size $s$ for the runs (a)–(d), where the avalanche size $s$ is defined as an amount of interfacial energy per unit area which is released during an avalanche.[5] In this figure squares, triangles, open circles, and closed circles correspond to the runs (a)–(d), respectively. Although these data are rather scattered, the power-law behavior of $P(s; \dot{\gamma})$ is observed for every run and we cannot find any significant dependence on $\dot{\gamma}$. In Fig. 2 we plot the distribution $P(\tilde{s})$ of $\tilde{s} \equiv L^2 s$ for the runs (c) (open circles) and (L) (diamonds), where $L(= L_x = L_y = N_C^{1/2})$ is the system

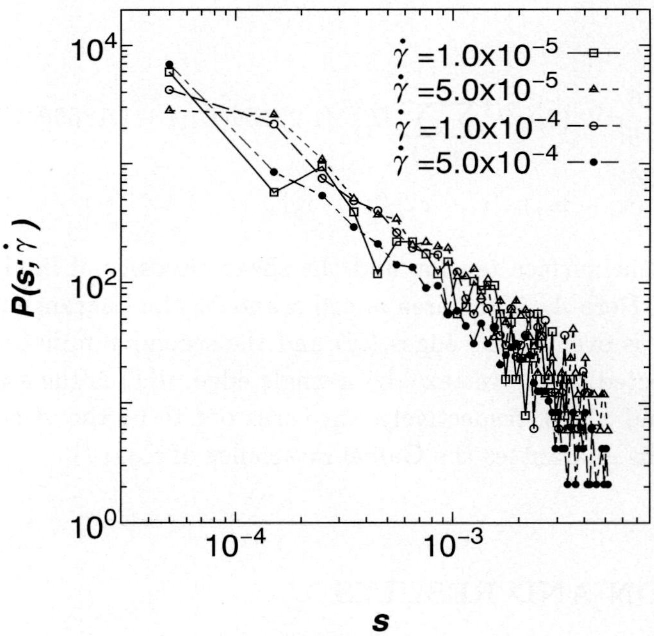


**Fig. 1** Probability density $P(s; \dot{\gamma})$ of avalanche size $s$ for various shear rates $\dot{\gamma}$. Squares, triangles, open circles, and closed circles correspond to $\dot{\gamma} = 1.0 \times 10^{-5}$, $5.0 \times 10^{-5}$, $1.0 \times 10^{-4}$, $5.0 \times 10^{-4}$, respectively. The total number of cells is fixed at 1071.

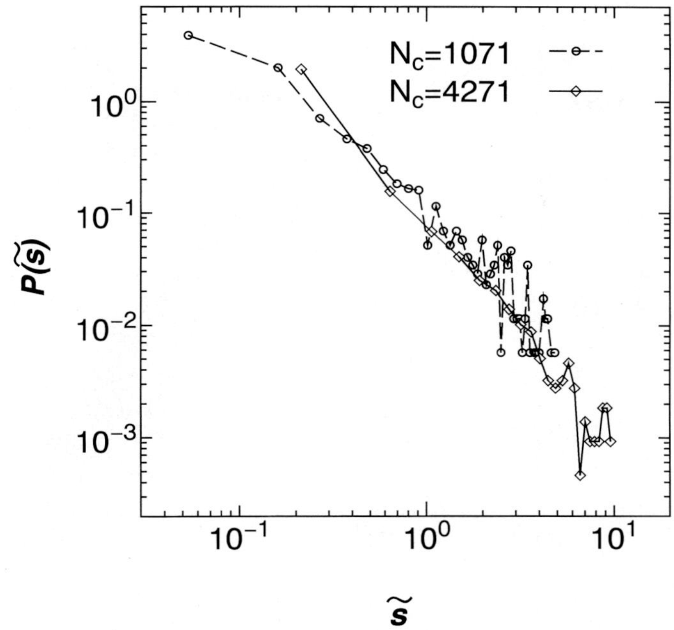


**Fig. 2** Probability density $P(\tilde{s})$ of the net total avalanche size $\tilde{s}$ for two different system sizes. The plots by open circles and diamonds are for the systems which contains 1071 and 4271 cells, respectively. The shear rate is fixed at $1.0 \times 10^{-4}$.

size and $N_C$ is the total number of cells. Here we have used the net total avalanche size $\tilde{s}$ instead of $s$ to compare the data with different system sizes.[5] These two data show almost the same power-law behavior.

Figure 3 shows the probability distribution $P(T; \dot{\gamma})$ of life-time or duration $T$ of an avalanche for the runs (a)–(d). The symbols in this figure are the same as in Fig. 1. It appears that there is a local maximum for each curve, especially for lower shear rates, and its peak position is shifted to smaller $T$ as $\dot{\gamma}$ increases. The appearance of these local maxima is probably the effect of finite system size. Indeed, for the run (L) the local maximum disappears (Fig. 4) and power-law behavior is observed.

Next, we investigate spatial structures of the flow accompanied with avalanches. The velocities of cells form a simple shear flow but it is disordered by avalanches which intermittently occur in large strain regime ($\dot{\gamma}t \gtrsim 1$).[5] In order to see the spatial structure of this disordered flow, we should consider the velocity correlation of the flow. Since there are some ambiguities in defining the velocity field, we introduce the quantity $\mathbf{j}(\mathbf{r})$ which is a function of position $\mathbf{r}(= (x, y))$ given by:

$$\mathbf{j}(\mathbf{r}) = \sum_{\alpha} \delta\mathbf{v}_\alpha \delta(\mathbf{r} - \mathbf{r}_\alpha) , \tag{8}$$

where $\delta\mathbf{v}_\alpha \equiv \mathbf{v}_\alpha - \bar{\mathbf{v}}(\mathbf{r}_\alpha)$ is the deviation of velocity of cell $\alpha$ from the simple shear flow $\bar{\mathbf{v}}(\mathbf{r}) \equiv (\dot{\gamma}y, 0)$ and $\mathbf{r}_\alpha$ and $\mathbf{v}_\alpha$ are the position and velocity of the center of cell $\alpha$, respectively. The structure factor $\langle \mathbf{j}_{-\mathbf{k}} \cdot \mathbf{j}_{\mathbf{k}} \rangle$ can include the velocity correlation and it also includes the density correlation of cells, where $\mathbf{j}_{\mathbf{k}}$ is the Fourier component of $\mathbf{j}(\mathbf{r})$ with wave vector $\mathbf{k}$ and $\langle \cdot \rangle$ means the ensemble average. In Figs. 5(a)–5(d), we show the structure factor of $\mathbf{j}(\mathbf{r})$ together with that of the density of cells $\rho(\mathbf{r}) \equiv \sum_\alpha \delta(\mathbf{r} - \mathbf{r}_\alpha)$ for the runs (a)–(d). The left and

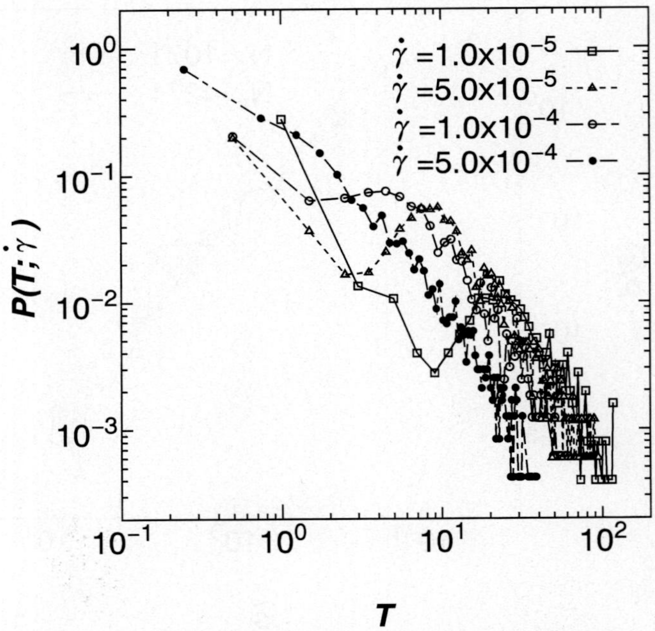


**Fig. 3** Probability density $P(T;\,\dot{\gamma})$ of life-time or duration $T$ of an avalanche for various shear rates $\dot{\gamma}$. The symbols are the same as in Fig. 1.

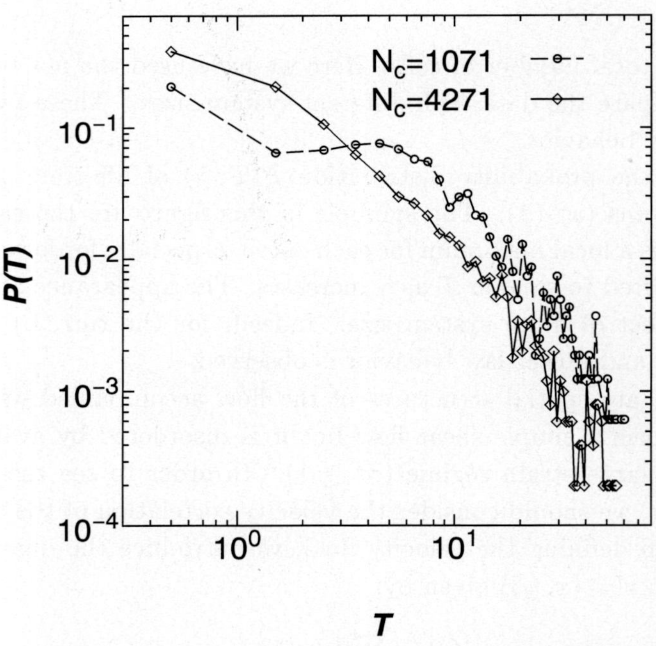


**Fig. 4** Probability density $P(T)$ of life-time or duration $T$ of an avalanche for two different system size. The symbols are the same as in Fig. 2.

right figures in Fig. 5 are the structure factors $\langle \rho_{-\mathbf{k}}\rho_{\mathbf{k}} \rangle$ and $\langle \mathbf{j}_{-\mathbf{k}} \cdot \mathbf{j}_{\mathbf{k}} \rangle$, respectively, where $\rho_{\mathbf{k}}$ is the Fourier component of $\rho(\mathbf{r})$. In each figure the intensity of the structure factor is expressed by the colors on the $\mathbf{k}$-plane, whose range is $-k_{\max} \le k_x \le k_{\max}$, $-k_{\max} \le k_y \le k_{\max}$, where $k_x(k_y)$ is the $x(y)$ component of $\mathbf{k}$ and $k_{\max} = 64k_{\min}$ with $k_{\min} \equiv 2\pi/L$. The intensity

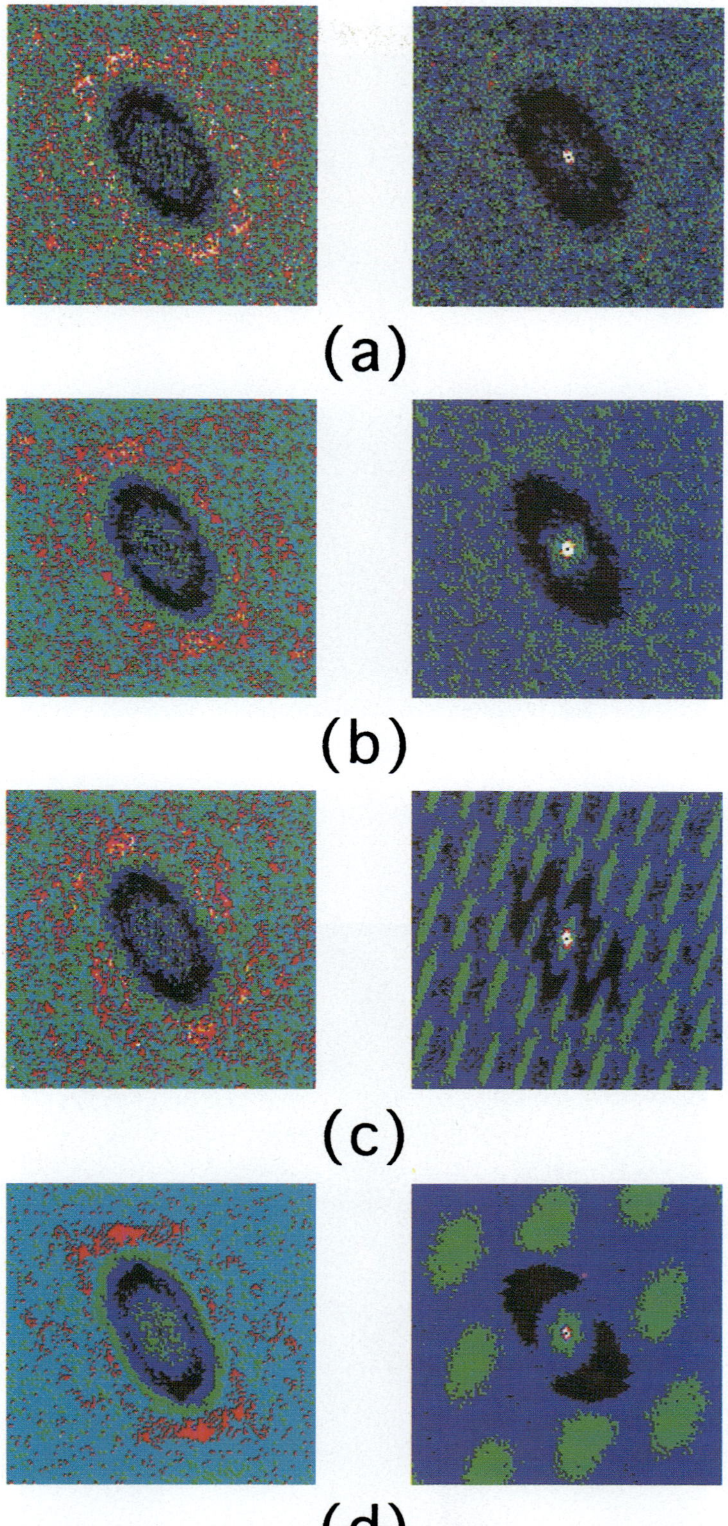



**Fig. 5** The structure factors of $\rho(\mathbf{r})$ and $\mathbf{j}(\mathbf{r})$ for the runs (a)–(d). Left and right figures are the structure factors $\langle \rho_{-\mathbf{k}} \rho_{\mathbf{k}} \rangle$ and $\langle \mathbf{j}_{-\mathbf{k}} \cdot \mathbf{j}_{\mathbf{k}} \rangle$, respectively. In each figure, the intensity of the structure factor is expressed by the colors on the $\mathbf{k}$-plane whose range is $-k_{\max} \le k_x \le k_{\max}$, $-k_{\max} \le k_y \le k_{\max}$ with $k_{\max} = 64 \cdot 2\pi/L$. The intensity increases in the following order of colors: black, blue, green, cyan, red, magenta, yellow, and white.

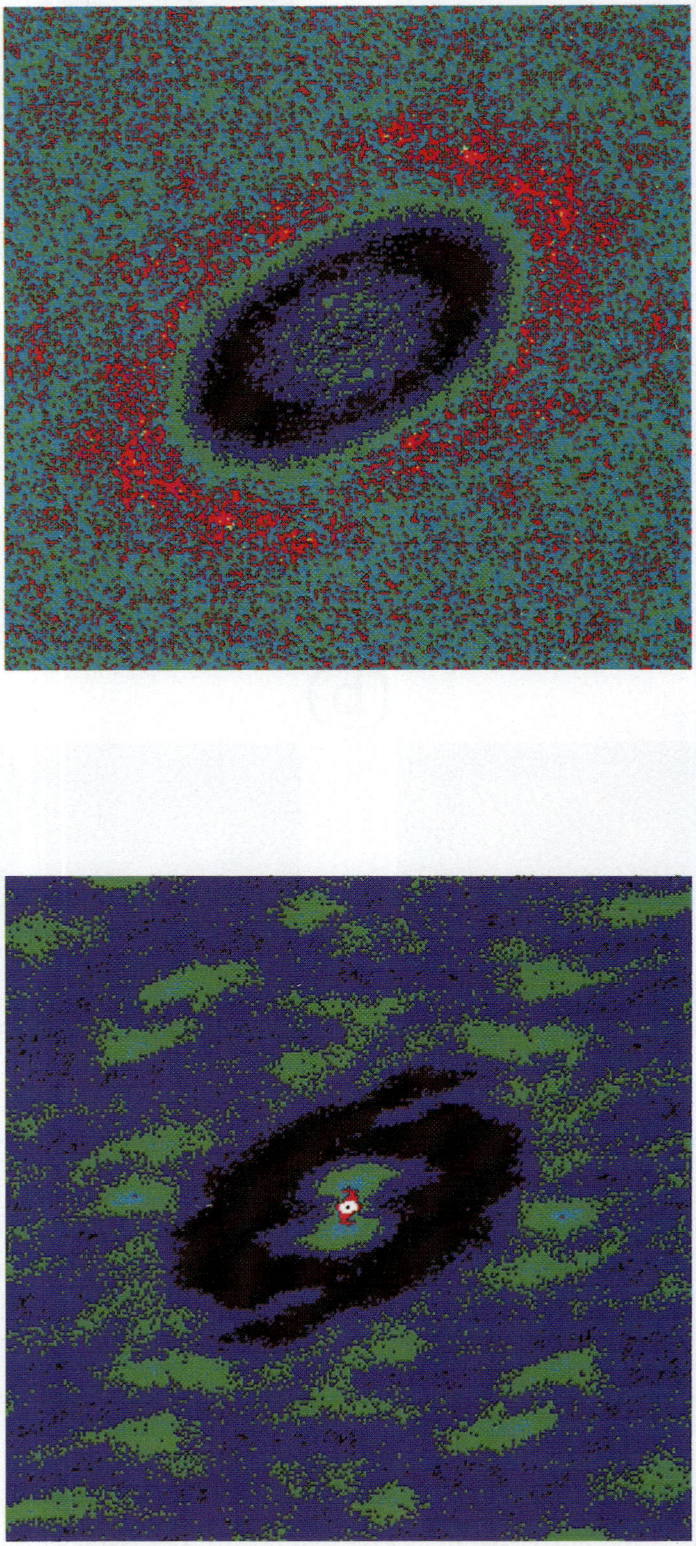

**Fig. 6** The structure factors of $\rho(\mathbf{r})$ (left) and $\mathbf{j}(\mathbf{r})$ (right) for the run (L). In each figure, the intensity of the structure factor is expressed by the colors on the $\mathbf{k}$-plane whose range is $-k_{\max} \leq k_x \leq k_{\max}$, $-k_{\max} \leq k_y \leq k_{\max}$ with $k_{\max} = 128 \cdot 2\pi/L$. The colors have the same meaning as in Fig. 5.

increases in the following order of colors: black, blue, green, cyan, red, magenta, yellow, and white. The average is taken over 100 snapshots which are arbitrarily chosen in the large strain regime for each run. From the figures of $\langle \rho_{-\mathbf{k}} \rho_{\mathbf{k}} \rangle$, we can see the distortion of the cellular pattern. The structure factors of $\mathbf{j}(\mathbf{r})$ for the runs (c) and (d) have the periodic structures, while for the runs (a) and (b), the periodic structure is not observed. This implies that a periodic array of vortex-like structure of the velocity field emerges under the boundary condition in Eq. (7) and a lower shear rate enhances a larger structure, although these periodic structures might be due to finite size effect. For the larger system [run (L)], such periodic structure is obscured as one can see in Fig. 6. As long as we concentrate on the region of small $|\mathbf{k}|$, we cannot observe any spot or peak of the intensity except for $\mathbf{k} = 0$, although the pattern is anisotropic. Therefore we expect that there exists no characteristic length in the velocity field at the limit $\dot{\gamma} \to 0$ and $L \to \infty$, i.e., the critical state is reached in this limit.

## 4. DISCUSSION

In this paper, we considered a model for the two-dimensional dry foam and found that the SOC-like behavior under externally applied shear strain increased steadily at small and constant rates. On the other hand, Hutzler et al.[6] found irregular SOC-like behavior in the stress–strain relation for two-dimensional wet foams under statically applied elongational strain, i.e., the strain is increased by small steps and the system is equilibrated after each step. This is connected with the second-order-like transition at the critical volume fraction 0.84 of the dispersed fluid, and hence appears to be different from SOC which is not related to the existence of a critical value of the control parameter of the system. Note that the static stress–strain relation of our model[14] shows that the behavior very similar to that of Ref. 6 in the dry limit. Thus the vertex model approximation is not responsible for the SOC-like behavior we have found. The case studied in this paper is perhaps more closely related to the earthquake models[2] in the sense that there is no control parameter that has to be adjusted and the stick–slip mechanism under externally applied strains increases steadily at small constant rates plays a role.

Finally, we comment on the experimental tests of our simulation results. Recently, Gopal and Durian[4] examined a soap froth with a volume fraction 0.92 under slowly increasing shear strain by diffusing-wave spectroscopy. They found that the autocorrelation function of the electric field amplitude of the transmitted light shows exponential decay in time, which contradicts large fluctuations expected from avalanches. Although the reason for this discrepancy is not understood at this time, we tend to attribute it to the small volume fraction. Hutzler et al.[6] showed in two dimensions that the mechanical behavior is extremely sensitive to the values of volume fraction near 1. In fact, the values of yield stress and yield strain drop roughly by an order of magnitude as the volume fraction decreases from 1 to 0.92. This does not literally apply to the three-dimensional experimental situation. However, similar singular behavior is expected in three dimensions. Furthermore, there are more ways to relieve stress in three dimensions than in two. On the other hand, it is possible that our simulation model lacks some essential ingredients that provide cut-offs to the power-law behavior. The extensions of our simulation model to three dimensions and to wet foams are a necessary but formidable task.

## ACKNOWLEDGMENTS

We would like to thank D. Durian for useful correspondence and for sending to us a preprint.[4] Thanks are also due to D. Weaire for sending a preprint of Ref. 6 before publication. This work is supported by a Grant-in-Aid for Encouragement of Young Scientists, from the Ministry of Education, Science and Culture, Japan.

## REFERENCES

1. P. Bak, C. Tang and K. Wiesenfeld, *Phys. Rev. Lett.* **59**, 381 (1987) and *Phys. Rev.* **A38**, 364 (1988).
2. J. M. Carlson and J. S. Langer, *Phys. Rev. Lett.* **62**, 2632 (1989) and *Phys. Rev.* **A40**, 6470 (1989).
3. For example, C. Domb and M. S. Green, eds. *Phase Transition and Critical Phenomena*, Vol. 6 (Academic Press, London, 1976).
4. A. D. Gopal and D. J. Durian, "Nonlinear bubble dynamics in a slowly driven foam," *Phys. Rev. Lett.* **75**, 2610 (1995).
5. T. Okuzono and K. Kawasaki, *Phys. Rev.* **E51**, 1246 (1995).
6. S. Hutzler, D. Weaire and F. Bolton, *Phil. Mag.* **B71**, 277 (1995).
7. A. M. Kraynik, *Ann. Rev. Fluid. Mech.* **20**, 325 (1988). A general review of foam rheology.
8. T. Okuzono and K. Kawasaki, *Trends in Stat. Phys.* **1**, 65 (1994).
9. L. W. Schwartz and H. M. Princen, *J. Colloid Interface Sci.* **118**, 201 (1987).
10. K. Kawasaki, T. Nagai and S. Ohta, in *Simulation and Theory of Evolving Microstructures*, eds. M. P. Anderson and A. D. Rollett (Minerals, Metals and Materials Society, Warrendale, PA, 1990); K. Kawasaki, T. Nagai, T. Okuzono and K. Fuchizaki, in *Modeling of Coarsening and Grain Growth*, eds. S. P. Marsh and C. S. Pande (Minerals, Metals and Materials Society, Warrendale, PA, 1993).
11. D. M. A. Buzza, C.-Y. D. Lu and M. E. Cates, *J. Phys. II* (France) **5**, 37 (1995).
12. K. Kawasaki and T. Okuzono, in *Dynamical Systems and Chaos*, Vol. 2: Physics, eds. Y. Aizawa, S. Saito and K. Shiraiwa (World Scientific, Singapore, 1995).
13. T. Okuzono, K. Kawasaki and T. Nagai, *J. Rheol.* (NY) **37**, 571 (1993).
14. K. Kawasaki, T. Okuzono and T. Nagai, *J. Mech. Behav. Mater.* **4**, 51 (1992).

Fractals, Vol. 4, No. 3 (1996) 349–354


# FRACTALS IN THE GLASS TRANSITION

ANTONIO CONIGLIO
*Dipartimento di Scienze Fisiche, Universitá di Napoli,*
*INFN and INFM sezione di Napoli,*
*Mostra d'Oltremare Pad. 19, 80125 Napoli, Italia*

**Abstract**

The static and fractal properties of the frustrated percolation model are investigated. This model, which contains frustration as an essential ingredient, displays glassy behavior at high density or low temperature and exhibits two transitions: a percolation transition at a temperature $T_p$ with critical exponents of the ferromagnetic $s = 1/2$ state Potts model, and a second transition at a lower temperature $T_g$ in the same universality class of the Ising spin glass model.

## 1. INTRODUCTION

One of the features of many complex systems is the appearance of many minima in the free energy, which allows the system to exist in one of many different states for a given set of external parameters. One mechanism leading to a landscape with many minima is based on the concept of frustration. In Spin Glasses[1] (SG), frustration occurs when the ferromagnetic and antiferromagnetic interactions are distributed in such a way that not all the spins can satisfy all the interactions simultaneously. In glass-forming systems[2] without underlying crystalline order, frustration arises because the molecules are intrinsically unable to form close-packed configurations at low temperature or high density, while for systems with underlying crystalline order, frustration arises when the local arrangement of molecules kinetically prevents all the molecules from reaching the crystalline state. In this paper we will describe the fractal properties of a generic model, called Frustrated Percolation[3] (FP), which contains geometric frustration as essential ingredient and has been proposed as a model for glassy systems.[4,5]

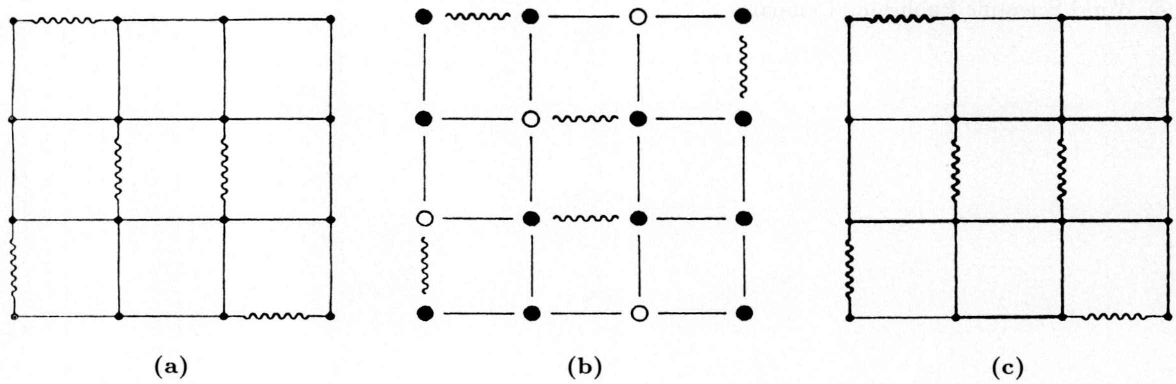

**Fig. 1** **(a)** First step to define the frustrated percolation model. The medium is prepared by assigning to each edge of the lattice a quenched "interaction" $\epsilon_{ij} = +1$ (straight line) or $-1$ (wavy line); **(b)** A configuration of particles (filled circles) and holes (empty circles) in the site-frustrated percolation model; **(c)** A bond configuration in the bond frustrated percolation model. Present bonds are represented by (bold) lines.

## 2. SITE-FRUSTRATED PERCOLATION

Let us consider a regular lattice and randomly assign to each edge of the lattice a quenched "interaction" $\epsilon_{ij} = \pm 1$ with equal probability. A loop is considered frustrated if the product of the signs along the loop is $-1$. We now randomly introduce particles on the vertices of this lattice, with the constraint that frustrated loops cannot be fully occupied [Figs. 1(a) and 1(b)]. That is, for a given number of particles, we assign in the partition function zero weight to all configurations of particles which contain at least one fully occupied frustrated loop, and the same constant nonzero weight to all other configurations. Frustration therefore induces defects or holes in the system, since a frustrated loop must necessarily contain at least one hole.

The dramatic effect of frustration on particle motion can be seen by allowing the particles to diffuse by exchanging their positions with a nearest-neighbor hole such that no frustrated loops become completely occupied. At low particle densities, motion is not inhibited by frustration because of the abundance of holes. However, at high densities, a given particle can diffuse through the system only by a large scale, cooperative rearrangement of many particles. Numerical results on a two-dimensional system show that at high density (or low temperature), the system exhibits complex dynamics with a two-step relaxation similar to that found in glass-forming systems.[6]

## 3. BOND-FRUSTRATED PERCOLATION

In the bond version of this model, we first prepare the lattice as before, by randomly assigning to each edge a quenched interaction $\epsilon_{ij} = \pm 1$ with equal probability. We then randomly introduce bonds onto the edges regardless of the sign $\epsilon_{ij}$, with the only constraint that a configuration of bonds which contains at least one fully bonded frustrated loop is not allowed [Fig. 1(c)]. That is, for a given number of bonds, we assign zero weight to all configurations of bonds which contain at least one fully occupied frustrated loop, and the same nonzero weight to all other configurations.

For a given set of interactions $\{\epsilon_{ij}\}$, the partition function is given by:

$$Z\{\epsilon_{ij}\} = \sum_C e^{\beta\mu_b n(C)}\,, \tag{1}$$

where $\mu_b$ is the chemical potential controlling the bond density, $\beta = 1/k_BT$, $T$ is the temperature, $k_B$ is Boltzmann's constant, $n(C)$ is the number of bonds in the bond configuration $C$, and the sum is performed over all allowed bond configurations $C$. In the following, we consider the chemical potential fixed and use the temperature as independent variable.

The partition function for the site version of this model is written in the same way provided that configurations of particles are considered instead of bonds.

## 4. STATIC PROPERTIES AND FRACTAL STRUCTURE OF THE FP MODEL

We have studied the static properties of the bond FP model via a renormalization group calculation[7] on a hierarchical lattice, using a Hamiltonian formalism which will be introduced in the next section. The results have been confirmed by Monte Carlo simulation on the square lattice for both bond and site FP[6,8] and we expect the result to be valid for any dimension. We find two critical points. The first critical point occurs at a percolation temperature $T_p$ and is in the same universality class of the ferromagnetic 1/2-state Potts model (as opposed to standard percolation[9] which is described by the ferromagnetic 1-state Potts model), while the second critical point at $T_g$ is in the same universality class as the Ising spin glass transition. Each critical point is characterized by a diverging length, associated with the quantities:

$$p_{ij} = p_{ij}^+ + p_{ij}^-\,, \tag{2}$$

and

$$g_{ij} = p_{ij}^+ - p_{ij}^-\,. \tag{3}$$

Here $p_{ij}^+$ ($p_{ij}^-$) is the probability that:

(1) Sites $i$ and $j$ are connected by at least one path of bonds.
(2) The product over all the signs $\epsilon_{mn}$ along the path connecting $i$ and $j$ is $+1$ ($-1$).

The length $\xi_p$ associated with the pair connectedness function $\overline{p_{ij}}$ (the bar represents the average over all possible interaction configurations $\{\epsilon_{ij}\}$), diverges at the percolation temperature $T_p$, with an exponent $\nu$, while the mean cluster size diverges with an exponent $\gamma$ identical to the corresponding critical exponents of the ferromagnetic 1/2-state Potts model. The fractal dimension of the percolating cluster is given by $D = 1/2(\gamma/\nu + d)$. As in random percolation, the fractal structure of the percolating cluster is made of links and blobs. However, the fractal dimension of the links or red bonds $D_R$ is not equal to $1/\nu$ but is a new independent exponent. For $d = 2$ we have[10] $\gamma = 3.25$, $\nu = 1.786$, $D = 1.91$, $D_R = 0.83$, $D_H = 1.787$, where $D_H$ is the fractal dimension of the hull of the percolating cluster.

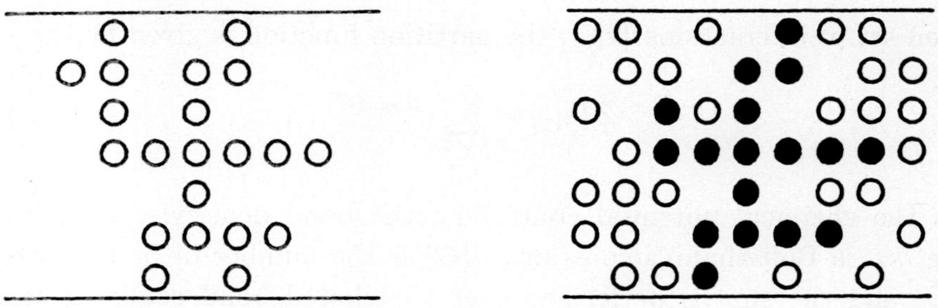

(a) (b)

**Fig. 2** (a) Schematic representation of a spanning cluster of particles; (b) A spanning cluster of quasifrozen particles (filled circles) in a sea of unfrozen particles (empty circles).

The second length $\xi$ associated with $\overline{g_{ij}^2}$ diverges at the lower transition temperature $T_g$. What is the physical meaning of the second diverging length? The second transition, like the quantum percolation transition,[11] occurs at a temperature lower than the usual percolation transition, due to the interference of paths with different phases. However, for a given pair of sites $i$ and $j$, this interference effect vanishes if there is a frozen path of bonds or sites between them. Since a path can never be completely frozen, the major contribution to $g_{ij}$ is due to those configurations in which $i$ and $j$ are connected by quasifrozen sites or bonds.[12] Therefore, $|g_{ij}|$ roughly coincides with the probability that $i$ and $j$ belong to the same quasifrozen cluster, and the length $\xi$ associated with $\overline{g_{ij}^2}$ roughly represents the linear dimension of these clusters. Since $\xi$ diverges at $T_g$, the quasifrozen clusters percolate at $T_g$ (Fig. 2).

The picture which emerges from the frustrated percolation model is that as the temperature is lowered towards the percolation transition $T_p$, the dynamics becomes slower but the bonds (or particles in the case of site FP) are still free to move and rearrange themselves. We interprete $T_p$ as the temperature below which the complex dynamics starts to appear, such as stretched exponential relaxations, due to the appearance of quasifrozen clusters. Below $T_p$, these clusters grow until they span the system at $T_g$, which is interpreted as the ideal glass transition (Fig. 2).

The presence of a percolation transition well above the glass transition has recently been discovered by Tomida and Egami[2] in a molecular-dynamics simulation of monoatomic liquids.

## 5. HAMILTONIAN FORMALISM FOR THE SITE AND BOND FRUSTRATED PERCOLATION

The bond FP model can be obtained using the following Hamiltonian formalism:

$$-H = \frac{q^J}{2} \sum_{\langle ij \rangle} \delta_{\sigma_i \sigma_j} (\epsilon_{ij} S_i S_j + 1) , \tag{4}$$

where each site contains both an Ising variable $S_i = \pm 1$ and an $s$-state Potts variable $\sigma_i = 1, \ldots, s$, with $s = q/2$. As before, the $\epsilon_{ij} = \pm 1$ are randomly distributed quenched variables. For a given set of interactions $\{\epsilon_{ij}\}$, the partition function for this Hamiltonian can be expressed as[13]:

$$Z\{\epsilon_{ij}\} = \sum_{\{S_i,\sigma_i\}} e^{-\beta H} = \sum_{C} e^{\beta\mu_b n(C)} q^{N(C)}\,, \tag{5}$$

where the second sum is over all the allowed bond configurations $C$ of the bond FP model, $n(C)$ and $N(C)$ are the number of bonds and the number of clusters, respectively, in the configuration $C$, $\beta\mu_b = \ln(e^{q\beta J} - 1)$.

When there is no frustration, Eq. (5) reproduces the Fortuin–Kasteleyn[14] cluster formulation of the partition function of the $q$-state Potts model. By taking the limit $q \to 1$ and $q \to 2$ in Eq. (5), we obtain the partition function of the bond FP model [Eq. (1)], and of the spin glass model, respectively.

As for $q = 1$ also for $q = 2$ (spin glass) we find two transitions. The percolation transition at $T_p$ now is in the same universality class as random percolation.[7,15] The clusters which diverge at $T_p$ in the spin glass are analogous to the clusters which diverge at the critical temperature in the Ising model.[16] However, because of the frustration caused by the competing interactions, the correlations are not carried by these clusters as they are in the Ising model, and hence these clusters do not diverge at the thermodynamic transition in the spin glass.[17]

The second transition coincides with the spin glass transition $T_g(q = 2) = T_{SG}$. According to the argument given in Sec. 4, we expect a second percolation transition at the spin glass transition $T_{SG}$ — this time a percolating cluster of spins connected by quasifrozen bonds. Indeed recent numerical simulations seem to support this picture.[12,18]

The general Hamiltonian needed to describe the site-bond frustrated percolation problem is obtained by diluting Eq. (4) with lattice gas variable:

$$-H = \frac{qJ}{2} \sum_{ij} [\delta_{\sigma_i\sigma_j}(\epsilon_{ij} S_i S_j + 1) - 2] n_i n_j - \mu \sum_i n_i\,, \tag{6}$$

where $n_i = 1$ if site $i$ is occupied and $n_i = 0$ if site $i$ is empty, $\mu$ is the chemical potential for a particle and $J$ is related to the chemical potential for a bond $\mu_b$ through $\beta\mu_b = \ln(e^{q\beta J} - 1)$.

In the limit $q \to 1$, the Hamiltonian [Eq. (6)] reproduces site-bond frustrated percolation. As a particular case in the limit $\mu \to \infty$ ($\mu_b \to \infty$), we obtain the frustrated bond (site) percolation model.

## 6. CONCLUSIONS

In conclusion, we have introduced a model whose essential ingredient — frustration — results in a constraint in configuration space and eventual breaking of ergodicity. The model exhibits two transitions, the first occurring at a temperature $T_p$ and the second occurring at a lower temperature $T_g$. $T_p$ coincides with the temperature below which a cluster of sites connected by bonds first spans the system. As the temperature is lowered below $T_p$, portions of these clusters become quasifrozen even though the system is still in equilibrium, resulting in complex dynamics such as two-step relaxation behavior and an apparent breaking of ergodicity at short time scales. As the temperature is lowered further, we expect quasifrozen clusters of sites to percolate at $T_g$, resulting in diverging time scales and breaking of ergodicity. This lower transition temperature is thus interpreted as the ideal glass transition.

## ACKNOWLEDGMENTS

I wish to thank A. Angell, V. Cataudella, A. de Candia, F. di Liberto, S. C. Glotzer, M. Nicodemi, S. Scarpetta and U. Pezzella for useful discussions.

## REFERENCES

1. K. Binder and A. P. Young, *Rev. Mod. Phys.* **58**, 801 (1986).
2. T. Tomida and T. Egami, *Phys. Rev.* **B52**, 3290 (1995).
3. T. Coniglio, in *Correlation and Connectivity*, eds. H. E. Stanley and N. Ostrowsky (Kluwer Acad. Publ., 1990); A. Coniglio, F. di Liberto, G. Monroy and F. Peruggi, *Phys. Rev.* **B44**, 12605 (1991).
4. A. Coniglio, *Il Nuovo Cimento* **D16**, 1027 (1994).
5. S. G. Glotzer and A. Coniglio, *Computational Material Science* **215**, 1 (1995).
6. S. Scarpetta, Thesis, University of Naples (1995).
7. U. Pezzella and A. Coniglio, to be published.
8. A. de Candia, V. Cataudella and A. Coniglio, in preparation.
9. D. Stauffer and A. Aharony, *Introduction to Percolation Theory*, 2nd ed. (Taylor and Francis, London, 1985).
10. A. Coniglio, *Phys. Rev. Lett.* **62**, 3054 (1989).
11. Y. Meir, A. Aharony and A. B. Harris, *Europhys. Lett.* **10**, 275 (1989) and references therein.
12. P. H. Poole, N. Jan, S. C. Glotzer and A. Coniglio, in preparation.
13. A. Coniglio, *J. de Physique IV Colloque C1, suppl. J. de Physique II* **3**, 1 (1993); V. Cataudella, A. Coniglio, L. de Arcangelis and F. di Liberto, *Physica* **A92**, 167 (1993).
14. C. M. Fortuin and P. W. Kasteleyn, *Physica* (Utrecht) **57**, 536 (1972).
15. L. de Arcangelis, A. Coniglio and F. Peruggi, *Europhys. Lett.* **14**, 515 (1991).
16. A. Coniglio and W. Klein *J. Phys.* **A13**, 2775 (1980).
17. For a general cluster formulation in which new clusters are defined so that the difference between $T_p$ and $T_c$ in frustrated systems is systematically reduced, see V. Cataudella, G. Franzese, M. Nicodemi, A. Scala and A. Coniglio, *Phys. Rev. Lett.* **72**, 1541 (1994).
18. D. Stauffer, *J. Phys.* **A26**, L525 (1993). M. Stevens, M. Cleary and D. Stauffer, *Physica* **A208**, 1 (1994); T. S. Ray and N. Jan, *J. Phys. I* (France) **3**, 2125, 1993.

Fractals, Vol. 4, No. 3 (1996) 355–367


# LOCALIZATION OF FRACTONS, RANDOM WALKS AND LINEAR POLYMERS IN PERCOLATION

A. BUNDE*,†, J. DRÄGER*,‡, S. HAVLIN*,†, H. E. ROMAN* and St. RUSS§

*Institut für Theoretische Physik, Justus-Liebig-Universität Giessen, D-35392 Giessen, Germany

†Minerva Center and Department of Physics, Bar-Ilan University, Ramat Gan, Israel

‡I. Institut für Theoretische Physik, Universität Hamburg, D-20355 Hamburg, Germany

§Laboratoire de Physique de la Matière Condensée, Ecole Polytechnique, F-91128 Palaiseau, France

## Abstract

We review analytical and numerical results for the vibrational amplitudes of localized excitations, the probability distribution of random walks and the distribution of linear polymers (modeled by self-avoiding walks of $N$ steps) on percolation structures at criticality. Our numerical results show that the fluctuations of these quantities, at fixed shortest-path distance ("chemical length") $\ell$ from the center of localization, are considerably smaller than at fixed Euclidean distance $r$ from the center. Using this fact, we derive via convolutional integrals explicit expressions for the averaged functions in $r$-space, and show analytically and numerically that three different localization regimes occur. In the short-distance regime, remarkably, the averages show a universal spatial decay behavior, with the same exponent for both fractons and random walks, while in the asymptotic regime, the averages depend explicitly on the number of configurations considered.

## 1. INTRODUCTION

It is well known that in random fractal structures vibrational excitations can be localized, i.e., the mean vibrational amplitudes of the vibrational excitations, called fractons, decay with increasing distance $r$ from a localization center.[1–9] The knowledge of the localization behavior on fractals is relevant for a large number of both experimental and theoretical issues.[10–14] Closely related to the localization behavior of fractons[1,2,7–9,15] is the way how the mean probability densities of random walks and self-avoiding walks on random fractal structures decreases with increasing distance $r$ from the starting point.[1,2,15–21] The probability density of a random walker is important for all physical processes that involve diffusing particles, e.g., in chemical reactions between diffusing reactants,[15] while Self Avoiding Walks (SAW)[22] are generally accepted for modeling linear polymers in good solvents. Most of the work on linear polymers in disordered media has so far concentrated on the problem of how the mean end-to-end distance of a polymer chain depends on the number of monomers $N$ of the polymer.[23,24] From a microscopic point of view, more important is the configurational averaged distribution function, which is defined as the probability that a polymer of $N$ monomers has an end-to-end distance $r$. The mean end-to-end distance is only the first moment of the distribution and contains only part of the information.

In this paper, we review the results of the vibrational excitations and of the probability distributions of random walks and self-avoiding walks on random fractal structures that we obtained recently.[4,5,19,21] We find that the *asymptotic* regimes of the relevant localization functions describing the three localization phenomena are divided into two parts.[25–27] There exists a crossover distance $r_\times(N_{\rm av})$ that increases logarithmically with the number $N_{\rm av}$ of configurations taken into account in the averages $\langle\,\rangle$. In general, the localization functions decay, for large distances $r$, from the center of localization, according to $\exp[-{\rm const}\, r^{d_\Psi}]$. The localization exponent $d_\Psi$ differs below and above $r_\times(N_{\rm av})$. Below $r_\times(N_{\rm av})$, we have $d_\Psi = d_w/(d_w - 1)$ (random walks), $d_\Psi = 1$ (fractons) and $d_\Psi = 1/(1 - \nu_r)$ (polymers), while above $r_\times(N_{\rm av})$, $d_\Psi = d_{\rm min} d_w/(d_w - d_{\rm min})$ (random walks), $d_\Psi = d_{\rm min}$ (fractons) and $d_\Psi = d_{\rm min}/(1 - \nu_r d_{\rm min})$ (polymers). Here, $d_w$ describes how the root-mean-square displacement $\langle R(t)\rangle \sim t^{1/d_w}$ of a random walker scales with time $t$ and $d_{\rm min}$ is the fractal dimension of the shortest path between two points at Euclidean distance $r$ on the fractal $\langle \ell(r)\rangle \sim r^{d_{\rm min}}$. For SAW's of $N$ steps, the exponent $\nu_r$ describes how the mean end-to-end distance $\langle r(N)\rangle$ of the linear chains scales with $N$, $\langle r(N)\rangle \sim N^{\nu_r}$.

In the short-distance regime, random walks and fractons behave in the same way and decay with the same exponent $g$, which is a purely structural quantity, describing the distribution of the minimum path distance $\ell$ between two points at fixed Euclidean distance $r$.[26] Since random walks and vibrational excitations require a slightly different treatment than polymer chains, we divide this paper into 2 parts. We start with a discussion of fractons and random walks.

## 2. FRACTONS AND RANDOM WALKS

To treat both localization phenomena simultaneously, we introduce a function $\psi_i(r)$, which stands either for the amplitude $|u_i(r, \omega)|$ of the displacement of a vibrating particle on a site $i$ at distance $r$ from the localization center, or for the probability $p_i(r, t)$ to find a random walker (after $t$ time steps) on a site $i$ at distance $r$ from its starting point on a given configuration of the random fractal substrate.

Our basic assumption is that the shortest-path distance ("chemical length") $\ell$ rather than the Euclidean distance $r$ between two points on the fractal is the relevant physical length in the problem, such that the fluctuations of the $\psi_i(\ell)$ on sites $i$ at fixed chemical distance $\ell$ from the localization center are small compared with the fluctuations of the $\psi_i(r)$ at fixed Euclidean distance $r$ from the localization center, for both the same and different configurations. According to Refs. 5 and 19, we can approximately neglect the fluctuations in $\ell$-space and write:

$$\psi_i(\ell) \cong \Psi(\ell) \sim \exp[-(\ell/\xi_\ell)^v]\,. \tag{1}$$

For fractons, we have $\xi_\ell \equiv \xi_\ell(\omega) \sim \omega^{-d_w/2d_{\min}}$ and $v = 1$. For random walks, $\xi_\ell$ is proportional to the mean chemical distance $\langle \ell(t) \rangle \sim t^{d_{\min}/d_w}$ and $v = d_w/(d_w - d_{\min})$. Assumption (1) is trivial for random walks on linear fractal structures ($v = 2$) or for localized vibrations on a linear fractal with a small impurity mass located at the origin ($v = 1$). Figure 1 illustrates that the assumption is also reasonable for random walks on percolation structures. In the figure we show the distribution of the $p_i(\ell, t)$ and $p_i(r, t)$, for fixed time $t = 1000$, $\ell = 150$ and $r = 70$, averaged over $5 \times 10^3$ configurations. The figure shows clearly that the distribution of the $p_i$ for fixed $\ell$ is narrow compared with the logarithmically broad distribution of the $p_i$ for fixed $r$. Indeed, from assumption (1) one can derive the functional form of the histogram for the $p_i$ in $r$-space (full line in the figure), which fully coincides with the result from the computer simulations.[15] Qualitatively similar results were obtained for the fracton amplitudes in percolation structures.[5]

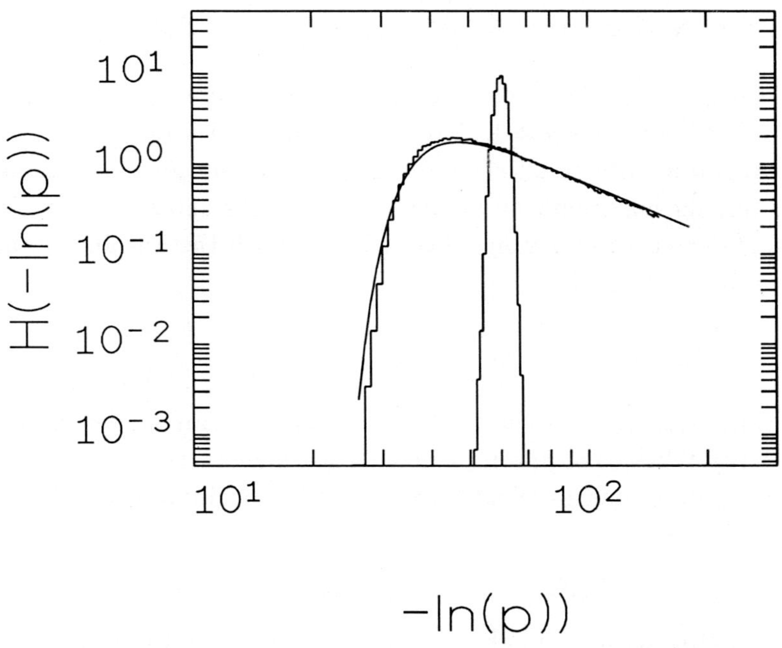


**Fig. 1** Plot of the histogram $H(-\ln p)$ vs. $-\ln(p)$ for $r = 70$ and $t = 1000$ (broad curve) for two dimensional percolation clusters at criticality. The smooth line represents the theoretical prediction.[15] The narrow curve represents the histogram in $\ell$-space, for $\ell = 150$ and $t = 1000$. The large fluctuations in $r$-space are in marked contrast to the very narrow fluctuations in $\ell$ space. For obtaining the histograms, $p_i(\ell, t)$ and $p_i(r, t)$ were enumerated on $5 \cdot 10^3$ percolation clusters.

To obtain the mean localization functions, one first averages $\psi_i(r)$ over all $N_r$ sites $i$ at distance $r$ from the localization center for a single configuration of the fractal substrate, according to Ref. 15:

$$\Psi(r) = \frac{1}{N_r}\sum_{i=1}^{N_r} \psi_i(r)\,. \tag{2}$$

Assuming that among the $N_r$ sites, $N_\ell(r)$ sites are at chemical distance $\ell$ from the center, and employing Eq. (1), we can write[25]:

$$\Psi(r) = \frac{1}{N_r}\sum_{\ell=\ell_{\min}(r)}^{\infty} N_\ell(r)\Psi(\ell)\,,$$

where $\ell_{\min}(r)$ is the shortest chemical length on the fractal reaching the distance $r$ from the center. Averaging over $N_{\rm av}$ configurations and replacing the sum by an integral, yields the localization function of interest,

$$\langle\Psi(r)\rangle_{N_{\rm av}} = \int_{\ell_{\min}(r,N_{\rm av})}^{\infty} \phi(\ell\,|\,r;\,N_{\rm av})\Psi(\ell)d\ell\,, \tag{3}$$

where $\phi(\ell\,|\,r;\,N_{\rm av})$ is the average of $N_\ell(r)/N_r$ over $N_{\rm av}$ configurations for a range $d\ell$ near $\ell$, $\ell_{\min}(r,\,N_{\rm av})$ is defined as the shortest chemical length on the fractal, among the $N_{\rm av}$ configurations, reaching the distance $r$ from the center, and $\phi(\ell\,|\,r;\,N_{\rm av}) = 0$ for $\ell < \ell_{\min}(r,\,N_{\rm av})$.

Equation (3) reveals that averaging the localization function involves both a simple arithmetic average $(\phi(\ell\,|\,r;\,N_{\rm av}))$ that is independent of $N_{\rm av}$ for large $N_{\rm av}$ and a minimization procedure $(\ell_{\min}(r,\,N_{\rm av}))$. Naturally, the minimum value $\ell_{\min}(r,\,N_{\rm av})$ depends strongly on $N_{\rm av}$, and this causes, as we shall see below, the breakdown of self-averaging and scaling at large distances $r$. The average $\phi(\ell\,|\,r;\,N_{\rm av})$ can be approximately written as[4] (see also[1,15,21,28,29]):

$$\phi(\ell\,|\,r;\,N_{\rm av}) \equiv \phi(\ell\,|\,r) = \frac{C_1}{\ell}\left(\frac{r}{\ell^{1/d_{\min}}}\right)^g \exp\left[-C_2\left(\frac{r}{\ell^{1/d_{\min}}}\right)^{\tilde\delta}\right], \tag{4}$$

with $\tilde\delta = d_{\min}/(d_{\min}-1)$. For percolation clusters, where $d_{\min} \cong 1.13$ in $d = 2$ and $d_{\min} \cong 1.37$ in $d = 3$,[15] one has $g \cong 1.35$ in $d = 2$ and $g \cong 1.5$ in $d = 3$,[21] for not too small $r/\ell^{1/d_{\min}}$. For asymptotically small $r/\ell^{1/d_{\min}} \ll 1$, the values for $g$ are slightly lower.[29] For fixed $r$, $\phi(\ell\,|\,r)$ has its maximum at $\ell_{\max}(r) \cong r^{d_{\min}}$. As shown in Fig. 2(a), the cut-off value $\ell_{\min}(r,\,N_{\rm av})$ decreases monotonically with $N_{\rm av}$ until the absolute minimum $\ell_{\min} \equiv r$ is reached, according to Ref. 25:

$$\ell_{\min}(r,\,N_{\rm av}) = \begin{cases} r, & r \ll r_c(N_{\rm av})\,, \\ \alpha_{\min}(N_{\rm av})r^{d_{\min}}, & r \gg r_c(N_{\rm av})\,. \end{cases} \tag{5}$$

The crossover value $r_c(N_{\rm av})$ and the prefactor $\alpha_{\min}(N_{\rm av})$ can be determined analytically: since the probability $W(N_{\rm av})$ of finding a percolation structure at the critical concentration $p_c$ with $\ell = r = r_c$ in a square- or s.c. lattice with coordination number $z$ is $W(N_{\rm av}) = 1/N_{\rm av} = z \cdot p_c^{r_c}$, we have:

$$r_c(N_{\rm av}) = (\ln z + \ln N_{\rm av})/\ln(1/p_c)\,. \tag{6}$$

The unknown prefactor $\alpha_{\min}(N_{\rm av})$ in Eq. (5) can be determined from the scaling assumption, $\ell_{\min}(r,\,N_{\rm av}) = r_c g(r/r_c)$. In order to satisfy Eq. (5), we must require $g(x) = x$ for $x \ll 1$ and $g(x) = g_1 x^{d_{\min}}$ for $x \gg 1$. This yields:

$$\alpha_{\min}(N_{\rm av}) = g_1 r_c^{1-d_{\min}}\,. \tag{7}$$

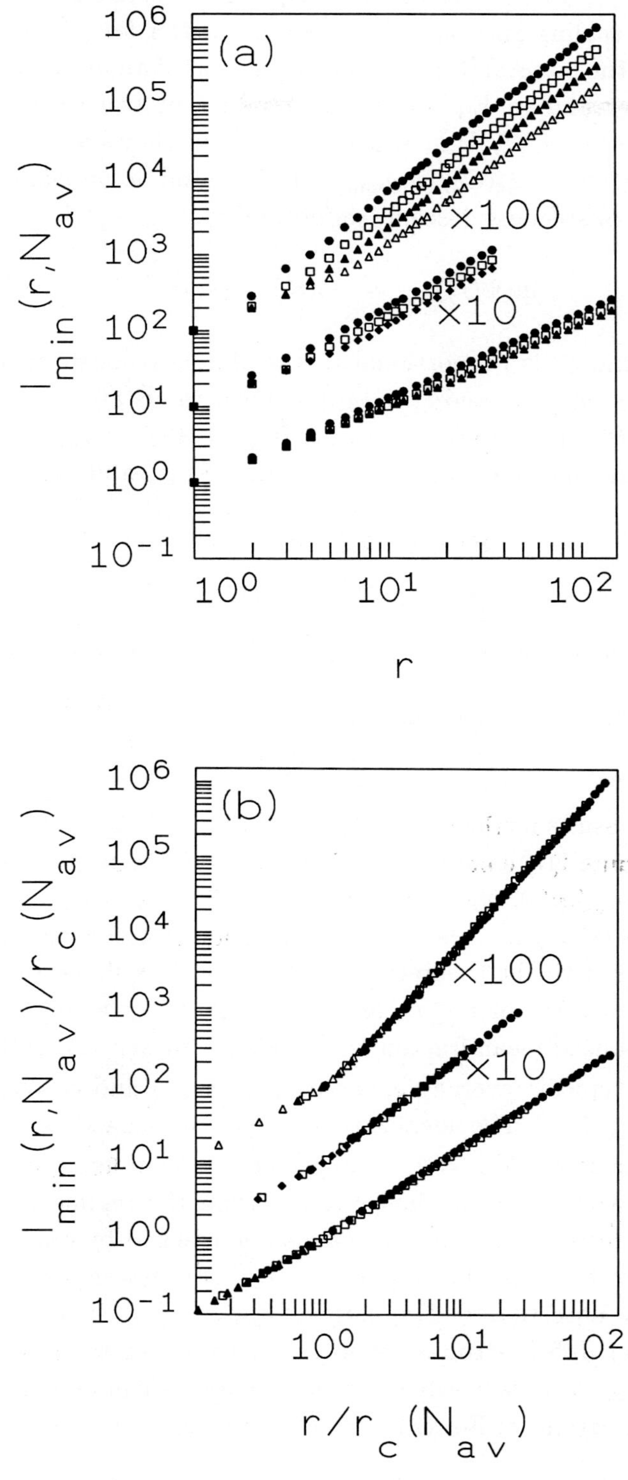


**Fig. 2** **(a)** The minimum distance $\ell_{\min}(r, N_{\rm av})$ vs. $r$ for: RW-structures on the s.c.-lattice [$N_{\rm av} = 1$ (full circle), 5 (square), 50 (full triangle) and 10 000 (triangle)] (upper set of points, shifted by a factor 100 for clarity); bond percolation on the simple cubic lattice [$N_{\rm av} = 1$ (full circle), 10 (square) and 1000 (full diamond)] (intermediate set of points, shifted by a factor 10 for clarity); and site percolation on the square lattice [$N_{\rm av} = 1$ (full circle), 100 (square) and 250 000 (full triangle) (lower set of points)]. For $N_{\rm av} < 10^4$, averages have been performed over typically 100 sets of $N_{\rm av}$ configurations. **(b)** Scale plot of $\ell_{\min}/r_c$ vs. $r/r_c$ for the same structures and the same $N_{\rm av}$ values shown in **(a)**. Here, the same shifts as in **(a)** have been used (after Ref. 25).

Figure 2(b) shows $\ell_{\min}/r_c$ vs. $r/r_c$ for the same $N_{\rm av}$ values as in Fig. 2(a). The data collapse supports strongly the scaling ansatz, and shows that $g_1 \cong 1$ for percolation.

Using Eqs. (4)–(7), the integral [Eq. (3)] can be calculated analytically. For $\ell_{\min}(r, N_{\rm av}) < \ell < \ell_{\max}(r)$, the integrand in Eq. (3) is the product of two exponential functions $\Psi \cdot \phi \sim \exp[-C_2(r/\ell^{1/d_{\min}})^{\tilde{\delta}} - (\ell/\xi_\ell)^v] \equiv \exp[-\eta(\ell)]$, which shows a steep maximum at $\ell^* = \xi_r^{d_{\min}}(r/\xi_r)^{u/v}$, with $\xi_r^{d_{\min}} = \xi_\ell(C_2/(v(d_{\min}-1)))^{1/v}$ and $u = vd_{\min}/(1+v(d_{\min}-1))$. Applying the method of steepest descent, we obtain[5,19]:

$$\ln\langle\Psi(r)\rangle_{N_{\rm av}} \sim -\eta(\ell^*) \sim -\left(\frac{r}{\xi_r}\right)^u . \tag{8}$$

For fractons, $u = 1$ and $\xi_r$ is proportional to the standard localization length $\lambda(\omega)$. For random walks, $u = d_w/(d_w - 1)$ and $\xi_r$ is proportional to $\langle R(t)\rangle$.

By definition, Eq. (8) holds only for $\ell_{\min}(r, N_{\rm av}) < \ell^* < \ell_{\max}(r)$, and this restriction determines the $r$-regime $r_1 < r < r_\times(N_{\rm av})$ where Eq. (8) is valid. We find:

$$r_1 = \xi_r[(g + d_{\min})/C_2\tilde{\delta}] , \tag{9}$$

$$r_\times(N_{\rm av}) \cong \xi_r r_c^{1/u}(N_{\rm av}) . \tag{10}$$

For $r > r_\times(N_{\rm av})$, the integrand in Eq. (3) is peaked sharply at $\ell = \ell_{\min}(r, N_{\rm av})$, and

$$\ln\langle\Psi(r)\rangle_{N_{\rm av}} \sim -(\ell_{\min}(r, N_{\rm av})/\xi_\ell)^v \sim -r_c(N_{\rm av})^{v(1-d_{\min})}(r/\xi_r)^{vd_{\min}} , \quad r > r_\times(N_{\rm av}) . \tag{11}$$

Figure 3 shows our results for the asymptotic regime $r > r_1$ for site percolation clusters on the square lattice. Figure 3(a) compares the logarithm of the normalized mean probability density $(-\ln[\langle p(r, t)\rangle_{N_{\rm av}}/\langle p(0, t)\rangle_{N_{\rm av}}])$ of random walks for several $N_{\rm av}$ values with the theoretical predictions (straight lines). The $N_{\rm av}$ dependent crossover $r_\times(N_{\rm av})$ of the data is clearly seen and agrees with the theoretical predictions. The slopes of the lines correspond to our predictions $d_\psi = 1.53$, for $r < r_\times$ and $d_\psi = 1.86$, for $r > r_\times$. Figure 3(b) shows the logarithm of the normalized mean fracton amplitude $(-\ln\langle|u(r, \omega)|/|u(0, \omega)\rangle)$ as a function of $r/\xi_\ell^{1/d_{\min}}$, for the "typical" average introduced in Ref. 5 and corresponding to the case $N_{\rm av} = 1$ in Eqs. (5)–(11).[25] The localization length in $\ell$-space $\xi_\ell \sim 29$ was also found in Ref. 5. The crossover $r_\times(N_{\rm av} = 1)$ is expected to occur for $r_\times(1)/\xi_\ell^{1/d_{\min}} \cong 8.7$ and separates the regime with $d_\psi = 1$, when $r < r_\times$, from the regime with $d_\psi = d_{\min}$, when $r > r_\times$. However, the latter regime $r > r_\times$ cannot be reached by our present data.

Next we discuss the short-distance regime $r < r_1$.[26] To determine the integral in Eq. (3) in an efficient way we note that $r < r_1$ implies $\ell_{\max}(r) < \xi_\ell \simeq r_1^{d_{\min}}$. Hence for $\ell < \xi_\ell$, the behavior of $\phi(\ell\,|\,r)$ differs strongly from the behavior of $\Psi(\ell)$: while $\phi(\ell\,|\,r)$ shows a steep maximum at $\ell_{\max}$, $\Psi(\ell)$ is nearly constant. Hence we can assume that to a very good approximation, the integrand of Eq. (3) can be written as,

$$\phi(\ell\,|\,r)\Psi(\ell) \cong \begin{cases} \phi(\ell\,|\,r)\langle\Psi(0)\rangle , & \ell \le \xi_0 , \\ 0 , & \ell > \xi_0 . \end{cases} \tag{12}$$

The cutoff-length $\xi_0$ is determined by the condition:

$$\int_0^\infty \Psi(\ell)^\tau \ell^{d_\ell - 1} d\ell = \int_0^{\xi_0} \langle\Psi(0)\rangle^\tau \ell^{d_\ell - 1} d\ell , \tag{13}$$

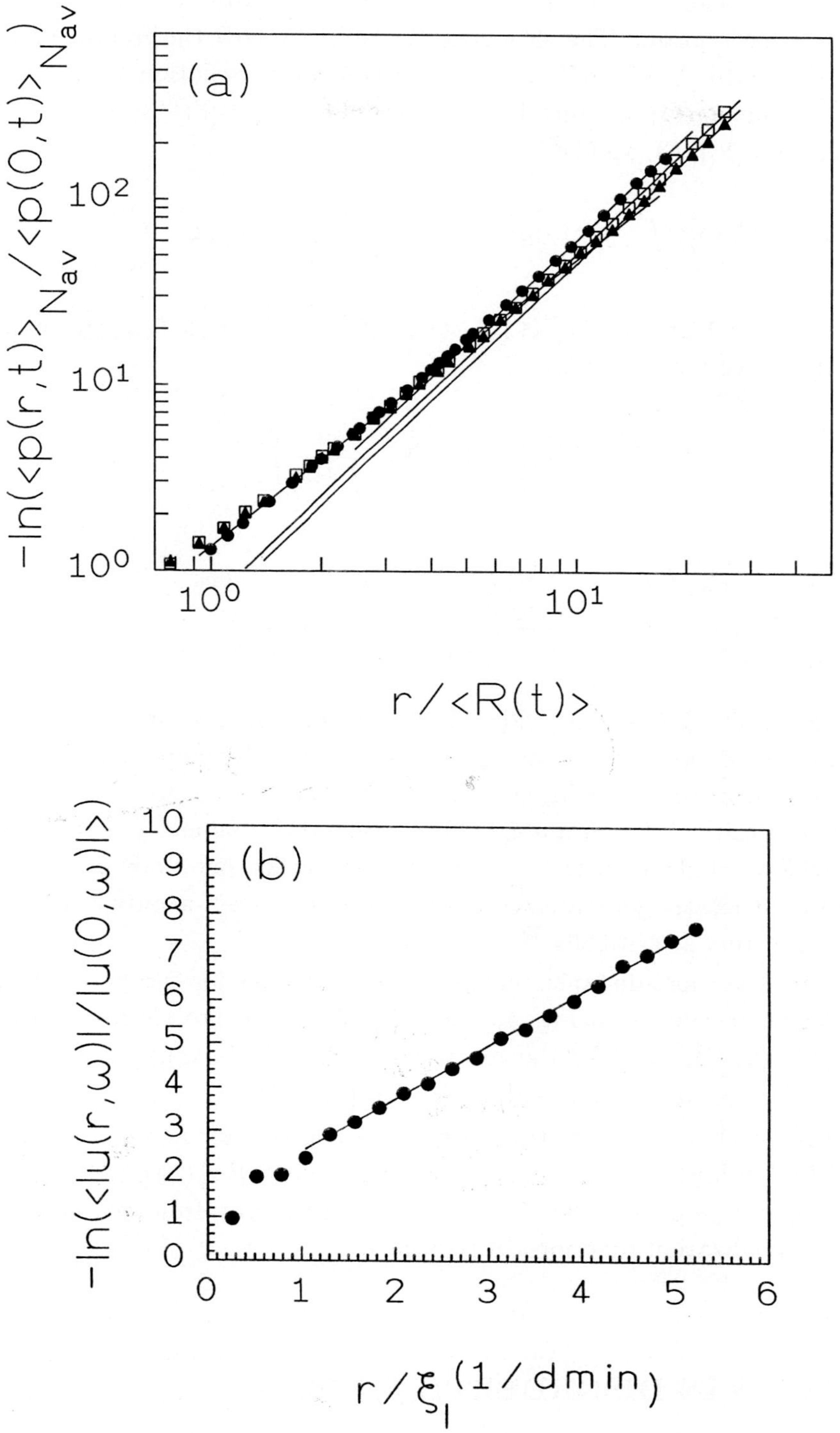


**Fig. 3** **(a)** Plot of $-\ln(\langle p(r,t)\rangle_{N_{av}}/\langle p(0,t)\rangle_{N_{av}})$ for random walks vs. $r/\langle R(t)\rangle$ for site percolation clusters on the square lattice [$t = 400$, $N_{av} = 5$ (square) and 50 (full triangle)] compared with the typical probability density (full circle) corresponding to the case $N_{av} = 1$ (see Ref. 25); $\langle R(t)\rangle$ is the r.m.s. displacement, and the straight lines have the theoretical slopes $d_\psi = 1.53$, for $r < r_\times$ and $d_\psi = 1.86$, for $r > r_\times$ (after Ref. 25). **(b)** Plot of $-\ln\langle|u(r,\omega)|/|u(0,\omega)|\rangle$ vs. $r/\xi_\ell^{1/d_{min}}$ for the "typical" average, corresponding to the case $N_{av} = 1$ (see Ref. 25), of a fracton ensemble of frequencies $\omega \cong 0.1f/M$, generated on site percolation clusters at criticality on a square lattice. The different clusters contained between 15 000 and 30 000 particles. Details of the calculations as well as the determination of $\xi_\ell \sim 29$ can be found in Ref. 5. The straight line is shown as a guide to the eye.

which for random walks ($\tau = 1$) is the normalization conditon, and for fractons ($\tau = 2$) reflects energy conservation. The exponent $d_\ell$ describes how the fractal mass scales with $\ell$, $\langle M(\ell)\rangle \sim \ell^{d_\ell}$, with $d_\ell \cong 1.678$ and $d_\ell \cong 1.885$ for percolation in $d = 2$ and $d = 3$, respectively.[15] Inserting Eq. (1) into Eq. (13), we obtain $\xi_0 = \Gamma(\frac{d_\ell}{v}+1)^{1/d_\ell}\tau^{-1/v}\xi_\ell$. Substituting Eq. (12) into Eq. (3) yields[26]:

$$\langle\Psi(r)\rangle \cong \langle\Psi(0)\rangle \int_0^{\xi_0} \phi(\ell\,|\,r)d\ell = \langle\Psi(0)\rangle\left(\int_0^{\infty} \phi(\ell\,|\,r)d\ell - \int_{\xi_0}^{\infty} \phi(\ell\,|\,r)d\ell\right). \tag{14}$$

Since $\int_0^\infty \phi(\ell\,|\,r)d\ell = 1$ and $\phi(\ell\,|\,r) \cong (C_1/\ell)(r/\ell^{1/d_{\min}})^g$ for $\ell \gg \ell_{\max}$, we finally obtain in the short distance regime ($r < r_1$):

$$\frac{\langle\Psi(r)\rangle}{\langle\Psi(0)\rangle} = 1 - A\left(\frac{r}{\xi_\ell^{1/d_{\min}}}\right)^g, \tag{15a}$$

with

$$A = \frac{C_1 d_{\min}}{g}\left(\frac{\tau^{1/v}}{\Gamma\left(\frac{d_\ell}{v}+1\right)^{1/d_\ell}}\right)^{g/d_{\min}}. \tag{15b}$$

According to Eq. (15), the decay of $\langle\Psi(r)\rangle$ is solely characterized by the substrate geometry represented by the structural exponent $g$ in Eq. (4).[26] This exponent depends explicitly on the dimension $d$ of the embedding space. It can be shown, however, that Eq. (15a) is valid only below a critical dimension $d_c$, above which the exponent $g$ is replaced by another exponent $g_c$ which is independent of $d$. For random walks, one finds $g_c = d_w$, with $d_c = 6$ on linear fractal structures generated by random walks and self-avoiding walks, and $d_c = 10$ on percolation clusters at criticality.[30]

The results for the short-distance regime $r < r_1$, obtained for site percolation clusters on the square lattice, are shown in Fig. 4, where (a) the mean probability density of random walks $\langle p(r,\,t)\rangle/\langle p(0,\,t)\rangle$, and (b) the mean amplitudes $\langle|u(r,\,\omega)|\rangle/\langle|u(0,\,\omega)|\rangle$ of localized vibrational excitations are plotted as functions of $(r/\xi_\ell^{1/d_{\min}})^g$.

The figures show that our predictions are in full *quantitative* agreement with the numerical results. The full lines represent our theoretical results, Eq. (15), with no fit parameters involved. The values of $\xi_\ell$ have been determined numerically from the decay of the corresponding mean localization functions in $\ell$-space.

## 3. POLYMERS IN DISORDERED MEDIA

Next we study the probability distributions of linear polymers modeled by $N$-step self-avoiding random walks in disordered structures, again both in $r$- and in $\ell$-space. In our calculations, SAW's of $N$ steps ($N$ monomers) are generated by the enrichment method (for details see Ref. 27), and as in the foregoing section, the disordered structure is modeled by the incipient percolation cluster on a square lattice.

Due to the self-avoiding constraint, however, SAW's cannot explore the whole percolation cluster but only the percolation backbone. Thus, in contrast to the definitions in Sec. 2, the distribution functions studied here are normalized on the percolation backbone instead

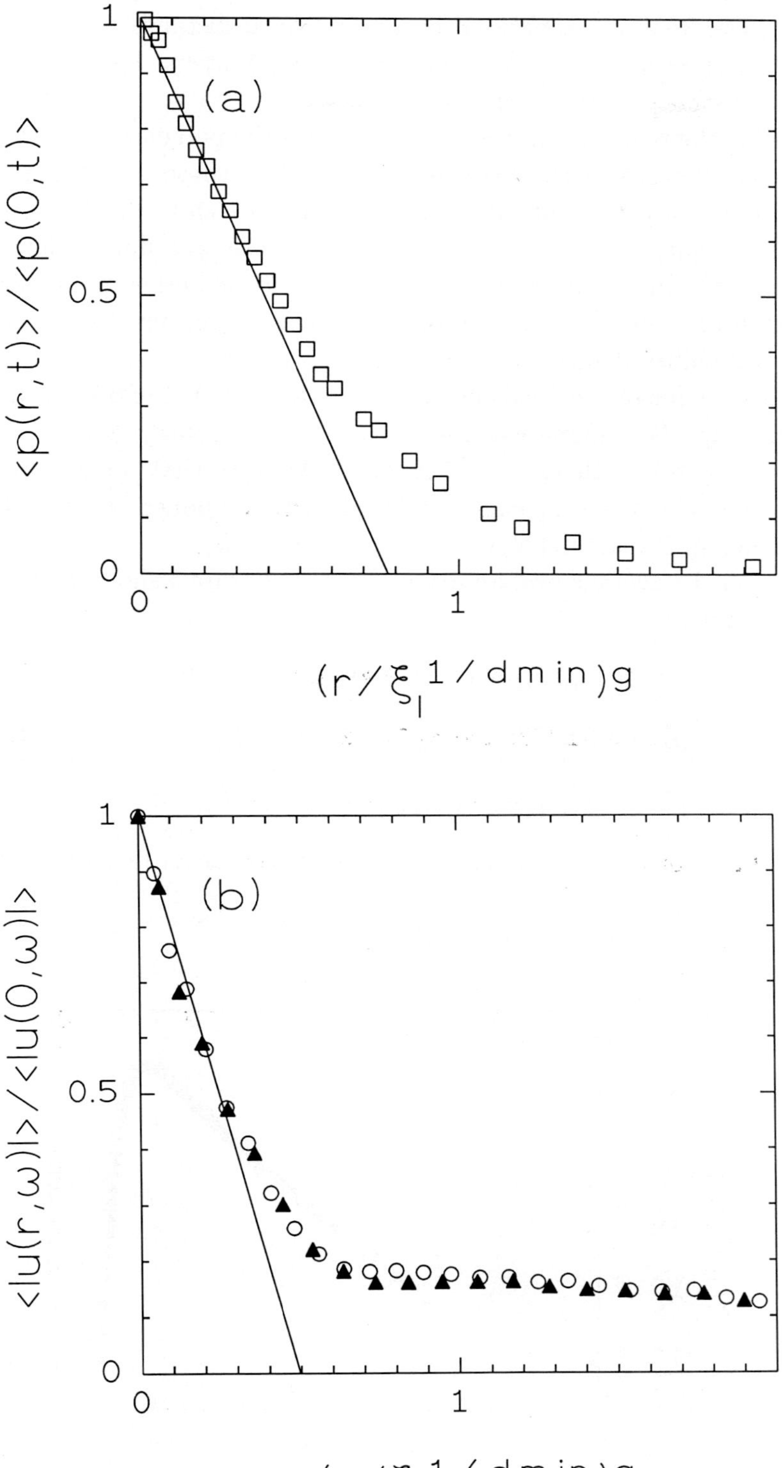


**Fig. 4** **(a)** The normalized mean probability density $\langle p(r,t)\rangle/\langle p(0,t)\rangle$ of random walks vs. $(r/\xi_\ell^{1/d_{\min}})^g$ for site percolation clusters on the square lattice ($g \cong 1.35$, $t = 10\,000$, $\xi_\ell \cong 40$, $N_{\rm av} = 100$). **(b)** The normalized mean amplitudes $\langle|u(r,\omega)|\rangle/\langle|u(0,\omega)|\rangle$ for fractons vs. $(r/\xi_\ell^{1/d_{\min}})^g$ for site percolation clusters on the square lattice ($g \cong 1.35$, $N_{\rm av} \cong 50$) for two frequencies: $\omega^2 \cong 0.005 f/M$ ($\xi_\ell \cong 21$) (circles) and $\omega^2 \cong 0.009 f/M$ ($\xi_\ell \cong 17$) (full triangles). The full lines in the plots represent the theoretical predictions, Eq. (15), without any fit parameters (after Ref. 26).

of on the entire cluster. In $\ell$-space, we evaluate the probability $P(\ell, N)$ that after $N$-steps, the endpoints of the SAW are separated by the chemical distance $\ell$. Correspondingly, the quantity $P(r, N)$ represents the probability that after $N$-steps the endpoints of the SAW are at Euclidean distance $r$ from each other.

Since for SAW's, the relevant disordered substrate is the percolation backbone, we expect that the exponents describing the backbone rather than those describing the percolation structure will enter the probability distributions $P(\ell, N)$ and $P(r, N)$. In spite of these differences we find that, similar to the random walk case, the fluctuations in $\ell$-space are significantly smaller than the fluctuations in $r$-space and we derive, in analogy to the treatment of random walks and fractons, an analytical expression for the asymptotic behavior of the average probability density in $r$-space $\langle P(r, N)\rangle$.

First we discuss the mean chemical end-to-end distance $\langle \ell(N)\rangle$ of SAW's on $2d$ percolation clusters at criticality. Our numerical results (see Ref. 27) suggest that $\langle \ell(N)\rangle$ follows a power law $\langle \ell(N)\rangle \sim N^{\nu_\ell}$ with $\nu_\ell = 0.87 \pm 0.02$. We note that for $d \geq 6$, it is expected that $\nu_\ell = 1$.[27] Since $\ell \sim r^{d_{\min}}$, $\nu_\ell$ can be related to the exponent $\nu_r$ describing the mean end-to-end distance in $r$-space $\langle r(N)\rangle \sim N^{\nu_r}$ by $\nu_r = \nu_\ell / d_{\min}$.

Next we focus on the average probability $\langle P(\ell, N)\rangle$. Figure 5 shows $\ell\langle P(\ell, N)\rangle$ vs. $\ell/N^{\nu_\ell}$ and suggests that for $\ell/\xi_\ell > 0.28$,

$$\langle P(\ell, N)\rangle \sim \ell^{-1}(\ell/\xi_\ell)^{g'_\ell} \exp[-(\ell/\xi_\ell)^{\delta_\ell}], \tag{16a}$$

with $\delta_\ell = 1/(1-\nu_\ell)$, $g'_\ell = 3.0 \pm 0.2$, and $\xi_\ell \cong 1.27 N^{\nu_\ell}$. For $\ell/\xi_\ell < 0.28$, the data suggest that:

$$\langle P(\ell, N)\rangle \sim \ell^{-1}(\ell/\xi_\ell)^{g_\ell}, \tag{16b}$$

with $g_\ell = 2.5 \pm 0.2$. Notice that $\langle P(\ell, N)\rangle$ is normalized according to $\int d\ell \langle P(\ell, N)\rangle = 1$.

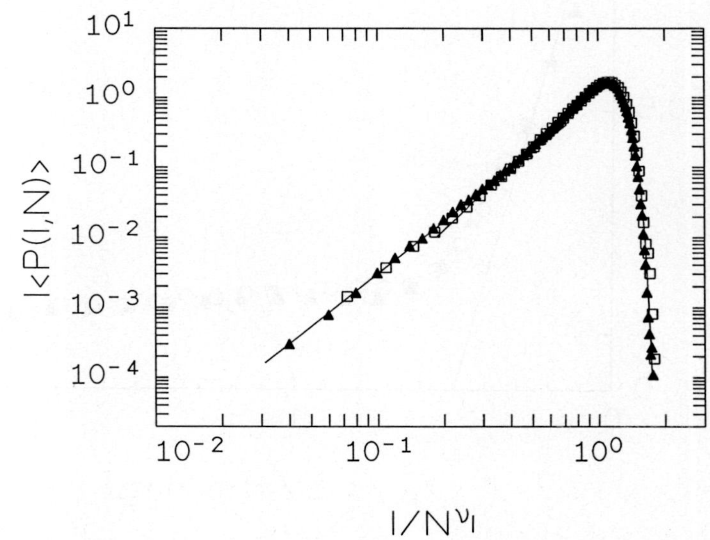


**Fig. 5** Plot of $\ell\langle P(\ell, N)\rangle$ vs. $\ell/N^{\nu_\ell}$ on a double logarithmic graph, for SAW's on two-dimensional percolation clusters at criticality grown on a square lattice up to 400 chemical shells. The circles represent the numerical data for $N = 100$ and the triangles for $N = 200$. The dashed line represents the functional form of Eq. (16a) with $\nu_\ell = 0.87 \pm 0.02$, $\delta_\ell = 1/(1-\nu_\ell)$, $g'_\ell = 3.0 \pm 0.2$, and $\xi_\ell = 1.27\ N^{\nu_\ell}$. For $\ell/N^{\nu_\ell} < 0.35$, a smaller value of the exponent $g_\ell$ represents the data better and we find $g_\ell = 2.5 \pm 0.2$ [Eq. (16b)] (after Ref. 27).

We have recently shown that for self-avoiding walks in $2d$ percolation clusters at criticality, the fluctuations in $\ell$-space are much smaller than in $r$-space.[27] According to this result, we can proceed in a similar way as in Sec. 2. We define the probability $p_i(r, N)$ that after $N$ steps the SAW has arrived at a backbone site $i$ at distance $r$ from the origin of the SAW, on a given configuration. Assuming that there are $N_r^B$ sites on the backbone at distance $r$ in the considered configuration, the probability $P(r, N)$ for that configuration is now given by [cf. Eq.(2)]:

$$P(r, N) = \sum_{i=1}^{N_r^B} p_i(r, N) . \tag{17}$$

Among the $N_r^B$ sites, there are $N_\ell^B(r)$ sites at chemical distance $\ell$ from the center, and we can write:

$$\begin{aligned} P(r, N) &= \sum_{\ell=\ell_{\text{min}}^B(r)}^{\infty} N_\ell^B(r) p_i(\ell, N) , \\ &= \sum_{\ell=\ell_{\text{min}}^B(r)}^{\infty} \frac{N_\ell^B(r)}{N_\ell^B} P(\ell, N) . \end{aligned} \tag{18}$$

In deriving the second equality, we have assumed that there are $N_\ell^B$ sites at chemical distance $\ell$ from the center and that $p_i(\ell, N) \simeq P(\ell, N)/N_\ell^B$. This is justified, since the fluctuations of the $p_i(\ell, N)$ are small. By definition, $\ell_{\text{min}}^B(r)$ is the length of the shortest chemical path between the center and any site $i$ on the backbone at fixed distance $r$ in the considered configuration. Assuming further that, similarly to the random-walk case, the fluctuations of $P(\ell, N)$ between different configurations are small, we can write $P(\ell, N) \simeq \langle P(\ell, N)\rangle$. Averaging over $N_{\text{av}}$ configurations and replacing the sum in Eq. (18) by an integral, yields the desired relation between $\langle P(\ell, N)\rangle$ and $\langle P(r, N)\rangle_{N_{\text{av}}}$,

$$\langle P(r, N)\rangle_{N_{\text{av}}} = \int_{\ell_{\text{min}}^B(r, N_{\text{av}})}^{\infty} \phi^B(r\,|\,\ell; N_{\text{av}}) P(\ell, N) d\ell , \tag{19}$$

where $\phi^B(r\,|\,\ell; N_{\text{av}}) \equiv \phi^B(r\,|\,\ell) \equiv N_\ell^B(r)/N_\ell^B$ is the probability that two sites on the backbone separated by chemical distance $\ell$ are at Euclidean distance $r$. We like to note that $\phi^B(r\,|\,\ell)$ is not fully analogous to $\phi(\ell\,|\,r)$ in Eq. (4). While $\phi^B(r\,|\,\ell)$ is normalized in $r$-space, $\int_0^\infty \phi^B(r\,|\,\ell)dr = 1$, $\phi(\ell\,|\,r)$ is normalized in $\ell$-space, $\int_0^\infty \phi(\ell\,|\,r)d\ell = 1$. Recently, it has been shown that $\phi^B(r\,|\,\ell)$ can be written in a form analogous to Eq. (4)[31]:

$$\phi^B(r\,|\,\ell) = \frac{C_1^B}{r} \left(\frac{r}{\ell^{1/d_{\text{min}}}}\right)^{g_B} \exp\left[-C_2^B \left(\frac{r}{\ell^{1/d_{\text{min}}}}\right)^{\tilde{\delta}}\right] , \tag{20}$$

with $\tilde{\delta} = d_{\text{min}}/(d_{\text{min}} - 1)$. For fixed $r$, $\phi^B(r\,|\,\ell)$ has its maximum at $\ell_{\text{max}}^B(r) \cong r^{d_{\text{min}}}$.

The lower integration limit in Eq. (19), $\ell_{\text{min}}^B = \ell_{\text{min}}^B(r, N_{\text{av}})$, is that of the backbone, which is larger than $\ell_{\text{min}}$ for the entire percolation cluster, but still obeys an equation similar to Eq. (5) with $\alpha_{\text{min}}(N_{\text{av}})$ replaced by $\alpha_{\text{min}}^B(N_{\text{av}}) \sim r_c^B(N_{\text{av}})^{1-d_{\text{min}}}$ and $r_c^B(N_{\text{av}}) \sim \ln(N_{\text{av}})$.

Equation (19) has the same functional form as Eq. (3) and can be solved in exactly the same way as described above for random walks and fractons. Following the procedure outlined in Sec. 2 [Eqs. (8)–(11)], we find:

$$\ln\langle P(r, N)\rangle_{N_{\text{av}}} \sim \begin{cases} -(r/\xi_r)^{\delta_r} , & r_1^B < r < r_\times^B(N_{\text{av}}) , \\ -r_c^B(N_{\text{av}})^{\delta_\ell(1-d_{\text{min}})}(r/\xi_r)^{\delta_\ell d_{\text{min}}} , & r > r_\times^B(N_{\text{av}}) , \end{cases} \tag{21}$$

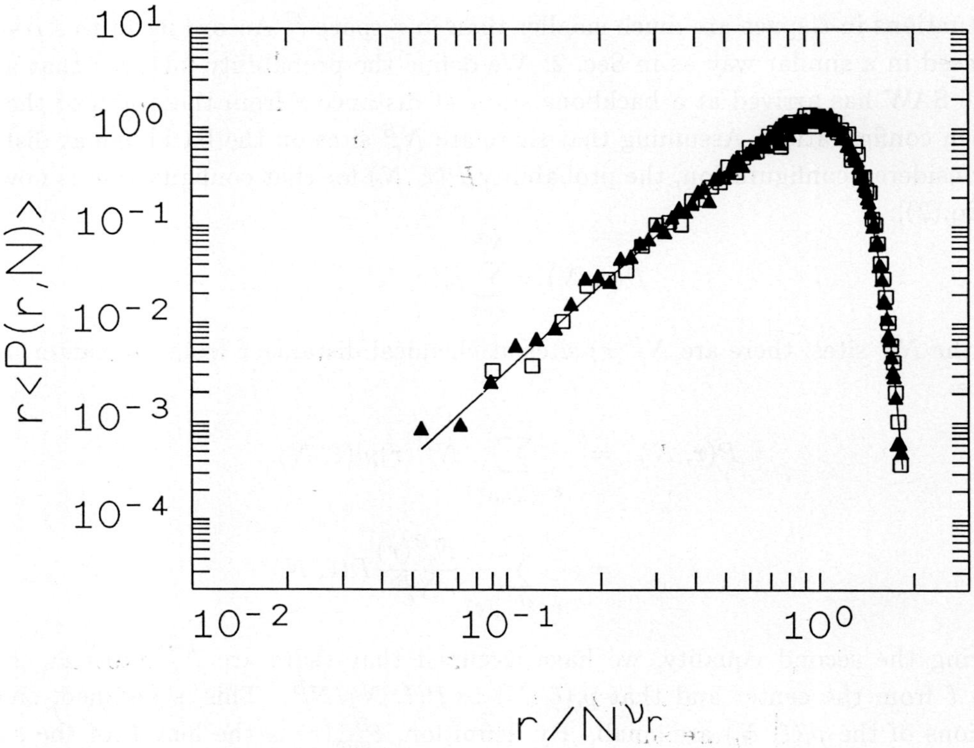


**Fig. 6** Plot of $r\langle P(r,\, N)\rangle$ vs. $r/N^{\nu_r}$ for the SAWs considered in Fig. 5, on a double logarithmic graph. The squares represent the numerical data for $N = 100$ and the triangles for $N = 200$. The line represents the analytical function, Eq. (22), which fits the data for all $r/N^{\nu_r}$, with $\nu_r = 0.76 \pm 0.02$, $\delta_r = 1/(1-\nu_r)$, $g_r = 2.9 \pm 0.2$, and $\xi_r = 1.04\ N^{\nu_r}$ (after Ref. 27).

with $\delta_r = 1/(1-\nu_r)$, $\xi_r^{d_{\min}} = \xi_\ell[C_2^B/(\delta_\ell(d_{\min}-1))]^{(1-\nu_\ell)}$, and the limits $r_1^B$ and $r_\times^B$ correspond to $r_1$ and $r_\times$ from Eqs. (9) and (10). Equation (21) represents only the exponential part of $\langle P(r,\, N)\rangle_{N_{\rm av}}$, which describes the asymptotic regime $r > r_1^B$. For $r < r_1^B$, we expect that $\langle P(r,\, N)\rangle_{N_{\rm av}}$ displays power-law dependence in the scaling variable $r/\xi_r \sim r/N^{\nu_r}$. Hence, in order to describe both the short-distance and the intermediate-distance regime, we infer the more general form:

$$\langle P(r,\, N)\rangle \sim r^{-1}(r/\xi_r)^{g_r}\exp[-(r/\xi_r)^{\delta_r}] \quad r < r_\times^B(N_{\rm av})\,. \tag{22}$$

Figure 6 shows $r\langle P(r,\, N)\rangle$ vs. $r/N^{\nu_r}$. The figure supports our scaling result [Eq. (22)] with $g_r = 2.9 \pm 0.2$. Since $\langle P(r = 1,\, N)\rangle = \langle P(\ell = 1,\, N)\rangle$, the exponents $g_\ell$ and $g_r$ describing the small $\ell$- and $r$-behavior in Eqs. (16b) and (22) can be related to each other by:

$$g_r = g_\ell d_{\min}\,. \tag{23}$$

Our numerical results yield $g_\ell = 2.5 \pm 0.2$ and $g_r = 2.9 \pm 0.2$, in good agreement with this prediction.

The exponent $\delta_\ell d_{\min} \cong 8.7$ for $r > r_\times^B(N_{\rm av})$ in Eq. (21) is much larger than the exponent $\delta_r \cong 4.3$ in Eq. (21).[27] This enhanced decay, however, cannot be seen in the figure, since $r_\times^B(N_{\rm av}) > r_\times(N_{\rm av}) \simeq 2\xi_r$, for $N_{\rm av} = 1$ and we can only reach $r$ values below $r \simeq 2\xi_r$ with our data.

## ACKNOWLEDGMENTS

We wish to thank the Deutsche Forschungsgemeinschaft and the Alexander von Humboldt-Stiftung for financial support.

## REFERENCES

1. S. Havlin and D. Ben-Avraham, *Adv. in Phys.* **36**, 695 (1987).
2. A. B. Harris and A. Aharony, *Europhys. Lett.* **4**, 1355 (1987); A. Aharony and A. B. Harris, *Physica* **A163**, 38 (1990); *ibid.*, *Physica* **A191**, 365 (1992).
3. P. de Vries, H. de Raedt and A. Lagendijk, *Phys. Rev. Lett.* **62**, 2515 (1989).
4. H. E. Roman, St. Russ and A. Bunde, *Phys. Rev. Lett.* **66**, 1643 (1991).
5. A. Bunde, H. E. Roman, St. Russ, A. Aharony and A. B. Harris, *Phys. Rev. Lett.* **69**, 3189 (1992); St. Russ, Ph.D. thesis, Hamburg (1992).
6. C. Lambert and G. D. Hughes, *Phys. Rev. Lett.* **66**, 1074 (1991).
7. Y. A. Levy and B. Souillard, *Europhys. Lett.* **4**, 233 (1987).
8. T. Nakayama, K. Yakubo and R. L. Orbach, *Rev. Mod. Phys.* **66**, 381 (1994).
9. S. Alexander and R. L. Orbach, *J. Phys. Lett.* **43**, L625 (1982).
10. Y. Tsujimi, E. Courtens, J. Pelous and R. Vacher, *Phys. Rev. Lett.* **60**, 2757 (1988).
11. M. Montagna, O. Pilla, G. Viliani, V. Mazzacurati, G. Ruoco and G. Signorelli, *Phys. Rev. Lett.* **65**, 1136 (1990).
12. E. Stoll, M. Kolb and E. Courtens, *Phys. Rev. Lett.* **68**, 2472 (1992).
13. S. Alexander, *Phys. Rev.* **B40**, 7953 (1989); S. Alexander, E. Courtens, R. Vacher, *Physica* **A195**, 286 (1993).
14. St. Russ and B. Sapoval, *Phys. Rev. Lett.* **73**, 1570 (1994).
15. A. Bunde and S. Havlin, eds., *Fractals and Disordered Systems*, 2nd ed. (Springer Verlag, Heidelberg, 1996).
16. J. P. Bouchaud and A. Georges, *Phys. Rep.* **195**, 127 (1990).
17. K. W. Kehr and R. Kutner, *Physica* **A110**, 535 (1982); S. Havlin et al., *J. Phys.* **A18**, L719 (1985).
18. B. O'Shaughnessy and I. Procaccia, *Phys. Rev. Lett.* **54**, 455 (1985).
19. A. Bunde, S. Havlin and H. E. Roman, *Phys. Rev.* **A42**, 6274 (1990).
20. E. Eisenberg, S. Havlin and G. H. Weiss, *Phys. Rev. Lett.* **72**, 2827 (1994).
21. A. Bunde and J. Dräger, *Phil. Mag.* **B71** 721 (1995).
22. P. G. de Gennes, *Scaling Concepts in Polymer Physics* (Cornell University Press, Ithaca, 1979).
23. H. Nakanishi, in *Annual Reviews of Computational Physics*, ed. by D. Stauffer (World Scientific, Singapore, 1994).
24. K. Barat and B. K. Chakrabarti, preprint (1995).
25. A. Bunde and J. Dräger, *Phys. Rev.* **E52**, 53 (1995).
26. J. Dräger, St. Russ and A. Bunde, *Europhys. Lett.* **31**, 425 (1995).
27. H. E. Roman, J. Dräger, A. Bunde, S. Havlin and D. Stauffer, *Phys. Rev.* **E52**, 6303 (1995).
28. U. A. Neumann and S. Havlin, *J. Stat. Phys.* **52**, 203 (1988).
29. H. E. Roman, *Phys. Rev.* **E51**, 5422 (1995).
30. S. Rabinovich, H. E. Roman, S. Havlin and A. Bunde, *Phys. Rev.* **E** (submitted); H. E. Roman, *Phys. Rev.* **E** (submitted).
31. M. Porto, A. Bunde, S. Havlin and H. E. Roman, preprint (1995).

Fractals, Vol. 4, No. 3 (1996) 369–375


# ANOMALOUS DIFFUSION IN THE KICKED HARPER SYSTEM

SATOSHI SAITÔ, YASUYUKI NOMURA* and YOSHI H. ICHIKAWA†
*School of Engineering, Nagoya University, Nagoya 464-01, Japan*
**Department of Electronics and Information Engineering, Fukui National College of Technology, Sabae 916, Japan*
*†College of Engineering, Chubu University, Kasugai 487, Japan*

**Abstract**

Stochastic property of the Hamiltonian dynamical systems exhibits anomalous behavior owing to intermittent processes between regular orbits and chaotic motion. In particular, the accelerator modes give rise to the anomalous enhancement of the diffusion process, which could be attributed to temporal fractal property. In order to explore the effect of the accelerator modes on the stochastic diffusion, the kicked Harper system is investigated in detail. It is found that the numerical observation of the diffusion coefficient shows good agreement with the analytical expression obtained by the characteristic function method, except in the domains where the accelerator modes exist. It is interesting to note that, unlike the case of the standard map, the multi-period, multi-step accelerator modes are responsible for much of the profound effect of the enhancement of diffusion even in the domain of larger stochastic parameters.

## 1. INTRODUCTION

Stochastic property of chaotic orbits is the main subject which explains the relation of the nonlinear dynamics and statistical mechanics. In the Hamiltonian chaotic system, even in the global stochastic system, there has been shown anomaly in the long-time behavior of the statistical properties, which is caused by the integral orbit called accelerator modes.

We anticipate that such anomalous behavior could be attributed to the temporal fractal property of the system.

A well-known model for characterizing this mechanism is the kicked rotator model, which has been well studied in the analysis of the Hamiltonian stochasticity. Meiss et al.[1] obtained the theoretical expression for the momentum diffusion coefficient by applying the Fourier path integral method. They carried out renormalization calculation for the principal terms of the characteristic function. Ichikawa et al.[2] found that the renormalized diffusion coefficient is valid even for the small stochastic parameter values except for the region where the accelerator modes exist.

Since the kicked Harper system[3] has periodic perturbation both in the momentum and the coordinate directions, it is anticipated that the system will exhibit interesting stochastic behavior.

## 2. FUNDAMENTAL PROPERTIES OF THE KICKED HARPER MAP

The kicked Harper map (KHM) is expressed by the discretized mapping:

$$\begin{aligned} p_{n+1} &= p_n + \frac{A}{2\pi}\sin 2\pi q_n \\ q_{n+1} &= q_n - \frac{L}{2\pi}\sin 2\pi p_{n+1}\,, \end{aligned} \qquad (\text{mod } 1) \tag{1}$$

where $n$ denotes the discrete time, and $q$ and $p$ denote the dimensionless generalized position and momentum, respectively. $A$ and $L$ are the positive stochastic parameters. KHM depends on these two parameters; however, we concentrate for the sake of simplicity in our study on the special case of $A = L$. In the case of $A = L$, KHM coincides with the map which describes the stochastic web.[4] The phase portrait at $A = L = 1.5$ is shown in Fig. 1.

Since this map possesses the periodicity in both $p$ and $q$ with period 1, it admits the existence of the accelerator modes. Stable fundamental accelerator mode exists when the condition:

$$\frac{A}{2\pi}\sin 2\pi q = m\,, \quad -\frac{L}{2\pi}\sin 2\pi p = l \quad (m,\ l : \text{integers}) \tag{2}$$

is satisfied. Simultaneously, the linear stability condition:

$$0 < AL\cos 2\pi q \cos 2\pi p < 4 \tag{3}$$

must hold. If the initial point is on the fundamental accelerator mode, the momentum at time $T$ is $p_T = p_0 + mT$. In the case of $A = L$, the stability condition for the fundamental accelrator mode is:

$$2\pi l' < A < \sqrt{2\pi^2(m^2+l^2) + 2\sqrt{\pi^4(m^2-l^2)^2+4}}\,, \tag{4}$$

where $\max\{|m|,\ |l|\} = l'$.

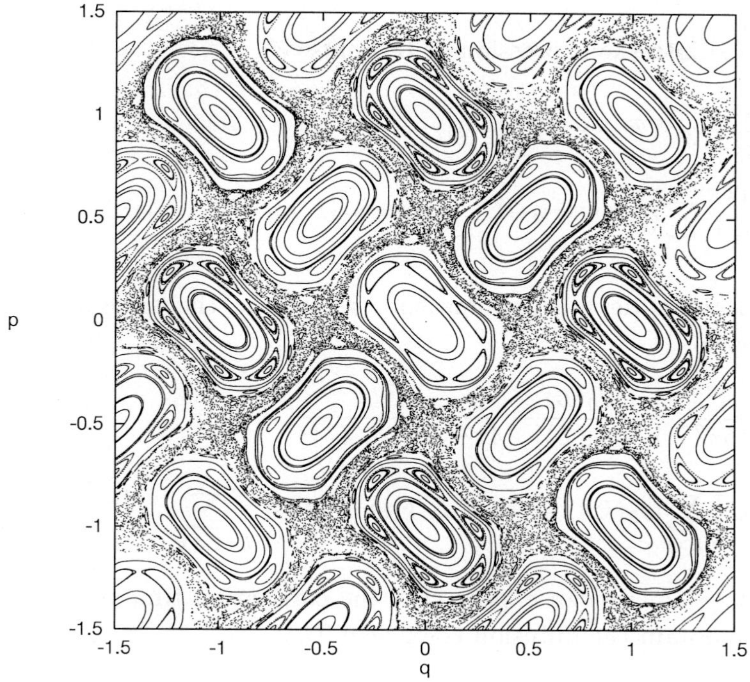


**Fig. 1** Phase portrait of the kicked Harper map at $A = L = 1.5$. The island structure has fractal property.

When the parameter values exceed the critical values, the last KAM curve is destroyed and the system shows the global stochasticity. In this situation, the system shows diffusive motion.

## 3. STATISTICAL PROPERTIES OF THE KICKED HARPER MAP

In this section, we investigate the diffusion process in the momentum space and analyze it by means of force-correlation.

We first define an acceleration of momentum (the force) in one step of time $\Delta p_i$:

$$\Delta p_i = p_{i+1} - p_i = \frac{A}{2\pi} \sin 2\pi q_i = -\frac{iA}{4\pi}\{e^{2\pi i q_i} - e^{-2\pi i q_i}\}. \tag{5}$$

Using $\Delta p_i$, we define a force-correlation $C_k$ by:

$$C_k \equiv \sum_i \langle \Delta p_i \Delta p_{i+k} \rangle , \tag{6}$$

where $\langle\ \rangle$ is an ensemble average of orbits. The diffusion process in the momentum space is characterized by the diffusion coefficient $D$,

$$D \equiv \left\langle \lim_{T\to\infty} \frac{1}{2T}(p_T - p_0)^2 \right\rangle = \frac{1}{2}C_0 + \lim_{T\to\infty} \sum_{j=1}^{T} \left(1 - \frac{j}{T}\right) C_j . \tag{7}$$

Inserting Eq. (5) into the expression of $D$ yields:

$$\begin{aligned} D = &-\frac{A^2}{32\pi^2}\langle\langle e^{4\pi i q_i} + e^{-4\pi i q_i} - 2\rangle\rangle \\ &- \frac{A^2}{16\pi^2}\sum_{j=1}^{\infty}\langle\langle e^{2\pi i(q_i+q_{i+j})} - e^{2\pi i(q_i-q_{i+j})} - e^{-2\pi i(q_i-q_{i+j})} + e^{-2\pi i(q_i+q_{i+j})}\rangle\rangle . \end{aligned} \tag{8}$$

Using the characteristic function method, we obtain the analytical expression of the diffusion coefficient:

$$\begin{aligned} D = &\frac{A^2}{16\pi^2}\cdot\frac{1+J_0(L)}{1-J_0(L)} - \frac{A^2}{4\pi^2\{1-J_0(L)\}^2} \\ &\times \left\{ \sum_{k\geq 1}[J_2(kA) - (-1)^k J_0(kA)]J_k^2(L) + \sum_{k=2,4,6,\ldots} \frac{J_1^2(kA)J_k^2(L)}{1-J_0(kA)} \right\} . \end{aligned} \tag{9}$$

In the limit $A \to \infty$, we can expect that the correlation time is short, i.e., $C_k = 0$ for $|k| > 0$. Thus, we obtain the diffusion coefficient $D_q$:

$$D_q = \frac{1}{2}C_0 = \frac{1}{2}\langle(\Delta p_0)^2\rangle = \frac{A^2}{8\pi^2}\langle \sin^2 2\pi q_0\rangle = \frac{A^2}{16\pi^2} \tag{10}$$

which is the value predicted by the quasi-linear theory.

In Eq. (9) for $A \leq L$, in the limit $L \to \infty D$, $D = D_q$. For $A > L$ in the limit of $A \to \infty$, however, the diffusion coefficient $D$ is reduced to:

$$D_q' = \frac{A^2}{16\pi^2}\cdot\frac{1+J_0(L)}{1-J_0(L)} . \tag{11}$$

Therefore, we take the quasi-linear diffusion coefficient as Eq. (11).

## 4. NUMERICAL EXPERIMENT

Now, let us compare our expression for the diffusion coefficient with the results of numerical observation.

Figure 2 illustrates the diffusion coefficient in the case of $A = L$. The horizontal axis stands for parameter $A$ and the vertical axis $D/D_q'$. The solid line is the theoretical value for $D$ given by Eq. (9) and the squares represent the numerical results. We observe that at larger values of $A$, the diffusion coefficient converges to the quasi-linear expression (11). The analytical result yields the oscillations about the quasi-linear term $D_q'$. These oscillations stand for the correction term of the correlation.

Overall agreement between the theoretical prediction and the numerical observation is extremely good in Fig. 2. We observe, however, an anomalous enhancement due to the fundamental accelerator mode. Furthermore, we note that derivation from the theoretical prediction occurs in the vicinity of $A = \pi$ and $A = 3\pi$ in Fig. 2. This anomaly is not directly related to the fundamental accelerator mode. In order to investigate this, we carry

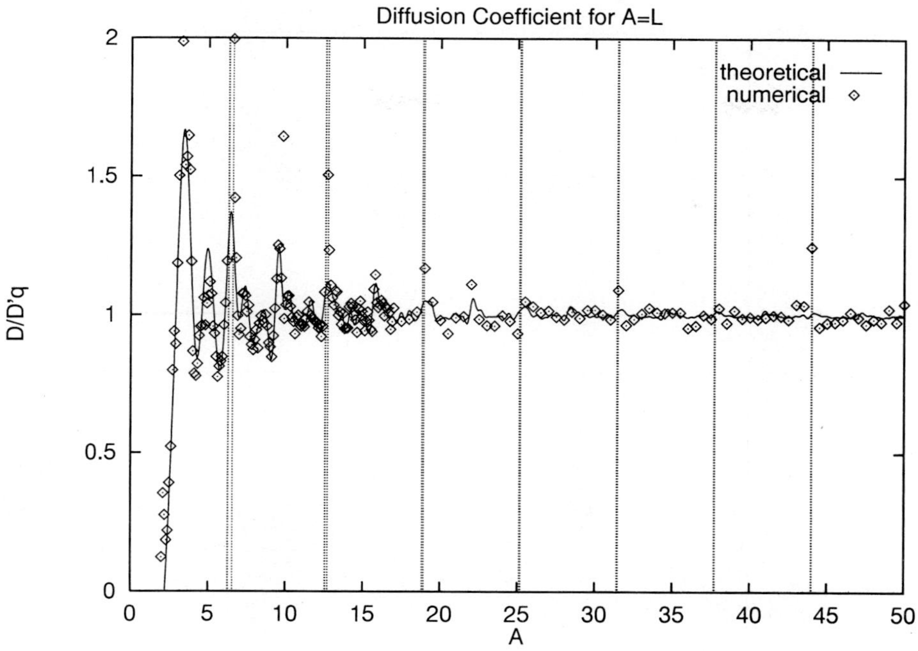


**Fig. 2** Comparison of numerical (points) and theoretical (solid line) diffusion coefficient in the case of $A = L$. The narrow region of $A$ between two vertical dotted lines represents a domain for the stable fundamental accelerator mode.

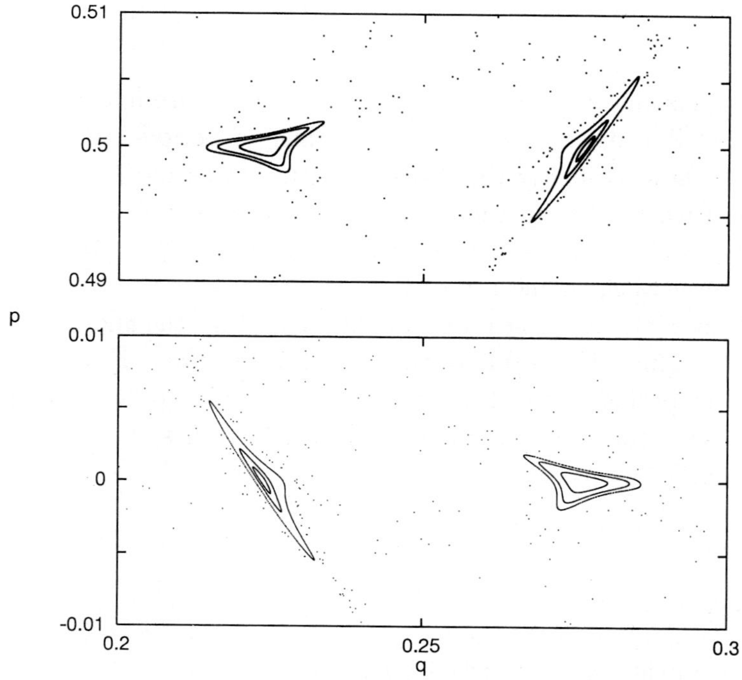


**Fig. 3** The phase portrait of the period-2 step-1 accelerator mode at $A = L = 3.185$. The upper part of this figure stands for the evolution of the lower part in one time step. The orbit starting at $(q, p) = (0.225, 0)$ jumps up to $p = 0.5$ and returns to original point, while the orbit starting at $(q, p) = (0.275, 0)$ repeats the same behavior.

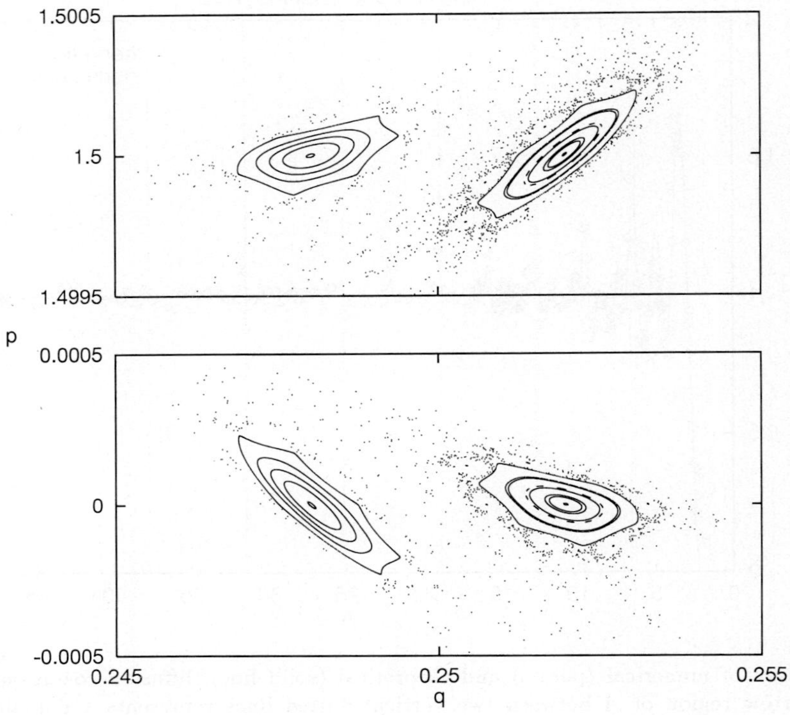


**Fig. 4** The phase portrait of the period-2 step-3 accelerator mode at $A = L = 9.4255$. The upper part of this figure stands for the evolution of the lower part in one time step. The orbit starting at $(q, p) = (0.248, 0)$ jumps up to $p = 1.5$ and returns to original point, while the orbit starting at $(q, p) = (0.252, 0)$ repeats the same behavior.

out a careful examination of the case of $A = L = 3.185$. Through the detailed analysis of the momentum distribution, we find that there exists a period-2 step-1 accelerator mode at $A = L = \pi$. We show explicitly the island structure of period-2 step-1 accelerator mode in Fig. 3. Let us turn to the enhancement peak in the region of $A = L = 3\pi$. We find that there exists a period-2 step-3 accelerator mode at $A = L = 3\pi$. We show the island structure of this accelerator mode in Fig. 4.

From the analysis of the two cases, we can anticipate that the anomalous enhancement of diffusion at $A = L = (2n+1)\pi$ is due to the period-2 step-$(2n+1)$ accelerator mode, while other modes such as period-3 step-1 accelerator mode will manifest its effect in the region of $A = L \simeq 2/3\pi$, as in the case of the standard map.[2] Studies of this aspect is reserved for future investigation.

## 5. SUMMARY

We have investigated the classical kicked Harper system.

We have examined the chaotic motion from the statistical point of view. We obtained the analytical expression for the diffusion coefficient by carrying out the renormalization for the principal term of the characteristic function. By comparing the renormalized diffusion coefficient with the numerical observation, we found that at the larger parameter values the diffusion converged to the value predicted by the quasi-linear term in Eq. (11). On the other hand, we observed an anomalous enhancement of the diffusion in the stable region of the fundamental accelerator modes. We have identified that period-2 step-1 and period-2 step-3 accelerator mode give rise to the enhancement in diffusion in the region of $A = L = \pi$ and $A = L = 3\pi$, respectively. Referring to our previous analysis on the standard map, we expect that other multi-period multi-step accelerator modes will give rise to anomalous enhancement of diffusion in the smaller values of stochastic parameter $A$ and $L$.

## REFERENCES


1. J. D. Meiss, J. R. Cary, C. Grebogi, J. D. Crawford, A. N. Kaufman and H. D. I. Abarbanel, *Physica* **D6**, 375 (1983).
2. Y. H. Ichikawa, T. Kamimura and T. Hatori, *Physica* **D29**, 247 (1987).
3. P. Leboeuf, J. Kurchan, M. Feingold and D. P. Arovas, *Phys. Rev. Lett.* **65**, 3076 (1990).
4. G. M. Zaslavskiĭ, M. Yu. Zakharov, R. Z. Sagdeev, D. A. Usikov and A. A. Chernikov, *Sov. Phys. JETP* **64**, 294 (1986).

Fractals, Vol. 4, No. 3 (1996) 377–384


# PERCOLATED CRACKS IN A SOLID CONTAINING MANY INITIAL DEFECTS

SABUROU NISHIUMA*
*Shiga Bunka College, Yokaichi, Shiga 527, Japan*

YUTAKA HASEGAWA and SASUKE MIYAZIMA†
*Department of Engineering Physics, Chubu University, Kasugai, Aichi 487, Japan*

## Abstract

Hydrogen-induced fracture of steel is characterized by the formation of internal voids caused by hydrogen precipitation at an inclusion-matrix interface, followed by the formation of microcrack array under the superposed action of internal hydrogen pressure and external forces. The propagation of the hydrogen-induced fracture is considerably random and the fracture develops by stepwise linking of the microcracks.

Crack growth in a solid containing many initial defects is studied by Monte Carlo simulation on a square lattice. Each initial defect is assumed to have two crack tips like the Griffith crack and percolated crack patterns in homogeneous medium are investigated. The effect of the density of the initial defects and the lattice size on percolation characteristics is also studied.

*E-mail: luh100@duaic.doshisha.ac.jp
†E-mail: miyazima@isc.chubu.ac.jp

## 1. INTRODUCTION

Hydrogen embrittlement of steel is defined as one of the important phenomena that causes decrease of ductility and strength under a hydrogen environment, such as the steel manufacturing process, welding, plating, acid cleaning and corrosion.[1,2]

The mechanism of hydrogen-induced cracking in a wide variety of steels and nonferrous metals has been a topic of extensive experimental and theoretical research in engineering and material sciences.[3–15]

We presented a fracture model describing the patterns of crack growth in the situation where microcracks initiate from randomly distributed defects and link up to macrocracks as observed in the hydrogen embrittlement crack in steel.[16] By analyzing time-dependent cluster size distribution of cracks and its dynamic scaling, it was found that the crack growth

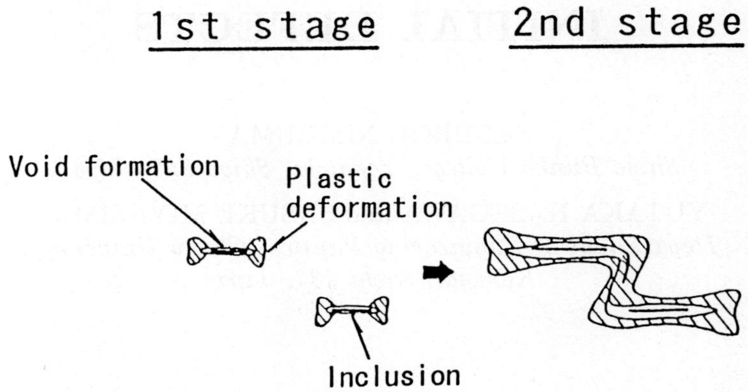


**Fig. 1** Coalescence process of microcracks around inclusions in steel.[17]

**Fig. 2** Metallographic section through fractured stainless steel spring.[14]

process has a fractal structure. In the previous paper, each defect is assumed to have one crack tip and allowed to grow from the crack tip. In this paper, each defect is assumed to have two crack tips like the Griffith crack and be able to grow from the two crack tips. The growing crack tip is selected randomly in each time step.

Figure 1 shows a schematic diagram of the coalescence process of microcracks around inclusions in a steel.[17] The microcrack grows from plastically deformed crack tip region and links to an adjacent microcrack. Figure 2 shows a metallographic section through fractured stainless steel spring.[14] Failures associated with hydrogen-assisted cracking are discontinuous crack growth.

## 2. FRACTURE MODEL

A fracture model is assumed to be isotropic on a finite medium of two dimensional lattice. Initial defects are randomly distributed on the square lattice points. The initial defects correspond to inclusion-matrix interface, microvoids, precipitates, impurities or dislocations in the steel. Hydrogen transported to and enriched at the initial defects can enhance void nucleation or growth, leading to microcrack coalescence and macroscopic fracture. In this model, a crack starts to grow from an initial defect to an adjacent lattice point. Each crack has two growing points (i.e., two crack tips). A growing crack and growing crack tip is

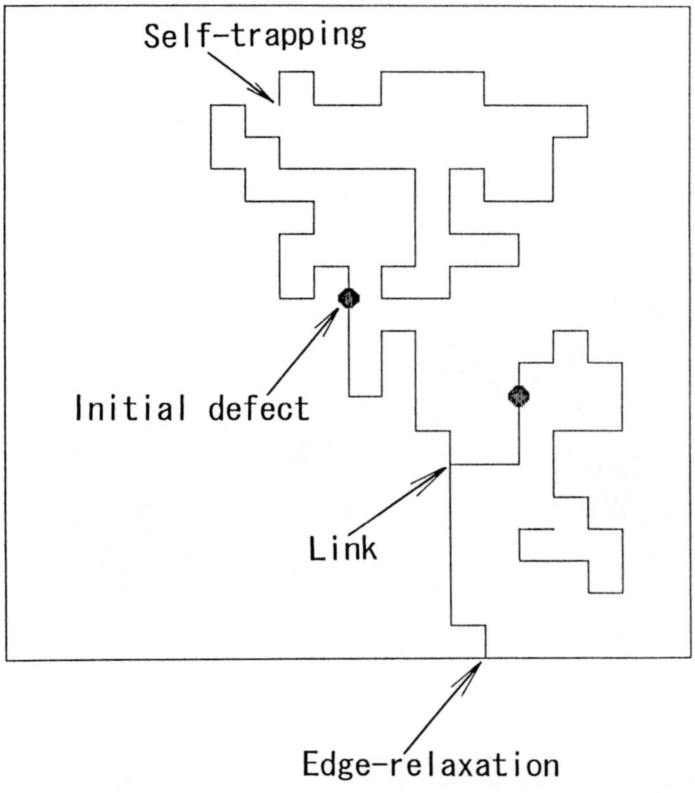


**Fig. 3** Crack patterns of 20 × 20 of the lattice plane size and density of initial defects $\rho_{id} = 0.00454$. The closed circles show the lattice points of initial defects.

selected randomly in each time step (one unit of time step is increased when one bond is broken, i.e., the cluster size increases by one). A crack can grow only from the growing lattice point and the growth of the crack is allowed to any adjacent lattice points with an equal probability, but which is inhibited from crossing over its own trace of growing microcrack (self-avoiding).

Usually the cracks cannot grow indefinitely in an actual material by stress relaxation at the tip of a crack and lowering of hydrogen pressure in the crack cavity. The conditions for cessation of crack growth in this model are as follows:

(1) A growth point of a crack reaches an adjacent crack (Link).
(2) A growth point is trapped in its own cluster (Self-trapping).
(3) A growth point reaches an edge of the lattice (Edge-relaxation).

Crack patterns in a 20 × 20 square lattice are shown in Fig. 3. The closed circles in the figure show the lattice points of randomly distributed initial defects and the solid lines, the trace of crack growth. The number of initial defects is 2 and the density of the initial defects is $\rho = 0.00454$.

## 3. PERCOLATED CRACKS

The clusters continue to grow in the same manner in a self-avoiding random walk until they satisfy the cessation conditions as described in Sec. 2, and the crack path may run from one of the edges to another of the lattice.

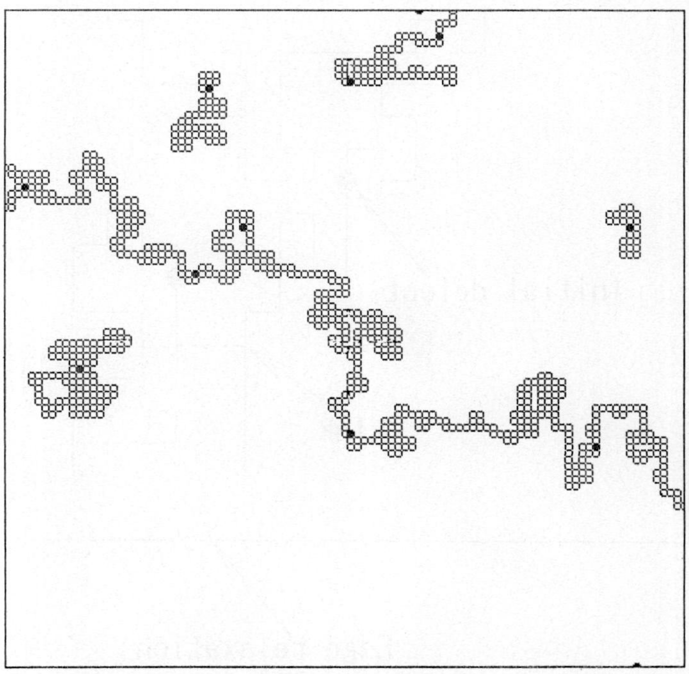

**Fig. 4** Percolated crack patterns of 100 × 100 of the lattice plane size and density of initial defects $\rho_{id} = 0.00118$. The closed circles show the lattice points of initial defects.

We define "percolated crack" as a crack path which runs from one of the edges to the opposite edge of the square lattice, i.e., from $x = 0$ to $x = L$ of the lattice, or from $y = 0$ to $y = L$, where $L$ is the size of square lattice. This percolated crack corresponds to mechanical breakdown of actual materials.

Figure 4 shows a percolated crack pattern of a lattice of 100 × 100. The closed circles show the initial defects and the density of the initial defects is $\rho_{id} = 0.00118$. One crack path runs from $x = 0$ to $x = L$.

Figure 5 shows a percolated crack pattern of lattice of 120 × 120. The closed circles show initial defects and the density of the initial defects is $\rho_{id} = 0.00128$. One crack path runs from $y = 0$ to $y = L$.

Figure 6 shows a percolated crack pattern of larger lattice of 200 × 200. The density of the initial defects is $\rho_{id} = 0.00495$. In this case, a crack path runs both from $x = 0$ to $x = L$ and $y = 0$ to $y = L$.

We define "percolation ratio" $R$ as follows:

$$R = \frac{N_{pc}}{N}$$

where $N$ is the number of iterated calculations under a given system size and density of initial defects and $N_{pc}$ is the number of frequency that a percolated crack appears.

Figure 7 shows a relation between the density of the initial defects and the percolation ratio for two different system sizes. It is clear that a percolated crack appears by increasing the density of the initial defects and a threshold value of the density of the initial defects exists. The percolation ratio increases with the density of the initial defects; however, the percolation ratio depends on the system size $L$. The percolation ratio decreases as the system size increases.

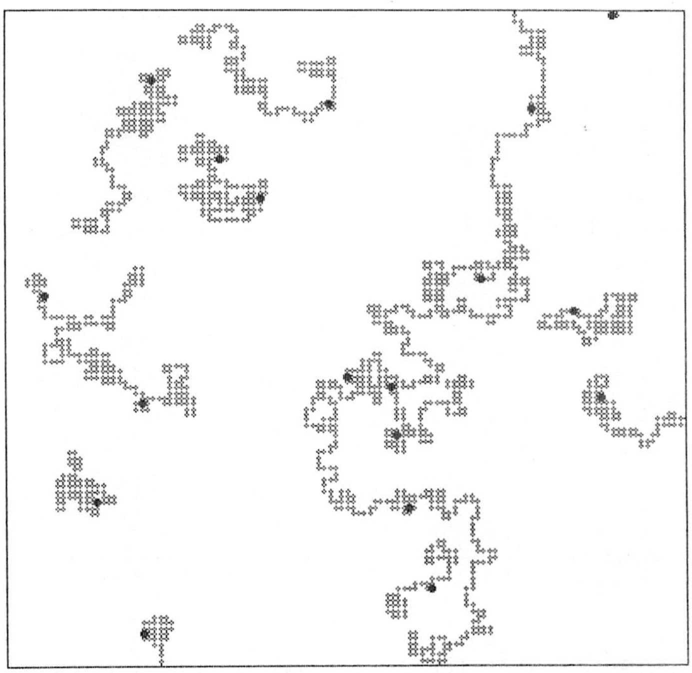

**Fig. 5** Percolated crack patterns of 120 × 120 of the lattice plane size and density of initial defects $\rho_{id} = 0.00123$. The closed circles show the lattice points of initial defects.

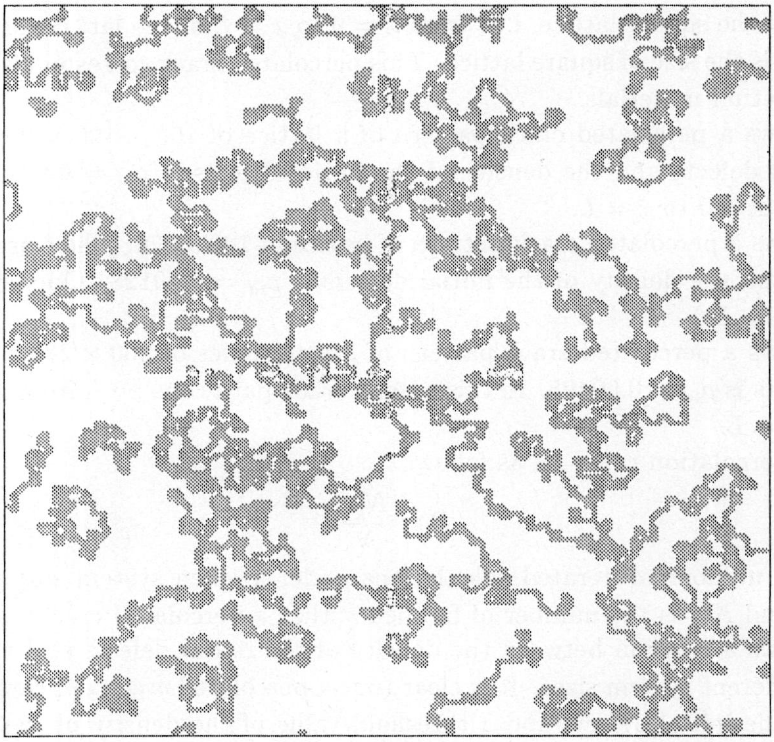

**Fig. 6** Percolated crack patterns of 200 × 200 of the lattice plane size and density of initial defects $\rho_{id} = 0.00495$.

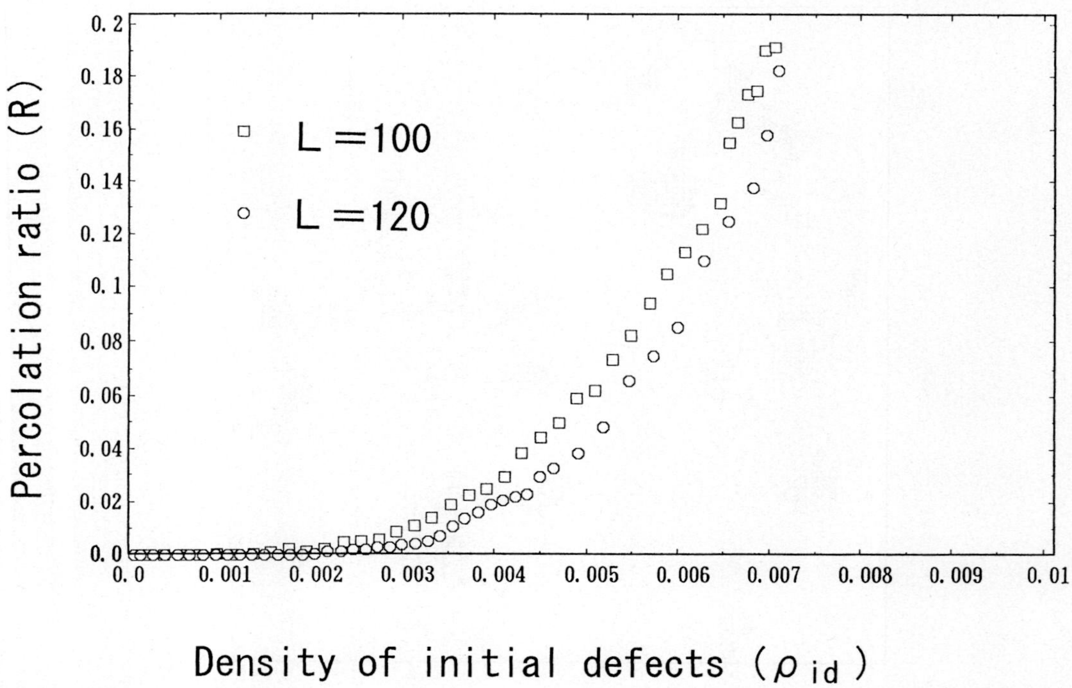


**Fig. 7** Relation between density of initial defects and percolation ratio.

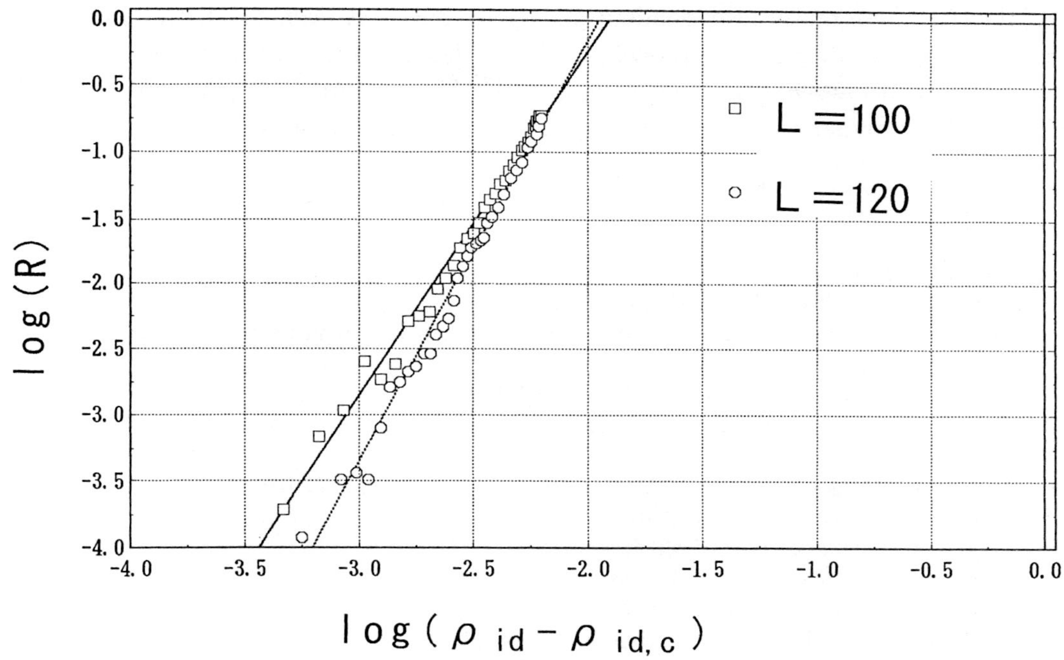


**Fig. 8** Log-log plot of density of initial defects vs. percolation ratio.

Figure 8 shows a log-log plot of the density of the initial defects vs. the percolation ratio for two different system sizes. The percolation ratio $R$ can be represented by a power law:

$$R \sim (\rho_{id} - \rho_{id,c})^r$$

where $\rho_{id}$ is the density and $\rho_{id,c}$ is the critical density, respectively, of the initial defects. It is found that the critical density $\rho_{id,c}$ and the exponent $r$ depend on the system size ($\rho_{id,c} = 0.0007$, $r = 2.620$ at $L = 100$ and $\rho_{id,c} = 0.0008$, $r = 1.706$ at $L = 120$, respectively). The critical density increases with the system size and the exponent decreases as the system size increases.

## 4. CONCLUDING REMARKS

We presented here a fracture model describing the process of crack growth by hydrogen embrittlement and investigated slow-dynamics and delayed crack with the Monte Carlo method. In a previous paper,[16] we assumed that each initial defect has one crack tip (one crack tip assumption) and in this paper, we assumed that each initial defect has two crack tips like the Griffith crack (two crack tips assumption). It is found that the percolation characteristics of the above two models is significantly different. The critical density of the initial defects $\rho_{id,c}$ of the "two crack tips model" is smaller more than two order compared with that of the "one crack tip model".

Our fracture model is basically based on self-avoiding random walk. The process of actual hydrogen embrittlement crack growth is affected by the hydrogen concentration, stress field, shape and distribution of non-metallic inclusions, anisotropy of composite, dimensions of material, etc. These material-dependent properties have not been considered in the present work. Within our stochastic approach, material dependence may somehow be related to the growth probability of four directions and the distribution of initial defects (origins of crack growth). Further investigation is required to clarify the abovementioned problems.

## REFERENCES


1. F. de Kazinczy, *J. Iron Steel Inst.* **177** (1954).
2. S. Matsuyama, *Delayed Fracture* (in Japanese) (Nikkan Kougyou Shinbun, 1989).
3. N. J. Petch, *Phil. Mag.* **1** (1956).
4. W. F. Savage and E. S. Szekeres, *Weld. J.* **46**, 2 (1967).
5. J. K. Tien, A. W. Thompson, I. M. Bernstein and R. J. Richards, *Met. Trans.* **A7** (1976).
6. E. Smith, *Int. J. Fracture* **14** (1978).
7. F. Matsuda, H. Nakagawa and T. Matsumoto, *Trans. JWRI* **8** (1979).
8. J. P. Smith, *Met. Trans.* **A11** (1980).
9. R. A. Oriani, *Corrosion-NACE* **43**(7) (1987).
10. S. P. Lynch, *Acta. Metall.* **36**(10) (1988).
11. D. IL. Kwon and R. J. Asaro, *Acta. Metall. Mater.* **38**(8) (1990).
12. N. R. Moody and S. L. Robinson, *Res. Mech.* **30** (1990).
13. S. H. Chen, Y. Katz and W. W. Gerberich, *Phil. Mag.* **63**(1) (1991).
14. A. J. McEvily and I. L. May, *Mater. Charct.* **26** (1991).
15. Y. Tong and J. F. Knott, *Scr. Metall.* **26** (1992).
16. S. Nishiuma and S. Miyazima, *Fractals* **3**(1) (1995).
17. F. Terasaki, *Materia Japan* **16**(11) (1977) (in Japanese).

Fractals, Vol. 4, No. 3 (1996) 385–392


# WATER EROSION ON FRACTAL SURFACE

R. LI, H. TAKAYASU and H. INAOKA
*Graduate School of Information Sciences, Tohoku University, Aramaki, Aoba-ku, Sendai, 980-77, Japan*

**Abstract**

We study numerically the water erosion process under a variety of conditions. The water erosion model that we use leads to a universal exponent describing the fractal basin area distribution in the steady state.

## 1. INTRODUCTION

A number of geophysical structures like coastlines, mountains and river networks have been shown to possess fractal structure, as has been shown by numerical analysis of real topographical data.[1–3] Rivers have frequently been cited as typical examples of fractals in nature; distribution of drainage basin area is known to follow a power law. The fractal properties of the river networks are a reflection of the absence of characteristic length scale in the process of river formation; this kind of processes are often referred to as a self-organized criticality.[4,5]

A realistic model of water erosion, leading to the formation of river network, has been introduced by Takayasu and Inaoka in Ref. 2 (henceforth we refer to it as the water erosion model). While producing impressively realistic images of eroded landscape, the model leads to a power law basin area distribution, with an exponent $\beta = 0.42 \pm 0.01$, in overall agreement with the field data.

In the present article, we report a numerical study of the water erosion (in the framework of an extended water erosion model) taking place under widely varying conditions. The

E-mail: takayasu@fractal.is.tohoku.ac.jp

value of the exponent governing the power-law distribution of drainage basin area changes from case to case by no more than the simulation error; we conclude that it is universal and estimate it as $\beta = 0.44 \pm 0.03$.

## 2. THE WATER EROSION MODEL

The water erosion model involves two major steps: determination of the water flow and taking into account the effect of erosion. For a given landform $h(x, y)$, at each time step, the flow direction and the height of the water surface $h_w(x, y)$ are calculated for each site as follows:

(1) The direction of water flow is determined at each time step by fixing the lowest site $(x', y')$ among its nearest neighbors. If the elevation of the destination is lower, then the water drains into the neighboring site; otherwise, the water accumulates until the water level exceeds the destination height.

(2) Using the effect of erosion, the decrease in the height of the land surface is calculated by:

$$\delta h(x, y) = F(J(x, y))\{h_w(x, y) - h_w(x', y')\}, \tag{1}$$

where $J(x, y)$ denotes the water power which is defined as:

$$J(x, y) \equiv s(x, y)\{h_w(x, y) - h_w(x', y')\}. \tag{2}$$

Here, $s(x, y)$ refers to the water-flow intensity. At each time step, the new $s(x, y)$ is determined as the sum of $s$ from nearest neighbors and a constant which corresponds to the water from rainfall.

The function $F$ controls the erosion speed and we assume the following functional form:

$$F(J) = CJ^a/(1 + J^a), \tag{3}$$

where $C$ and $a$ take positive constants. In the water erosion model, the effect of accumulation of eroded soil is neglected.

The boundary condition, if not mentioned otherwise, is given as an infinite wall on one side of the system, and an infinitely deep well on the other side. The remaining two sides are connected by a periodic boundary condition.

It is known that by these evolution rules, a nearly flat initial landscape evolves into a rough landscape having a stable river pattern.[2] The distribution of drainage basin area is found to follow as a power law:

$$P(\geq s) \sim s^{-\alpha}, \qquad \alpha = 0.42 \pm 0.01. \tag{4}$$

## 3. STABILITY OF THE BASIN-AREA DISTRIBUTION

### 3.1 Initial Conditions

The most widely applied method of generating fractal landscapes has been introduced by Voss, and is based on simulating the Fractional Brownian Motion by applying a random

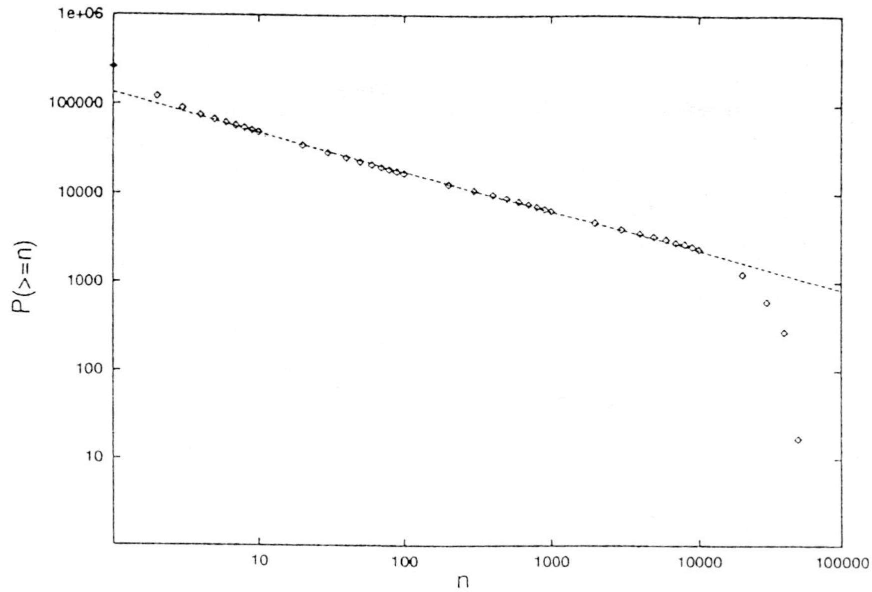


**Fig. 1** The cumulative distribution of drainage basin area starting from a fractional Brownian motion initial condition, with $H = 0.7$.

midpoint displacement algorithm in two dimensions.[5] Here we use it to generate a randomized initial condition for the water erosion model in order to check the stability of the steady state basin area distribution.

Figure 1 shows the cumulative distribution of drainage basin area, obtained starting from a landscape generated for $H = 0.7$ ($H = 0.7$ is most commonly found in natural landscapes[5]), given in a log-log plot. Apart from several points corresponding to big and small basin areas, the data collapse on a straight line, with the slope giving the exponent of the distribution. The exponents, obtained for different $H$ values at varying system sizes, are presented in Fig. 2. In the big system size limit, the exponents converge to similar values. We attribute the difference to random deviations and conclude that the exponent is unaffected by the fractal initial condition.

The case of an initial condition given as a plane put aslant has also been considered, and has led to the same conclusion. This case is of interest, since the evolution on initial stages is similar to the one in the so-called Scheidegger's model,[7] which is known to give the exponent $\alpha = 1/3$ theoretically.[6] The basin area distribution at different stages is given in Fig. 3. One can see that while, initially, the distribution corresponds to the one found theoretically for the Scheidegger's model, in the stationary state the exponent value characteristic for the water erosion model is recovered.

### 3.2 Distributed Erosion Rate

To simulate the inhomogeneity of the real soil, we introduce a spatial distribution of the erosion rates. The erosion function $F$ (Eq. 3) is controlled by the parameter $C$. We study the following cases:

(1) Two distinct $C$ values; the $C$ value in half of the system is different from the other half.
(2) The two distinct $C$ values are distributed according to a power law.

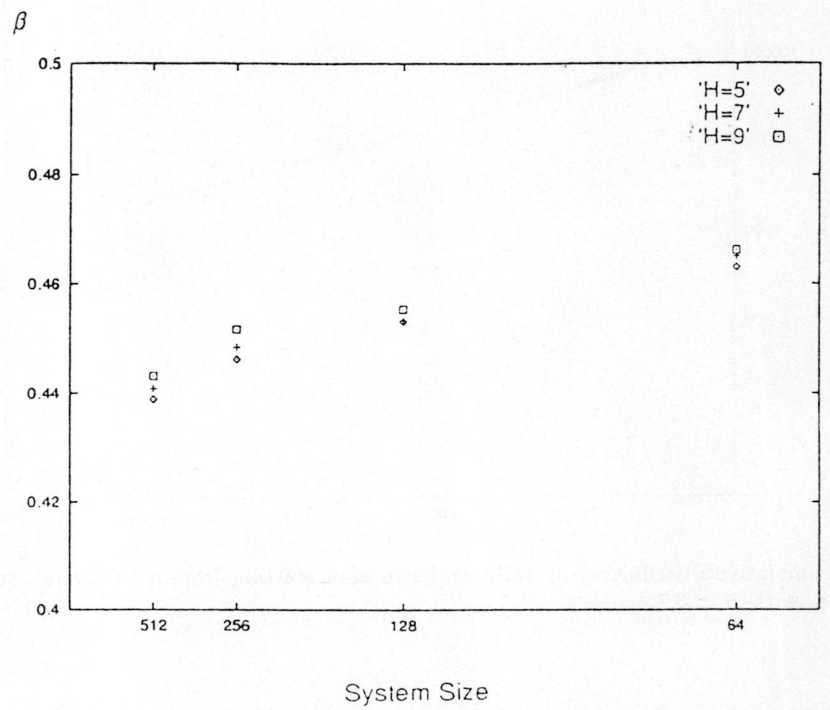


**Fig. 2** **System size dependence of the exponents for the drainage basin area distribution; initial condition given by fractional Brownian motion with different values of $H$.**

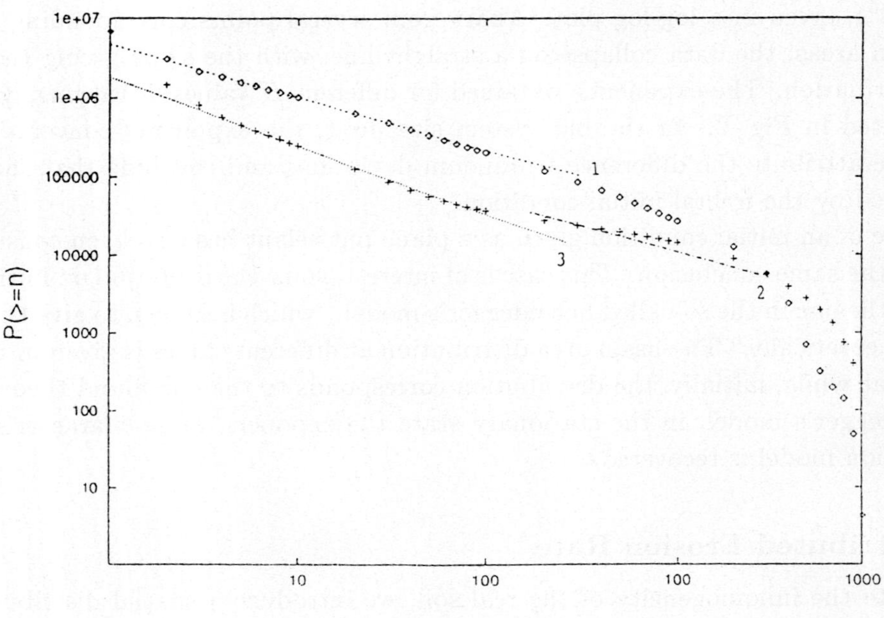


**Fig. 3** **The cumulative distribution of drainage basin area starting from a plain put aslant. Lines (1) and (2) have a slope $-1/3$ (the theoretical value for the exponent in the Scheidegger's model), line (3) is drawn with a slope $-0.44$.**

(3) Spatial correlations in the distribution of $C$ values are introduced by setting $C(x, y)$ according to a fractional Brownian motion, again generated using the Voss's midpoint displacement algorithm.

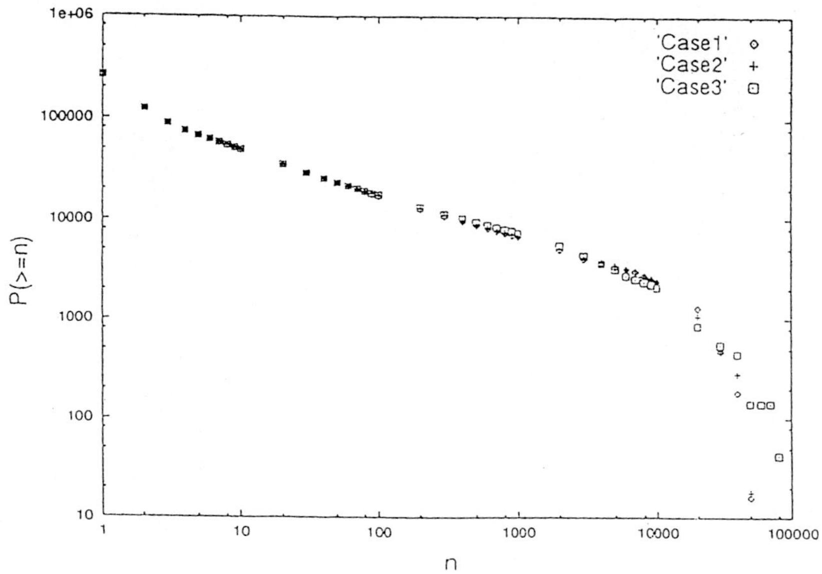


**Fig. 4** The cumulative distribution of drainage basin area for the 3 cases of the distributed erosion rate, given in the text of the article. (1) In half of the system $C = 0.3$; in the other half $C = 0.6$. (2) The two distinct $C$ values are set according to power law. (3) $C$ is obtained by fBm with $H = 0.7$.

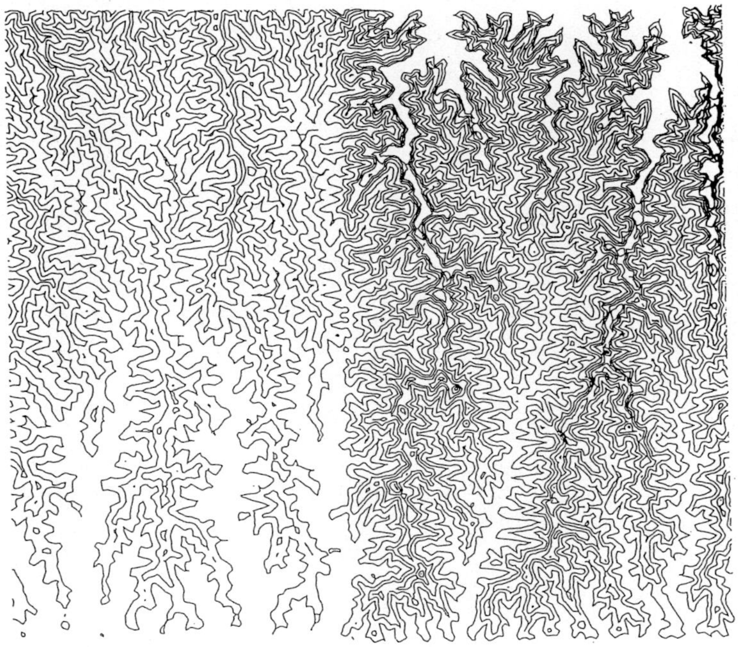

(a)

**Fig. 5** Contour plots for the landscapes obtained in conditions described in the caption for Fig. 4. **(a)** Case (1) of the Fig. 4. **(b)** Case (2). **(c)** Case (3).

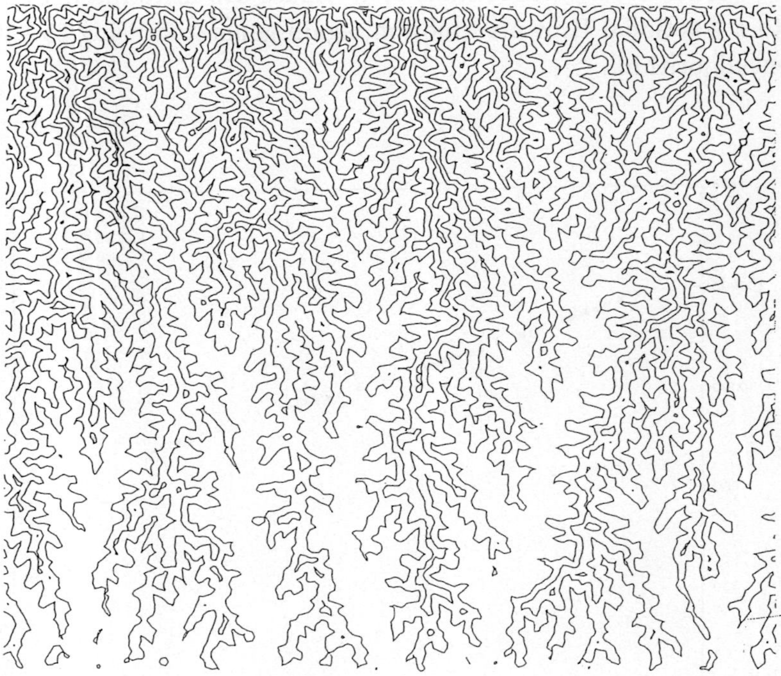

(b)

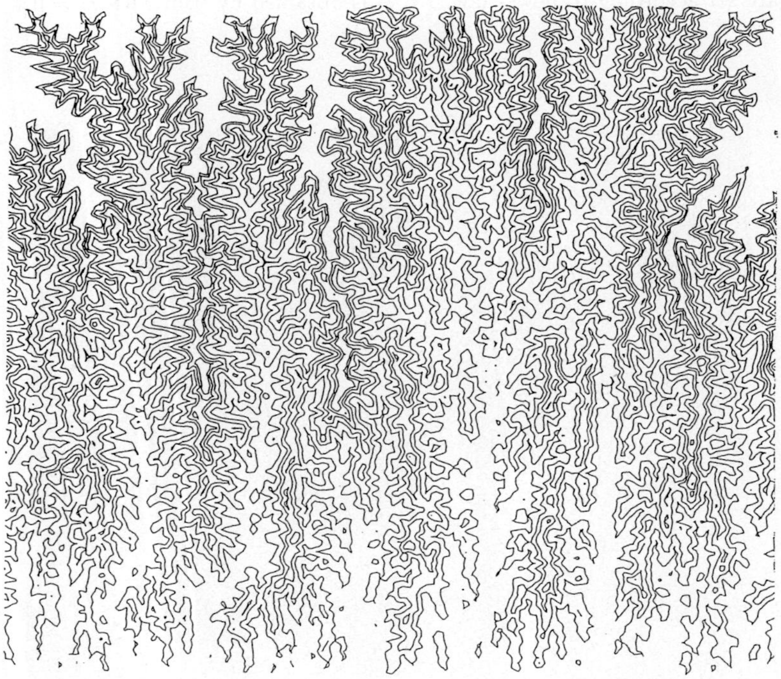

(c)

**Fig. 5** (*Continued*)

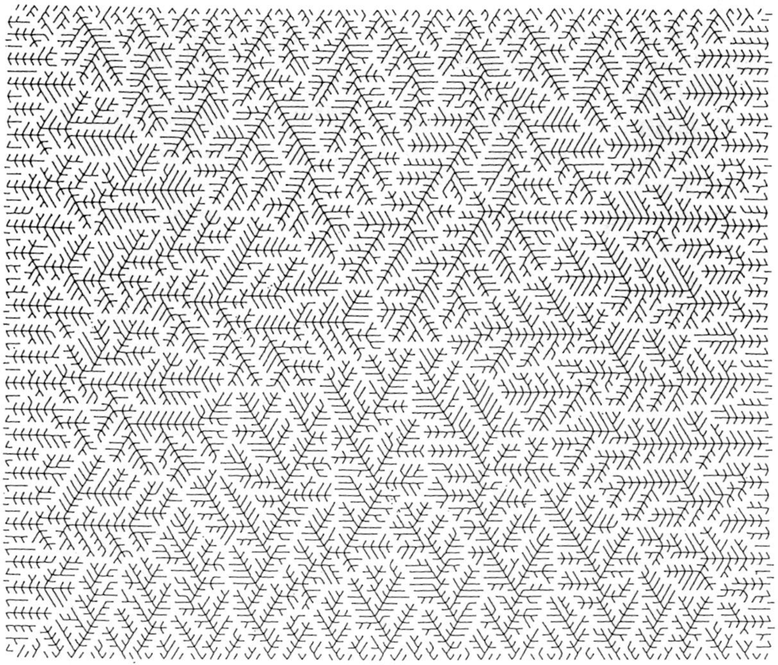

**Fig. 6** The river pattern obtained for the boundary condition given as a well on all four sides.

The results are shown in Fig. 4. The $\beta$ values, within an error of the simulation, are the same in all three cases. It is interesting to note that the appearance of the landscapes can vary quite significantly (see contour plots in Fig. 5). This kind of behavior is rather typical in systems with universality; processes looking quite differently at short distances lead to the same critical exponents, reflecting the similarity of their behavior on a large scale.

### 3.3 Boundary Condition

To complete our study of the stability of the basin area distribution, we address the problem of the boundary condition. We modify the boundary condition by setting a well on all four sides. The resulting river pattern is given in Fig. 6. Not surprisingly, the exponent value in this case is in agreement with the cases discussed above, $\beta = 0.43 \pm 0.02$.

## 4. CONCLUSION

We have shown that variation of initial and boundary conditions, as well as the introduction of an inhomogeneous distribution of erosion rate, does not lead to a change in the exponent describing the steady state basin area distribution, which remains universal.

## ACKNOWLEDGMENT

Thanks are due to Dr. Alex Tretyakov for helpful discussions.

## REFERENCES

1. B. B. Mandelbrot, *The Fractal Geometry of Nature* (Freeman, San Francisco, 1982).
2. H. Takayasu and H. Inaoka, *Phys. Rev. Lett.* **68**, 966 (1992); H. Inaoka and H. Takayasu, *Phys. Rev.* **E49**, 899 (1993).
3. D. G. Tarboton, R. L. Bras and I. Rodriguez-Iturbe, *Water Resour. Res.* **24**, 1317 (1988).
4. T. Vicsek, *Fractal Growth Phenomena* (World Scientific, Singapore, 1989).
5. J. Feder, *Fractals* (Plenum, New York, 1988).
6. H. Takayasu, *Fractals in the Physical Sciences* (Manchester University Press, Manchester, 1990).
7. A. E. Scheidegger, *Bull. I. A. S. H.* **12**(1), 15 (1967).

Fractals, Vol. 4, No. 3 (1996) 393–399


# FRACTAL ANALYSES FOR A MODEL OF IMPACT FRAGMENTATION

HAJIME INAOKA* and HIDEKI TAKAYASU
*Graduate School of Information Sciences, Tohoku University, Sendai 980-77, Japan*

**Abstract**

We perform fractal analyses for a model of three-dimensional impact fragmentation. The fractal dimension of the whole crack configuration in the system is numerically estimated as $D = 2.3$.

## 1. INTRODUCTION

Impact fragmentation is a familiar phenomenon in our daily life and it has been widely studied for many years for its importance in various fields of sciences. For examples, there is a pioneering theoretical and experimental study by Gilvarry et al.[1,2] and a recent study by Oddershede et al.[3] In these studies, the size of the fragments produced by impact fragmentation shows a power-law distribution. However, the mechanism of the process has not been fully understood yet because of its complexity and no numerical model has successfully reproduced the power-law fragment size distribution.

We first proposed a model of the three-dimensional impact fragmentation.[4] The model produces a power-law fragment mass distribution which is consistent with the experiments. The power-law distribution is deeply connected with the concept of fractals.[5–8] In this paper, we present analyses of fractal properties in the model.

*Research Fellow of the Japan Society for the Promotion of Science.

## 2. MODEL DESCRIPTION

Our main hypothesis in constructing the model is existence of a plane-like density wave as a crack propagating front. We schematically present the situation in Fig. 1. When an object undergoes an impact on a side of the object, a plane-like density wave occurs and propagates with a constant velocity, $v_0$, from the hit side toward the opposite side. Crack propagation occurs only on the propagating density wave and no secondary fragmentation occurs behind the density wave. Namely, we can consider the crack dynamics only on the plane density wave. We also assume that no crack is newly created and bifurcates during the propagation, so, crack density monotonically decreases as the wave propagates.

The crack dynamics must be connected with stress and strain field surrounding the cracks. We can convert the three-dimensional equation of deformation by elasticity, which is based on an equation of motion with Hooke's law, into two-dimensional Poisson's equation observing on the propagating density wave. For a detailed derivation of the equation, see

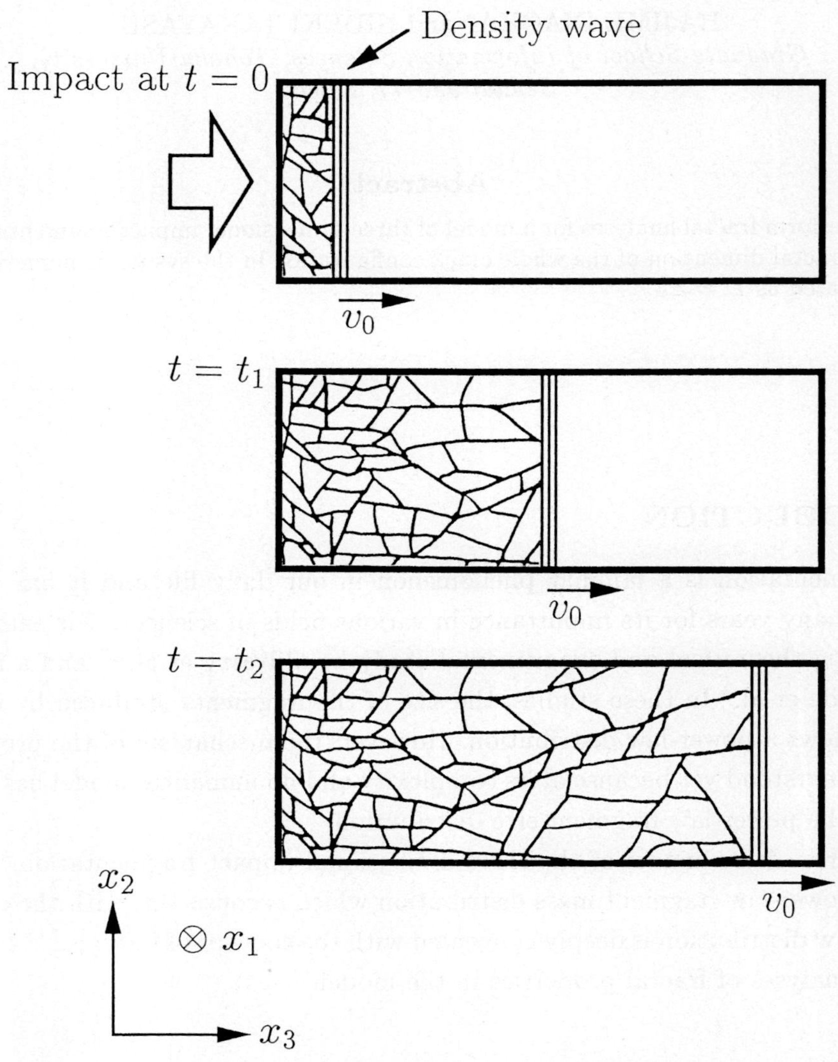


**Fig. 1** Schematic pictures of impact fracture process. An impact at $t = 0$ produces a density wave propagating in the $x_3$-direction with a constant velocity and it forms a crack propagating front.

Ref. 4. We set the propagating direction of the density wave parallel to the $x_3$ axis; then the equation is written as:

$$\nabla^2 \phi(x_1, x_2, t) = -\theta(x_1, x_2, v_0 t), \tag{1}$$

where $\phi$ is a scalar potential defined by two-dimensional displacement field, $\mathbf{u}$, on the density wave as:

$$\nabla \phi(x_1, x_2, t) = \mathbf{u}(x_1, x_2, t), \tag{2}$$

and $\theta$ denotes dilation field caused by the elasticity of the material. Note that the vectors and operators appearing in the equations are two-dimensional, namely, they are all perpendicular to the wave propagating direction. We assume that Eq. (1) is applicable to all non-crack part on the density wave and cracks determine the boundary condition of Eq. (1). The boundary condition is given as:

$$\phi(x_1, x_2, t) = 0, \tag{3}$$

on the cracks. The physical meaning of the condition is that the displacement near the cracks are perpendicular to the cracks.

Thus, we can treat the three-dimensional impact fracture process as a series of two-dimensional processes on the density wave. Note that a region which is encircled by the cracks in the model is not a fragment itself but a section of the fragment by the plane density wave.

The model is defined on a two-dimensional triangular lattice which represents the propagating plane density wave. Each lattice site $(x_1, x_2)$, has three variables; fragment number, $N_f(x_1, x_2, v_0 t)$, displacement potential, $\phi(x_1, x_2, t)$, and dilation field, $\theta(x_1, x_2, v_0 t)$. Hereafter, $v_0$ is set to be unity in the model; therefore, the time, $t$, and the depth $x_3 = v_0 t$ can be regarded as identical. The sites which have the same fragment number are regarded as belonging to the section of the same fragment. That is, there is a crack between any two adjacent sites having different fragment numbers. We introduce the displacement field, $\mathbf{u}(x_1, x_2, x_3)$, using Eq. (2). The displacement potential, $\phi(x_1, x_2, x_3)$, is determined by solving Eq. (1) under the boundary condition, Eq. (3), on the cracks. The randomness of the materials of the fractured objects is introduced through the dilation field, $\theta(x_1, x_2, x_3)$. We calculate the magnitude of the displacement, $|\mathbf{u}(x_1, x_2, x_3)|$, on the crack from both sides of the fragment sections. They are then compared and the crack is assumed to move by one site toward the site having the smaller value of $|\mathbf{u}(x_1, x_2, x_3)|$.

We start from the initial condition that each site in the system belongs to the different fragment, which physically means that the surface hit by an impact is completely fractured. The edges of the whole system are connected by the periodic boundary condition.

Starting from the initial condition, the time evolution of the system is performed by repeating the following procedures.

(1) For each site $(x_1, x_2)$, we assign a random number distributing uniformly in the interval $[1.0, 1.5)$ to the value of dilation field, $\theta(x_1, x_2, x_3)$.
(2) The displacement potential, $\phi(x_1, x_2, x_3)$, is solved by the relaxation method.
(3) The sites which are at the edges of the fragment sections are extracted from the system and make a set of sites named $A$.

(4) For each site $(x_1, x_2) \in A$, we make a set of the neighboring sites, $U$, which consists of the site $(x_1, x_2)$ itself and the six nearest neighbors of the site $(x_1, x_2)$. Then we find out the site $(x_1', x_2', x_3)$ which has the largest value of the displacement potential, $\phi$, in the set of $U \cap A$. The fragment number $N_f(x_1, x_2, x_3)$ is renewed by $N_f(x_1', x_2', x_3)$.

These procedures make one time-step and are repeated with $x_3$ increased by 1.

## 3. SIMULATION RESULTS

Examples of the fragments produced by this model is shown in Fig. 2. The figures can be regarded as the fragment sections parallel [Fig. 2(a)] and perpendicular [Fig. 2(b)] to the density wave. We extract the largest fragment section from the figures and analyze the fractal dimension of the sections of the fragment periphery by the box-counting method. The results are shown in Fig. 3. The fractal dimension of the fragment periphery sections parallel and perpendicular to the density wave are both observed to be very close to 1.0. These results confirms that the fractal dimension of the periphery of a fragment is 2 and it is an ordinary surface. Namely, each fragment does not have fractal properties in this model. However, we can recognize a kind of self-similarity from Fig. 2(b) as follows. We consider a rectangular part on Fig. 2(b) whose upside is placed just on the upside of the figure. The upside of Fig. 2(b) corresponds to the section of the hit surface of the fractured object. We

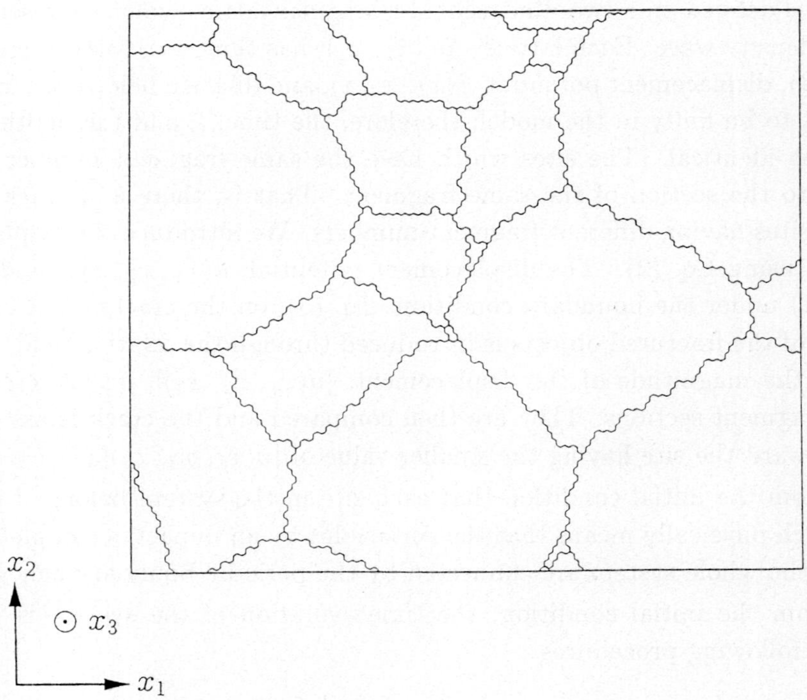


(a)

**Fig. 2** **(a)** A snapshot of the system of size 128 × 128. A part of the size 80 × 80 at 25 time-steps is presented. The solid lines show cracks. **(b)** A section perpendicular to Fig. 2(a) is presented. The width is 80 lattice size and it takes 64 time-steps. The time is shown in the direction from the top to the bottom in the figure.

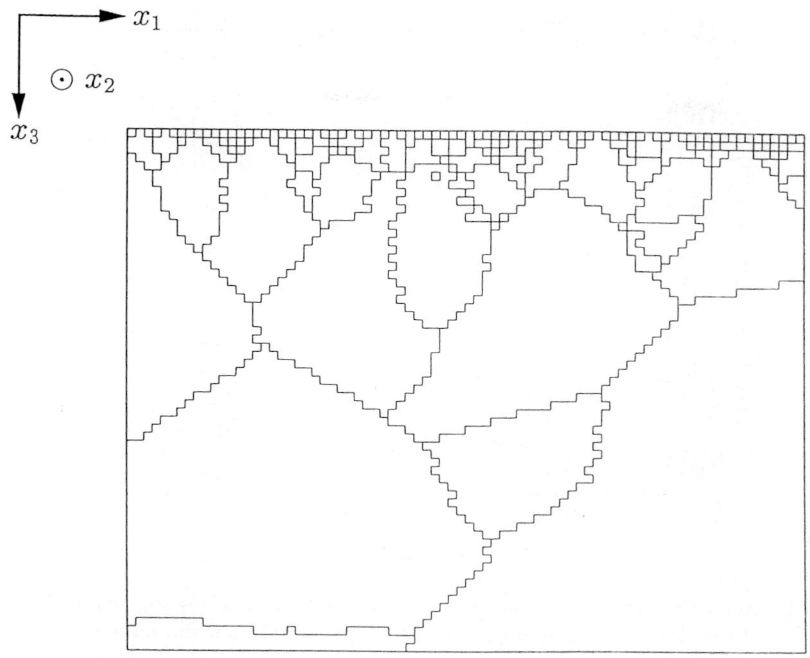


**(b)**

**Fig. 2** (*Continued*)

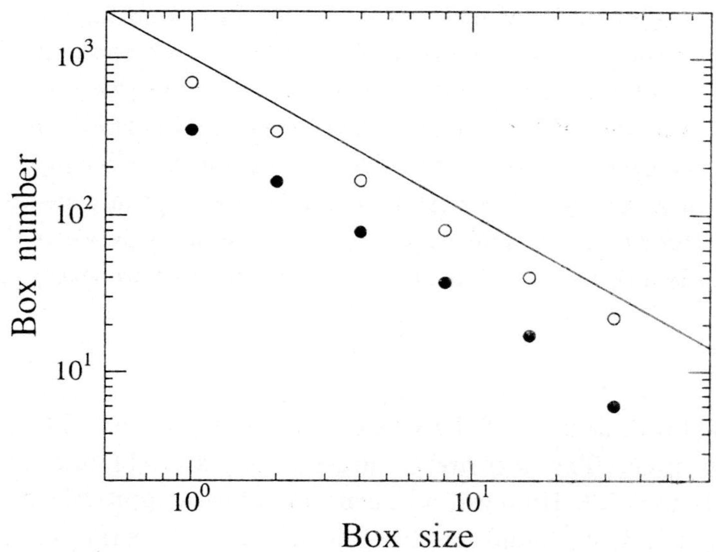


**Fig. 3** Results by the box-counting method applied to the section of fragment periphery by planes parallel (∘) and perpendicular (•) to the density wave. The solid line shows a slope of $D = 1.0$.

expand the rectangular part properly; the expanded configuration of the fragment sections then appears to have a similarity to the original Fig. 2(b). This fact implies that the fragments in the system are similar in shape to each other and that they construct a self-similar structure as a whole though each of them does not have fractal properties.

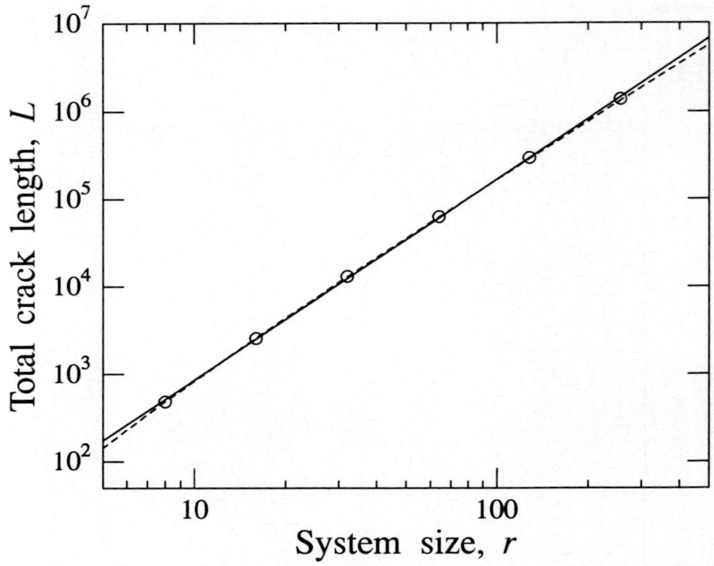


**Fig. 4** Relation between the system size and the total crack length in log-log scale. The solid line shows a relation $L \propto r^{2.3}$ and the dashed line shows the results from the dimension analysis $L \propto r^2 \ln r$.

Obviously, the system has an anisotropy due to the wave propagation. Because of the anisotropy, the self-similarity of the system requires a unique way of observation, i.e., only a rectangular parallelepiped region of the system, which has a part of the hit surface as one of its plane, can be similar to the whole system. So, the box-counting method is not suitable for the estimation of the fractal dimension of the whole crack configuration. To estimate the fractal dimension of the whole configuration of the cracks, we perform simulations of models of various sizes. We set the size of the model as $r$ and perform the simulation in $r/2$ time-steps so that the systems represent similar rectangular parallelepiped fractured object of each other. We measure the crack length on the lattice at each time-step and sum them up throughout all the evolution time-steps. Then the relation between the system size, $r$, and the total crack length in the whole system, $L$, is expected to satisfy a power-law as:

$$L \propto r^D \,, \tag{4}$$

where $D$ is the fractal dimension of the whole crack configuration. The result is shown in Fig. 4 on a log-log scale. The dots are apparently on a straight line, indicating that the fractal dimension is $D = 2.3$. However, a dimension analysis approach to the problem gives another result. In Ref. 4, we found that the mean fragment section area, $\langle s(x_3) \rangle$, defined as:

$$\langle s(x_3) \rangle \equiv \frac{\sum_i s_i(x_3)^2}{\sum_i s_i(x_3)} \,, \tag{5}$$

shows power-law expansion as:

$$\langle s(x_3) \rangle \propto x_3^2 \,, \tag{6}$$

where $i$ and $s_i$ denote the fragment number and the section area of the fragment numbered $i$, respectively. By Eq. (6), the number of the fragment sections, $N$, and the total crack length on the density wave, $l$, decrease as, $N \propto x_3^{-2}$ and $l \propto N \cdot x_3 \propto x_3^{-1}$, respectively. Since all the fragments in our model have their roots in the hit surface of the fractured object, the section number, $N$, should have the relation with the dimension of the object, $r$, as, $N \propto r^2$. Then the total crack length in the whole system, $L$, scales as:

$$L \propto \int_{\varepsilon}^{\frac{r}{2}} r^2 x_3^{-1} dx_3 \propto r^2 \ln r \,, \tag{7}$$

where $\varepsilon$ denotes the lattice constant. The result is also shown in Fig. 4 by a dashed line and it also seems to fit the dots. The difference between $r^{2.3}$ and $r^2 \ln r$ is very subtle in the observed range and we have to perform larger scale simulations to clarify it. However, it is difficult to perform them at present because of the lack of our computer ability.

## 4. SUMMARY

In this paper, we analyzed the fractal properties of the periphery of the identical fragment and the whole crack configuration. The whole crack configuration numerically shows fractality whose dimension is estimated as $D = 2.3$, though the periphery of the identical fragment does not show fractal properties.

The fact that the whole crack configuration shows a nontrivial scaling as $r^{2.3}$ or $r^2 \ln r$ has a physically significant meaning. Assuming that the crack surface propagation of unit area requires constant energy dissipation, the fragmentation of an object whose linear dimension is $r$ requires energy proportional to the nontrivial scaling. This relation might be confirmed experimentally.

## ACKNOWLEDGMENT

This work is partly supported by Japan Society for the Promotion of Science and Grant-in-Aid for Scientific Research from the Ministry of Education, Science and Culture.

## REFERENCES

1. J. J. Gilvarry, *J. Appl. Phys.* **32**, 391 (1961).
2. J. J. Gilvarry and B. H. Bergstrom, *J. Appl. Phys.* **32**, 400 (1961).
3. L. Oddershede, P. Dimon and J. Bohr, *Phys. Rev. Lett.* **71**, 3107 (1993).
4. H. Inaoka and H. Takayasu, *Physica A*, in press.
5. B. B. Mandelbrot, *The Fractal Geometry of Nature* (Freeman, San Francisco, 1982).
6. J. Feder, *Fractals* (Plenum, New York, 1988).
7. T. Vicsek, *Fractal Growth Phenomena* (World Scientific, Singapore, 1989).
8. H. Takayasu, *Fractals in the Physical Sciences* (Manchester University Press, Manchester, 1990).

Fractals, Vol. 4, No. 3 (1996) 401–406


# DOMAIN WALLS IN THE TRANSVERSE FIELD ISING SPIN GLASSES

MAI SUAN LI[*,†] and MAREK CIEPLAK[*]
[*]*Institute of Physics, Polish Academy of Sciences, 02-668 Warsaw, Poland*
[†]*Faculty of Engineering and Design,*
*Kyoto Institute of Technology, Sakyo-ku, Kyoto 606, Japan*

## Abstract

Interfaces in Ising spin glasses in the transverse field are studied by the local mean field method. The quantum fluctuations are found to make the length of domain walls shorter and smoother and to reduce the fractal dimensionality of the wall. The domain wall are almost as narrow as in the classical Ising case.

Recently, there has been a growing interest in the study of interfaces with non-trivial geometry.[1] In particular, the interfaces in classical Ising spin glasses have been shown to be fractal[2–4] and in 2- and 3-dimensional systems the fractal dimensionalities, $D_f$, are estimated to be equal to $1.27 \pm 0.04$ and $2.57 \pm 0.05$, respectively. It should be noted that in the strong disorder limit[5] the $D_f$'s are lower and equal to about 1.2 and 2.5, respectively. Here, we are interested in interfaces in quantum spin glasses and ask how quantum effects affect them. Specifically, we consider the model of a 2-dimensional Ising spin glass in a transverse field and at zero temperature. The corresponding Hamiltonian is given by Kopeć et al.[6]:

$$\mathcal{H} = -\sum_{\langle ij \rangle} J_{ij} \sigma_i^z \sigma_j^z - \Gamma \sum_i \sigma_i^x , \tag{1}$$

where $\sigma_i^x$ and $\sigma_i^z$ are the Pauli matrices for a spin at site $i$. The distribution of random nearest-neighbor couplings $J_{ij}$ is taken to be Gaussian with zero mean and dispersion $J$.

This model is believed to mimic the physics of proton glasses[7] and of some magnetic systems in the external field.[8]

Our studies are based on the local mean field approach[9] which allows one to determine the interfaces in a straightforward fashion. More importantly, this approach can easily be implemented in quantum situations. Although within the local mean field approximation one can never reach the true ground state, this method has turned out[4] to yield a fractal dimensionality of metastable interfaces in 2-dimensional Ising spin glass, which is consistent with the numerically exact transfer matrix estimates.[2,3]

Here, we will demonstrate that the quantum fluctuations make the interfaces shorter and smoother. The corresponding $D_f$ decreases monotonically with increase in $\Gamma$ and becomes equal to 1 above a critical value, $\Gamma_c$, of the transverse field. In the paramagnetic phase, the (longitudinal) local moments become equal to zero and the interfaces cannot be defined.

The fundamental simplification of the local mean field approach is the factorization of the density matrix into single site terms.[10] This leads to the following expressions for the $z$- and $x$-components of the local magnetic moments:

$$m_{zi} = \frac{H_i}{\sqrt{H_i^2 + \Gamma^2}} \tanh(\sqrt{H_i^2 + \Gamma^2}/k_B T) , \tag{2}$$

$$m_{xi} = \partial F/\partial \Gamma = \frac{\Gamma}{\sqrt{H_i^2 + \Gamma^2}} \tanh(\sqrt{H_i^2 + \Gamma^2}/k_B T) . \tag{3}$$

Here $H_i = \sum_j J_{ij} m_{zj}$, where $j$ is a nearest neighboring site to site $i$. We consider systems of size $L \times L$. At a given temperature we start from some trial spin configuration, typically obtained by slow cooling, and solve Eqs. (2) and (3) iteratively by proceeding from spin to spin sequentially. The convergence is assumed to be achieved if:

$$\max |m_{i,z,x}^{(n+1)} - m_{i,z,x}^{n}| < 10^{-8} , \tag{4}$$

where $n$ is the order of iteration. We study spin configurations which have been obtained on cooling the system to $T = 0$.

On application of the transverse field, the system becomes magnetized in the direction of the field, whereas it is either spin glassy or paramagnetic in the $z$ direction[8] depending on whether $\Gamma < \Gamma_c$ or $\Gamma > \Gamma_c$. The critical value $\Gamma_c(T = 0)$ above which the spin glass phase disappears at $T = 0$ is calculated for various $L$ and is extrapolated to $L = \infty$, as shown in Fig. 1. We find that $\Gamma_c/J = 4.05 \pm 0.02$ in this 2D system.

In order to study the interfaces, we let spins at boundaries of one direction to be aligned in the $z$-direction and the periodic boundary condition is applied in the remaining directions. Once a local energy minimum is found, the spins at one boundary are twisted antiparallel to the $z$-direction. The spin configurations obtained by solving Eqs. (2) and (3) in two cases are denoted by $\{m_i^P\}$ and $\{m_i^{AP}\}$, respectively. The interface is defined as a boundary separating the region with $m_{iz}^P \cdot m_{iz}^{AP} > 0$, from the region with $m_{iz}^P \cdot m_{iz}^{AP} < 0$. From Eq. (3), it is clear that $m_{ix}$ are always positive and no interface is generated in the transverse spin components.

In the classical limit, the $z$-spin components at $T = 0$ are of unit magnitude. This is no longer the case for $\Gamma \neq 0$, when $|m_{iz}|$ at $T = 0$ varies from site to site, as shown in Fig. 2. The profile, $P$, which is appropriate to characterize the width of the domain wall, is then given by:

$$P(i) = \frac{|m_{iz}^P - m_{iz}^{AP}|}{|m_{iz}^P| + |m_{iz}^{AP}|} . \tag{5}$$

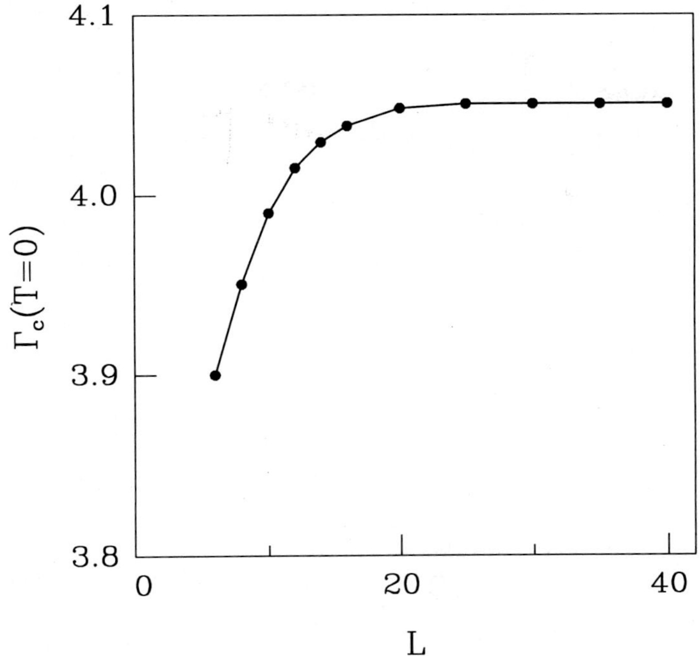


**Fig. 1** The dependence of $\Gamma_c(T = 0)$ on $L$ for the 2D systems. $\Gamma_c(T = 0)$ saturates at the value of $(4.05 \pm 0.02)J$.

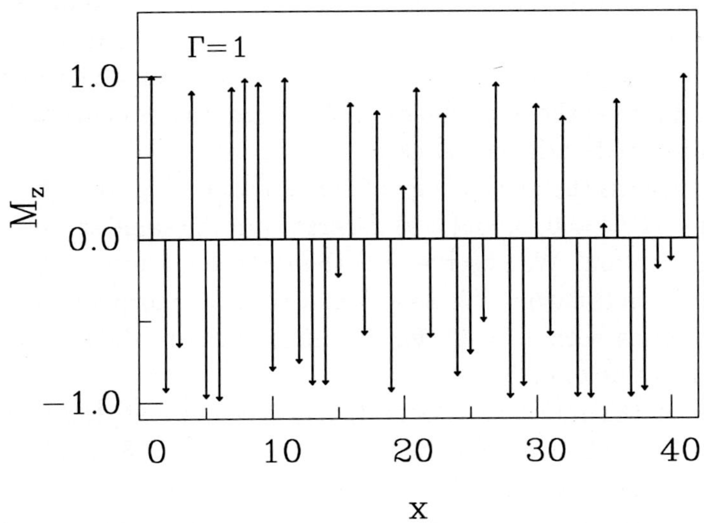


**Fig. 2** Typical values of $m_{iz}^P$ across the interface at $T = 0$ for $\Gamma = J$ ($L = 40$). In the classical case $|m_{iz}^{P,AP}|$ are equal to 1 everywhere.

The typical 2D profile for the $z$-component of spins is shown in Fig. 3 where $L = 20$ and 40, and $\Gamma = J$. In the classical case, the interface separates the phase characterized by $P(i) = 1$ from the phase with $P(i) = 0$. In the quantum case, one has two phases: one with $P(i) = 1$ and the other with $P(i) < 1$. The crossover region from $P(i) = 1$ to $P(i) = 0$, as one can see from Fig. 3, is of 2 lattice constants. This result remains valid for larger $L$. The

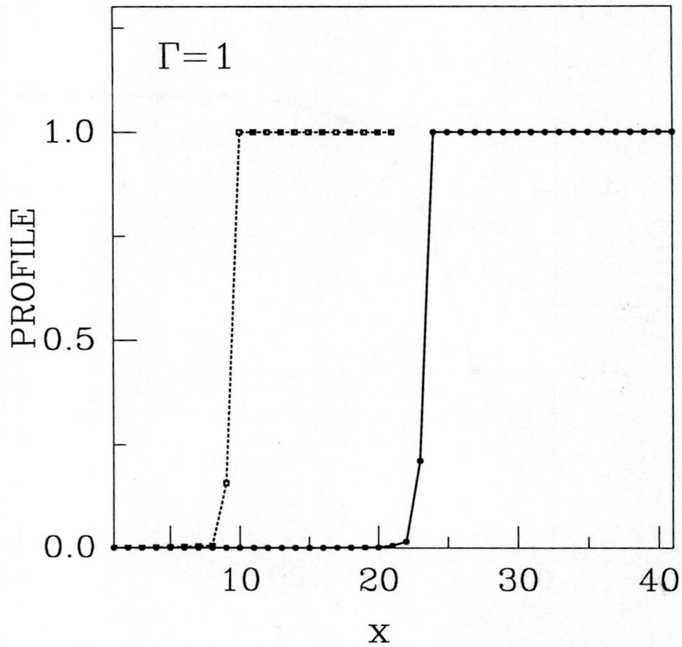


**Fig. 3** Typical profile $P(i)$ for $z$-component for D = 2, $\Gamma = J$, $L = 20$ (dotted line) and 40 (solid line). Clearly, the width of the crossover region from $P(i) = 1$ to $P(i) = 0$ is of 2 lattice constants (in the classical case the corresponding width is equal to zero). $x$ denotes the direction across the interface.

width of the domain wall becomes somewhat larger when $\Gamma$ approaches $\Gamma_c$. It is expected that an exact solution would generate interfaces with widths that diverge at $\Gamma_c$. Thus the local mean field method generates interfaces of finite width but this width is bounded. This conclusion is also valid for the one-dimensional system, the domain wall width of which is, however, larger compared with the 2D case.

We now turn to discuss the morphology of domain walls. Figure 4 shows the interfaces for different values of $\Gamma$ ($\Gamma/J = 0$, 3, and 3.9) generated in the same sample and with the same starting spin configuration. We observe that the quantum fluctuations make the length of the interfaces shorter and flatter. We have found that the morphology of the domain walls of 2D quantum systems with $L = 40$, for example, starts to differ from the classical ones when $\Gamma$ exceeds the value of $0.25J$.

The smoothing of the interface is reflected in the gradual decrease of the fractal dimensionality with $\Gamma$, as shown in Fig. 5 for $\Gamma/J = 0$, 3, and 3.9. The number of samples considered in the simulations ranges from 11 000 for $L = 4$ to 600 for $L = 40$. The averaging over samples is found to be more important than the averaging over starting spin configurations, so we generate one interface per sample. The best fit dives $D_f = 1.27$, 1.24, and 1.16 (all $\pm 0.04$) for $\Gamma/J = 0$, 3, and 3.9, respectively. The $\Gamma = 0$ value is consistent with the previous results.[2–4] Our results, although obtained for metastable states, were determined by studying system sizes which were much larger than in the transfer matrix studies, where $L$ of up to 12 was considered. It is possible, however, that the $D_f$ we obtained for a finite $\Gamma$ is merely a crossover value, valid up to some finite length scale. This point needs to be clarified by other methods.

It is known that the SG phase is chaotic.[2] The chaotic behavior is characterized by the Lyapunov exponent $\zeta$. In the case of Ising spins $D_f = 2\zeta + y$, where $y$ is the $T = 0$ scaling

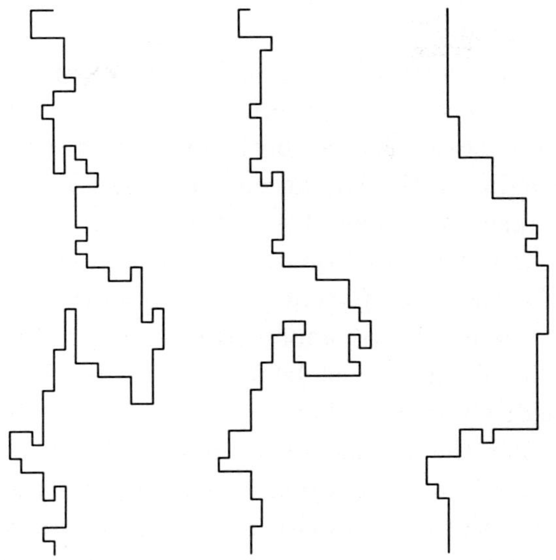


**Fig. 4** The interfaces of 2D systems ($L = 40$) for $\Gamma/J = 0$, 3, and 3.9. Their length gets shorter with increasing of $\Gamma$. The overhangs are present in all cases.

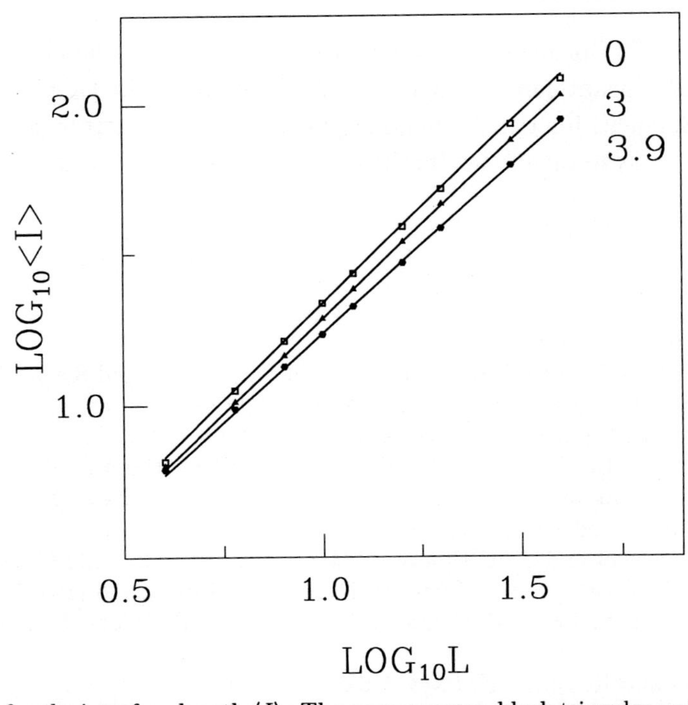


**Fig. 5** Scaling plot for the interface length $\langle I \rangle$. The open squares, black triangles and hexagons correspond to $\Gamma = 0$, $3J$ and $3.9J$ for 2D, respectively. The values of $\Gamma/J$ are also shown next to the curves. Straight lines are the best fits. For $\Gamma/J = 0$, 3, and 3.9, we have $D_f$ of 1.27, 1.24, and 1.16 ($\pm 0.04$), respectively.

exponent. The value of $\zeta$ (and consequently of $D_f$) is found to become smaller in the critical region.[11] In analogy with this classical case, one can expect that in the region very close to the $T = 0$ quantum phase transition, $D_f$ may have a value different from that for $\Gamma < \Gamma_c$.

Study of this question requires a determination of $\Gamma_c$ with high accuracy which is hard to reach in the local mean field approximation.

We now discuss the morphology of the interfaces when the periodic and antiperiodic boundary conditions are applied to the $x$-component of the spins. In this case, the periodic boundary condition means that at both boundaries $m_{ix}$ are kept at 1. For the antiperiodic boundary condition, the $m_{ix}$ on one boundary are equal to 1, whereas on the opposite boundary they take on a value of $-1$. The $m_{iz}$ on the boundaries are, therefore, equal to 0 for both two boundary conditions. It follows, from Eq. (2), that $P(i) = 0$ and no interface is created for the $z$-component. One can show that in the local mean field approximation, the interface for the $x$-component is a straight line without any roughness. This fact may be understood since $m_{ix}$ is defined via $m_{iz}$ by Eq. (3).

In this paper, we have demonstrated that the morphology of the domain walls is affected by the quantum fluctuations: the interfaces become smoother and flatter. Another interesting problem is to determine how the roughness of weakly disordered Ising ferromagnets[12] becomes affected by application of a transverse field. Our preliminary studies show that roughness, already in the classical limit, is much more sensitive to the nonequilibrium aspects of the local mean field method than is the fractal dimensionality, and it is hard to get any reliable estimates. This problem should be studied by some other methods.

## ACKNOWLEDGMENTS

The discussions with P. Gornicki are appreciated. This research has been supported by the Polish Agency KBN (grant number 2 P302 127 07). M. S. Li thanks the Japan Society for Promotion of Sciences for the financial support and the generous allocation of computing time on the Sun system at the Institute of Theoretical Physics, Cologne University, Germany.

## REFERENCES

1. F. Family and T. Vicsek, eds., *Dynamics of Fractal Surfaces* (World Scientific, Singapore, 1991).
2. A. J. Bray and M. A. Moore, *Phys. Rev. Lett.* **58**, 57 (1987).
3. M. Cieplak and J. R. Banavar, *J. Phys.* **A23**, 4385 (1990).
4. M. Cieplak and M. S. Li, *Fractal* **2**, 481 (1994); the result for the fractal dimensionality in three dimensions has been quoted as $2.57 \pm 0.07$ — our reduced error bars given in the current paper are based on an increased statistics.
5. M. Cieplak, A. Maritan and J. R. Banavar, *Phys. Rev. Lett.* **72**, 2322 (1994).
6. T. K. Kopeć, K. D. Usadel and G. Büttner, *Phys. Rev.* **B39**, 12418 (1989); H. Rieger and A. P. Young, *Phys. Rev. Lett.* **72**, 4141 (1994); M. Guo, R. N. Bhatt and D. Huse, *Phys. Rev. Lett.* **72**, 4144 (1994).
7. R. Pirc, B. Tadić and R. Blinc, *Z. Phys.* **B61**, 69 (1985).
8. W. Wu, D. Bitko, T. F. Rosenbaum and G. Aeppli, *Phys. Rev. Lett.* **71**, 1919 (1993).
9. C. M. Soukoulis, G. S. Grest and K. Levin, *Phys. Rev. Lett.* **48**, 1756 (1982).
10. M. S. Li, R. A. Hassan and M. Cieplak, *J. Phys.: Condens. Matter* **6**, 2595 (1994).
11. M. Nifle and H. J. Hilhorst, *Phys. Rev. Lett.* **68**, 2992 (1992); M. S. Li and M. Cieplak, *Physica* **A197**, 5022 (1993).
12. D. A. Huse and C. L. Henley, *Phys. Rev. Lett.* **54**, 2708 (1985).

Fractals, Vol. 4, No. 3 (1996) 407–414


# DYNAMICAL BEHAVIOR OF ACOUSTICALLY COUPLED CHAOS OSCILLATORS

Y. HARADA, K. MASUDA and A. OGAWA
*Department of Applied Physics, Faculty of Engineering, Fukui University, Bunkyo 3-9-1, Fukui, Japan*

**Abstract**

We have found, for the first time, that Acoustically Coupling Chaos Oscillator (ACCO) also exhibits a chaotic attractor called the "double scroll Chua's attractor" by using the modified Chua's circuit. When each ACCO is in an oscillating periodic state, the interaction of sound waves with acoustic coupling of two ACCOs can also result in the appearance of chaotic sound waves.

## 1. INTRODUCTION

It has been generally accepted that Chua's circuit is a very suitable subject for the study of dynamical chaos by means of both laboratory experiments and computer simulations.[1] Chua's circuit[2] is the simplest autonomous circuit that can exhibit bifurcation and chaos. It has been studied extensively and is one of the very few circuits in which a formal proof of the existence of chaos has been accomplished. Moreover, theoretical and simulated behavior of this circuit can be accurately reproduced experimentally. The chaotic nature of double scroll Chua's attractor was proved in Chua et al.[3] and Silva[4] by establishing the existence of a homoclinic loop of the saddle-focus at the origin for some values of parameters and by applying the Shil'nikov theorem.

In this paper, we report the experimental results and the implementation details of acoustically coupled chaos oscillator by using Chua's circuit. We have shown that the acoustically coupled chaos oscillator produces chaotic signals whose attractor is double scroll.

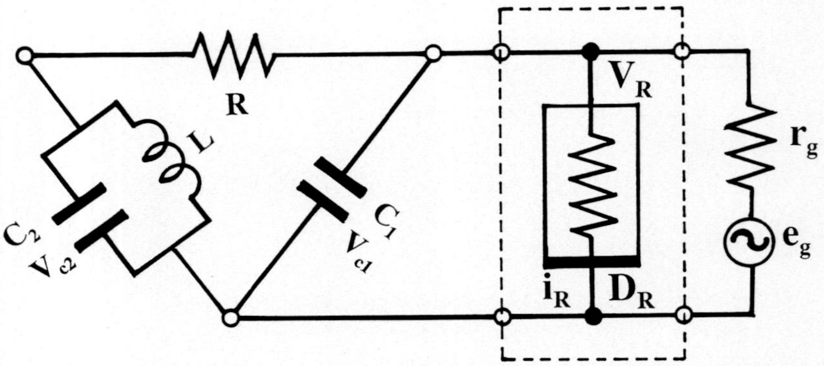


**Fig. 1** Construction of our chaos oscillator. The dashed line box is a nonlinear resistor. This part is called as Chua's diode ($D_R$). We choose a piecewise-linear function for nonlineality which induces period doubling route to chaos. $C_1$ and $C_2$ are 0.00068 $\mu$F and 0.0068 $\mu$F, respectively. $L$ is 3.3 mH. These values are fixed in our all experiments. $R$ is the control parameter in our oscillator.

Chua's circuit is a nonlinear electronic circuit that is the object of much scientific research activities.[1] The circuit contains four linear elements and a nonlinear resistor which can be built using op-amps as shown in Fig. 1. The state equation of Chua's oscillator are as follows:

$$\frac{dv_1}{dt} = \frac{1}{C_1}\{G(v_2 - v_1) - f(v_1)\}, \tag{1}$$

$$\frac{dv_2}{dt} = \frac{1}{C_2}\{G(v_1 - v_2) + i_3\}, \tag{2}$$

$$\frac{dv_3}{dt} = -\frac{1}{L}(v_2 + R_0 i_3), \tag{3}$$

where

$$G = \frac{1}{R}, \tag{4}$$

and

$$f(v_1) = G_b v_1 + \frac{1}{2}(G_a - G_b)\{|v_1 + E| - |v_1 - E|\} \tag{5}$$

is the $v-i$ characteristic of the nonlinear resistor $D_R$ called Chua's diode, with a slope equal to $G_a$ in the inner region and $G_b$ in the outer region. A typical $v-i$ characteristic of $D_R$ is the piecewise-linear function as done by Chua et al. By choosing appropriate values for $G_a$, $G_b$ and $E$, any continuous three-segment odd-symmetric piecewise-linear $v-i$ characteristic for Chua's diode can be specified. We choose appropriate values for $G_a$, $G_b$ and $E$ which induce period doubling bifurcation route to chaos with double scroll attractors.

## 2. CHAOS OSCILLATORS — CHUA'S CIRCUIT

In order to generate the acoustic coupling chaos oscillator, we chose the same frequency band from that of the acoustic transducer as discussed in the next section. In this study, adopted nonlinear characteristic was piecewise-linear functions. In our experimental set-up, the resistor value $R$ was varied as a control parameter and the following values were fixed: $C_1 = 0.00068\ \mu F$, $C_2 = 0.0068\ \mu F$, $L = 3.3$ mH, $E = 1\ V$ and $G_a < G_b < 0$. Although any circuit which realizes the nonlinear $v - i$ characteristic Chua's diode can be used experimentally, we used two diodes and one op-amp.

Decreasing the value of $R$, we observed time waveforms (time series) of $V_{c1}(t)$ and $V_{c2}(t)$, attractors projected on the $V_{c1} - V_{c2}$ plane and power spectrum of $V_{c1}(t)$. Until $R = 1.416\ k\Omega$, the attractor was point attractor, i.e., stable equilibrium point. At a threshold value of $R = 1.416\ k\Omega$, an equilibrium point lost stability and stable periodic attractor (limit cycle) emerged through an Hopf bifurcation. Further decreasing of $R$ led to the period doubling bifurcation. We observed period two and period four behavior clearly at $R = 1.388\ k\Omega$ and $R = 1.380\ k\Omega$, respectively. From $R = 1.374\ k\Omega$ to $R = 1.334\ k\Omega$, we observed double scroll attractor which is the typical attractor in the Chua's circuit. The power spectrum of $V_{c1}$ is continuous for the wide range of frequency. This double scroll attractor is constructed by connection of two screw attractors which are symmetric around the original point. Homoclinicity in the double scroll attractor is the origin of the chaotic behavior.[1]

Examples of time waveforms of $V_{c1}$ and $V_{c2}$, the double scroll attractor and the power spectrum of $V_{c1}$ are shown in Figs. 2, 3 and 4, respectively. From these results, chaotic behavior is confirmed.

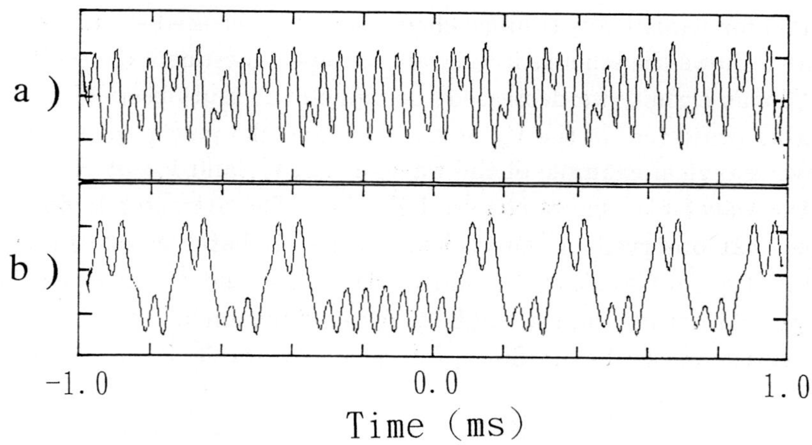


**Fig. 2** Time waveforms of: **(a)** $V_{c1}$ and **(b)** $V_{c2}$ ($R = 1.334\ k\Omega$). Chaotic behaviors are found in them. Different waveforms between them are due to the difference in capacitances between $C_1$ and $C_2$.

**Fig. 3** The attractor projected on a $V_{c1} - V_{c2}$ plane ($R = 1.334\ k\Omega$). This attractor is double scroll attractor which is a typical one in the Chua's circuit ($R = 1.334\ k\Omega$). This attractor is constructed through the connection of two screw attractors.

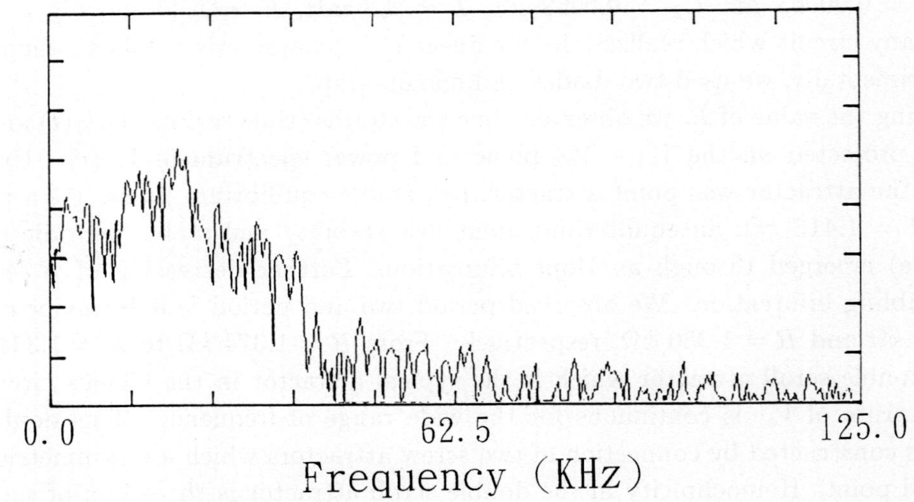


**Fig. 4** Power spectrum of $V_{c1}$. This continuous spectrum is chaotic signal ($R = 1.334\ k\Omega$).

## 3. ACOUSTICALLY COUPLED CHAOS OSCILLATORS

As a model system for acoustically coupled chaos oscillator, we used an ultrasonic transducer MA23L3. Figure 5 shows the schematic diagram of these experimental circuits. As was done in the case of Chua's circuit, we decreased the value of $R$ and observed waveforms of $V_{c1}$ and $V_{c2}$, attractors projected on the $V_{c1} - V_{c2}$ plane and the power spectrum of $V_{c1}$.

Figure 6 shows a typical example of the waveform of $V_{c1}$ and $V_{c2}$ at $R = 1.424\ k\Omega$. This time series of the waveforms shows chaotic behavior. The attractor is double scroll type which resembles that observed in the typical Chua's oscillator as shown in Fig. 7. The double scrolls in a typical Chua's circuit have holes at $P^+$ and $P^-$ in which two coexisting homoclinic loops originating from a saddle-focus equilibrium point located at the origin of an odd-symmetric vector field.[3] Since $P^+$ is an "unstable focus" when restricted to

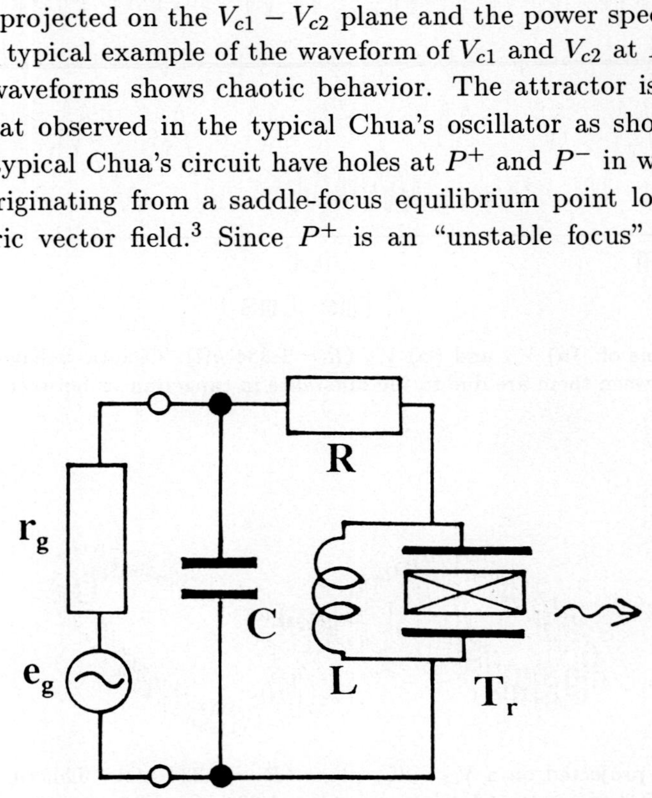


**Fig. 5** Our acoustic chaos oscillator which utilize ultrasonic transducer MA23L3.

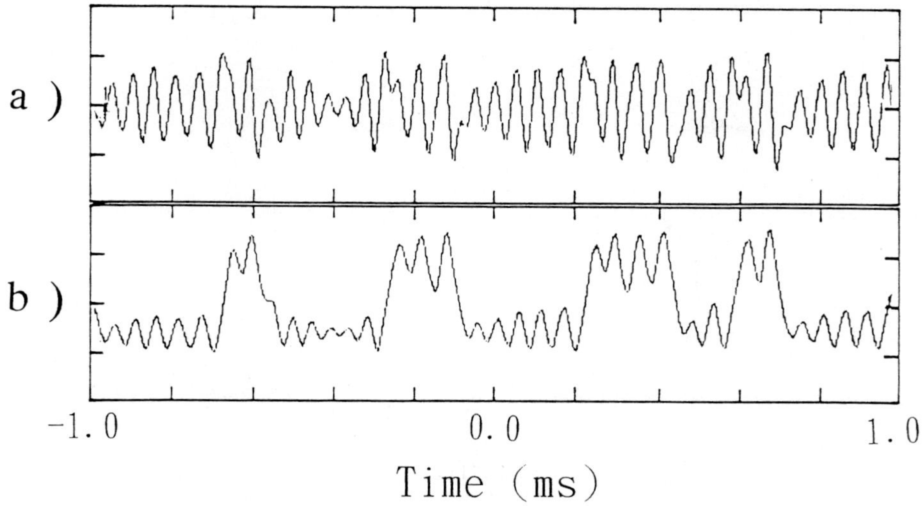


**Fig. 6** Chaotic time waveform in the acoustic chaos oscillator. **(a)** $V_{c1}$ and **(b)** $V_{c2}$ ($R = 1.424\ k\Omega$).

**Fig. 7** Hole-filling Double scroll attractor. The center of scrolls fluctuate in time due to two peaks in the band width of the transducer MA23L3 ($R = 1.424\ k\Omega$).

the eigenspace, it follows that the resulting double scroll will not have a hole and henceforth called a hole-filling orbit. The double scroll shown in Fig. 7 corresponds to a case in hole-filling heteroclinic orbit. The behavior of attractors from a periodic orbit to double scroll attractor for the case of Chua's circuit is clearly different from the hole-filling heteroclinic chaotic orbit. We confirmed that the ACCO exhibits a hole-filling heteroclinic chaotic behavior. The power spectrum closely resembles that in the case of the Chua's circuit and it shows a continuous spectrum as shown in Fig. 8. The behavior of attractors is, limit cycle → screw attractor → double scroll attractor, without period doubling bifurcation. We can conclude that our acoustic oscillator is able to generate chaotic signals. This fact leads to the possibility of the large number of ACCO system.

As a prototype of such coupled systems, we constructed one pair of acoustic chaos oscillators as shown in Fig. 9. As an example, we investigated a coupling between two limit cycles. Each oscillator's $R$ was set at a value which generates simple periodic signal (Fig. 10). If two oscillators were coupled by acoustic periodic waves, both waveforms of $V_{c1}$ became chaotic as in Fig. 11. This chaotic behavior was also recognized by the continuous power spectrum and calculated Lyapunov exponent.[5] When we shut the coupling, limit cycle oscillations were recovered. Therefore chaotic behavior was clearly induced by the acoustic coupling.

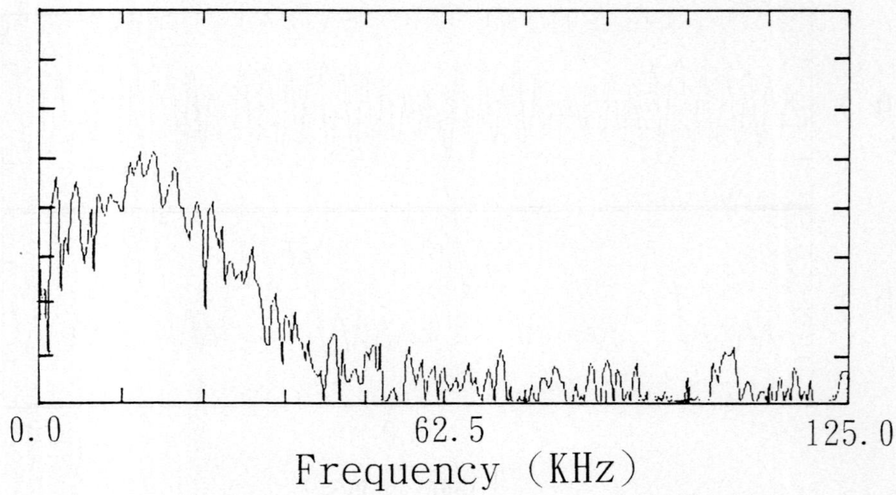


**Fig. 8** Power spectrum of $V_{c1}$. Continuous spectrum typical in chaotic signals is apparent ($R$=1.424 $k\Omega$).

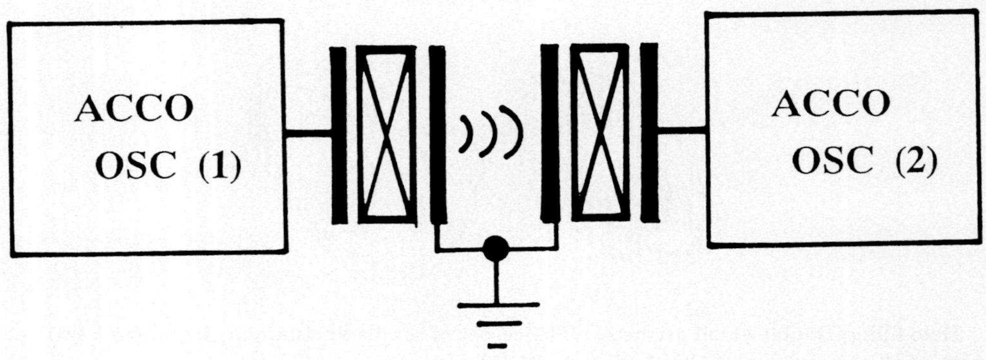


**Fig. 9** Schematic figures of Acoustically Coupled Chaos Oscillator (ACCO). Two chaos oscillators are coupled by ultrasonic sound.

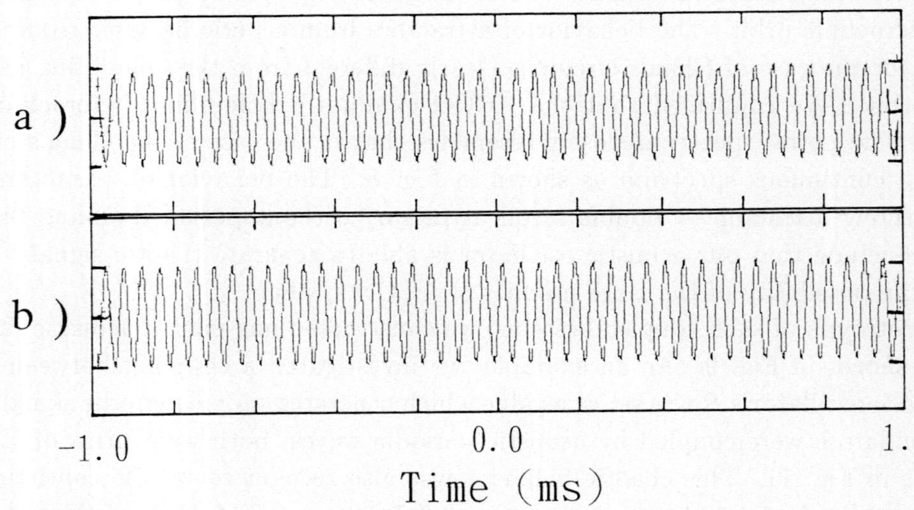


**Fig. 10** Time waveforms of $V_{c1}$ in **(a)** oscillator 1 and **(b)** oscillator 2 without coupling between two oscillators. Simple periodic states are realized.

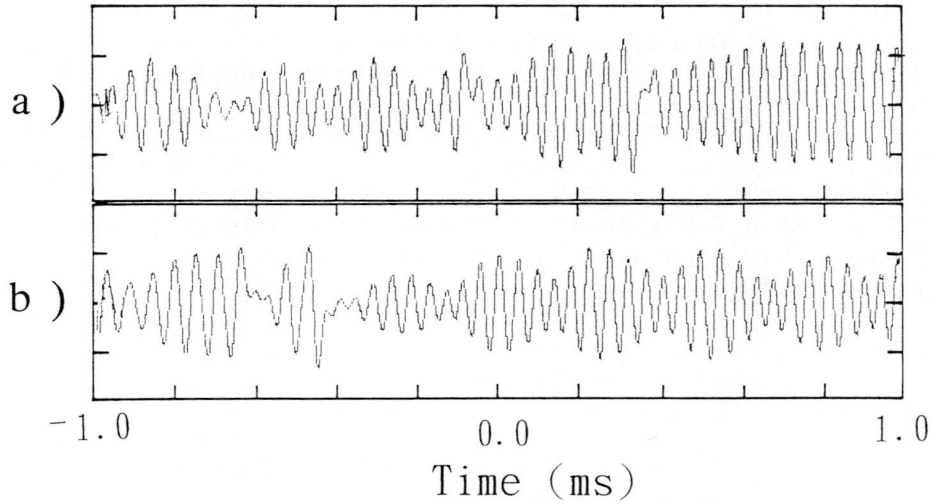


**Fig. 11** Time waveforms of $V_{c1}$ in **(a)** oscillator 1 and **(b)** oscillator 2 with acoustic coupling. In both oscillators chaotic signals are realized. If we shut the coupling physically, simple periodic waveforms are recovered for both oscillators.

As in other types of coupling, we are investigating the case of the periodic-chaos and chaos-chaos couplings now.

## 4. CONCLUSIONS

In this study, we have realized an acoustic chaos oscillator and its coupling utilizing transducers MA23L3 on Chua's circuit. A variety of interesting features were confirmed as follows:

(1) The route to chaos in ACCO is, limit cycle → screw attractor → double scroll attractor.
(2) Acoustic oscillator can exhibit chaotic behavior and the attractor is hole-filling double scroll attractor, due to the electro-mechanical effect.
(3) Due to the acoustic interactions between two ACCOs, two periodic oscillations change to chaotic behaviors.

Our ACCO would lead to wide applications such as electro-mechanical devices, chaos communications[6] and a typical model of large numbers of globally coupled chaos oscillators.[7]

## ACKNOWLEDGMENTS

The authors would like to thank Mr. H. Uesaka for the technical assistance and Mr. Y. Okagawa for the help of experiments. We acknowledge Mr. J. Inoue of MURATA MFG. CO., LTD. for preparing transducers (MA23L3).

## REFERENCES


1. L. P. Shil'nikov, *Int. J. Bifurcation and Chaos* **3**, 489 (1994).
2. L. O. Chua, C. W. Wu, A. S. Huang and G. Q. Zhong, *IEEE Trans. Circuits and Systems I* **40**, 732 (1993).
3. L. O. Chua, M. Komuro and T. Matsumoto, *IEEE Trans. Circuit Syst. CAS* **33**, 1072 (1986).
4. C. P. Silva, in *Chua's Circuit: A Paradigm for Chaos*, ed. R. N. Madan (Singapore, World Scientific, 1993), p. 671.
5. A. Wolf, J. B. Swift, H. L. Swinny and J. A. Vastano, *Physica* **D16**, 285 (1985).
6. K. M. Cuomo and A. V. Oppenheim, *Phys. Rev. Lett.* **71**, 65 (1993).
7. K. Kaneko, *Physica* **D34**, 1 (1989).

Fractals, Vol. 4, No. 3 (1996) 415–425


# CAN STATISTICAL PHYSICS CONTRIBUTE TO THE SCIENCE OF ECONOMICS?

MICHAEL H. R. STANLEY*, LUÍS A. N. AMARAL*, SERGEY V. BULDYREV*, SHLOMO HAVLIN*, HEIKO LESCHHORN*, PHILLIPP MAASS*, MICHAEL A. SALINGER†, and H. EUGENE STANLEY*

**Center of Polymer Studies and Department of Physics, Boston University, Boston, MA 02215, USA*

*†School of Management, Boston University, Boston, MA 02215, USA*

## Abstract

In recent years, a breakthrough in statistical physics has occurred. Simply put, statistical physicists have determined that physical systems which consist of a large number of interacting particles obey universal laws that are independent of the microscopic details. This progress was mainly due to the development of scaling theory. Since economic systems also consist of a large number of interacting units, it is plausible that scaling theory can be applied to economics. To test this possibility we study the dynamics of firm size. This may help to build a more complete characterization of the nature and processes behind firm growth.

To date, the study of firm dynamics has primarily focused on whether small firms on average have higher growth rates than large firms. To a lesser extent, attention has been placed on the relationship between firm size and variation in growth rate. Our research goes beyond these questions by looking at the relationship between numerous firm characteristics and the entire distribution of growth rates. Thus, it may provide a better understanding of the mechanisms behind firm dynamics. In contrast to previous studies, this research analyzes data over many time scales, instead of just a single time interval.

From a scientific standpoint, this work could be useful because it will affect the formulation of firm modeling — one of the basic building blocks of all economic analysis. In addition, this work will have practical applications. For example, there are Federal policies that are designed to encourage small businesses. While such policies might be justified on grounds other than their contribution to growth, any systematic difference in the growth rates of small and large firms might be relevant for evaluating such policies. Also, there has traditionally been a concern that an excessive amount of economic activity might become concentrated in a small number of firms. A more detailed understanding of the firm growth process will provide evidence for whether such concerns have any scientific foundation.

## 1. INTRODUCTION

The objective of our recent research has been to use methods from statistical physics to provide a more complete description of the dynamics of firm growth. Economists have already studied the static distribution of firms[1–5] and the relationship between firm size on the one hand, and the mean and standard deviations of growth rates on the other.[2,6–8] A key exercise in statistical physics is to examine the distributions of observable features of complex dynamic systems at different scales. In studying turbulent fluids, for example, statistical physicists examine the evolution of the velocity distribution over time. In studying the dynamics of firms, two natural scales to study are size and time. From the standpoint of statistical physics, the relationship between size on the one hand, and the mean and standard deviations of the growth rates on the other is not a complete representation of firm dynamics. Rather, a complete description would include how the entire shape of the distribution varies with size. Similarly, every existing study of firm growth has examined growth rates over a single time scale. A complete description of the dynamics would include how the entire distribution of growth rates evolves over different time scales.

The theory of the firm is fundamental to economic analysis; yet, the classical model of the firm, in which the firm is viewed simply as a production function, is inadequate for many applications in economics. There is an older literature about alternative theories of the firm,[9–11] but the classical model has survived as the dominant one.[12] Recently, however, it has been recognized that more sophisticated models are needed for many applications, most notably in the areas of corporate finance, labor economics, law and economics, industrial organization, and management strategy. As a result, there has been a recent revival in research on the nature of the firm.[13–18] Much of this work has been theoretical. The objective of this research is to provide a set of empirical findings with which future theories of the firm must be consistent. There is a tradition in economics of reconciling theories of the firm with statistical analysis of firm size and growth. For example, both Lucas'[19] seminal paper on the distribution of firm size and Nelson and Winter's[15] work on firm and economic growth made use of Gibrat's Law.

Our results should have potential applications in many areas of economics. For example, ever since the seminal work of Solow,[20] it has been generally accepted that technical change (as opposed to capital accumulation) is the primary source of economic growth. There are a number of hypotheses about the relationship between firm size and efficiency in doing research and development (R&D).[21,22] A better understanding of the relationship between size and firm growth can provide insights into which (if any) of these hypotheses is correct. Another application is the growing literature on the role of internal capital markets in allocating funds for investment.[23–26] One view holds that internally generated funds are irrelevant for investment decisions because funds for profitable projects can be raised on the capital markets. The alternative view is that there are significant costs to external forms of finance so that the availability of internally generated funds does affect investment decisions. However, there are competing hypotheses about the implications of the importance of the availability of internally generated funds. Under the free cash flow hypothesis,[27] the availability of funds gives managers the freedom to pursue their own objectives at the expense of shareholders. The alternative is that the availability of funds makes it possible to finance profitable investments. An examination of the entire distribution of growth rates as a function of firm characteristics (including but not limited to size) can shed light on the relative importance of each of these hypotheses.

Before discussing the specifics of this research, we should note that this work represents a collaboration between scholars from two different disciplines: statistical physics and economics. The model of statistical physicists exploring topics in economics in daily collaboration with economists has great potential. The methodologies of economics and statistical physics have evolved in different ways and viewing economics through the lens of statistical physics, we can provide a fresh perspective. Currently, there are several preliminary results which, if confirmed by more extensive work, could be of significance.[5,28,29]

## 2. SCALING THEORY

Recently, statistical physicists have formulated scale invariance theory, also known as scaling theory. This has led to a breakthrough in our understanding of complex physical and chemical systems that consist of a large number of interacting particles.[30–33]

The main concept of scaling theory is that the properties of a many-particle system under certain conditions will become scale invariant, i.e., the behavior of a system at short distances or times is similar to the behavior of a system for long distances or times. Thus a unique scaling function can describe the whole range of phenomena. This scaling function does not depend on most of the microscopic details of the system, such as inter-molecular interactions or the structure of the molecules, but only on a few fundamental properties. In statistical physics, this phenomena is called universality. Thus, knowledge of the properties of a model system at short times or short length scales can be used to predict the behavior of a real system at large times and large length scales.

Since economic systems consist of a large number of interacting units, it is plausible that they might be amenable to scaling analysis. In fact, a recent study determined that the fluctuations in the S&P 500 index exhibited scaling behavior.[29] Using 1.5 million records at 1 minute intervals over 6 years of trading, Mantegna and Stanley determined that fluctuations on a one minute time interval were reflected in 10-minute, 100-minute, and 1000-minute intervals.[29] The distribution of index returns fits a Lévy (similar to a Pareto) distribution with an exponential drop off in the tails, as shown in Fig. 1. These scaling properties mean that viewing stock market returns on one minute intervals provides insight on 1000-minute intervals. Unlike other models of the stock market, e.g., ARCH/GARCH models,[34,35] Mantegna and Stanley demonstrated that even the crash of October 1987 — an event that occurs only once every several million minutes — can be characterized by the perturbations that obey scaling theory.

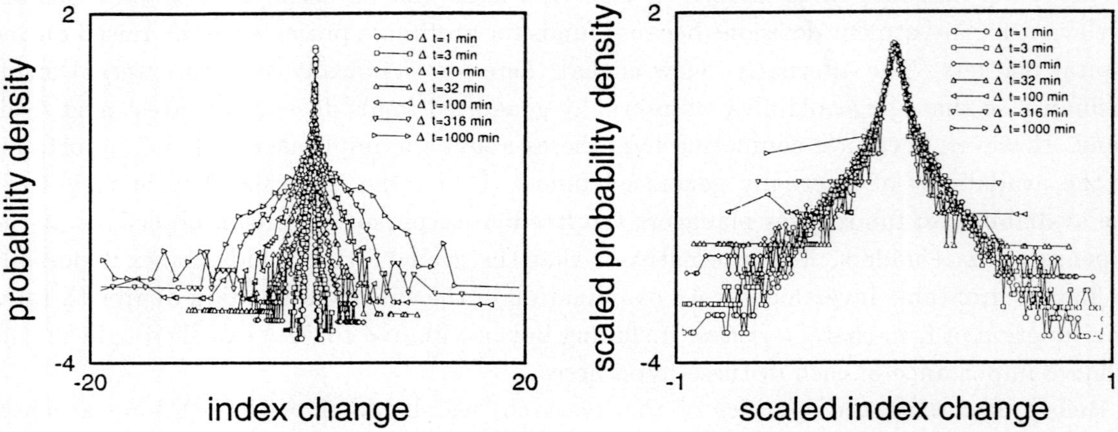


**Fig. 1** (a) Probability distribution of the S&P 500 index variations observed at time intervals $\Delta t$ which range from 1 to 1000 minutes. By increasing $\Delta t$, spreading of the probability distribution is observed. (b) By scaling the plot in (a), all the data collapses on the $\Delta t = 1$ min. distribution. The points outside the average behavior define the noise level of that specific distribution.

## 3. SIZE DISTRIBUTION OF FIRMS

### 3.1 Review of Recent Results

Gibrat[1] suggested a simple stochastic dynamic model of firm growth that predicts the distribution of firm size to be log-normal. Later research by Simon and Bonini[3] suggested that the distribution of the largest firms is a power law, resulting in an upper tail of the overall distribution that is too thick relative to the log-normal. Their result has to be viewed under the restrictions of the time. For example, for a comprehensive data set, they used the Fortune 500, which is only a list of the largest firms. Using both a more complete database, *Compustat* (which includes all publicly-traded US manufacturing firms), and techniques only recently developed in statistical physics, we confirmed that the log-normal fits the data reasonably well. However, in contrast to previous findings, we found that the upper tail of the distribution is too thin relative to the log-normal, not too fat.[5] In all likelihood, we could not have made this discovery without the use of "Zipf plots" — a technique of statistical physics used to highlight the upper tail of a distribution.

In Fig. 2(a), we show the histogram of the data fit by a log-normal curve. From this graph, it is not possible to discern any systematic deviations in the upper tail. However, Fig. 2(b) utilizes the "Zipf plot" and the deviation becomes remarkably pronounced.

### 3.2 Open Questions

Our results suggest that the sales of very large firms grow by an entirely different mechanism than the sales of smaller firms. Much work remains to be done to test the above ideas. A few of the questions that we seek to answer are:

- Does the distribution of other characteristics, such as the number of employees, exhibit a similar shape?

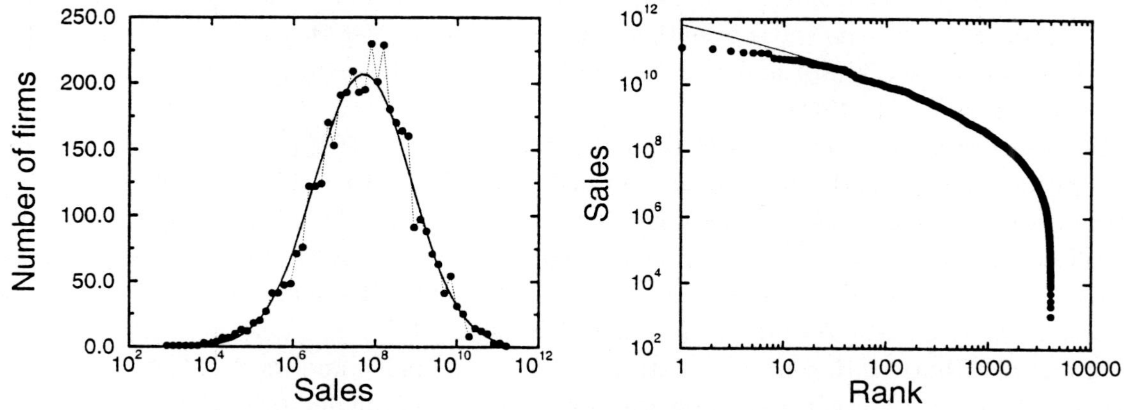


**Fig. 2** **(a)** Histogram showing the number of firms as a function of 1993 sales. The data are for the 4071 *Compustat* firms with standard industrial classification codes 2000-3999. The value of the sales are binned in powers of $2^{1/2}$. The solid curve is a log-normal fit to the data using the mean of the log of sales and the standard deviation of the log of sales as fitting parameters. From the histogram, the deviations in the upper tail are not noticeable. **(b)** The data are a Zipf plot (double logarithmic plot of sales vs. rank) for the same sample as in (a). The solid line is a predicted Zipf plot from the log-normal fit shown in (a). From the Zipf plot, the deviations in the upper tail are easily noticed.

- Are the distributions of firms outside the manufacturing sector similar to the distribution in Fig. 2?
- How does the distribution in Fig. 2 appear for different countries? Do the distributions for different economies obey the scaling theory?

## 4. GROWTH RATE DISTRIBUTION

### 4.1 Review of Recent Results

In statistical physics, it is common to explore the underlying mechanisms of a complex dynamic process by studying the time evolution of distribution functions. Using the *Compustat* database over the time period 1975–1991, our results suggest:

(i) The distribution of annual logarithmic growth rates for firms with approximately the same size displays an exponential form.
(ii) The fluctuations in the growth rates — measured by the width (or standard deviation) of this distribution — decrease exponentially with the logarithm of firm size.

We made plots on a double logarithmic scale of the probability density function of the real growth rates in sales from 1990 to 1991. Independent of the initial sales, the curves display a simple "tent-shaped" form. Hence the distribution is not Gaussian — as expected from the Gibrat hypothesis[1] — but rather is exponential, with the probability density $p$ obeying the following relation:

$$p(r_1|s_0) = \frac{1}{\sqrt{2}\sigma_1(s_0)} \exp\left(-\frac{\sqrt{2}|r_1 - \bar{r}_1(s_0)|}{\sigma_1(s_0)}\right), \tag{1}$$

where $s_0$ is the log of the initial year sales and $r_1$ is the logarithm of the growth rate, defined as the ratio of next years' sales over this year's (the initial year's) sales. We calculated the average growth rate $\bar{r}_1(s_0)$ and the standard deviation $\sigma_1(s_0)$ by fitting the data set to Eq. (1).

We also tested the robustness of the exponential function of Eq. (1) for other one-year intervals. We find that the data for *each* annual interval from 1975–1991 fit well into Eq. (1), with only small variations in the parameters $\bar{r}_1(s_0)$ and $\sigma_1(s_0)$. To improve the statistics, we therefore calculated a new histogram by averaging all the data from the 16 annual intervals in the database. The data, converting sales to 1987 constant dollars and combining, now scatter much less and the shape is well described by Eq. (1).

In any distribution, it is crucial to study its shape, its mean and its standard deviations. One notices that the larger firms grow less rapidly than the smaller ones. We quantitatively analyze the relation of the mean growth rate over one-year, $\bar{r}_1(s_0)$, and initial sales, $s_0$. The data suggests that while there is a slight trend toward slower growth in large firms, the mean growth rate is relatively constant for relatively large firms.

We also study the dependence on $s_0$ of $\sigma_1(s_0)$ which decreases with increasing $s_0$. We quantitatively analyze this feature by plotting $\sigma_1(s_0)$ vs. $s_0$, and we find that the standard deviation is well approximated over more than 7 orders of magnitude (from sales of less than $10^4$ dollars up to sales of more than $10^{11}$ dollars) by the law:

$$\sigma_1(s_0) = a \exp(-\beta s_0), \tag{2}$$

where $a \simeq 6.66$ and $\beta = 0.15 \pm 0.01$. In physical systems, it is not unusual to find such simple relationships holding over a wide range, but such a relation for economic data is striking and merits further analysis.

In order to further check the robustness of our findings, we performed a parallel analysis for the number of employees, which is also given in the *Compustat* database for each successive year. We find that the analogs of $p(r_1|s_0)$ and $\sigma_1(s_0)$ behave in the same fashion. For the standard deviation of the number of employees, we find that the data are linear over roughly 5 orders of magnitude, from firms with only 10 employees to firms with almost $10^6$ employees. The slope $\beta = 0.16 \pm 0.01$ is the same, within error bars, as found for the sales.

In statistical physics, scaling phenomena of the sort that we have uncovered in the sales and employee distribution functions are sometimes represented graphically by plotting a suitably scaled dependent variable as a function of a suitably scaled independent variable. If scaling holds, then the data for a wide range of parameter values are said to "collapse" upon a *single* curve. To test the present data for such data collapse, we show in Fig. 3 the scaled probability density as a function of the scaled growth rates of sales and employees. The data indeed are found to collapse upon a single curve. This collapse suggests that the scaling theory might be applicable to the system of firms. The high degree of similarity in the behavior of sales and the number of employees points to the existence of large mutual correlations between sales and the number of employees. We hope to quantitatively determine these correlations in our future work.

## 4.2 Open Questions

We plan to complete and extend our study in order to get a comprehensive picture of the firm dynamics described in terms of conditional distribution functions. Because age has

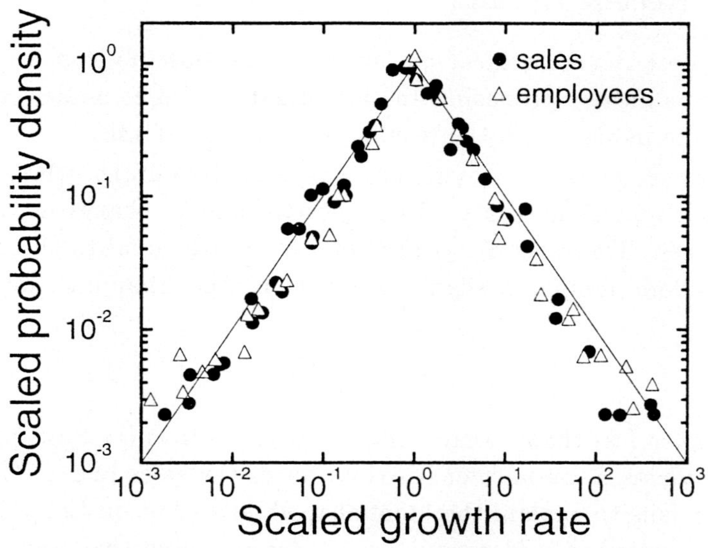


**Fig. 3** Scaled probability density $p_{scal} \equiv 2^{1/2}\sigma_1(s_0)p(r_1|s_0)$ as a function of the scaled growth rate $r_{scal} \equiv 2^{1/2}|r_1 - \bar{r}_1(s_0)|/\sigma_1(s_0)$. Also we show the analogous scaled quantities for the number of employees. All the data collapse upon the universal curve $p_{scal} = \exp(-|r_{scal}|)$ as predicted by Eq. (2).

played prominently in the literature on firm dynamics,[8,14] we propose to conduct a similar study for the dependence of the growth distribution (of sales or employment) on the firm's age. Also, the change of the distribution functions over longer growth intervals than one year will be investigated and other important quantities considered, as for example the assets of the firms. The aim of these future studies is to address the following questions:

- Does the distribution of growth rates (of sales, employment, or assets) change more rapidly in time for younger than for older firms? To what extent is this influenced by the relatively smaller size of young firms?
- Is the distribution of growth rates outside the manufacturing sector similar?
- Does the shape of the distribution functions change for shorter and larger time intervals $\Delta t$? Is this shape then still independent of the initial conditions and the quantity considered, i.e., does scaling hold as exemplified in Fig. 3? What are the characteristic time scales over which the changes occur? Are the values for these time scales universal or unique to particular sectors?
- Does the exponent $\beta$ change when different markets are studied? When different economies are studied?
- What model describes the time evolution of the growth distributions?

## 5. MODELING

A cornerstone of modern statistical physics has been the interpretation and explanation of data analysis using simple models. While these models do not precisely describe the mechanisms of a system, they do provide insight into "why" certain things occur in a complex system.

## 5.1 Review of Recent Results

In this section we describe some ideas on how we could possibly contribute to the understanding of firm dynamics by proposing and investigating simple models and by comparing the predictions of the models with the results of our data analysis.

The first attempt to model firm dynamics was made by Gibrat, who hypothesized that firm growth rate is independent of size.[1] He suggested that the change in sales was a random multiplicative process: The sales $S_{t+\Delta t}$ at time $t + \Delta t$ are equal to the sales $S_t$ at time $t$ multiplied by a random number $\varepsilon_t$ slightly smaller or larger than one, so:

$$S_{t+\Delta t} = S_t \varepsilon_t \,, \tag{3}$$

where $\Delta t$ can be much less than 1 year. After many iterations (i.e., after a long time), this stochastic process yields a *log-normal* distribution of the sales and of the annual growth rates. Therefore, it fails to explain the "tent shaped" distribution Eq. (1), obtained in our data analysis. We can try to understand Eq. (1) by assuming that successive values of $S_t$ are correlated. Each of the firms has a tendency to maintain its initial value $S_0$, since it cannot adapt to large changes in demand in a short time $\Delta t$. The easiest way to incorporate these correlations is to consider a constant "back-drift," i.e.,

$$\frac{S_{t+\Delta t}}{S_t} = \begin{cases} k\varepsilon_t, & \text{for } S_t < S_0 \,, \\ \frac{1}{k}\varepsilon_t, & \text{for } S_t > S_0 \,, \end{cases} \tag{4}$$

where $k$ is a constant larger than one. One can show that this simple process indeed yields the "tent shaped" distribution given in Eq. (1) with a width proportional to $\ln k$.

One might expect that for sufficiently long times, the distribution of growth rates approaches the distribution of the sales. Since the sales distribution can be approximated by a log-normal distribution (with the exception of very large firms), we must seek to find a model which yields a "tent shaped" distribution for small times but a log-normal distribution for large times.

A simple example of such a model can be easily constructed. We assume that each firm can be characterized by three properties: one static property and two dynamic ones. The static property is the initial sales, $S_0$, which is initially the same for all firms. The dynamic properties are the firm's sales and its optimal value of sales. The optimal value may be considered to be the value at which a firm maximizes its profits. There are only two rules for the model. First, the optimal value, which initially equals $S_0$, follows a random, multiplicative process, as suggested by Gibrat.[1] Second, the sales of the firm follow the dynamics suggested by Eq. (4) with one caveat. Instead of the "back-drift" attracting the sales toward initial sales, $S_0$, the sales will be attracted toward the optimal value. The growth of the optimal value is much *slower* than the growth of the sales.

Following these rules, the model produces several interesting results that correspond to the analysis in Sec. 4. Growth rates over short time intervals are characterized by a "tent-shape" distribution, Eq. (1). In contrast, growth rates over long time intervals are distributed according to a log-normal distribution.

Unfortunately, the simplification to only 2 rules appears incomplete. Although the model fits the results of our preliminary data analysis, the distribution made by the model has a

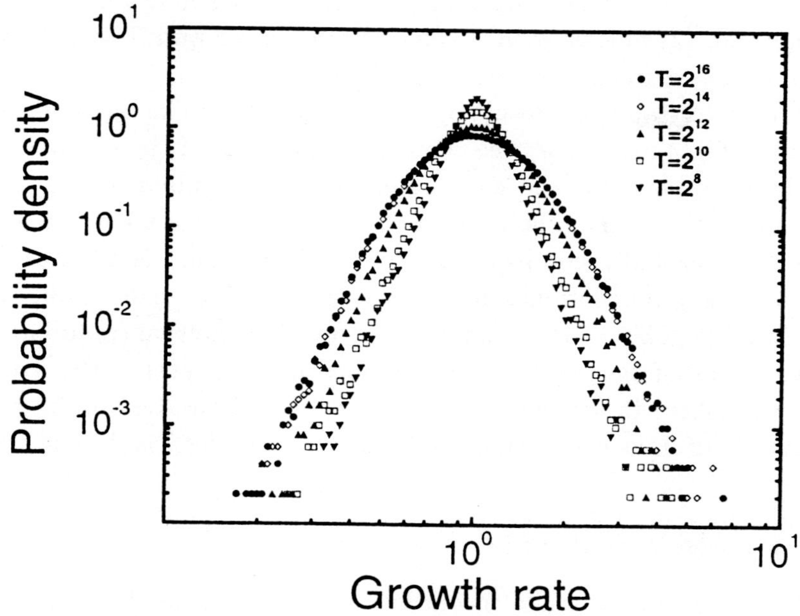


**Fig. 4** Probability density of the growth rate after various computer time-steps $T$. For small times $T$, the distribution is "tent-shaped", but it approaches a stable log-normal distribution for larger $T$.

width that increases with time as $t^{1/2}$. However, a careful consideration of the data suggests that there is little growth in the width of the distribution over time.

Thus, we must change the model to describe the data better. It appears that the mean growth rate slightly decreases with increasing value of sales. Incorporating a linear decrease into the model, we get for long time intervals $T$ a *stable* log-normal distribution, as shown in Fig. 4.

Up to now, we have neglected the interactions between firms, which is in many cases a reasonable approximation; for instance, to describe the size distribution of all publicly traded firms. However, for other situations such an extreme simplification is not justifiable: e.g., if we consider firms in the same industry, we obviously have to take into account the interactions and correlations between them.

The results obtained from the studies of correlations will help us to build more realistic models of firm dynamics. For example, by understanding how the growth rates of a firm are correlated from year to year we will be able to determine the time scale for the reaction of a firm to changes in its environment.

## 5.2 Open Questions

### 5.2.1 *Relation between initial size and standard deviation of growth rates*

Given that a simple relation between firm size and the standard deviation of the growth rate holds over such a wide range, that relation merits explanation. One extreme hypothesis is that modern firms consist of multiple, independent business units and that large firms consist of more units than do small firms. Based on our results, however, we can reject this hypothesis. To see this point, assume that the sales of a firm $S$ are given by: $S = \sum_{i=1}^{N} \xi_i$, where $\xi_i$ represents the sales of business unit $i$. If the $\xi_i$'s are independently and identically distributed, then Eq. (2) follows with $\beta = 0.5$. However, we find $\beta = 0.15$, which indicates the presence of strong correlations among the firm's units.

We can try to understand this result by considering a tree-like hierarchical organization of a typical firm. The root of the tree represents the head of the company, whose policy is passed to the level beneath, and so on, until finally the units in the lowest level take action. Let us first assume that the head of the firm suggests a policy that could result in changing the sales of each division by an amount $\delta$. If this policy is propagated through the hierarchy without any modifications, then $\beta = 0$ follows. More realistically, each unit is not only influenced by the policy of the head but also by other internal (random) factors of that unit. Then the policy of the head can be modified at each level of the firm management with a small probability. In this case a value $0 < \beta < 1/2$ can be obtained, depending on the average number of links connecting the levels. This analogy is developed elsewhere.[36]

### 5.2.2 *Entries, mergers, and exits*

In many of the models under study in our team, it is assumed that the number of firms acting on a given market remains constant in time. This simplification is quite useful, since it allows a better understanding of the remaining factors at play of the dynamics of firm growth. However, a more realistic approach to the problems must inevitably include the possibility of entry of new firms, the merger of existing firms, and the exit of firms from the market.

## REFERENCES


1. R. Gibrat, *Les Inégalités Economiques* (Sirey, Paris, 1933).
2. P. E. Hart and S. J. Prais, *J. Roy. Stat. Soc. Series* **A119**, 150–191 (1956).
3. H. A. Simon and C. P. Bonini, *Am. Econ. Rev.* **48**, 607–617 (1958).
4. Y. Ijiri and H. A. Simon, *Skew Distributions and the Sizes of Business Firms* (North Holland, Amsterdam, 1976).
5. M. H. R. Stanley, S. V. Buldyrev, S. Havlin, R. N. Mantegna, M. A. Salinger and H. E. Stanley, "Zipf plots and the size distribution of firm," *Economics Letters* **49**, 453–457 (1995).
6. S. Hymer and P. Pashigian, *J. Pol. Econ.* **52**, 556–569 (1962).
7. B. H. Hall, *J. Indust. Econ.* **35**, 583–606 (1987).
8. D. S. Evans, *J. Pol. Econ.* **95**, 657–674 (1987).
9. W. Baumol, *Business Behavior, Value, and Growth* (MacMillan, New York, 1959).
10. R. Cyert and J. March, *A Behavioral Theory of the Firm* (Prentice-Hall, Englewood Cliffs, New Jersey, 1963).
11. H. A. Simon, *Am. Econ. Rev.* **68**, 1 (1978).
12. F. Machlup, *Am. Econ. Rev.* **57**, 1 (1967).
13. M. C. Jensen and M. H. Meckling, *J. Fin. Econ.* **3**, 305–360 (1976).
14. B. Jovanovic, *Econometrica* **50**, 649–670 (1982).
15. R. R. Nelson and S. G. Winter, *An Evolving Theory of Economic Change* (Harvard University Press, Cambridge, Massachusetts, 1982).
16. B. R. Holmstrom and J. Tirole, in *Handbook of Industrial Organization* Vol. 1, eds. R. Schmalensee and R. Willig (North Holland, Amsterdam, 1989), pp. 61–133.
17. O. E. Williamson, in *Handbook of Industrial Organization* Vol. 1, eds. R. Schmalensee and R. Willig (North Holland, Amsterdam, 1989), pp. 135–182.
18. P. Milgrom and J. Roberts, *Economics, Organization, and Management* (Prentice-Hall, Englewood Cliffs, New Jersey, 1992).
19. R. Lucas, *Bell J. Econ.* **9**, 508–523 (1978).

20. R. Solow, *Review Econ. Stat.* **39**, 312 (1957).
21. M. I. Kamien and N. L. Schwartz, *Market Structure and Innovation* (Cambridge University Press, Cambridge, 1982).
22. W. M. Cohen and R. C. Levin, in *Handbook of Industrial Organization* Vol. 2, eds. R. Schmalensee and R. Willig (North Holland, Amsterdam, 1989), p. 1059.
23. S. R. Fazzari, R. G. Hubbard and B. Petersen, *Microeconomics Brookings Papers on Economic Activity*, 141 (1988).
24. S. C. Myers and N. Majluf, *J. Fin. Econ.* **13**, 147 (1984).
25. R. G. Hubbard, (ed.), *Asymmetric Information, Corporate Finance and Investment* (University of Chicago Press, Chicago, 1990).
26. K. A. Froot, D. S. Scharfstein and J. C. Stein, *Harvard Business Review*, 91 (1994).
27. M. C. Jensen, *Am. Econ. Rev.* **76**, 323 (1986).
28. M. H. R. Stanley, L. A. N. Amaral, S. V. Buldyrev, S. Havlin, H. Leschhorn, P. Maass, M. A. Salinger and H. E. Stanley, "Scaling behavior in the growth of companies," *Nature* **379**, 804–806 (1996).
29. R. N. Mantegna and H. E. Stanley, *Nature* **376**, 46–49 (1995).
30. H. E. Stanley, *Introduction to Phase Transitions and Critical Phenomena* (Oxford University Press, New York, 1971).
31. A. Bunde and S. Havlin (eds.), *Fractals in Science* (Springer-Verlag, Berlin, 1994).
32. A. L. Barabasi and H. E. Stanley, *Fractal Concepts in Surface Growth* (Cambridge University Press, Cambridge, 1995).
33. S. V. Buldyrev, P. Garik, S. Milosevic, H. E. Stanley, E. F. Taylor and P. A. Trunfio, *Fractals in Science* (Springer Verlag, Berlin and New York, 1994).
34. R. F. Engle, *Econometrica* **50**, 987–1007 (1982).
35. T. Bollerslev, R. Y. Chou and K. F. Kroner, *J. Economet.* **52**, 5–59 (1992).
36. See article of H. E. Stanley et al. in this volume.
37. S. V. Buldyrev, L. A. N. Amaral, S. Havlin, H. Leschhorn, P. Maass, M. A. Salinger, H. E. Stanley and M. H. R. Stanley, "Scaling behavior in economics: Empirical results and modeling," (preprint).

Fractals, Vol. 4, No. 3 (1996) 427–451


# SCALING AND UNIVERSALITY IN LIVING SYSTEMS

H. E. STANLEY*, L. A. N. AMARAL*, S. V. BULDYREV*,
A. L. GOLDBERGER†, S. HAVLIN*,§, B. T. HYMAN‡, H. LESCHHORN*,
P. MAASS*, H. A. MAKSE*, C.-K. PENG*,†, M. A. SALINGER¶,
M. H. R. STANLEY* and G. M. VISWANATHAN*

*Center for Polymer Studies and Department of Physics,
Boston University, Boston, MA 02215

†Cardiovascular Division, Harvard Medical School,
Beth Israel Hospital, Boston, MA 02215 USA

‡Neurology Service, Massachusetts General Hospital,
Boston, MA 02114 USA

§Minerva Center and Department of Physics,
Bar-Ilan University, Ramat Gan, Israel

¶School of Management, Boston University,
Boston, MA 02215, USA

## 1. INTRODUCTION

The twin pillars of scaling and universality support much of our current conceptualization on the general subject of just how complex systems formed of interacting subunits should actually behave.[1] These pillars were constructed a quarter century ago by scientists interested in the behavior of a system near its critical point. Progress was made possible by a remarkable combination of experiment and phenomenological theory. To help conceptualize the problem, the Ising model came to play a key role, and indeed most of the ideas that emerged — save one — were tested on this model system.[2]

The Ising model is defined as a set of classical spins localized on the sites of a lattice. Each spin is a one-dimensional object that can point either "up" or "down." The ferromagnetic

Ising interaction is particularly simple: If two neighboring spins are parallel (both up or both down), then there is a negative contribution to the energy. Hence the lowest energy of the entire system will be the configuration in which all the spins in the system are parallel. The Ising model can be regarded as a crude model of a ferromagnet if we think of the classical spins as representing the constituent microscopic moments comprising the ferromagnet.

Studies of the Ising model reveal a remarkable feature. If one tunes a "control parameter" — the temperature $T$ — then one finds that at a certain critical value $T_c$ spins remarkably far apart have orientations that are strongly correlated. Such correlations do not lend themselves to a ready explanation. The reason is that normally in physics modeling "you get out what you put in". In the case of the Ising model, we "put in" an interaction that extends a finite distance. Magically, we "get out" a correlation that spreads an *infinite* distance. How does this happen?

Our intuition tells us that the correlation $C(r)$ between subunits separated by a distance $r$ should decay exponentially with $r$ — for the same reason the value of money stored in ones mattress decays exponentially with time (each year it loses a constant fraction of its worth). Thus $C(r) \sim e^{-r/\xi}$, where $\xi$ is termed the correlation length — the characteristic length scale above which the correlation function is negligibly small. Experiments and also calculations on mathematical models confirm that correlations do indeed decay exponentially so long as the system is not exactly *at* its critical point, where the rapid exponential decay turns into a long-range power-law decay of the form $C(r) \sim r^{-\eta}$, where $\eta$ is called a critical exponent.[1] If correlations decay with a power-law form, we say the system is "scale free" since there is no characteristic scale associated with a simple power law.

Critical exponents, such as $\eta$, are found empirically to depend most strongly on the system dimension and on the general symmetry properties of the constituent subunits, and not on other details of the system under investigation. We understand power-law decay as arising primarily from the multiplicity of interaction paths that connect two spins in dimensions larger than one.[3] Exact enumeration methods take into account exactly the contributions of such paths — up to a maximum length that depends on the strength of the computer used and the patience of the investigator. To obtain quantitative results, the hierarchy of exact results is extrapolated to infinite order. In some sense, although the correlation along each path *decreases* exponentially with the length of the path, the number of such paths *increases* exponentially. The "gently-decaying" power-law correlation emerges as the victor in this competition between the two warring exponential effects.

Cyril Domb, Michael Fisher, and their colleagues pioneered such exact enumeration approaches in the 1970s. Among the results emerging from their efforts was a complete calculation of the scaling equation of state of the Ising model in a magnetic field.[4] This equation of state was recently tested by the experiments of Back and co-workers.[2] A remarkable feature of their work is that the system they study does not perfectly mirror the conditions of the Ising model definition, yet the scaling equation of state that they measure conforms almost perfectly to the calculated result.[4] How can that be?

The quick answer is one word: "universality". The word universality connotes the fact that quite disparate systems behave in a remarkably similar fashion near their respective critical points — simply because near their critical points what matters most is not the details of the microscopic interactions but rather the nature of the "paths along which order is propagated". The experiments in question are essentially two-dimensional, and the component "magnetic moments" have sufficient anisotropy in their interactions that when acting collectively they behave as one-dimensional classical spins. Hence the experimental system mirrors a two-dimensional Ising model!

At one time, it was imagined that the "scale-free" case was relevant to only a fairly narrow slice of physical phenomena.[1] However, the range of systems that apparently display power-law correlations has increased dramatically in recent years, ranging from base pair correlations in DNA,[5] lung inflation[6] and interbeat intervals of the human heart[7] to complex systems involving large numbers of interacting subunits that display "free will," such as govern city growth[8] and even economics.[9] The principle of universality seems to be reflected in the empirical fact that quite different systems can have remarkably similar critical exponents — perhaps the "interaction paths" between the constituent subunits dominate the observed cooperative behavior more than the detailed properties of the subunits themselves.

The purpose of this paper is to describe a few examples of recent progress in applying concepts of modern statistical physics to systems of relevance to biology, medicine, economics, and urban growth.

## 2. SCALING IN BIOLOGY AND MEDICINE: A BRIEF SURVEY

In the last decade, it was realized that some biological systems have no characteristic length or time scale, i.e., they have fractal — or, more generally, self-affine — properties.[10–12] However, the fractal properties in different biological systems, have quite different nature, origin, and appearance. In some cases, it is the geometrical shape of a biological object itself that exhibits obvious fractal features, while in other cases the fractal properties are largely "hidden" and can only be perceived if data are studied as a function of time or mapped onto a graph in some special way. After an appropriate mapping, such a graph may resemble a mountain landscape, with jagged ridges of all length scales from very small bumps to enormous peaks (see Fig. 1). Mathematically, these landscapes can be quantified in terms of fractal concepts such as self-affinity.

In contrast to compact objects, fractal objects have a very large *surface* area. In fact, they are composed almost entirely of "surface." This observation explains why fractals are ubiquitous in biology, in which surface phenomena[12] are of crucial importance.

Lungs exemplify this feature (see Fig. 2). The surface area of a human lung is as large as a tennis court. The mammalian lung is made up of self-similar branches with many length scales, which is the defining attribute of a fractal surface. The efficiency of the lung is enhanced by this fractal property, since with each breath oxygen and carbon dioxide have to be exchanged at the lung surface. The structure of the bronchial tree has been quantitatively analyzed using fractal concepts.[11–14] In particular, fractal geometry could explain the power-law decay of the average diameter of the bronchial tube with the generation number, in contrast to the classical model which predicts an exponential decay.[15]

Not only the geometry of the respiratory tree but also the time-dependent features of inspiration, is described by fractal geometry. Specifically, Suki et al.[14] studied airway opening in isolated dog lungs. They found that during constant flow inflations, the lung volume changes in discrete jumps (Fig. 2), and that the probability distribution function of the relative size $x$ of the jumps, $\Pi(x)$, and that of the time intervals $t$ between these jumps, $\Pi(t)$, follow a power law over nearly two decades of $x$ and $t$ with exponents of 1.8 and 2.7,

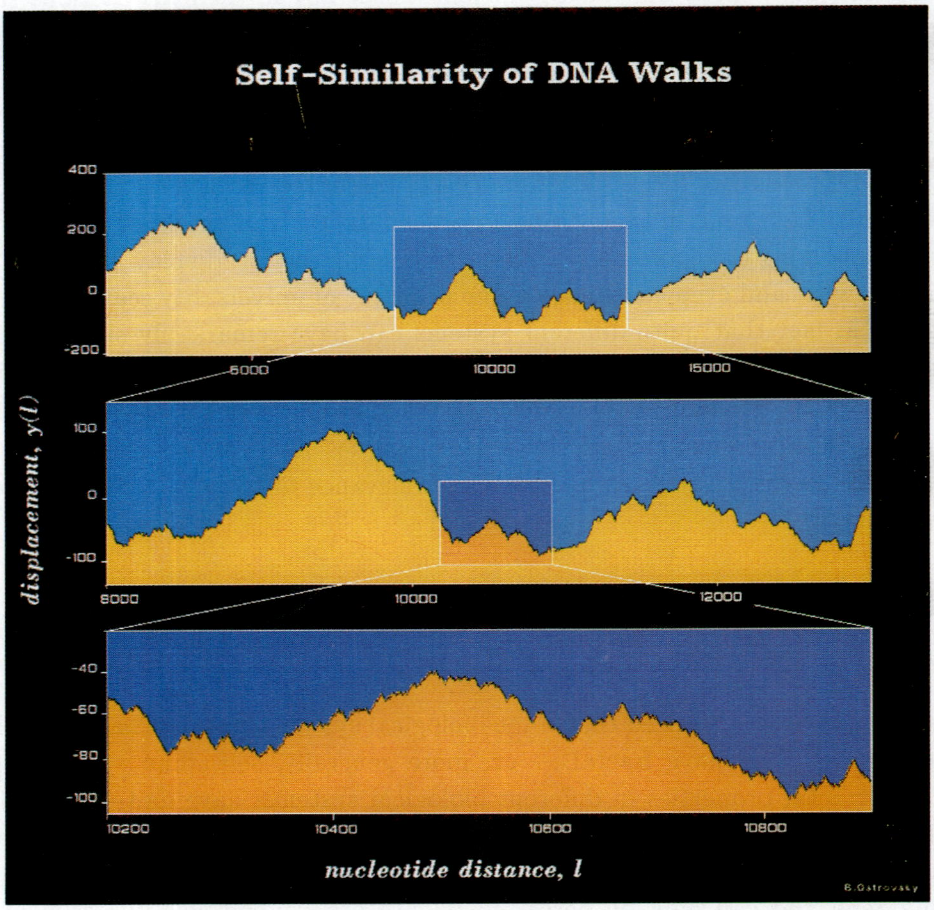


**Fig. 1** The DNA walk representation for the rat embryonic skeletal myosin heavy chain gene ($\alpha = 0.63$). At the top the entire sequence is shown. In the middle the solid box shown in the top is magnified. At the bottom the solid box shown in the middle is magnified. The statistical self-similarity of these plots is consistent with the existence of a scale-free or fractal phenomenon which we call a fractal landscape. Note that one must magnify the segment by different factors along the $\ell$ (horizontal) direction and the $y$ (vertical) direction; since $F$ has the same units (dimension) as $y$, these magnification factors $M_\ell$ and $M_y$ (along $\ell$ and $y$ directions respectively) are related to the scaling exponent $\alpha$ by the simple relation $\alpha = \log(M_y)/\log(M_\ell)$ [e.g., from top to middle, $\log(M_y)/\log(M_\ell) = \log(2.07)/\log(3.2) = 0.63$].

respectively. To interpret these findings, they developed a branching airway model in which airways, labeled $ij$, are closed with a uniform distribution of opening threshold pressures $P$. When the "airway opening" pressure $P_{\text{ao}}$ exceeds $P_{ij}$ of an airway, that airway opens along with one or both of its daughter branches if $P_{ij} < P_{\text{ao}}$ for the daughters. Thus, the model predicts "avalanches" of airway openings with a wide distribution of sizes, and the statistics of the jumps agree with those $\Pi(x)$ and $\Pi(t)$ measured experimentally. They concluded that power-law distributions, arising from avalanches triggered by threshold phenomena, govern the recruitment of terminal airspaces.

A second example is the arterial system which delivers oxygen and nutrients to all the cells of the body. For this purpose blood vessels must have fractal properties.[16,17] The diameter distribution of blood vessels ranging from capillaries to arteries follows a power-law distribution which is one of the main characteristics of fractals. Sernetz et al.[18] have

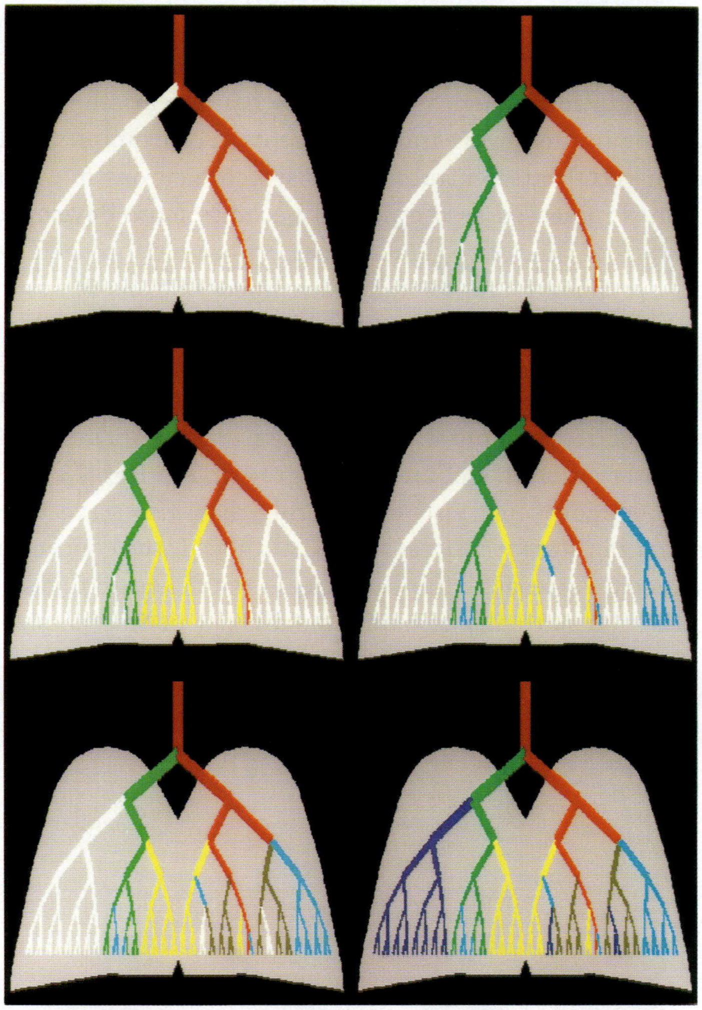

**Fig. 2** The dynamic mechanism responsible for filling the lung involves "avalanches" or "bursts" of air that occur in all sizes — instead of an exponential distribution, one finds a power law distribution.[14] The underlying cause of this scale-free distribution of avalanches is the fact that every airway in the lung has its own threshold below which it is not inflated. Shown here is a diagram of the development of avalanches in the airways during airway opening. At first, almost all airways whose threshold value is smaller than the external pressure (red) are closed. Then the airway opening pressure increases until a second threshold is exceeded, and as a result all airways further up the tree whose thresholds are smaller become inflated (green). The airway opening pressure is successively increased until third, fourth, and fifth thresholds are exceeded (yellow, brown, and blue). The last threshold to be exceeded results in filling the airways colored violet; we notice that this last avalanche opens up over 25% of the total lung volume, thereby significantly increasing the total surface area available for gas exchange. After Ref. 14.

studied the branching patterns of arterial kidney vessels. They analyzed the mass-radius relation and found that it can be characterized by fractal geometry, with fractal dimensions between 2.0 and 2.5. Similarly, the branching of trees and other plants, as well as root systems have a fractal nature.[19] Moreover, the size distribution of plant-supported insects was found to be related to the fractal distribution of the leaves.[20]

One of the most remarkable examples of a fractal object is the surface of a cauliflower, where every little head is an "almost" exact reduced copy of the whole head formed by intersecting Fibonacci spirals of smaller heads, which in turn consist of spirals of smaller and smaller heads, up to the fifth order of hierarchy. West and Goldberger were first to describe such a "Fibonacci fractal" in the human lung.[11] (For a general review of fractals in physiology and medicine see also Goldberger, Rigney, and West.[11])

Considerable interest in the biological community has also arisen from the possibility that neuron shape can be quantified using fractal concepts. For example, Smith et al.[21] studied the fractal features of vertebrate central nervous system neurons in culture and found that the fractal dimension is increased as the neuron becomes more developed. Caserta et al.[22] showed that the shapes of quasi-two-dimensional retinal neurons can be characterized by a fractal dimension $d_f$. They found for fully developed neurons *in vivo*, $d_f = 1.68 \pm 0.15$, and suggest that the growth mechanism for neurite outgrowth bears a direct analogy with the growth model called *diffusion limited aggregation* (DLA), (see Fig. 3). The branching pattern of retinal vessels in a developed human eye is also similar to DLA.[17] The fractal dimension was estimated to be about 1.7, in good agreement with DLA for the case of two dimensions. For an alternative model for retinal growth, see Ref. 23.

The DLA-type model governing viscous fingering may also serve to resolve the age-old paradox, "*Why doesn't the stomach digest itself?*"[24] Indeed, the concentration of hydrochloric acid in the mammalian stomach after each meal is sufficient to digest the stomach itself, yet the gastric epithelium normally remains undamaged in this harsh environment. One protective factor is gastric mucus, a viscous secretion of specialized cells, which forms a protective layer and acts as a diffusion barrier to acid. Bicarbonate ion secreted by the gastric epithelium is trapped in the mucus gel, establishing a gradient from pH 1–2 at the

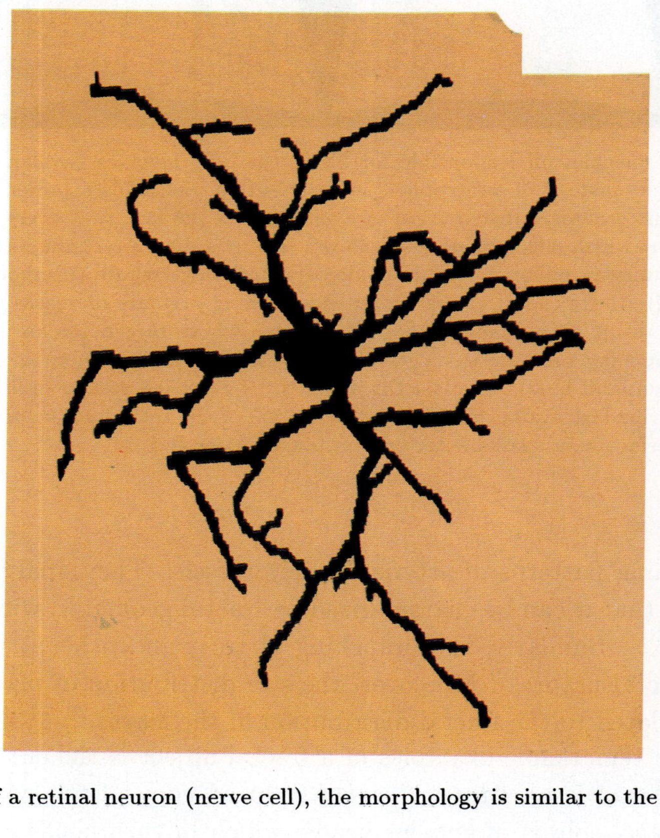

**Fig. 3** Photograph of a retinal neuron (nerve cell), the morphology is similar to the DLA archetype. After Ref. 22.

**Fig. 4** Viscous fingers reflect the complex interface that develops when one fluid is pumped through another of higher viscosity. Shown is the formation of such viscous fingers or channels when hydrochloric acid is injected into solutions of gastric mucin. These channels may confine the acid and direct it to the lumen, thus protecting the gastric mucosa from acidification and ulceration; when the gastric glands contract, acid is ejected under high enough pressure to form viscous fingers. After Ref. 24.

lumen to pH 6–7 at the cell surface. The puzzle, then, is how hydrochloric acid, secreted at the base of gastric glands by specialized parietal cells, traverses the mucus layer to reach the lumen without acidifying the mucus layer. Bhaskar et al.[24] resolved this puzzle using experiments that demonstrate the possibility that the flow of hydrochloric acid through mucus involves viscous fingering — the phenomenon that occurs when a fluid of lower viscosity is injected into a more viscous one (see Fig. 4). Specifically, Bhaskar et al. demonstrated that injection of hydrochloric acid through solutions of pig gastric mucin produces fingering patterns which are strongly dependent on pH, mucin concentration, and acid flow rate. Above pH 4, discrete fingers are observed, while below pH 4, hydrochloric acid neither penetrates the mucin solution nor forms fingers. These *in vitro* results suggest that hydrochloric acid secreted by the gastric gland can penetrate the mucus gel layer (pH 5–7) through narrow fingers, whereas hydrochloric acid in the lumen (pH 2) is prevented from diffusing back to the epithelium by the high viscosity of gastric mucus gel on the luminal side.

Yet another example of DLA-type growth is bacterial colony spread on the surface of agar (gel with nutrient) plates[25] (see Fig. 5). Vicsek et al.[26] studied bacterial colony growth on

**Fig. 5** A typical example of DLA-like colony patterns incubated at 35°C for three weeks after inoculation on the surface of agar plates containing initially 1 g/$\ell$ of peptone as nutrient. This pattern has a fractal dimension of $d_f \cong 1.72$. After Matsushita and Fujikawa.[25]

a strip geometry which results in a self-affine surface (see Fig. 13.19 in Ref. 27). They calculated the roughness exponent $\alpha$ for this surface and found $\alpha = 0.78 \pm 0.07$. The interfacial pattern formation of the growth of bacterial colonies was studied systematically by Ben-Jacob et al.[28] They demonstrated that bacterial colonies can develop a pattern similar to morphologies in diffusion-limited growth observed in solidification and electrochemical deposition. These include fractal growth, dense-branching growth, compact growth, dendritic growth and chiral growth. The results indicate that the interplay between the micro-level (individual bacterium) and the macro-level (the colony) plays a major role in selecting the observed morphologies similar to those found in nonliving systems.

Another example of fractal interface appears in ecology, in the problem of the territory covered by $N$ diffusing particles[29] (see Fig. 6). As seen from the figure, the territory initially grows with the shape of a disk with a relatively smooth surface until it reaches a certain size, at which point the surface becomes increasingly rough. This phenomenon may have been observed by Skellam[30] who plotted contours delineating the advance of the muskrat population and noted that initially the contours were smooth but at later times they became rough (see Fig. 1 in Ref. 30).

Other biological contexts in which fractal scaling seems to be relevant are the relation between brain size and body weight,[31] between bone diameter and bone length,[32] between muscle force and muscle mass,[32] and between an organism's size and its rate of producing energy and consuming food.[33]

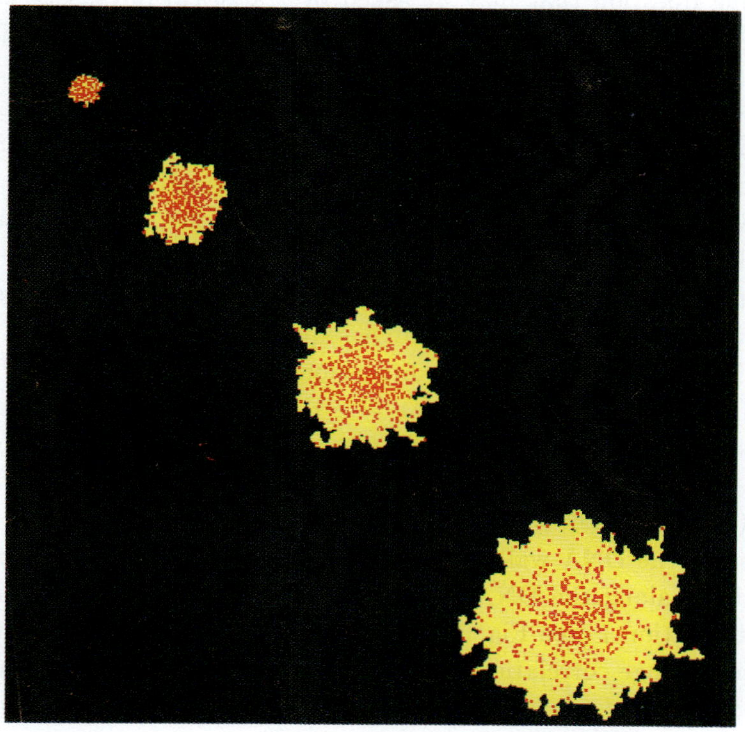

**Fig. 6** Snapshots at successive times of the territory covered by $N$ random walkers for the case $N = 500$ for a sequence of times. Note the roughening of the disc surface as time increases. The roughening is characteristic of the experimental findings for the diffusive spread of a population.[30] After Ref. 29, courtesy of P. Trunfio.

## 3. SCALING IN DNA BASE PAIR SEQUENCES

The role of genomic DNA sequences in coding for protein structure is well known.[34] The human genome contains information for approximately 100 000 different proteins, which define all inheritable features of an individual. The genomic sequence is likely the most sophisticated information database created by nature through the dynamic process of evolution. Equally remarkable is the precise transformation of information (duplication, decoding, etc.) that occurs in a relatively short time interval.

The building blocks for coding this information are called *nucleotides.* Each nucleotide contains a phosphate group, a deoxyribose sugar moiety and either a *purine* or a *pyrimidine base.* Two purines and two pyrimidines are found in DNA. The two purines are adenine (A) and guanine (G); the two pyrimidines are cytosine (C) and thymine (T).

In the genomes of high eukaryotic organisms only a small portion of the total genome length is used for protein coding (as low as 3% in the human genome). The segments of the chromosomal DNA that are spliced out during the formation of a mature mRNA are called *introns* (for intervening sequences). The coding sequences are called *exons* (for expressive sequences).

The role of introns and intergenomic sequences constituting large portions of the genome remains unknown. Furthermore, only a few quantitative methods are currently available for analyzing information which is possibly encrypted in the noncoding part of the genome.

One interesting question that may be asked by statistical physicists would be whether the sequence of the nucleotides A, C, G, and T behaves like a one-dimensional "ideal gas", where the fluctuations of density of certain particles obey Gaussian law, or if there exist long range correlations in nucleotide content (as in the vicinity of a critical point). These result in domains of all sizes with different nucleotide concentrations. Such domains of various sizes were known for a long time but their origin and statistical properties remain unexplained. A natural language to describe heterogeneous DNA structure is long-range correlation analysis, borrowed from the theory of critical phenomena.[1]

In order to study the scale-invariant long-range correlations of a DNA sequence, we first introduced a graphical representation of DNA sequences, which we term a *fractal landscape* or *DNA walk*.[5] For the conventional one-dimensional random walk model,[35] a walker moves either "up" $[u(i) = +1]$ or "down" $[u(i) = -1]$ one unit length for each step $i$ of the walk. For the case of an uncorrelated walk, the direction of each step is independent of the previous steps. For the case of a correlated random walk, the direction of each step depends on the history ("memory") of the walker.

One definition of the DNA walk is that the walker steps "up" if a pyrimidine (C or T) occurs at position $i$ along the DNA chain, while the walker steps "down" if a purine (A or G) occurs at position $i$ (see Fig. 1). The question we asked was whether such a walk displays only short-range correlations (as in a Markov chain) or long-range correlations (as in critical phenomena and other scale-free "fractal" phenomena — see Fig. 1). A different type of DNA walk was introduced earlier by Azbel.[36]

There have also been attempts to map DNA sequence onto multi-dimensional DNA walks.[37,38] However, recent work[39] indicates that the original purine-pyrimidine rule provides the most robust results, probably due to the purine-pyrimidine chemical complementarity.

An important statistical quantity characterizing any walk is the root mean square fluctuation $F(\ell)$ about the average of the displacement of a quantity $\Delta y(\ell)$ defined by $\Delta y(\ell) \equiv y(\ell_0 + \ell) - y(\ell_0)$, where

$$y(\ell) \equiv \sum_{i=1}^{\ell} u(i)\,. \tag{1}$$

If there is no characteristic length (i.e., if the correlations were "infinite-range"), then fluctuations will also be described by a power law

$$F(\ell) \sim \ell^{\alpha} \tag{2}$$

with $\alpha \neq 1/2$. The exponent $\alpha$ is the self similarity parameter mentioned above and therefore is directly related to long-range correlations in the sequence.

The fact that data for intron-containing and intergenic (i.e., noncoding) sequences are linear on this double logarithmic plot confirms that $F(\ell) \sim \ell^{\alpha}$. A least-squares fit produces a straight line with slope $\alpha$ substantially larger than the prediction for an uncorrelated walk, $\alpha = 1/2$, thus providing direct experimental evidence for the presence of long-range correlations.

Figure 7 shows the DFA exponent $\alpha(\ell)$ for the nine sequenced chromosomes of *Saccharomyces cerevisiae* using the purine-pyrimidine rule and the hydrogen bond energy rule. Note that although the landscapes look quite different, the LRC exponent $\alpha(\ell)$ is very similar for different chromosomes. For $\ell < 1000$ bp the different chromosomes have almost

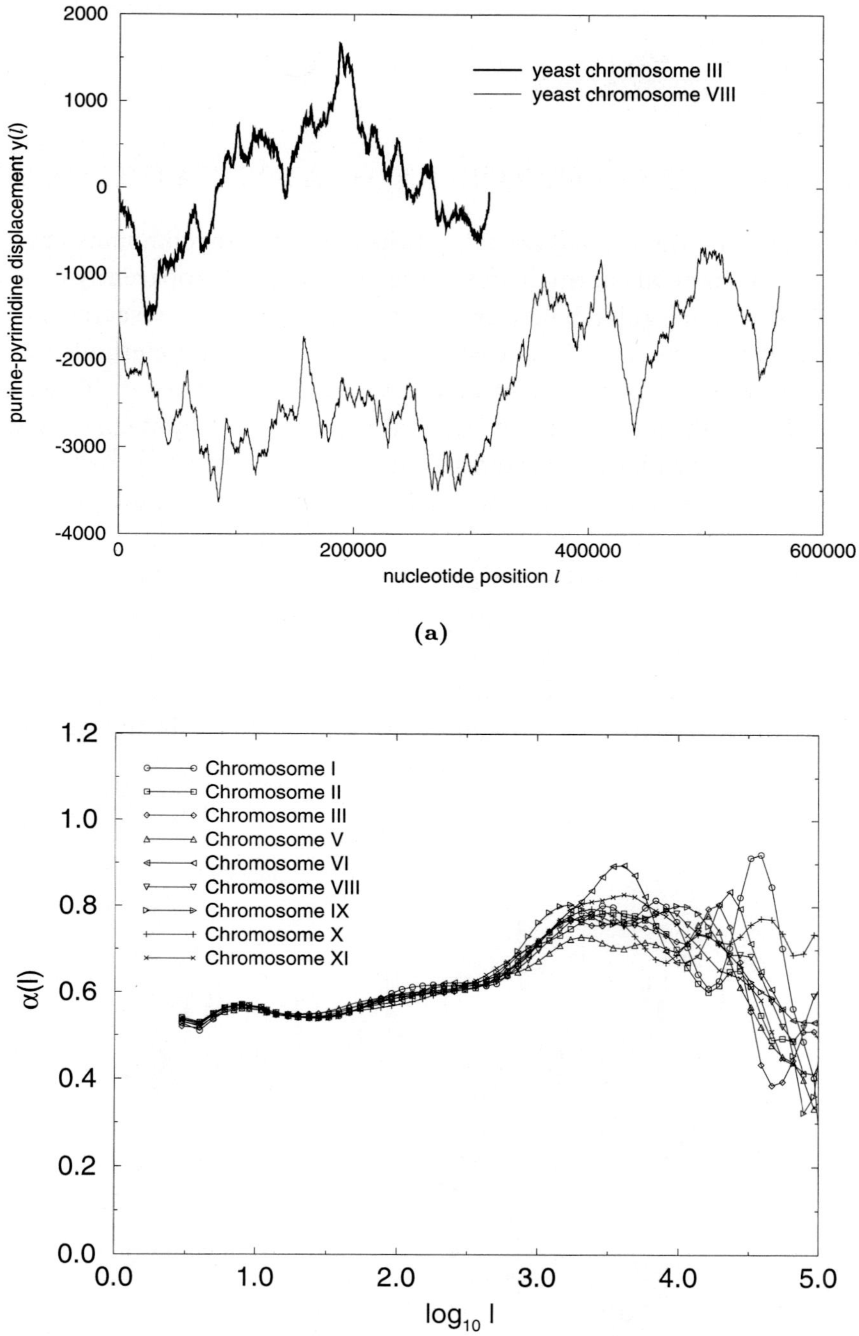


**(a)**



**(b)**

**Fig. 7** **(a)** DNA walk for yeast chromosomes III and VIII. **(b)** Local exponent $\alpha(\ell)$ measured on length scale $\ell$. Note that even though the 2 chromosomes have dramatically different landscapes, the $\alpha(\ell)$ functions are similar.

identical $\alpha(\ell)$. This similarity indicates that the correlation properties of the different chromosomes are almost the same for $\ell < 1000$ bp. Note also how the first couple of peaks in $\alpha(\ell)$ roughly coincide for the different chromosomes in Fig. 7(b). This indicates that the nine chromosomes have similar patch sizes, because peaks in $\alpha(\ell)$ correspond to characteristic patch sizes.

## 4. SCALING IN NONSTATIONARY HEARTBEAT TIME SERIES

Clinicians often describe the normal activity of the heart as "regular sinus rhythm." But in fact cardiac interbeat intervals normally fluctuate in a complex, apparently erratic manner[41] [Fig. 8(a)]. This highly irregular behavior has recently motivated researchers[42,43] to apply time series analyses that derive from statistical physics, especially methods for the study of critical phenomena where fluctuations at all length (time) scales occur. These studies show that under healthy conditions, interbeat interval time series exhibit long-range power-law correlation reminiscent of physical systems near a critical point.[1,44] Furthermore, certain disease states may be accompanied by alterations in this scale-invariant (fractal) correlation property.

The work in Ref. 7 is based on the digitized electrocardiograms of beat-to-beat heart rate fluctuations recorded with an ambulatory (Holter) monitor. The time series obtained by plotting the sequential intervals between beat $i$ and beat $i+1$, denoted by $B(i)$, typically reveals a complex type of variability [Fig. 8(a)]. The mechanism underlying such fluctuations appears to be related primarily to countervailing neuroautonomic inputs. The nonlinear

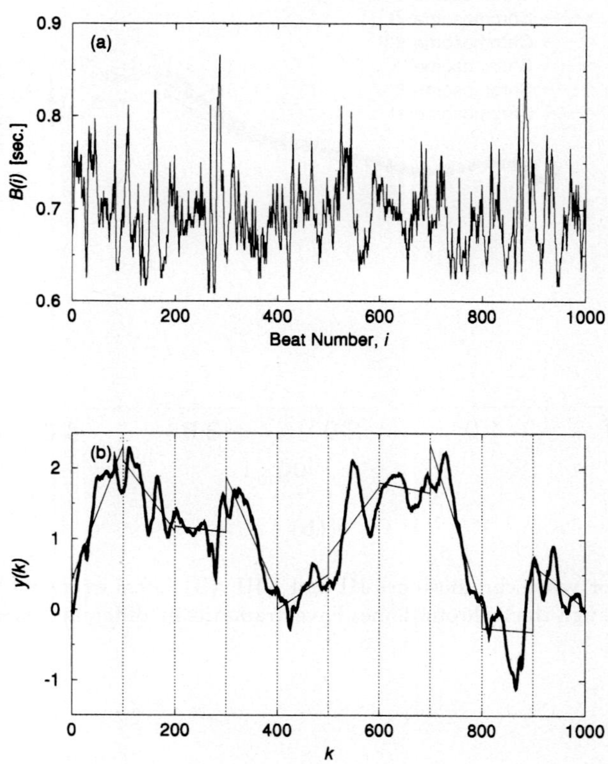


**Fig. 8** **(a)** The interbeat interval time series $B(i)$ of 1000 beat. **(b)** The integrated time series: $y(k) = \sum_{i=1}^{k}[B(i) - B_{\mathrm{av}}]$, where $B(i)$ is the interbeat interval shown in (a). The vertical dotted lines indicate box of size $n = 100$, the solid straight lines segments are the estimated "trend" in each box by least-squares fit.

interaction (competition) between the two branches of the autonomic nervous system is the postulated mechanism for the type of erratic heart rate variability recorded in healthy subjects.[41,45]

An immediate problem facing researchers applying time series analysis to interbeat interval data is that the heartbeat time series is often highly non-stationary. To overcome this problem, we introduced a modified fluctuation analysis of a random walk — termed *detrended fluctuation analysis* (DFA)[46] — to the analysis of physiological data. The advantages of DFA over conventional methods (e.g. spectral analysis and Hurst analysis) are that it permits the detection of long-range correlations embedded in a seemingly nonstationary time series, and also avoids the spurious detection of apparent long-range correlations that are an artifact of non-stationarity.

To illustrate the DFA algorithm, we use the interbeat time series shown in Fig. 8(a) as an example. Briefly, the interbeat interval time series (of total length $N$) is first integrated, $y(k) = \sum_{i=1}^{k} [B(i) - B_{\rm av}]$, where $B(i)$ is the $i$th interbeat interval and $B_{\rm av}$ is the average interbeat interval. Next the integrated time series is divided into boxes of equal length, $n$. In each box of length $n$, a least squares line was fit to the data (representing the *trend* in that box) [Fig. 8(a)]. The $y$ coordinate of the straight line segments is denoted by $y_n(k)$. Next we detrend the integrated time series, $y(k)$, by subtracting the local trend, $y_n(k)$, in each box. The root-mean-square fluctuation of this integrated and detrended time series is calculated by

$$F(n) = \sqrt{\frac{1}{N}\sum_{k=1}^{N} [y(k) - y_n(k)]^2}\,. \tag{3}$$

Figure 9 compares the DFA analysis of representative 24 hour interbeat interval time series of a healthy subject (○) and a patient with congestive heart failure (△). Notice that for large time scales (asymptotic behavior), the healthy subject shows almost perfect power-law scaling over two decades ($20 \leq n \leq 10\,000$) with $\alpha = 1$ (i.e., $1/f$ noise), while for

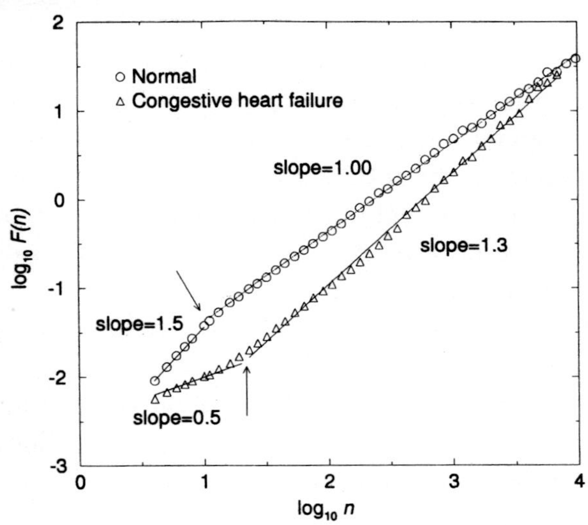


**Fig. 9** Plot of log $F(n)$ vs. log $n$ for two very long interbeat interval time series ($\sim$ 24 hr). The circles are for a healthy subject while the triangles are from a subject of congestive heart failure. Arrows indicate "crossover" points in scaling.

the pathologic data set $\alpha \approx 1.3$ (closer to Brownian noise). This result is consistent with our previous finding that there is a significant difference in the long-range scaling behavior between healthy and diseased states.[42,43]

## 5. SIZE DISTRIBUTION OF PLAQUE IN THE ALZHEIMER BRAIN

The recent revolution in the molecular biology of disease has led to the development of molecular epidemiology, in which a population of individuals with a disease is subdivided according to molecular characteristics.

It is unknown how genetic alterations are related to the fundamental pathological alterations of Alzheimer's disease (AD), specifically the aggregation of peptide to form senile plaques (SP). There is a need to define the phenotypic correlations of these subgroups in order to forge a link between molecular epidemiology and pathophysiology.

In a recent work,[47] we considered 40 AD and 7 elderly, demented trisomy 21 cases were studied. Fifty micrometer thick sections of the temporal lobe were immunostained with antibody 1,[48] a monoclonal antibody directed against $A'$, and a Bioquant image analysis system was used to semiautomatically assess SP number and cross sectional area,[49] from which the size distribution of plaques was calculated. A key point is to study in detail the size distribution histograms $P(V)$, where $P(V)\Delta V$ is the fraction of SP having cross-sectional area in the range $A + \Delta A$. In general, a Gaussian (i.e., a normal distribution) for the $P(V)$ histogram is usually found when $V$ is proportional to the sum of several parameters, each of which has independent fluctuations about its mean. In contrast, alog-normal distribution for the histogram occurs when $V$ is proportional to the product of several random parameters,[50–52] each of which has independent fluctuations about its mean. In a log normal distribution, $\log(V)$ replaces $V$, so that $P(\log(V))$ histogram is Gaussian. Consideration of analogous problems in statistical physics 21 and of the biological issues influencing SP growth lead to three possible models for the rate of increase in volume $V$ of SP:

(i) $dV/dt = k(t)$, where $k(t)$ is a parameter that changes with time in a random fashion around a positive average value until plaque size reaches a steady state. In this case, the histogram $P(V)$, giving the number of SP of volume $V$, is a Gaussian.

(ii) $dV/dt = k(t)S$, where $S$ is the total surface area. For spheres, $S$ is proportional to $r^2$, while volume $V$ is proportional to $r^3$; here $r$ denotes the typical radius. Hence one can re-write the equation in terms of $r$ as $dr/dt = k(t)$. The solution of this equation produces a histogram that is Gaussian in the variable $r$.

(iii) $dV/dt = k(t)V$, i.e., the rate is proportional to the volume. In this case, the histogram $P(V)$ is dramatically skewed to the right. However if we change variables from $V$ to $x = \log V$, then $dx/dt = k(t)$ and the distribution is Gaussian in this new variable and is thus log-normal.[51,53] The distribution of the cross sectional area $A$, $P(A)$ rather than $P(V)$, is plotted because $A$ was directly measured.

In accord with previous observations, we found that the profile of plaque sizes when plotted as a function of $A$ along a linear axis is not Gaussian, but instead is skewed dramatically[49,54,55] [see Fig. 10(a)]. We have now analyzed these observations using

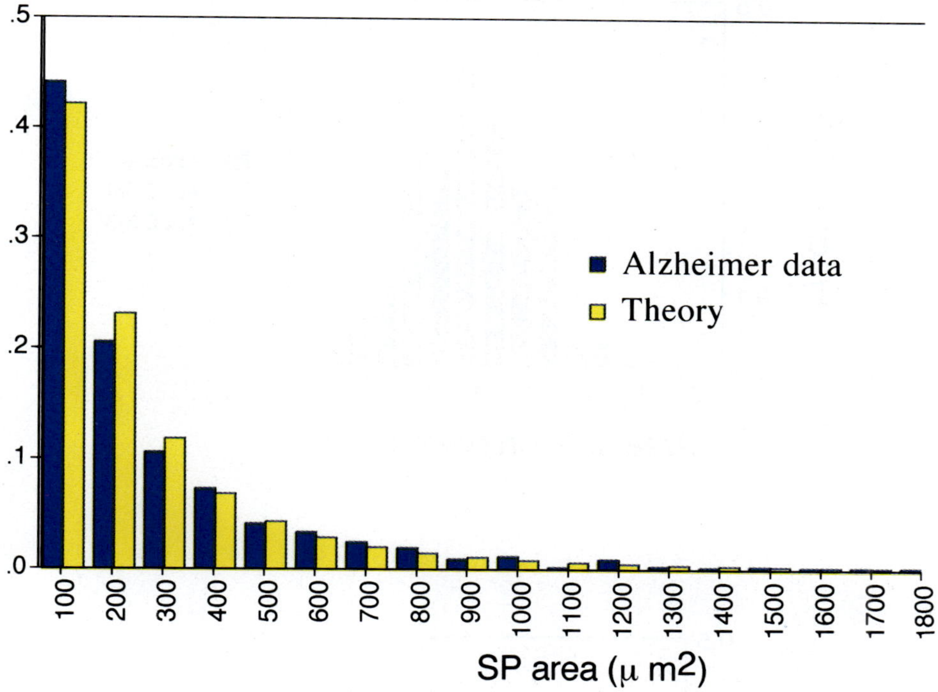


**(a)**

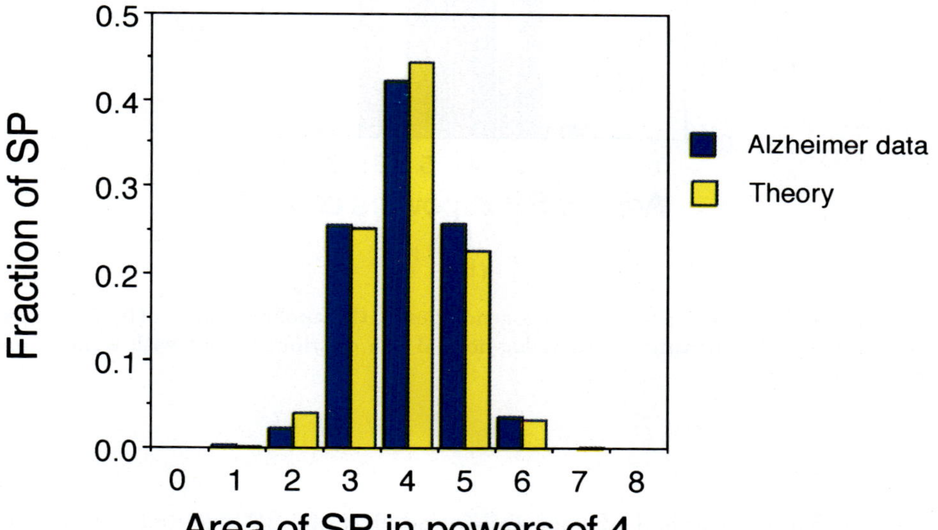


**(b)**

**Fig. 10** Senile plaque size profile distribution histograms of an AD patient. A 50 micrometer thick section of the temporal lobe was immunostained with anti-A' monoclonal antibody 10D5. In this case, six 780 micrometer by the depth of the cortex traverses of the inferior bank of the superior temporal sulcus located approximately 1 cm from the crest of the middle temporal gyrus were analyzed. Plaque size profile (closed bars) is **(a)** plotted arithmetically in bins of 100 micrometers squared, or **(b)** as a log 4 function of plaque size (in bins of powers of 4). A curve generated from a theoretical log-normal distribution is shown as open bars in (b) to illustrate how well this distribution fits the data.

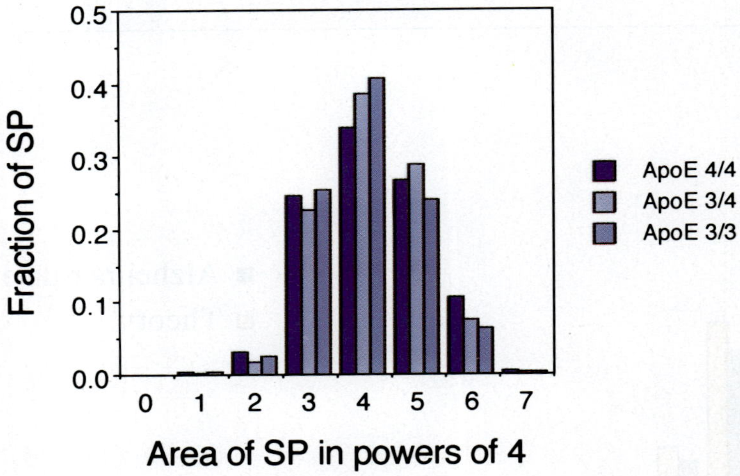


**(a)**

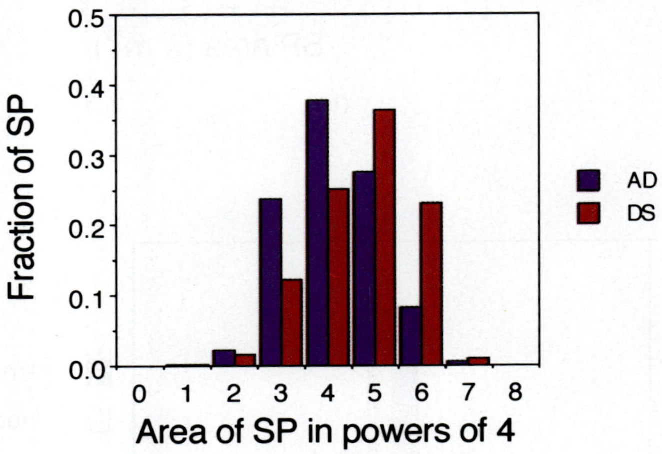


**(b)**

**Fig. 11** **(a)** same as Fig. 9(b) except for all 3 genotypes of the apolipoprotein. **(b)** Comparison of AD with Down syndrome (DS), showing the same log-normal size distribution but with a mean roughly four times larger.

concepts derived from statistical physics and found that the SP size distribution profile can be approximated by a log-normal distribution. Indeed, when the distribution is plotted as a function of log $A$, it has the shape of a Gaussian [see Fig. 10(b)].

Note that for roughly spherical objects, a log normal distribution of cross sections yields a log normal distribution of the volumes. Thus our data support models which result in a log normal distribution, and are inconsistent with models that predict a Gaussian distribution. The observation that SP size distribution is consistent and fits the log-normal law allows quantitative comparison across different subgroups of AD patients defined by molecular characteristics (see Fig. 11).

## 6. SCALING IN URBAN GROWTH PATTERNS

Predicting urban growth is important for the challenge it presents to theoretical frameworks for cluster dynamics.[56–58] Recently, the model of diffusion limited aggregation (DLA)[59] has been applied to describe urban growth,[56] and results in tree-like dendritic structures which have a core or "central business district" (CBD). The DLA model predicts that there exists only one large fractal cluster that is almost perfectly screened from incoming "development units" (people, capital, resources, etc.), so that almost all the cluster growth occurs in the extreme peripheral tips. In a recent work,[8] an alternative model to DLA that better describes the morphology and the area distribution of systems of cities, as well as the scaling of the urban perimeter of individual cities, has been developed. The results agree both qualitatively and quantitatively with actual urban data. The resulting growth morphology can be understood in terms of the effects of interactions among the constituent units forming a urban region, and can be modeled using the correlated percolation model in the presence of a gradient.

In the model one takes into account the following points:

(i) Urban data on the population density $\rho(r)$ of actual urban systems are known to conform to the relation[60] $\rho(r) = \rho_0 e^{-\lambda r}$, where $r$ is the radial distance from the CBD situated at the core, and $\lambda$ is the density gradient. Therefore, in our model the development units are positioned with an occupancy probability $p(r) \equiv \rho(r)/\rho_0$ that behaves in the same fashion as is known experimentally.

(ii) In actual urban systems, the development units are not positioned at *random*. Rather, there exist *correlations* arising from the fact that when a development unit is located in a given place, the probability of adjacent development units increases naturally — i.e., each site is not independently occupied by a development unit, but is occupied with a probability that depends on the occupancy of the neighborhood.

In order to quantify these ideas, we consider the *correlated* percolation model.[61–63] In the limit where correlations are so small as to be negligible,[64] a site at position $r$ is occupied if the occupancy variable $u(r)$ is smaller than the occupation probability $p(r)$; the variables $u(r)$ are uncorrelated random numbers. To introduce correlation among the variables, we convolute the uncorrelated variables $u(r)$ with a suitable power law kernel,[63] and define a new set of random variables $\eta(r)$ with long-range power-law correlations that decay as $r^{-\alpha}$, where $r \equiv |r|$. The assumption of power-law interactions is motivated by the fact that the "decision" for a development unit to be placed in a given location decays gradually with the distance from an occupied neighborhood. The correlation exponent $\alpha$ is the only parameter to be determined by empirical observations.

To discuss the morphology of a system of cities generated in the present model, we show in Fig. 12 our simulations of correlated urban systems for a fixed value of the density gradient $\lambda$, and for different degree of correlations. The correlations have the effect of agglomerating the units around a urban area. In the simulated systems the largest city is situated in the core, which is regarded as the attractive center of the city, and is surrounded by small clusters or "towns". The correlated clusters are nearly compact near their centers and become less compact near their boundaries, in qualitative agreement with empirical data on actual large cities such as Berlin, Paris and London (see Refs. 56 and 65).

So far, we have argued how correlations between occupancy probabilities can account for the irregular morphology of towns in a urban system. As can be seen in Fig. 13, the

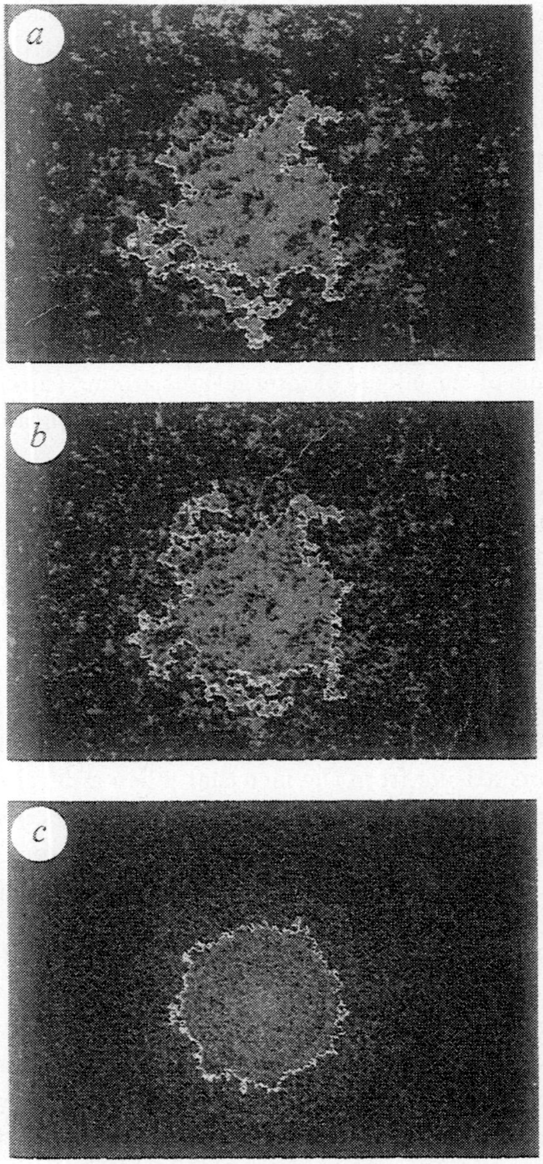


**Fig. 12** Simulations of urban systems for different degree of correlations. Here, the urban areas are red, and the external perimeter or urban boundary of the largest cluster connected to the CBD is light green. In all the figures, we fix the value of the density gradient to be $\lambda = 0.009$. **(a)** and **(b)** Two different examples of interactive systems of cities for correlation exponents $\alpha = 0.6$ and $\alpha = 1.4$, respectively. The development units are positioned with a probability that decays exponentially with the distance from the core. The units are located not randomly as in percolation, but rather in a correlated fashion depending on the neighboring occupied areas. The correlations are parametrized by the exponent $\alpha$. The strongly correlated case corresponds to small $\alpha$ ($\alpha \to 0$). When $\alpha > d$, where $d$ is the spatial dimension of the substrate lattice ($d = 2$ in our case), we recover the uncorrelated case. Notice the tendency to more compact clusters as we increase the degree of correlations ($\alpha \to 0$). **(c)** As a zeroth order approximation, one might imagine the morphology predicted in the extreme limit whereby development units are positioned at *random*, rather than in the correlated way of Figs. 12(a) and 12(b). The results for this crude approximation of a non-interactive (uncorrelated) system of cities clearly display a drastically different morphology than found from data on real cities [such as shown in Fig. 13(a)]. The non-interactive limit looks unrealistic in comparison with real cities, for the lack of interactions creates a urban area characterized by many small towns spread loosely around the core.

a b

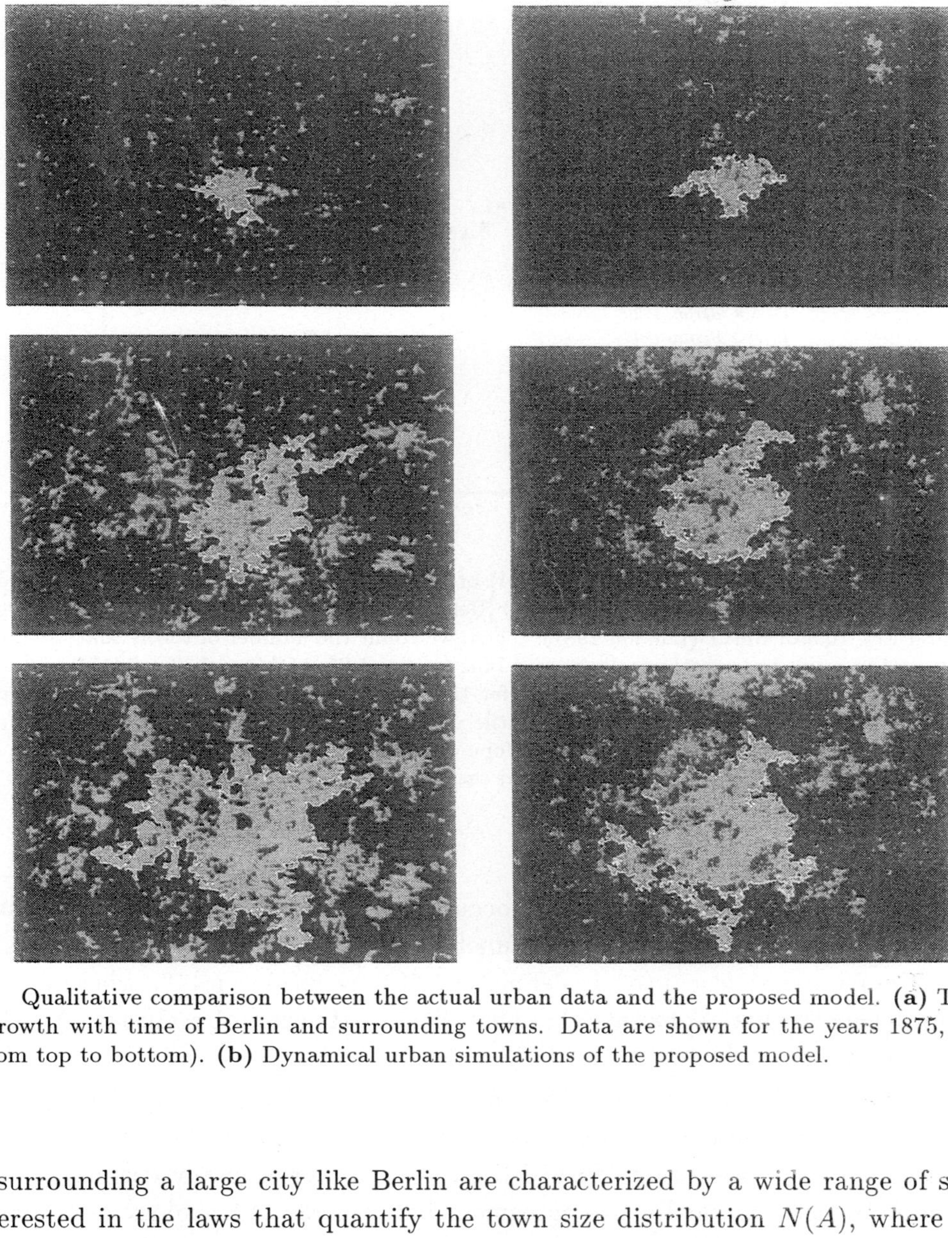

**Fig. 13** Qualitative comparison between the actual urban data and the proposed model. **(a)** Three steps of the growth with time of Berlin and surrounding towns. Data are shown for the years 1875, 1920, and 1945 (from top to bottom). **(b)** Dynamical urban simulations of the proposed model.

towns surrounding a large city like Berlin are characterized by a wide range of sizes. We are interested in the laws that quantify the town size distribution $N(A)$, where $A$ is the area occupied by a given town or "mass" of the agglomeration, so we calculate the actual distribution of the areas of the urban settlements around Berlin and London, and find (Fig. 14) that for both cities, $N(A)$ follows a power-law.

This new result of a power law area distribution of towns, $N(A)$, can be understood in the context of our model. Insight into this distribution can be developed by first noting that the small clusters surrounding the largest cluster are all situated at distances $r$ from the CBD such that $p(r) < p_c$ or $r > r_f$. Therefore, we find $N(A)$, the cumulative area distribution of clusters of area $A$, to be

$$N(A) \equiv \int_0^{p_c} n(A, p)\, dp \sim A^{-(\tau + 1/d_f \nu)} . \tag{4}$$

Here, $n(A, p) \sim A^{-\tau} g(A/A_0)$ is defined as the average number of clusters containing $A$ sites for a given $p$ *at a fixed distance* $r$, and $\tau = 1 + 2/d_f$. Here, $A_0(r) \sim |p(r) - p_c|^{-d_f \nu}$

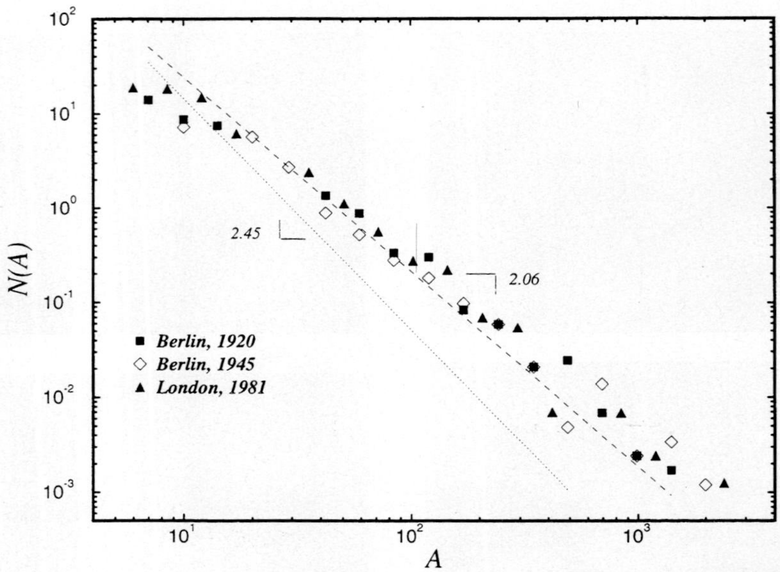


**Fig. 14** Log-log plot of the area distribution $N(A)$ of the actual towns around Berlin and London. We first digitize the empirical data of Fig. 4.1 of Ref. 10 [Berlin 1920 and 1945, shown in the last two panels of Fig. 2(a)], and Fig. 10.8 of Ref. 1 (London 1981). Then, we count the number of towns that are covered by $A$ sites, putting the result in logarithmically spaced bins (of size $1.2^k$, with $k = 1, 2, \ldots, 15$), and averaging over the size of the bin. A power-law is observed for the area distributions of both urban systems. The dotted line shows the predictions of our model for the uncorrelated case (slope $= 2.45$), while the dashed line gives results for the strongly correlated case (slope $= 2.06$). Note that the area distributions for both cities agree much better with the strongly correlated case ($\alpha \to 0$).

corresponds to the maximum typical area occupied by a cluster situated at a distance $r$ from the CBD, while $g(A/A_0)$ is a scaling function that decays rapidly (exponentially) for $A > A_0$. The exponent $\nu = \nu(\alpha)$ is defined by $\xi(r) \sim |p(r) - p_c|^{-\nu}$, where $\xi(r)$ is the connectedness length that represents the mean linear extension of a cluster at a distance $r > r_f$ from the CBD.

In Fig. 14 we also plot the power-law for the area distribution predicted by Eq. (4) along with the real data for Berlin and London. We find that the slopes of the plots for both cities are consistent with the prediction (dashed line) for the case of highly correlated systems. These results quantify the qualitative agreement between the morphology of actual urban areas and the strongly correlated urban systems obtained in our simulations.

## 7. SCALING IN ECONOMICS

Another quite unrelated phenomenon is the behavior of industrial firm sales or their employment. We have studied the dependence of the fluctuations of the annual firm growth rates on the initial size of the firm.[9] We computed the average annual fluctuation of size $\sigma(S_0)$ as a function of the initial value of size $S_0$. The remarkable linearity of the $\sigma(S_0)$ vs. $S_0$ function on a log-log scale over many orders of magnitude may indicate some universal law of economics that is applicable for small companies with sales of several thousands of dollars per year as well as for giants such as General Motors with hundred billion annual sales. We found that $\sigma(S_0)/S_0 \sim 1/S_0^{\beta}$, where $\beta \approx 0.2$ (Fig. 15).

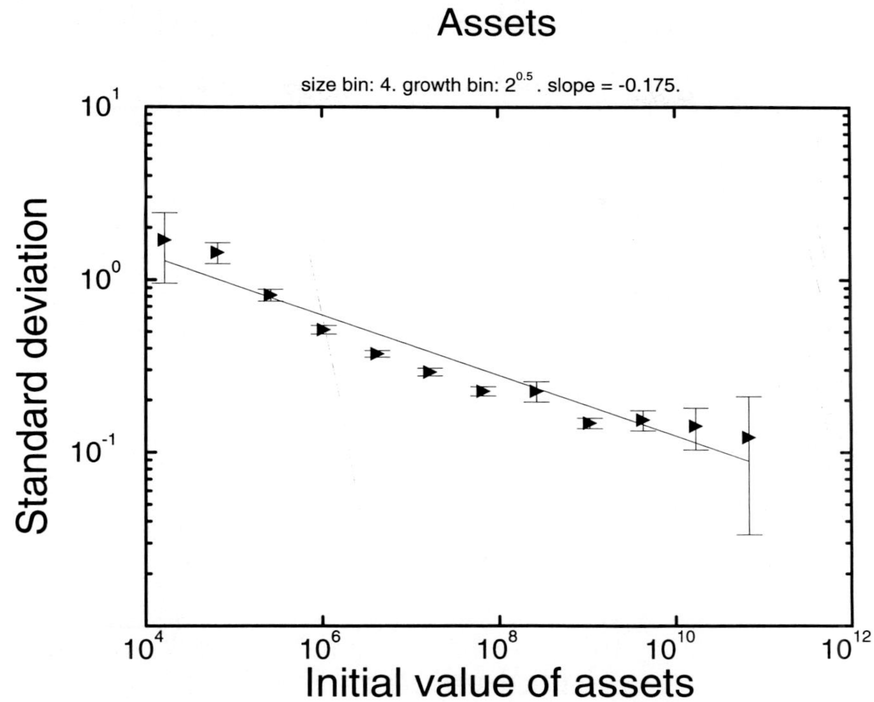


**(a)**

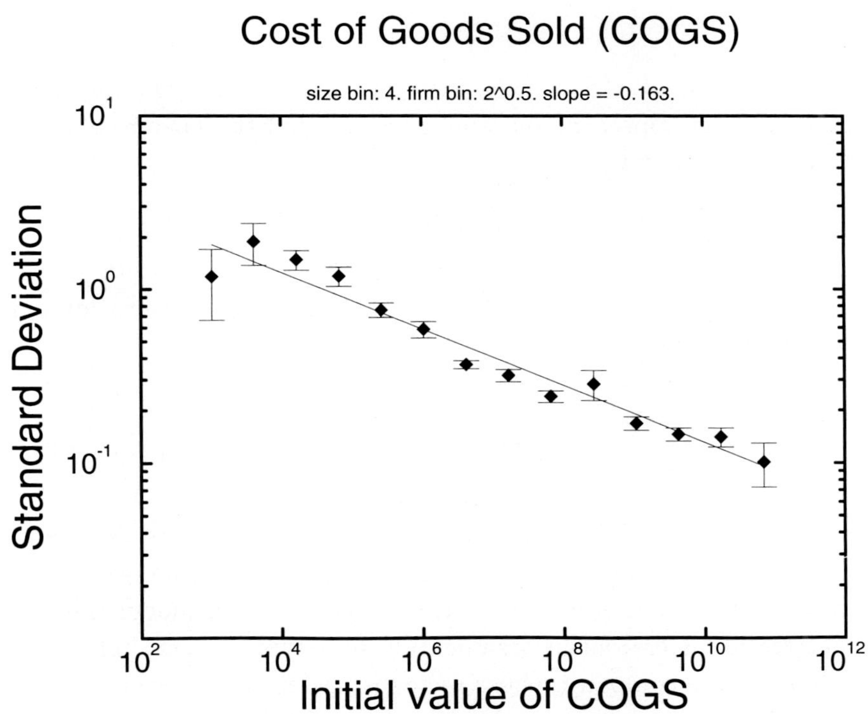


**(b)**

**Fig. 15** Power-law dependence of standard deviation $\sigma(S)/S$ on initial size $S$ for 3 quantities: **(a)** $S$ denotes assets (slope of $-0.18$), **(b)** $S$ denotes cost of goods sold (slope of $-0.16$) and **(c)** $S$ denotes property, plant, and equipment (slope of $-0.18$).

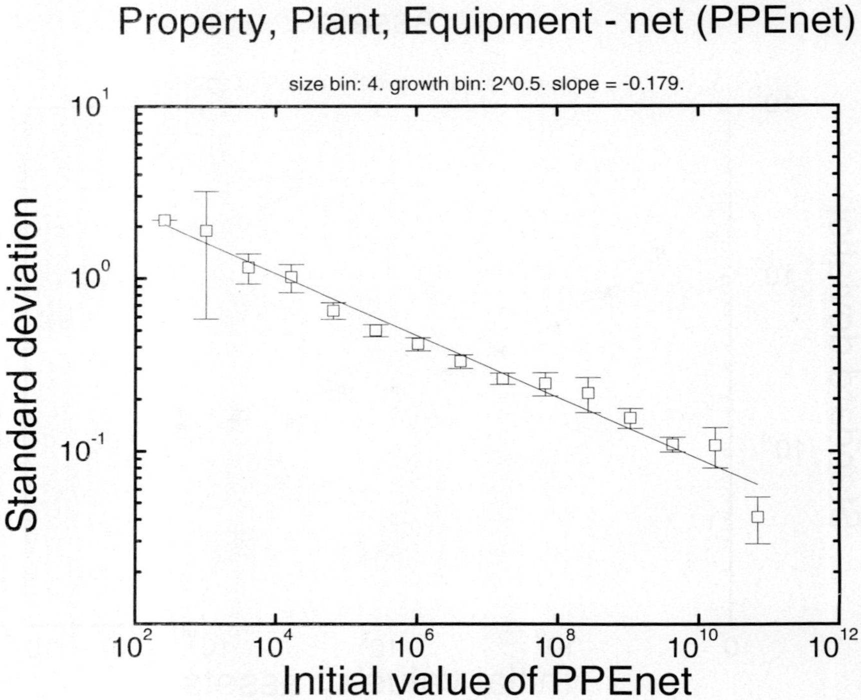


**(c)**

**Fig. 15** (*Continued*)

The power law increase of $\sigma(S_0)$ on $S_0$ may have its origin in the internal structure of each firm. In the simplest approach, one would assume that the sales $S_0$ of a given company result from $N$ independent units

$$S_0 = \sum_{i=1}^{N} \xi_i \,. \tag{5}$$

If the unit sales $\xi_i$ have a typical average $\langle \xi \rangle = S_0/N$ and an annual variation $u_i$ independent of $s_0$. Then the annual change in sales is

$$\Delta S = \sum_{i=1}^{N} u_i \,. \tag{6}$$

In analogy with a random walk, $\sigma(S_0)$ would grow as $\sqrt{N}$ or since $N$ is proportional to $S_0$ as $\sqrt{S_0}$, thus giving $\alpha = 1/2$. The much larger value of $\alpha$ that we find indicates the presence of strong correlations among the firm's units. We can model this phenomena by considering the tree-like hierarchical organization of a typical firm. The root of the tree represents the head of the company, whose policy is passed to the level beneath, and so on, until finally the units in the lowest level take action. Each of these units has an average sales value $\langle \xi \rangle = S_0/N$ and a corresponding typical fluctuation $u$. The number of links connecting the levels will vary from level to level, but there is a value $z$ which represents a certain average number of links. Then the number of units $N$ is equal to $z^k$, where $k$ is the number of levels.

What are the consequences of this simple model? Let us first assume that the head of the firm suggests a policy that could result in changing the sales of each unit by an amount $u$. If this policy is propagated through the hierarchy without any modifications, then the change in sales is simply $\Delta S = N_u = S_0 u/\langle\xi\rangle$. Accordingly, $\alpha = 1$.

More realistically, the policy of the head can be modified (undergo "mutation") at each level of the firm management with a small probability $p_m$. Hence the sales of the entire firm becomes a random variable with a standard deviation that can be explicitly computed using recursion relations among the levels of the tree. The result is $\sigma^2(S_0) = u^2[4zp_m(1-p_m)(y^k - z^k)/(y-z) + y^n]$, where $y = z^2(1-2p_m)^2$ and $k = \ln(S_0/\langle\xi\rangle)/\ln z$. For large $k$, the model predicts $\alpha = 1 - |\ln(1-2p_m)|/\ln(z)$ if $z > 1/(1-2p_m)^2$ and $\alpha = 1/2$ otherwise. Despite the model's simplicity, this result seems to be plausible. For example, if we require $\alpha = 0.8$, as suggested by our data, and choose $z = 3$, we obtain $p_m = 0.1$, which might be a reasonable situation.

The models proposed are very elementary, and show that simple mechanisms can provide some insight into our findings.

## ACKNOWLEDGMENTS

We are grateful to many individuals, including R. N. Mantegna, M. E. Matsa, S. M. Ossadnik, and F. Sciortino, for major contributions to those results reviewed here that represent collaborative research efforts. We also wish to thank C. Cantor, C. DeLisi, M. Frank-Kamenetskii, A. Yu. Grosberg, G. Huber, I. Labat, L. Liebovitch, G. S. Michaels, P. Munson, R. Nossal, R. Nussinov, R. D. Rosenberg, J. J. Schwartz, M. Schwartz, E. I. Shakhnovich, M. F. Shlesinger, N. Shworak, and E. N. Trifonov for valuable discussions. Partial support was provided by the National Science Foundation, National Institutes of Health (Human Genome Project), the G. Harold and Leila Y. Mathers Charitable Foundation, the National Heart, Lung and Blood Institute, the National Aeronautics and Space Administration, the Israel-USA Binational Science Foundation, Israel Academy of Sciences, and (to C-KP) by an NIH/NIMH Postdoctoral NRSA Fellowship.

## REFERENCES

1. H. E. Stanley, *Introduction to Phase Transitions and Critical Phenomena,* (Oxford Univ. Press, Oxford, 1971).
2. C. H. Back et al., *Nature* **378**, 597–600 (1995).
3. H. E. Stanley and N. Ostrowsky, eds., *Correlations and Connectivity: Geometric Aspects of Physics, Chemistry and Biology* (Kluwer, Dordrecht, 1990).
4. D. S. Gaunt and C. Domb, *J. Phys.* **C3**, 1442–1461 (1970).
5. C.-K. Peng, S. Buldyrev, A. Goldberger, S. Havlin, F. Sciortino, M. Simons and H. E. Stanley, *Nature* **356**, 168–171 (1992).
6. B. Suki, A.-L. Barabási, Z. Hantos, F. Peták and H. E. Stanley, *Nature* **368**, 615–618 (1994).
7. C. K. Peng, S. Havlin, H. E. Stanley and A. L. Goldberger, *Chaos* **5**, 82–87 (1995).
8. H. Makse, S. Havlin and H. E. Stanley, *Nature* **377**, 608–612 (1995).
9. M. H. R. Stanley, L. A. N. Amaral, S. V. Buldyrev, S. Havlin, H. Leschhorn, P. Maass, M. A. Salinger and H. E. Stanley, "Scaling Behavior in the Growth of Companies," *Nature* **379**, 804–806 (1996); M. H. R. Stanley, L. A. N. Amaral, S. V. Buldyrev, S. Havlin, H. Leschhorn, P. Maass, M. A. Salinger and H. E. Stanley, "Can Statistical Physics Contribute to the Science of Economics?" [Proc. International Conf. on "Future of Fractals"] *Fractals* **4**(3), 415–425 (1996).

10. B. B. Mandelbrot, *The Fractal Geometry of Nature* (W. H. Freeman, San Francisco, 1982).
11. B. J. West and A. L. Goldberger, *J. Appl. Physiol.*, **60**, 189 (1986); B. J. West and A. L. Goldberger, *Am. Sci.*, **75**, 354 (1987); A. L. Goldberger, B. J. West and Yale, *J. Biol. Med.* **60**, 421 (1987); A. L. Goldberger, D. R. Rigney and B. J. West, *Sci. Am.* **262**, 42 (1990); B. J. West and M. F. Shlesinger, *Am. Sci.* **78**, 40 (1990); B. J. West, *Fractal Physiology and Chaos in Medicine* (World Scientific, Singapore, 1990); B. J. West and W. Deering, *Phys. Reports* **254**, 1–100 (1994); J. B. Bassingthwaighte, L. S. Liebovitch and B. J. West, *Fractal Physiology* (Oxford University Press, New York, 1994).
12. A. Bunde and S. Havlin, eds., *Fractals and Disordered Systems* (Springer-Verlag, Berlin, 1991); A. Bunde and S. Havlin, eds., *Fractals in Science* (Springer-Verlag, Berlin, 1994); A.-L. Barabasi and H. E. Stanley, *Fractal Concepts in Surface Growth* (Cambridge University Press, Cambridge, 1995).
13. M. F. Shlesinger and B. J. West, *Phys. Rev. Lett.* **67**, 2106 (1991).
14. B. Suki, A.-L. Barabási, Z. Hantos, F. Peták and H. E. Stanley, *Nature* **368** 615–618 (1994).
15. E. R. Weibel and D. M. Gomez, *Science* **137**, 577 (1962).
16. A. A. Tsonis and P. A. Tsonis, *Perspectives in Biology and Medicine* **30**, 355 (1987).
17. F. Family, B. R. Masters and D. E. Platt, *Physica* **D38**, 98 (1989); B. R. Masters, F. Family and D. E. Platt, *Biophys. J. (Suppl.)* **55**, 575 (1989); B. R. Masters and D. E. Platt, *Invest. Ophthalmol. Vis. Sci. (Suppl.)* **30**, 391 (1989).
18. M. Sernetz, J. Wübbeke and P. Wlczek, *Physica* **A191**, 13 (1992).
19. H. Takayasu, *Fractals in the Physical Sciences* (Manchester University Press, Manchester, 1990).
20. D. R. Morse, *Nature* **314**, 731 (1985).
21. T. G. Smith, W. B. Marks, G. D. Lange, W. H. Sheriff Jr. and E. A. Neale, *J. Neuroscience Methods* **27**, 173 (1989).
22. H. E. Stanley, *Bull. Am. Phys. Soc.* **34**, 716 (1989); F. Caserta, H. E. Stanley, W. D. Eldred, G. Daccord, R. E. Hausman and J. Nittmann, *Phys. Rev. Lett.* **64**, 95 (1990); F. Caserta, R. E. Hausman, W. D. Eldred, H. E. Stanley and C. Kimmel, *Neurosci. Letters* **136**, 198 (1992); F. Caserta, W. D. Eldred, E. Fernandez, R. E. Hausman, L. R. Stanford, S. V. Buldyrev, S. Schwarzer and H. E. Stanley, *J. Neurosci. Methods* **56**, 133 (1995).
23. D. R. Kayser, L. K. Aberle and R. D. Pochy, *L. Lam, Physica* **A191**, 17 (1992).
24. K. R. Bhaskar, B. S. Turner, P. Garik, J. D. Bradley, R. Bansil, H. E. Stanley and J. T. LaMont, *Nature* **360**, 458 (1992).
25. T. Matsuyama, M. Sugawa and Y. Nakagawa, *FEMS Microb. Lett.* **61**, 243 (1989); H. Fujikawa and M. Matsushita, *J. Phys. Soc. Jpn.* **58**, 387 (1989); M. Matsushita and H. Fujikawa, *Physica* **A168**, 498 (1990).
26. T. Vicsek, M. Cserzö and V. K. Horváth, *Physica* **A167**, 315 (1990); S. Matsuura and S. Miyazima, *Physica* **A191**, 30 (1992).
27. T. Vicsek, *Fractal Growth Phenomena, Second Edition* (World Scientific, Singapore, 1992).
28. E. Ben-Jacob, H. Shmueli, O. Shochet and A. Tenenbaum, *Physica* **A187**, 378 (1992); E. Ben-Jacob, A. Tenenbaum, O. Shochet and O. Avidan, *Physica* **A202**, 1 (1994).
29. H. Larralde, P. Trunfio, S. Havlin, H. E. Stanley and G. H. Weiss, *Nature* **355**, 423 (1992); M. F. Shlesinger, *Nature* **355**, 396 (1992); H. Larralde, P. Trunfio, S. Havlin, H. E. Stanley and G. H. Weiss, *Phys. Rev.* **A45**, 7128 (1992); S. Havlin, H. Larralde, P. Trunfio, J. E. Kiefer, H. E. Stanley and G. H. Weiss, *Phys. Rev.* **A46**, R-1717 (1992); S. Havlin, M. Araujo, H. Larralde, A. Shehter, H. E. Stanley and P. Trunfio, *Nuovo Cimento* **16**, 1039–1051 (1995).
30. J. G. Skellam, *Biometrika* **38**, 196 (1951).
31. P. H. Harvey and J. R. Krebs, *Science* **249**, 140 (1990).
32. R. H. Peters, *The Ecological Implications of Body Size* (Cambridge University Press, Cambridge, 1983); J. E. I. Hokkanen, *J. Theor. Biol.* **499**, 499 (1956); A. A. Biewener, *Science* **250**, 1097 (1990).
33. M. Sernetz, B. Gelleri and J. Hofmann, *J. Theor. Biol.* **187**, 209 (1985); R. R. Strathmann, *Science* **250**, 1091 (1990).
34. J. D. Watson, M. Gilman, J. Witkowski and M. Zoller, *Recombinant DNA* (Scientific American Books, New York, 1992).
35. E. W. Montroll and M. F. Shlesinger, "The wonderful world of random walks," in *Nonequilibrium Phenomena II. From Stochastics to Hydrodynamics* eds. J. L. Lebowitz and E. W. Montroll, pp. 1–121 (North-Holland, Amsterdam 1984).

36. M. Ya. Azbel, *Phys. Rev. Lett.* **31**, (1973) 589.
37. W. Li and K. Kaneko, *Europhys. Lett.* **17**, (1992) 655.
38. C. L. Berthelsen, J. A. Glazier and M. H. Skolnick, *Phys. Rev.* **A45**, (1992) 8902.
39. S. M. Ossadnik, S. V. Buldyrev, A. L. Goldberger, S. Havlin, R. N. Mantegna, C.-K. Peng, M. Simons and H. E. Stanley, *Biophys. J.* **67**, (1994) 64.
40. C.-K. Peng, S. V. Buldyrev, S. Havlin, M. Simons, H. E. Stanley and A. L. Goldberger, *Phys. Rev.* **E49**, (1994) 1685.
41. A. L. Goldberger, D. R. Rigney and B. J. West, *Sci. Am.* **262**, 42–49 (1990).
42. C.-K. Peng, J. Mietus, J. M. Hausdorff, S. Havlin, H. E. Stanley and A. L. Goldberger, *Phys. Rev. Lett.* **70**, 1343–1346 (1993).
43. C.-K. Peng, S. V. Buldyrev, J. M. Hausdorff, S. Havlin, J. E. Mietus, M. Simons, H. E. Stanley and A. L. Goldberger, in *Fractals in Biology and Medicine*, T. F. Nonnenmacher, G. A. Losa and E. R. Weibel, eds. (Birkhaüser Verlag, Basel, 1994), pp. 55–65.
44. P. Bak, C. Tang and K. Wiesenfeld, *Phys. Rev. Lett.* **59**, 381–384 (1987).
45. M. N. Lévy, *Circ. Res.* **29**, 437–445 (1971).
46. C.-K. Peng, S. V. Buldyrev, S. Havlin, M. Simons, H. E. Stanley and A. L. Goldberger, *Phys. Rev.* **E49**, 1691–1695 (1994).
47. B. T. Hyman, H. L. West, G. W. Rebeck, S. V. Buldyrev, R. N. Mantegna, M. Ukleja, S. Havlin, and H. E. Stanley, *Proc. Natl. Acad. Sci. USA* **92**, 3586–3590 (1995).
48. B. T. Hyman, R. E. Tanzi, K. M. Marzloff, R. Barbour and D. Schenk, *J. Neuropath. Exp. Neurol.* **51**, 76–83 (1992).
49. B. T. Hyman, K. Marzloff and P. V. Arriagada, *J. Neuropath. Exp. Neurol.* **52**, 594–600 (1993).
50. E. C. Pielou, *An Introduction to Mathematical Ecology* (Wiley-Interscience, New York, 1969).
51. E. W. Montroll and M. F. Shlesinger, *Proc. Natl. Acad. Sci.* **79**, 3380–3383 (1982).
52. S. Redner, *Am. J. Phys.* **58**, 267–273 (1990).
53. C. G. Granqvist and R. A. Buhrman, *J. Appl. Phys.* **47**, 2219–2200 (1976).
54. M. Kawai, P. Cras and G. Perry, *Brain Res.* **592**, 278–282 (1992).
55. R. A. Armstrong, D. Myers and C. U. Smith, *Neurosci. Lett.* **127**, 223–226 (1991).
56. M. Batty and P. Longley, *Fractal Cities* (Academic Press, San Diego 1994).
57. L. Benguigui and M. Daoud, *Geog. Analy.* **23**, 362-368 (1991).
58. L. Benguigui, *Physica* **A219**, 13–26 (1995).
59. T. A. Witten and L. M. Sander, *Phys. Rev. Lett.* **47**, 1400–1403 (1981).
60. C. Clark, *J. R. Stat. Soc.* (Series A) **114**, 490–496 (1951).
61. A. Coniglio, C. Nappi, L. Russo and F. Peruggi, *J. Phys.* **A10**, 205–209 (1977).
62. A. Weinrib, *Phys. Rev.* **B29**, 387–395 (1984).
63. S. Prakash, S. Havlin, M. Schwartz and H. E. Stanley, *Phys. Rev.* **A46**, R1724–1727 (1992).
64. J.-F. Gouyet, *Physics and Fractal Structures* (Springer-Verlag, Berlin, 1995).
65. P. Frankhauser, *La Fractalité des Structures Urbaines* (Collection Villes, Anthropos, Paris, 1994).

Fractals, Vol. 4, No. 4 (1996) 453–462


# NUCLEAR RELAXATION IN IRREGULAR OR FRACTAL PORES

B. SAPOVAL, S. RUSS, J. P. KORB and D. PETIT
*Laboratoire de Physique de la Matière Condensée,*
*Ecole Polytechnique, C.N.R.S., 91128, Palaiseau, France*

Received August 10, 1995; Accepted November 13, 1995

**Abstract**

We apply a fractal description of pore surface irregularity to study the nuclear relaxation of a liquid confined in an irregular pore. Our calculation on a pre-fractal volume with infinite surface relaxation rate shows that the long time relaxation behavior is dominated by an exponential mode characterizing a free diffusive volume. At short time, the calculated magnetization follows closely the power-law behavior previously proposed by de Gennes. For finite surface relaxation rate, we extend our studies of the efficiency of exchange on fractal membranes to nuclear relaxation in the same geometry. We predict a decay involving two characteristic times. For very slow surface relaxation, there will exist exponential relaxation governed by the surface relaxation. These laws show that the pore surface irregularities may play an important role in nuclear relaxation.

## 1. INTRODUCTION

The role of confined geometry on the dynamics of liquids in porous media has recently been widely studied in view of its importance for a variety of technological processes including catalysis, chromatography, oil recovery, lubricants studies and foodstuffs properties. It is also important when considering the general problem of exchange dynamics of surface-active molecules between an irregular solid interface and a bulk liquid. Nuclear magnetic relaxation has proven useful to explore the motion of molecules in porous systems. It is particularly sensitive to confinement and has been used for the analysis of pore sizes in rocks[1] as well as pore

This paper was presented at the International Conference on "Future of Fractals" held at Aichi, Japan from 25–27 July 1995. It should have been published in Fractals, Vol. 4, No. 2 (Fractals Vol. 4, Nos. 2 & 3 being special issues of the journal devoted to "Future of Fractals"). The publisher wishes to apologise to the authors for the delayed publication of the paper in this issue.

structures in xerogels,[2] resins[3] and silica glasses.[4–6] The pulsed-field-gradient NMR has also given information on the surface to volume ratio[7,8] as well as the tortuosity[9] of porous media. The relaxation properties are modified by the fact that the spin-bearing molecules move in a restricted or confined space. Confinement may influence relaxation in several ways. First, there can exist a contribution to relaxation from the pore walls which may possess, for instance, paramagnetic impurities. Second, even without any relaxation on the walls, the random motion of the molecules which causes the spin-spin interaction to vary in time, may be affected by the fact that both translational[6] and reorientational molecular motions[10] are confined. In these studies the possible geometrical irregularity of the pores is not taken into account for two reasons: (1) the lack of a general frame to describe geometrical irregularity; and (2) the lack of analytical results describing the magnetization evolution in an irregular system.

The introduction of the concept of fractal geometry has been a breakthrough in the description of irregularity.[11] Our purpose here is not to show that one can really ascribe a definite fractal geometry to pores but to answer the following question: If the pores are strongly irregular, how will this modify nuclear relaxation? For this we consider fractal or prefractal pores of the type of geometry shown in Fig. 1. Note that here we consider surface fractals (pores bounded by a fractal surface) and not mass fractals[12] for which we have already proposed a relaxation method for measuring the fractal dimension.[13] Note also that we consider physical fractals which have a finite smaller cut-off $\ell$.

In this preliminary study we consider only two types of effects. First, we present a numerical study of nuclear relaxation in irregular cylindrical pores in the limiting situation where the surface relaxation rate is infinite. For this we use the time-dependent solution of the diffusion equation in irregular pores, using a numerical method which has been used to compute the eigenmodes and eigenfrequencies of the fractal drum.[14,15]

Second, we discuss the effect of a finite surface relaxation rate using what we know of the problem of Laplacian transfer to and across irregular interfaces. Examples are diffusion to and across fractal membranes or the response of porous fractal electrodes in electrochemistry.[16,17] In the same way that nuclear spins have to diffuse to the pore walls to hit paramagnetic impurities and relax, the ions in an

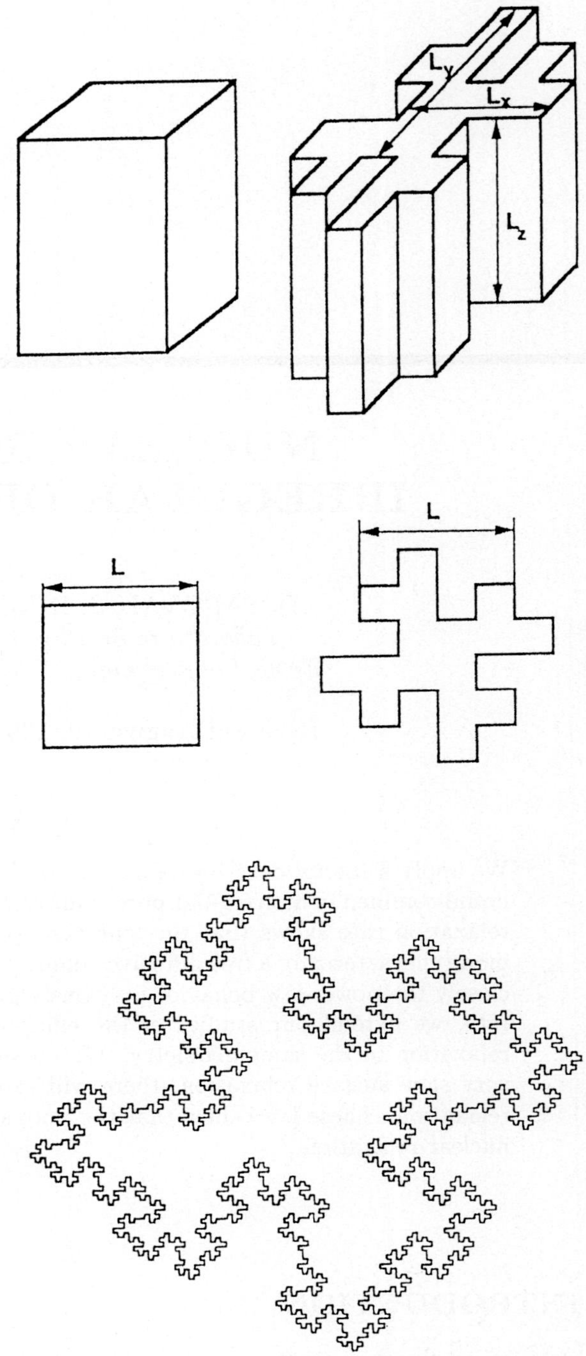


**Fig. 1** The geometries that we compare: (*Top*) We show a square pore on the left and an irregular pore on the right. In this paper we consider only "cylinders" of very large height $L_z \gg L_x = L_y = L$ with irregular basis. (*Bottom*) The geometry of the basis is shown for different stages of fractal generations $\nu$. We see the square initiator ($\nu = 0$) and the stage $\nu = 1$ and 3 of fractal iteration. The volume $L^2 L_z$ of the pore is conserved through the iteration process. The fractal dimension of the perimeter is $d_f = \ln 8/\ln 4 = 3/2$ and the fractal dimension of the pore surface is $d_f + 1 = 5/2$.

electrochemical cell have to be transported to the electrode to be reduced or oxidized. Correspondingly, oxygen molecules have to diffuse in air before hitting the membrane of our alveolae for oxidizing hemoglobin. What is common in these otherwise very different phenomena, is that the flux of magnetization towards a relaxing wall is governed by Fick's law in the same way as the flux of charge (electric current) or molecules (diffusion flux from Fick's law) is due to a field which satisfies Laplace equation in the steady state situation.[16–19]

The general problem that we address here is to find the time evolution of the total magnetic moment of an irregular cylindrical pore in which the bulk magnetization follows the equation:

$$\partial m/\partial t = D\Delta m - (m - m_0)/T_{1,B} \tag{1}$$

where $D$ is the diffusion coefficient of the spin-bearing particles, $T_{1,B}$ is the bulk spin-lattice relaxation time due to other relaxation mechanisms and $m_0$ is the equilibrium magnetization. In our case of a long pore, these equations are planar equations. Following Ref. 20, we change variables from $m(x, y, t)$ to $M(x, y, t)$ through the relation:

$$m(x, y, t) = M(x, y, t)\exp(-t/T_{1,B}) + m_0\,. \tag{2}$$

This transformation replaces the evolution equation [Eq. (1)] by a diffusion equation:

$$\partial M/\partial t = D\Delta M\,. \tag{3}$$

This equation describes the evolution of magnetization in the bulk but does not account for the influence of the surface. There are two ways to go further: First, one can write the relaxation boundary condition in the form:

$$D\partial M(x, y, t)/\partial n = a\mu_s M(x, y, t) \tag{4}$$

where $\mu_s$ represents the surface relaxation rate and $a$ is a length of the order of the thickness of a monolayer. This latter equation is the magnetization conservation equation. The Fick's flux at the boundary is equal to the relaxing flux on the surface. In principle, the problem is stated entirely by these equations. Equivalently, one can write:

$$M = \Lambda\partial M/\partial n \quad \text{with} \quad \Lambda = D/a\mu_s \tag{5}$$

In a form of arbitrary geometry, the evolution of the magnetization can be obtained by numerical solution of Eqs. (3, 4). In this formulation, the relaxation efficiency is assumed to be uniform on the surface.

It is interesting to discuss the order of magnitude of the parameters involved in these equations. We can base this discussion on our knowledge of the problem of fractal membranes or electrodes which work with similar boundary conditions. In the same way that molecules bounce many times before crossing a membrane of finite permeability, particles moving at random have to hit the relaxing walls many times before being relaxed if the surface relaxation is weak. From the study of the fractal membranes and electrodes, we know that the relative surface and bulk transport properties are characterized by the length $\Lambda$. Consider for instance a square relaxing surface of side $L$. It presents a spin "admittance" equal to $a\mu_s L^2$. But to reach this surface, the spins have to cross by diffusion a cube of side $L$, which presents a diffusion admittance equal to $LD$. These two admittances are equal when $L$ is equal to $\Lambda$. In an equivalent fashion, one can write Eq. (5) as $\partial M/M = \partial n/\Lambda$, which tells us that the typical range of variations in space of M is $\Lambda$.

For an irregular surface, the length $\Lambda$ characterizes the range of screening in the sense that any part of the pore wall with a perimeter larger or equal to $\Lambda$, can be considered in first approximation as totally absorbing for the magnetization. This is discussed in Refs. 16 to 19. We recall briefly how this length plays a role in our problem. Consider a part of a fractal surface with a diameter $L$ and smaller cut-off $\ell$. It presents an "admittance of relaxation" equal to $a\mu_s\ell^2(L/\ell)^{D_f}$. This means that the flux of magnetization that is lost per unit time, if the magnetization $m$ is uniform along the surface, is equal to $ma\mu_s\ell^2(L/\ell)^{D_f}$. But to reach such a surface, spins have to diffuse across a cube of side $L$ which has an admittance for diffusion equal to DL. These two admittances are equal for a diameter $L = L_{cg}$ given by:

$$ma\mu_s\ell^2(L_{cg}/\ell)^{D_f} = mDL_{cg} \quad \text{or}$$
$$L_{cg} = \ell(N/\ell)^{1/(D_f-1)}\,. \tag{6}$$

The diameter $L_{cg}$ is the diameter of a region such that the perimeter of a cut of the fractal surface in that region is equal to $\Lambda$. This follows from Eq. (6), which gives $\Lambda = \ell(L_{cg}/\ell)^{(D_f-1)}$. (It is known that a cut of a fractal surface of dimension $D_f$ is a fractal line with dimension $d_f = D_f - 1$). The subscript $cg$ stands for coarse-graining for the following reasons.

Regions of size smaller than $L_{cg}$ work uniformly for relaxation because they present a small surface admittance as compared with the admittance to reach them from the bulk. On the other hand, if we consider pore walls with size features bigger than $L_{cg}$ (in the previous sense), these pore walls can be considered as totally absorbing in the sense that the magnetization flux to reach them is limited by diffusion and not by the surface relaxation. These "coarse-grained" regions will then appear as absorbing for spins.

To relate these numbers to the case of porous materials, we suppose that the pore wall possesses relaxing sources (like paramagnetic impurities) in relative concentration equal to 1. We suppose that the considered spin-bearing molecule is moving randomly in the tridimensional space (gas or liquid). From time to time the molecule hits the wall and stays on it for some residence time $\tau_r$. Most often, the local magnetic field and/or the residence time are too small to cause a spin rotation of the order of $\pi$ at each interaction and we are in the regime of weak interactions. We call $\gamma$ the gyromagnetic ratio and $h_1$ the transverse local field experienced by the nuclear spin when it sits on the surface. In that case the maximum surface relaxation rate will be of the order of:

$$\mu_s \approx (\gamma h_1)^2 / 2\omega \tag{7}$$

in an external field in which the nuclear frequency is $\omega$[21]. Considering protons and a local field $h_1$ of about 3 teslas and ordinary NMR frequencies of the order of 300 MHz, one obtains a maximum surface rate of the order of $\mu_s \approx 3 \times 10^8$ sec$^{-1}$. Taking $a = 2 \times 10^{-8}$ cm and the diffusion coefficient of water $D = 2.5 \times 10^{-5}$ cm$^2$/sec, one obtains for the smallest feasible value of $\Lambda$ the number $\Lambda_{\min} \approx 10^{-5}$ cm. This value is really an extreme value because we have supposed that the pore walls are covered with paramagnetic impurities with concentration 1 and that the time $\tau_r$ is of the order of $1/\omega$. More realistic values of $\Lambda$ will be of the order of $10^{-3}$ cm or even larger.

We will consider in the following three relaxation regimes, depending on the value of $\Lambda$, as compared with the characteristic lengths which describe the real pore geometry. If $\Lambda$ is smaller or equal to $\ell$, the size of the building block of the structure, we will be in a situation of infinite surface relaxation which we call Diffusion Limited Relaxation (DLR). If $\ell < \Lambda < \ell(L/\ell)^{(D_f-1)}$, we will be in the finite surface relaxation case where a part of the decay is limited by surface relaxation and an inner part of the pore relaxes by diffusion as we will see. If the surface relaxation is very slow, the characteristic length $\Lambda$ exceeds the perimeter of a typical cut of the pore $\ell(L/\ell)^{(D_f-1)}$, where $L$ is the pore size. In that last case the relaxation will be surface limited.

## 2. DIFFUSION LIMITED RELAXATION

Although this situation is not the most common, we first discuss the case where infinite relaxation exists on the surface, i.e., a situation where $\mu_s$ can be considered as infinite or the walls totally absorbing. This means that $\Lambda$ is smaller or equal to $\ell$. The relaxation is then determined and limited by the diffusive motion towards the surface. In this case Eq. (4) imposes the value of $M$ on the surface to be equal to zero. This is the Dirichlet boundary condition and we have to solve the diffusion equation [Eq. (3)] with this boundary condition. This has already been done in the study of the vibrations for a fractal drum.[14,15] The general solution of Eq. (3) is of the form:

$$M(x, y, t) = \Sigma_n c_n \exp(-t/\tau_n) \Psi_n(x, y) \tag{8}$$

where $\Psi_n(x, y)$ is an eigenfunction of the eigenvalue equation:

$$\Delta \Psi_n(x, y) = -(D\tau_n)^{-1} \Psi_n(x, y) \,. \tag{9}$$

The set of all the solutions can be ordered with decreasing values of $\tau_n$, so that $\Psi_n$ is the eigenfunction associated with the $n$th mode, with:

$$\tau_0 > \tau_1 \geq \tau_2 \geq \cdots \geq \tau_n \geq \cdots \,. \tag{10}$$

The set of $\Psi_n$ forms an orthonormal basis and any function $M(x, y)$ defined on $V$ can be expanded as:

$$M(x, y) = \Sigma_n c_n \Psi_n(x, y) \,. \tag{11}$$

To compute the magnetization evolution we start at time $t = 0$, with a constant function $M_0(x, y, t = 0) = 1/\text{Vol.}$, where Vol. is the pore volume. Equation (11) states that, due to the hierarchy of the different time constant $\tau_n$, the coefficients of the linear expansion will tend to zero with different time constants. After some time the numerical evolution leads to a relaxation function $c_0 \Psi_0(x, y) \exp(-t/\tau_0)$.

This result is general. It shows that whatever the geometry, the long time magnetization time evolution due to diffusion from walls with infinite surface

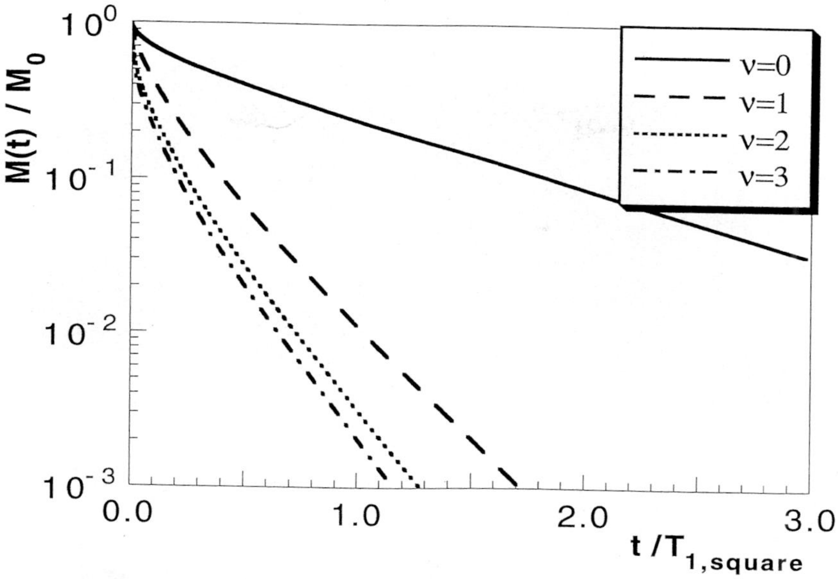


**Fig. 2** Magnetization decay in the case of Diffusion Limited Relaxation for pores of stage $\nu = 0$; 1; 2 and 3. One observes a nonexponential decay for short times. The unit of time is the diffusion limited relaxation time in the square pore. Non-exponential relaxation will dominate for highly irregular pores ($\nu = 3$). Even for $\nu = 1$, the exponential contribution governs about 10% of the total magnetization in our strongly irregular geometry.

relaxation rates, will be exponential with a relaxation time $T_1 = \tau_0$. This is verified in the numerical results that are presented in Fig. 2.

We first discuss the numerical method. For numerical efficiency, the diffusion equation [Eq. (3)] was discretized both in space and time, so that the computation took place on a square grid compatible with the boundary conditions. There are in principle three types of errors in this method: Systematic errors due to time and space discretization and numerical errors due to truncation of numbers in the computation. We will not repeat the discussion of the uncertainty which was given in Ref. 15, where it was shown that the method is very efficient. The average time for the computation of one time evolution is of the order of one hour for the discretized square 256*256 using a Sun Sparc 20 workstation. For a given geometry, the computing time grows linearly with the number of points inside the pore.

For the case of the square and owing to the same boundary condition, the fundamental eigenfunction of the continuous diffusion equation is of the form $\sin(\pi x/L)\,\sin(\pi y/L)$ which corresponds to an eigenvalue:

$$T_{1,\text{square}}^{-1} = 2\pi^2 D/L^2 \,. \tag{12}$$

In the following, the value [Eq. (12)] will be considered as our unit for exponential relaxation rate so that we will give our numerical results in the form of relative relaxation rates given by:

$$T_{1,r}^{-1} = T_1^{-1}/T_{1,\text{square}}^{-1} \,. \tag{13}$$

The unit relaxation time is then the relaxation time of an infinite square pore having the same volume as our irregular pores.

We work on a discretized lattice with $Z$ segments per side. In this case, there exists an exact solution of the eigenvalue problem:

$$T_1^{-1} = 4Z^2 D/L^2[1 - \cos(\pi/Z)] \tag{14}$$

which we can use to test the numerical accuracy of our computation. With our unit, the value of $T_{1,r}^{-1}$ (equal to 1 for the continuous diffusion equation), is predicted to be equal to 0.999 949 992 from Eq. (14), with $Z = 127$. The numerical result is also found to be equal to 0.999 949 992. The concordance of these 9 digits shows that our numerical procedure is indeed very efficient.

The time evolution shown in Fig. 2 can be discussed in the following manner. The long time relaxation due to diffusion towards relaxing surfaces is exponential as predicted. For the pores of generation $\nu = 1$, 2 and 3 we find, respectively, for the exponential relaxation time the relative values:

$$\begin{aligned} T_{1,r}^{-1} &= 3.306\,692\,35 \text{ using 63 sites per segment} \\ T_{1,r}^{-1} &= 4.121\,279\,88 \text{ using 15 sites per segment} \\ T_{1,r}^{-1} &= 4.410\,699\,84 \text{ using 3 sites per segment} \end{aligned} \tag{15}$$

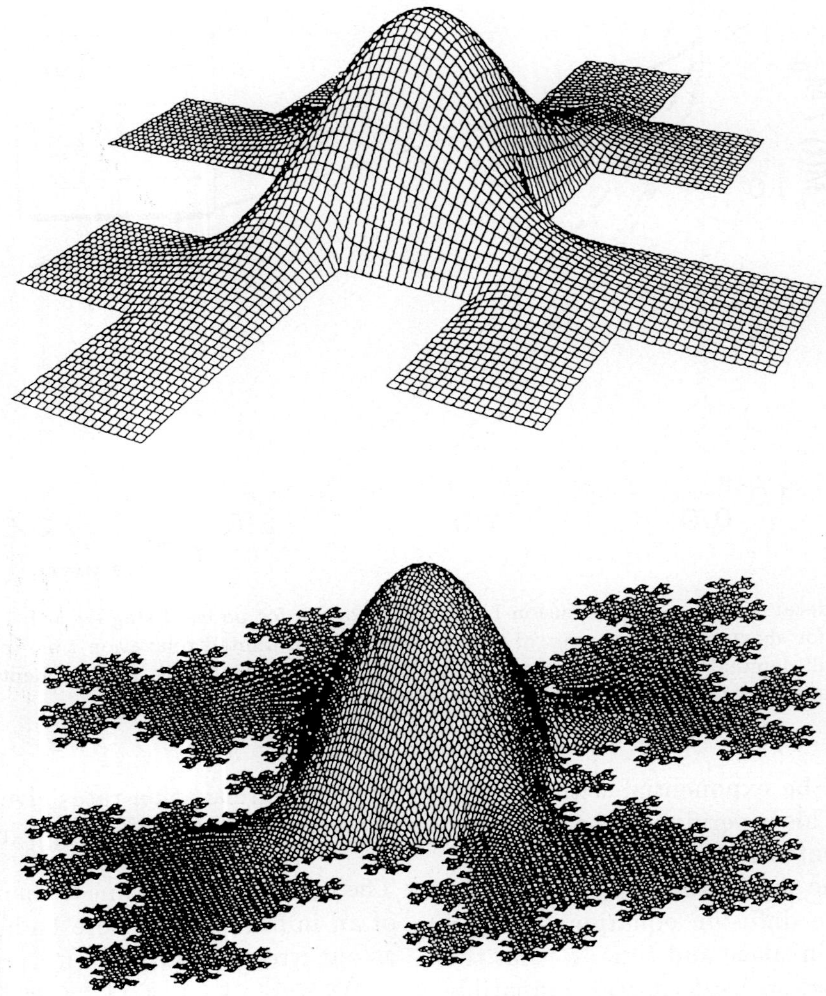

**Fig. 3** Space distribution of the magnetization at long time for D.L.R. in the cases $\nu = 1$ and 3. The decrease of the magnetization in the "bays" is indicative of the strong efficiency of the irregular walls to relax the regions of the pore which are surrounded by the geometrical irregularities. In a very crude approximation, the long time magnetization is confined in the inner square of the prefractal shown in Fig. 1 and the exponential relaxation regime governs only the fraction of the pore which is inside the free volume.

According to the formal equivalence which was explained in Refs. 14 and 15, the spatial distribution of the magnetization is identical to the distribution of the vibration amplitude for the fundamental state of the fractal drum. These fundamental states are shown in Fig. 3 for pores of generation $\nu = 1$ and $\nu = 3$. They exhibit no node line and are centered in the central part of the pore. One observes in the figure that the amplitude decays very strongly from the center towards the pore walls.

One can give the following simple argument for the existence of an extremely rapid decrease of the magnetization in smaller and smaller regions of the fractal geometry.[15] From the general mathematical results quoted above, the fundamental state has an exponential decay time $\tau_0$ of the order of $L^2/D$, where $L$ is the macroscopic size of the pore. This characteristic time is to be compared with a local diffusion time in a region of the drum with a much smaller size $\ell$. Locally, the decay time is of the order of $\ell^2/D$, much smaller than $\tau_0$. Because we know that the magnetization decays exponentially with a single time constant $\tau_0$ everywhere, there must exist a mechanism which compensates for the rapid absorption of spins by the nearby frontier. The compensating mechanism can only be the diffusion from a broader region. But to compensate for the rapid decay, the diffusion flux must be large and this is possible only if the gradient of the magnetization is large enough. In other words, the magnetization has to vary very rapidly in space. The fact that it is very small in the smaller geometrical features is a visual indication of the fractal efficiency to relax narrow regions.

There is a correspondence between the values [Eq. (15)] of the relaxation rates and the distributions shown in Fig. 3. The relaxation rates are of the order of relaxation rates for a square pore with a side of order $L/2$, namely, $T_{1,r}^{-1} = 4$. The long time distributions of Fig. 3 are approximately enclosed in an inner free diffusion volume, here with a size of order $L/2$, although in our case it contains only a quarter of the total number of the nuclei.

At short time the magnetization decay shown in Fig. 2 is not exponential. This can be understood qualitatively by figuring relaxation as the emission of magnetization "killers" which diffuse from the irregular surface. These killers will destroy the magnetization in the regions that they have invaded. At time $t$ the killers will have reached a distance $\ell_D \approx 2(Dt)^{1/2}$ from the surface. Then the region of the pore at distances smaller than $\ell_D$ from the walls will have lost its magnetization. This way to treat that question was proposed by de Gennes in 1982.[22] We then have to consider the "$\varepsilon$-neighboring" of the surface which is defined as the volume $V(\varepsilon)$ in the liquid, which is at a distance of the surface smaller than $\varepsilon$. The principle of such a neighboring (also named Minkovski half-sausage volume, see for instance Refs. 11, 12 or 16) is shown in Fig. 4. If the pore is a surface fractal with a fractal dimension $D_f$, and if the pore has an apparent size (diameter) $L$, the Minkovski-Bouligand "half-sausage" volume is of the order of:

$$V(\varepsilon) = (L/\varepsilon)^{D_f}\varepsilon^3 \tag{16}$$

If the pore is a cylinder of height $L_z$ as shown in Fig. 1, and an apparent size $L$, the Minkovski-Bouligand "half-sausage" volume is of the order of

$$V(\varepsilon) = L_z(L/\varepsilon)^{d_f}\varepsilon^2 \tag{17}$$

where $d_f$ is the fractal dimension of the horizontal cut, and the short time evolution of the magnetization will be given by:

$$(M_0 - M(t))/M_0 \sim L^{d_f-2}(Dt)^{(2-d_f)/2}\,. \tag{18}$$

Our numerical results are shown in Fig. 5. Apart from the very short times which correspond to the diffusion to the first or second row of our discretized sites, one observes power laws. For the square pore, the power exponent is equal to $1/2$ because we have a nonfractal pore, while for the fractal pore of generation 3, the exponent is close to $1/4$, being equal to $(2-d_f)/2$, with $d_f = 1.5$. At longer time when the diffusion length becomes comparable with the diffusion time to cross the cell (of the order of $L^2/D$), we will reach the long time exponential relaxation discussed above. One should note that for an arbitrary geometry, the same argument leads to:

$$(M_0 - M(t))/M_0 \sim V(\ell_D \approx 2(Dt)^{1/2}) \tag{19}$$

and that if one is able to realize this infinite surface rate condition, one would be able to study arbitrary morphologies through this measurement.

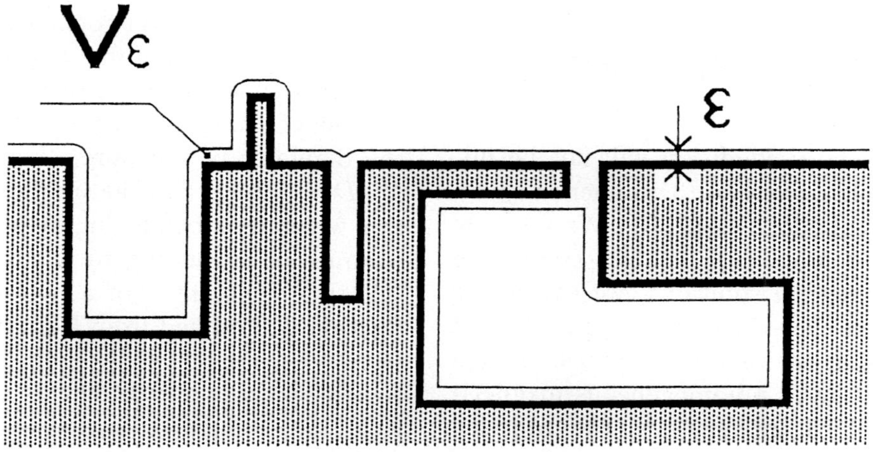


**Fig. 4** Definition of the Minkovski $\varepsilon$-neighboring. The Minkovski neighboring of a surface is the volume which is within a distance $\varepsilon$ of that surface. This zone is called the "half-saussage". For fractal surfaces, this volume is given by Eqs. (16) or (17) but the notion is not to be restricted to fractal surfaces.

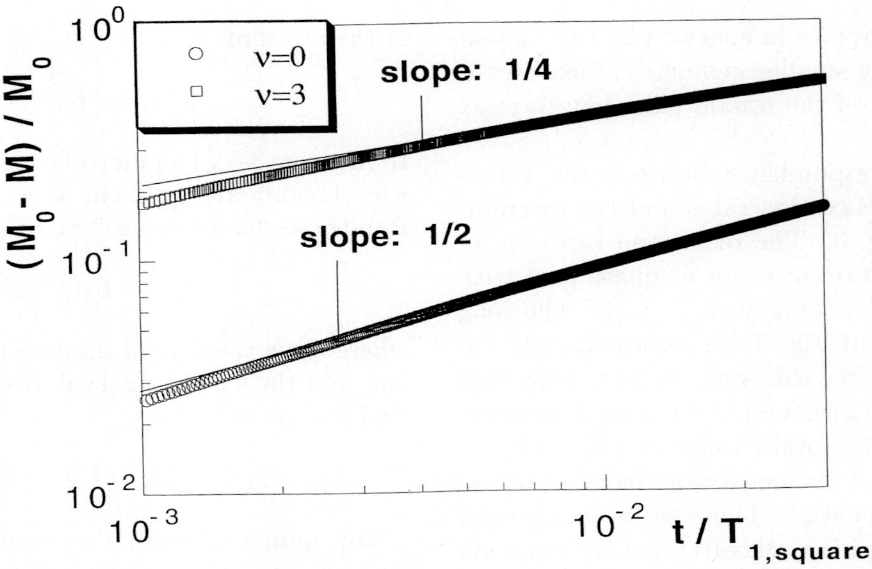


**Fig. 5** D.L.R. short time magnetization decay for the square and $\nu = 3$ pore. One observes power laws with exponent 1/2 in the case of the nonfractal square and 1/4 in the $\nu = 3$ fractal pore. This is in agreement with relation (18). The small discrepancy observed for the very short times is attributed to the effect of numerical discretization.

The above description of the infinite surface relaxation regime will apply to pores which have their smaller features larger than $\Lambda$. The long time exponential relaxation times [Eq. (14)] should be compared with $T_{1b}$ which has a value of 3.6 s for water. The times [Eq. (14)] are of the order of 3.6 s for $L$ of the order of 0.8 mm. The above mechanisms will then dominate for pores with $L$ of a few hundred microns, i.e., with an inner space of the order of 100 microns if the irregularity sizes are of the order of or larger than $\Lambda$.

## 3. CASE OF FINITE SURFACE RELAXATION

In most cases the above condition will not be fulfilled and one should solve Eq. (3) numerically using the boundary condition [Eq. (4)]. We are now in the situation $\ell < \Lambda < \ell(L/\ell)^{(D_f-1)}$. Under the present lack of numerical solutions, we can apply what we know from the corresponding problem of steady-state diffusion to fractal interfaces. If one starts at time $t = 0$ with a homogeneous magnetization, the spins which are very near the interface will relax rapidly while the spins in the center of the pore and far from the pore walls have to diffuse to the pore walls to be able to relax.

How do we decide if a spin is "near" or "far"? The critical distance which separates these regions is the coarse-graining length $L_{cg}$. From its very definition, the spins within this distance diffuse easily to the walls and are relaxed with a short time constant. So there will be a first-time decay with a characteristic time of the order of:

$$T_{\text{short}} \sim L_{cg}^2/D = \ell^2(\Lambda/\ell)^{2(D_f-1)}/D = \ell^{2(D_f-2)/(D_f-1)} D^{(3-D_f)/(D_f-1)} (a\mu_s)^{2/(1-D_f)} . \tag{20}$$

This time is a power law of the surface relaxation rate and the diffusion coefficient. For instance, in the case of relaxation due to paramagnetic impurities, this time will be a power law of the concentration of these impurities. It is independent of the diameter of the pore. For $D_f = 2$, $T_{\text{short}} = \Lambda^2/D = D/(a\mu_s)^2$. The fraction of the magnetization which is embedded in the Minkovski-Bouligand volume corresponding to $L_{cg}$ will be relaxed with this time constant. For a pore of diameter $L$, this fraction is obtained from Eq. (16):

$$(L_{cg}/L)^{(3-D_f)} . \tag{21}$$

For a cylindrical pore, the fraction of the magnetization which follows this rapid decay is from Eq. (17)

$$(L_{cg}/L)^{(2-d_f)} . \tag{22}$$

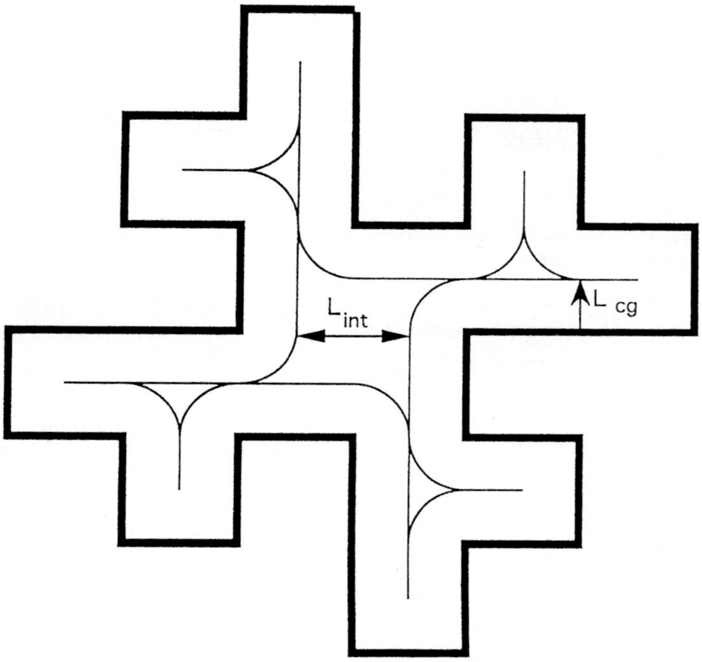


**Fig. 6** Scheme of the two relaxation regions in the case of finite surface relaxation. In the region within a distance of the order of $L_{cg}$ from the surface, a short time decay will be found with a characteristic time of the order of $L_{cg}^2/D$. The center part will exhibit a diffusion limited relaxation corresponding approximately to $L_{\rm int}$. The two zones are separated symbolically by a Minkovski frontier.

The other fraction of the magnetization, i.e., the fraction embedded in the volume inside the pore to the frontier of the Minkovski-Bouligand limit, will decay from diffusion only with a time constant of the order of those described above for this "free" volume. This situation is schematized in Fig. 6. The second time constant will be of the order of $L_{\rm int}^2/2\pi^2 D$, where $L_{\rm int}$ is of the order of the size of the free volume inside the Minkovski boundary related to the distance $L_{cg}$. This decay with two successive time constants will be observed when the coarse-graining length $L_{cg}$ is smaller than the pore diameter $L$. In particular, it can be observed when a pore of large diameter has a rough surface.

If $L_{cg} > L$, all the spins have easy diffusive access to the surface, which then works uniformly. The magnetization is uniform and the relaxation is limited by the surface. We reach a surface limiting regime with a single exponential relaxation time weighed by the relative volume of the surface layer $a\ell^2(L/\ell)^{D_f}$ to the volume of the pore volume $L^3$ and

$$T_{1s}^{-1} = a\mu_s \ell^{(2-D_f)} L^{(D_f-3)} \,. \tag{23}$$

## 4. CONCLUSION

To discuss the influence of irregular geometry on nuclear relaxation, we have used fractal geometry as a model of physical geometrical irregularity. We have shown that depending on the surface relaxation rates, there exist three possible relaxation behaviors in irregular pores. In the (nonfrequent) diffusion limited regime, the long time exponential relaxation is directly related to the size of the inner free volume of the pore, while the short time regime provides in principle a way to study the pore morphology. In the case of finite surface relaxation, we predict also a nonexponential decay. From this preliminary work, one can draw the general conclusion that the irregular nature of the pore geometry plays a role in the magnetization decay. In all three regimes either the fractal dimension intervenes or the size of the free inner volume does. More numerical work on various geometries particularly in the case of finite relaxation, is needed to confirm our predictions.

Thus far we have not studied the other effects of irregular confinement. For instance, in pulse gradient studies, if the gradient is not parallel to the cylindrical pore axis, the apparent diffusion will be found to be less effective because of the existence of a partial trapping of the magnetization within the bays. There will also exist an effect of the pore irregularity on the time dependence of the correlation functions and the translational spin-spin relaxation will be modified. These correlation functions can be obtained from a knowledge of the eigenstates of the diffusion equation with reflecting (Neumann) boundary conditions. This work is currently being done at the Ecole polytechnique.

## ACKNOWLEDGMENTS

This work was supported by D.R.E.T. under contract 93-2525A. One of us (S.R.) has benefited from the E.E.C. "Human Capital and Mobility" program. The Laboratoire de Physique de la Matière Condensée is "Unité de recherches associée du C.N.R.S. n°1254".

## REFERENCES

1. M. Lipsicas, J. R. Banavar and J. Willemsen, *Appl. Phys. Lett.* **48**, 1544 (1986).
2. D. P. Gallegos, D. M. Smith and C. J. Brinker, *J. Colloid Interface Sci.* **124**, 186 (1988).
3. J.-P. Korb, B. Sapoval, C. Chachaty and A. M. Tistchenko, *J. Phys. Chem.* **94**, 53 (1990); C. Chachaty, J.-P. Korb, J. R. C. Van der Maarel, W. Bras and P. Quinn, *Phys. Rev.* **B44**, 4778 (1991), C. Chachaty and J.-P. Korb, *J. Phys. Chem.* **95**, 6058 (1991).
4. F. d'Orazio, W. P. Halperin, K. Eguchi and T. Mizusaki, *J. Appl. Phys.* **65**, 742 (1989); W. P. Halperin, F. d'Orazio, S. Bhattacharja and J. C. Tarczon, in *Molecular Dynamics in Restricted Geometries*, ed. J. Klafter and J. M. Drake (Wiley, New York, 1989), p. 311.
5. G. Liu, Y. Li and J. Jonas, *J. Chem. Phys.* **95**, 6892 (1991).
6. J.-P. Korb, Shu Xu and J. Jonas, *J. Chem. Phys.* **98**, 2411 (1993); J.-P. Korb, A. Delville, G. Demeulenaere, P. Costa and J. Jonas, *J. Chem. Phys.* **101**, 7074 (1994).
7. P. P. Mitra, P. N. Sen, L. M. Schwartz and P. Le Doussal, *Phys. Rev. Lett.* **68**, 3555 (1992); L. L. Latour, P. P. Mitra, R. L. Kleinberg and C. H. Sotak, *J. Magn. Reson.* **A101**, 342 (1993).
8. P. Callaghan, *Bull. Ampere* **44**, N°178, 7 (1995).
9. L. L. Latour, R. L. Kleinberg, P. P. Mitra and C. H. Sotak, *J. Magn. Reson.* **A112**, 83 (1995).
10. J.-P. Korb, L. Malier, F. Cros, S. Xu and J. Jonas, in *Dynamics in Small Confining Systems II*, MRS Vol. 366, eds. J. Klafter and J. Drake (1995).
11. See for instance B. B. Mandelbrot, *The Fractal Geometry of Nature* (Freeman, San Francisco, 1982) or B. Sapoval, *Fractals* (Aditech, Paris, 1990).
12. See for instance P. Pfeiffer in *The Fractal Approach to Heterogeneous Chemistry*, ed. D. Avnir (John Wiley, Chistester, New York, 1989).
13. F. Devreux, J.-P. Boilot, F. Chaput and B. Sapoval, *Phys. Rev. Lett.* **65**, 614 (1990).
14. B. Sapoval, T. Gobron and A. Margolina, *Phys. Rev. Lett.* **67**, 2974 (1991).
15. B. Sapoval and T. Gobron, *Phys. Rev.* **E47**, 3013 (1993).
16. For a review, see for instance B. Sapoval, "Transport Across Irregular Interfaces: Fractal electrodes, fractal membranes and fractal catalysts" in *Fractals and Disordered Systems*, eds. A. Bunde and S. Havlin (Springer-Verlag, Heidelberg, 1995 and references therein), pp. 232–261.
17. R. Gutfraind and B. Sapoval, *Journal de Physique I* (France) **3**, 1801 (1993).
18. B. Sapoval, R. Gutfraind, P. Maekin, M. Keddam and H. Takenouti, *Phys. Rev.* **E48**, 3333 (1993); E. Chassaing and B. Sapoval, *J. Electrochem. Soc.* **141**, 2711 (1994).
19. B. Sapoval, *Phys. Rev. Lett.* **73**, 3314 (1994).
20. D. J. Wilkinson, D. L. John and L. Schwartz, *Phys. Rev.* **B44**, 4960 (1991).
21. See for instance A. Abragam, *The Principles of Nuclear Magnetism* (Clarendon Press, Oxford, 1961).
22. P. G. de Gennes, *C. R. Acad. Sci.* **295**, 1061 (1982).

Fractals, Vol. 4, No. 4 (1996) 463–468


# FRACTAL IMAGE ANALYSIS OF NATURAL SCENES AND MEDICAL IMAGES

TAKASHI SATO, MAKOTO MATSUOKA and HIDEKI TAKAYASU
*Graduate School of Information Sciences,*
*Tohoku University,*
*Aoba-ku, Sendai 980, Japan*

Received June 1, 1995; Accepted August 20, 1995

**Abstract**

We construct color map images of fractal dimension distribution from natural scenes and medical images by applying the box-counting method locally. The map images clearly show the difference between clouds and rocks, as well as between cancer parts and normal tissue in the colon. The method is simple and may be expected to be applicable to a real-time video-data processing.

## 1. INTRODUCTION

Fractal shapes have been characterized in many ways: fractal dimensions, scaling relations, power spectrum exponents, and area size distributions. In applications of fractal theory to image analysis, interesting results have been reported on the classification of textures in natural scenes and medical images as follows. Pentland[1] calculated the power spectra for the estimation of fractal dimension and found double maxima in the histogram of the fractal dimension in some natural scenes. He suggested using this evaluation to perform texture segmentation and classification. Peleg et al.[2] used the relationship between surface area size and resolutions to obtain the fractal dimension of texture image and then classify six natural textures. Chen et al.[3] used the averaged absolute intensity difference between pixel pairs as a function of scales (distances of pixel pairs) to calculate the Hurst coefficient in both normal and abnormal ultrasonic images of liver tissue and to distinguish them.

These works evidently show the effectiveness of fractal image analysis; however, the processing times are rather long to realize real-time processing. In our approach we apply the simplest method, the box-counting method, to binary digitized images in order to shorten the processing time. It is shown that the effectiveness is still comparable to the former methods.

## 2. THE METHOD

Original images are obtained from a royalty-free CD-ROM data base (a historigical picture) and image scanner (a natural scene). We separate them into red, green and blue components which have 256 gray levels for each pixel. In the following analyses, we use only green component images because they include more information about brightness and better contrast of original images than others.

We transform these gray level images into binary images in the pre-image processing to apply our fractal dimension calculation program. A binary image has only black and white pixels and has a much reduced amount of data compared with the original image. The pre-image processing is as follows. First, by applying the common gradient operator for edge enhancement[4] to a gray image, we obtain the edge enhanced image. These images still have gray levels for each pixel at this time. Then the edge-detected gray image is transformed into a binary image using a threshold method with a given threshold value. This value is chosen interactively to provide the best feature of edge-enhanced image.

A binary image which has $n$ pixels for the width and $m$ pixels for the height is divided into $32 \times 32$ pixel block areas with overlapping portions and then the fractal dimension is calculated for each block area locally using the box-counting method.[5] In each block area we take the log of $N(r)$ and $r$ and plot log $N(r)$ versus log $r$. Here $N(r)$ is the number of boxes needed to cover the image and $r$ is the size across the block. Then we use a method of least squares linear regression to estimate the slope of the resultant curve, which gives the local fractal dimension value $D$ over the chosen image area. In our method, the box size $r$ is varied by $2^x$ ($x = 0, 1, 2, 3$), that is, 1, 2, 4 and 8 pixels. This process is done repeatedly, with shifting of the $32 \times 32$ pixel block window by an increment of 1 pixel over the whole image. The obtained fractal dimension for each block is attributed to its center pixel. The fractal dimension values for the various pixels are classified into 10 levels ranging from 1.0 to 2.0, for convenience of display. In cases where the calculated values of the fractal dimension are out of the range, we do not display the pixels in the fractal dimension distribution map. With these processes, we construct a fractal dimension distribution color (or gray graduation) map image from the original image.

## 3. RESULTS

We first apply the method to a Sierpinski Gasket to check its validity. The theoretical value of this fractal figure is given as $D = \frac{\log 3}{\log 2} = 1.5849\ldots$. Figure 1(a) illustrates the original figure digitized with $320 \times 320$ pixels. The pre-image processing is not required in this case because this figure is originally a binary image. Figure 1(b) shows a log-log plot for an area selected by the rectangle in Fig. 1(a). The linear fit is quite good (correlation coefficient = 0.999) and the calculated fractal dimension value is 1.58, which is quite close to the theoretical value of this figure. After applying the above processes, we obtain a fractal dimension distribution map as shown in Fig. 1(c), in which each pixel value is a fractal dimension. As one can see from Fig. 1(c), most of the values of the fractal dimension lie between 1.5 and 1.6, proving the validity of our method.

Now we analyze a natural image including clouds and a rocky sea shore as an example. Figure 2(a) illustrates the original image digitized with $512 \times 256$ pixels. After the pre-image processing has been applied to the original image, we obtain Fig. 2(b). Applying our method to this binarized image, a map of the fractal dimension distribution is plotted as shown in Fig. 2(c). As we can see from Fig. 2(c), the fractal dimension values for the cloud parts lie mostly between 1.4 and 1.5, while most of the values of the fractal dimension for the rock parts lie between 1.8 and 1.9. This result clearly demonstrates that clouds and rocks in a natural scene can be classified by the fractal dimension calculated by our method.

Next we use colorectal cancer image as an example of medical images. Figure 3(a) shows the original image digitized with $256 \times 256$ pixels, which includes a normal tissue and a cancer tissue. The surface structure of the normal tissue (right down part of the figure) is rather regular. On the other hand, the surface structure of the cancer tissue (left

**(a)**

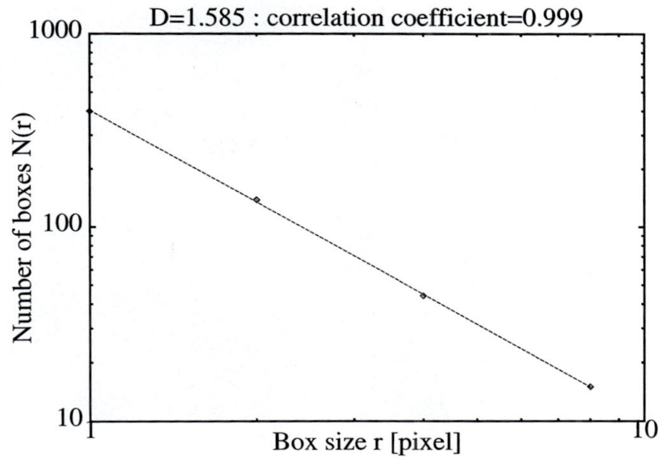


**(b)**

**(c)**

**Fig. 1** **(a)** A Sierpinski Gasket. **(b)** A log-log plot for an area marked by the rectangle in Fig. 1(a), shown with regression line. The slope is $-1.58$. **(c)** A map of fractal dimension distribution for Fig. 1(a) shown with color bar.

middle part of the figure) is complicated and the shape is irregular. Figure 3(b) shows the binarized image obtained by pre-image processing for this image. Sample regions are marked by rectangles with numbers at 1 and 2 in Fig. 3(a). The sizes of their regions are both $64 \times 64$ and the calculated fractal dimensions of selected cancer and normal tissues are 1.86 and 1.39, respectively. Figure 3(c) shows the color map image of the fractal dimension distribution which is $256 \times 256$ in size. As we can see from this figure, the cancer part and the normal tissue can be classified clearly. These results suggest the possibility of a quantitative approach in the diagnosis of colorectal cancer using our method.

## 4. DISCUSSION

In the above result for medical image analysis, the fractal dimension value of each region is obtained by the box-counting method. We also apply the fBm method[5] in order to compare the fractal dimension values between the two methods.

The calculation of the fBm method is much more complicated than the box-counting method. Actually, typical computation times (CPU times) by the fBm method are about 1000 times greater than the computation times by the box-counting method. The results of fractal analysis and their comparison are shown in Table 1. As shown in this table, the

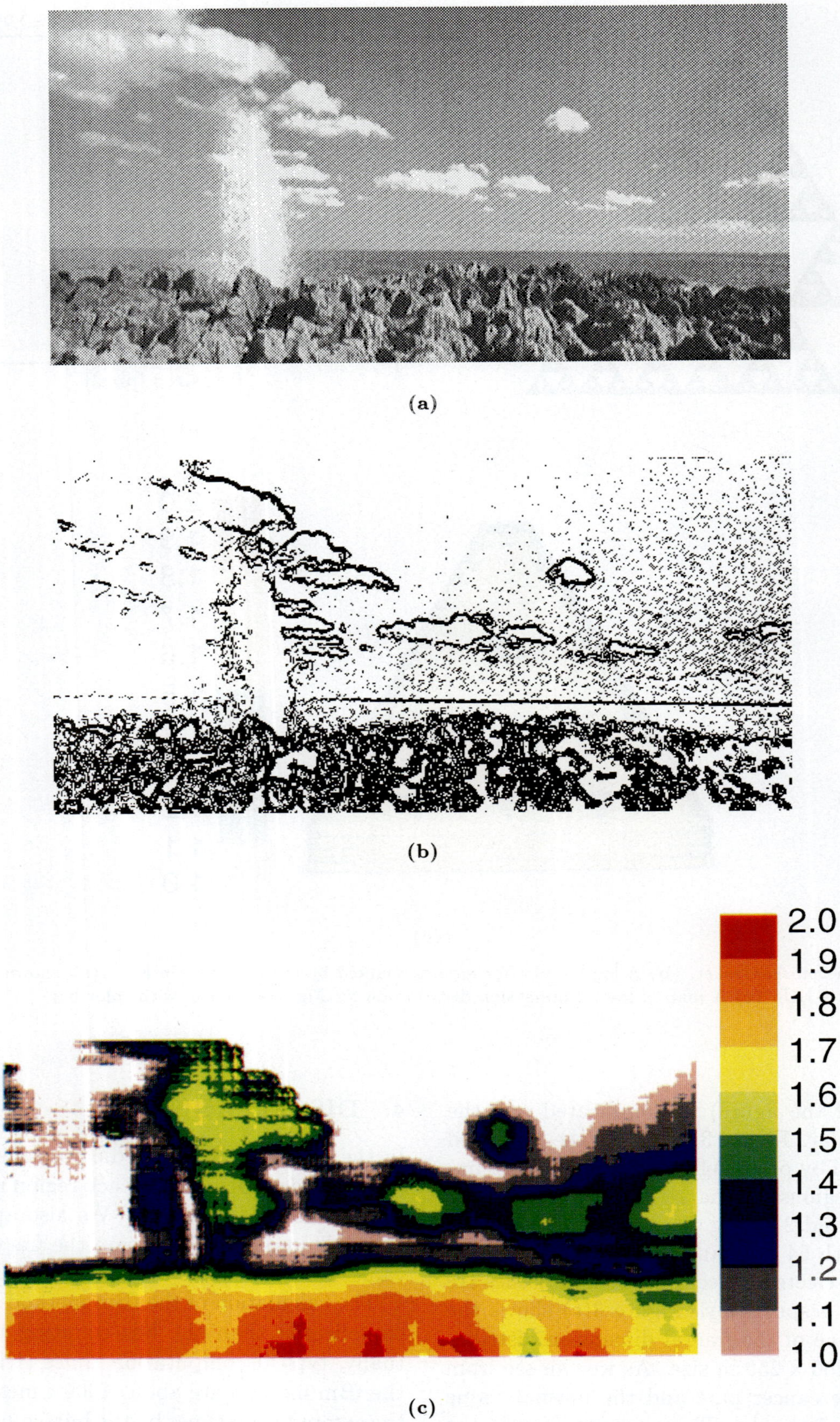


**Fig. 2** (a) A natural scene image. (b) An edge enhanced and then binarized image for Fig. 2(a). (c) A map of fractal dimension distribution for Fig. 2(a).

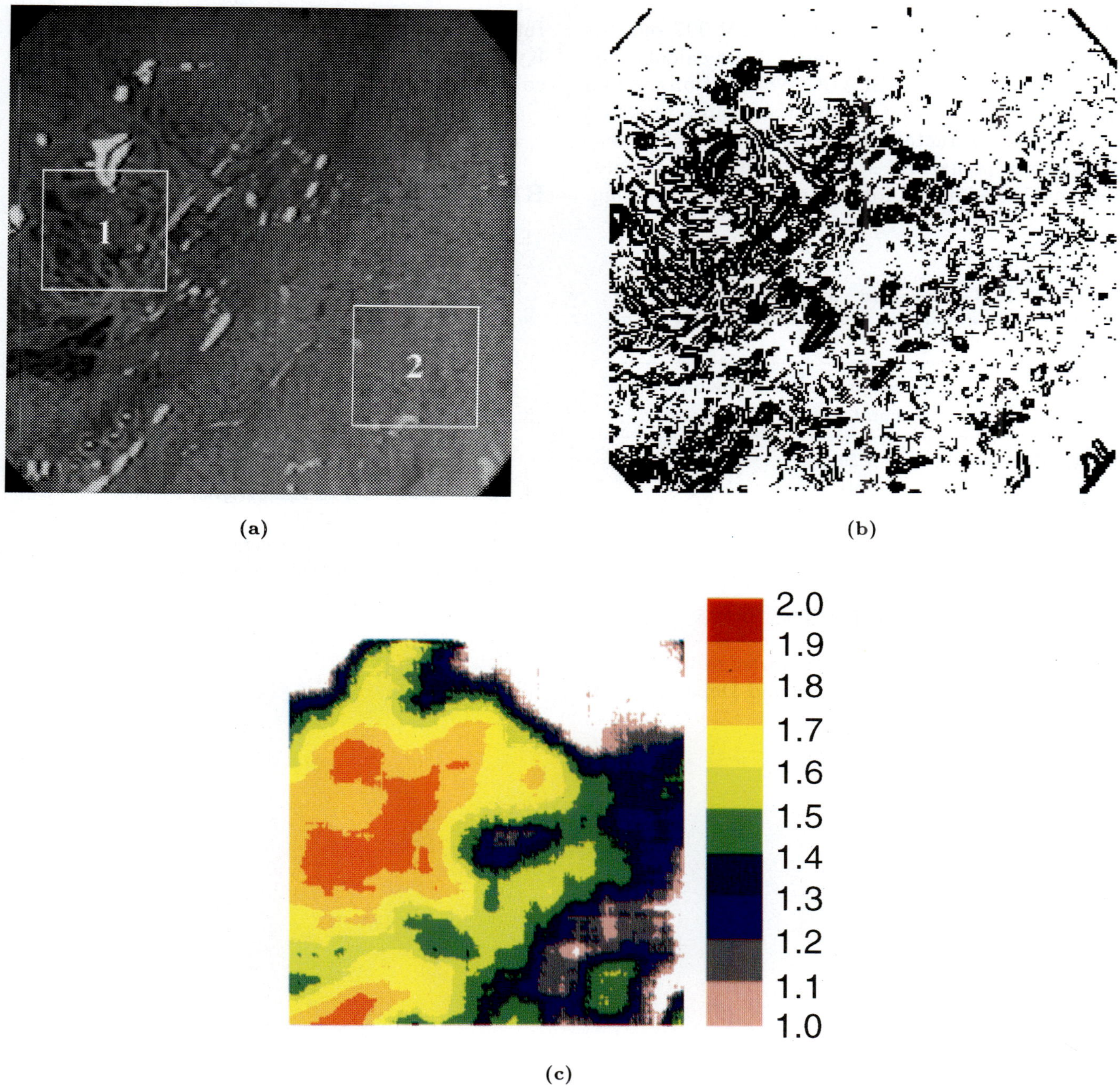


**Fig. 3** **(a)** A medical image including colorectal cancer (area 1) and normal tissue (area 2). **(b)** An edge enhanced and then binarized image for Fig. 3(a). **(c)** A map of fractal dimension distribution for Fig. 3(a).

fractal dimension values of cancer are apparently different from the values of normal tissue in both methods. Here, in the case of the fBm method, the fractal dimension is defined as a value within a range from 2.0 to 3.0. By subtracting 1.0 from the fractal dimension values by the fBm method, we can compare them with our values. As shown in the table, the values look consistent. We can see that using the binary data is enough to calculate the fractal dimensions of cancer and normal tissues and that the 256 gray levels data is not required. A reduction of the amount of data can cut down the calculation time and increase the process speed. In the calculations for a medical image of $256 \times 256$ pixels, with an area of size $32 \times 32$ pixels using the box-counting and fBm methods, the computation times

(CPU times) are 139 seconds and 151 997 seconds (almost two days), respectively. These calculations have been done by the SPARC Workstation with no optimization for speed. From the results reported here, our local fractal analysis can classify different image area in natural scenes and medical images, and may be expected for real-time data processing.

**Table 1 Comparison of fractal dimension values obtained by box-counting and fBm methods**

| | Local Fractal Dimension Value | |
|---|---|---|
| | **Normal Tissue** | **Cancer** |
| Box-Counting | 1.39 | 1.86 |
| fBm | 2.31 | 2.78 |

## ACKNOWLEDGMENTS

The authors would like to acknowledge the contributions of Dr. A. Iso of the Second Department of Internal Medicine, Kyoto First Red Cross Hospital, Kyoto, Japan, in providing us with and reading the colorectal cancer images used in this study.

## REFERENCES

1. A. Pentland, "Fractal-based description of natural scenes," *IEEE Trans. Pattern Anal. Machine Intell.*, Vol. PAMI-6, pp. 661–674 (1984).
2. S. Peleg, J. Naor, R. Hartley and D. Avnir, "Multiple resolution texture analysis and classification," *IEEE Trans. Pattern Anal. Machine Intell.*, Vol. PAMI-6, pp. 518–523 (1984).
3. C. Chen, J. Daponte and M. Fox, "Fractal feature analysis and classification in medical imaging," *IEEE Trans. Med. Imaging*, Vol. MI-8, pp. 133–142 (1989).
4. A. Rosenfeld and A. Kak, *Digital Picture Processing* (Academic Press, 1976).
5. B. Mandelbrot, *The Fractal Geometry of Nature*, (Freeman, San Francisco, CA, 1982).

Fractals, Vol. 4, No. 4 (1996) 469–475


# SOLVING THE TWO-DIMENSIONAL INVERSE FRACTAL PROBLEM WITH THE WAVELET TRANSFORM

ZBIGNIEW R. STRUZIK
*Department of Computer Science, University of Amsterdam*
*Kruislaan 403, 1098 SJ Amsterdam, The Netherlands*

Received June 1, 1995; Accepted June 15, 1995


### Abstract

The methodology of the solution to the inverse fractal problem with the wavelet transform[1,2] is extended to two-dimensional self-affine functions. Similar to the one-dimensional case, the two-dimensional wavelet maxima bifurcation representation used is derived from the continuous wavelet decomposition. It possesses translational and scale invariance necessary to reveal the invariance of the self-affine fractal. As many fractals are naturally defined on two-dimensions, this extension constitutes an important step towards solving the related inverse fractal problem for a variety of fractal types.


## 1. INTRODUCTION

The initial definition referred to the Hausdorff dimension as the principal feature of a fractal. Soon it became obvious that the world of fractals would be unnecessarily restricted if such a definition were maintained. It is the notion of some sort of similarity in scale which has loosely defined the fractal ever since. We adopted this point of view, and with the help of the wavelet transform in providing the amazing ability to unfold objects in scale, we were able to reveal the scale position similarity of self-affine fractals. A properly chosen wavelet transform of a fractal exhibits a characteristic regularity, consistent with the renormalization properties of the fractal. By unfolding a fractal in the scale domain, it becomes possible to investigate its invariance properties.

Indeed, fractals (in particular self-affine attractors) possess by definition invariance which must be identifiable in the form of some elementary transformations (for example, those used to create the fractal object). The problem of recovering these transformations from the fractal object is known as the Inverse Fractal Problem (IFP). The wavelet transform therefore provides the necessary means to make the IFP tractable. The crucial step in this approach consists of the possibility of verification of such a scale-space similarity in terms of invariance with respect to some iterative functions (maps). Contrary to the methods for solving the inverse fractal problem, which use a predefined class of Iterated Function System (IFS) functions[3] to approximate the invariance discussed, the approach we present aims at revealing the renormalization involved in the creation of the fractal, possibly bearing high relevance to the underlying physical phenomena.

We, therefore, proposed the hypothesis[1] that an object can be classified as a pre-fractal if there exists a solution to the related inverse fractal problem; in other words, if a set of construction rules can be found which, ultimately, would fully define the object within the scales where it is observed and which allows arbitrary extrapolation in the scale domain.

The goal of this paper is to extend the scope of the methodology developed in Refs. 1 and 2 to self-affine fractals on two-dimensional support. These can be created with $\mathbf{R}^3 \to \mathbf{R}^3$ iterated functions (maps). Like computer generated fractals, natural fractals have the property that they are created by the same phenomena acting on different scales. These phenomena may, however, show behavior ranging from purely stochastic to completely deterministic. Although the focus of this paper is on the deterministic limit, a more general application of the techniques is possible.

## 2. FRACTAL IFS FUNCTIONS IN TWO DIMENSIONS

In this section we will address relevant facts about self-affine functions over two-dimensional (or fractal $1 < D < 2$) support. Self-affinity is a concept inherent to fractal functions including self-similar sets of functions support as a special case. An *affine transformation* $S : \mathbf{R}^n \to \mathbf{R}^n$ is of the form:

$$S(x) = T(x) + b \tag{1}$$

where $T$ is a linear transformation on $\mathbf{R}^n$ (often represented by an $n \times n$ matrix) and $b$ is a vector in $\mathbf{R}^n$. Therefore, an affine transformation is in general a combination of translation, rotation, dilation and possibly reflection.

Of special relevance to the creation of fractal attractors is a *contraction* transformation $S$:

$$\begin{aligned} S : \mathbf{R}^n \to \mathbf{R}^n \quad & \text{is a contraction if there is a} \\ & \text{number } c \text{ and } 0 < c < 1 \\ \text{such that} \quad & |S(x) - S(y)| \le c|x - y| \\ & \text{for all } x, y \in \mathbf{R}^n . \end{aligned}$$

Affine transformations, having different contraction ratios in different directions, are to be distinguished from similarities which contract isotropically:

$$\text{if } |S(x) - S(y)| = c|x - y| \text{ for all } x,\, y \in \mathbf{R}^n$$

$$\text{then } S \text{ is a similarity} .$$

For $S_1, \ldots, S_N$ affine *contractions* on $\mathbf{R}^n$, by the standard result (see e.g. Ref. 4) the unique compact set $F$ invariant with respect to $S_n$ is guaranteed to exist and is termed a *self-affine set* or an *attractor* of $S_n$, $n = 1, \ldots, N$. Thus for the self-affine set $F$ obtained by:

$$\begin{aligned} F_k &= \left[ \bigcup_{n=1}^{N} S_n \right]^{\circ k} (B) \quad \text{for any } B \in \mathbf{R}^n ; \\ F &= \lim_{k \to \infty} F_k , \end{aligned} \tag{2}$$

where $\circ$ denotes the composition of transformations, so that:

$$(P_1 \circ P_2 \circ \cdots \circ P_k)(x) = P_1(P_2 \ldots (P_k(x)) \ldots)) ,$$

the following relation is satisfied:

$$F = S_n(F) \qquad \forall S_n,\ n = 1, \ldots, N . \tag{3}$$

The subsequent approximations $F_k$ of the set $F$ are called pre-fractals.

We will consider continuous contraction transformations $S_i : \mathbf{R}^3 \to \mathbf{R}^3$, in short to be called *maps*, chosen in such a way that the self-affine set they define is a functional mapping $f : \mathbf{R}^2 \to \mathbf{R}$.

Let $S_n (1 \le n \le N)$ be an affine transformation represented in matrix notation with respect to the coordinates $\{x_1,\, x_2,\, y\}$ by:

$$S_n \begin{pmatrix} x_1 \\ x_2 \\ y \end{pmatrix} = \begin{pmatrix} a_n & b_n & 0 \\ c_n & d_n & 0 \\ \gamma_{n1} & \gamma_{n2} & \alpha_n \end{pmatrix} \begin{pmatrix} x_1 \\ x_2 \\ y \end{pmatrix} + \begin{pmatrix} e_n \\ f_n \\ \delta_n \end{pmatrix} . \tag{4}$$

For the sub-transformation on the coordinates $\{x_1,\, x_2\}$ of the function $F(x_1,\, x_2)$:

$$S_{xn} \begin{pmatrix} x_1 \\ x_2 \end{pmatrix} = \begin{pmatrix} a_n & b_n \\ c_n & d_n \end{pmatrix} \begin{pmatrix} x_1 \\ x_2 \end{pmatrix} + \begin{pmatrix} e_n \\ f_n \end{pmatrix} \tag{5}$$

which describes an arbitrary affine transformation in $\mathbf{R}^2$, we will require:

$$0 < \sigma_n^{-1} = \begin{vmatrix} a_n & b_n \\ c_n & d_n \end{vmatrix} < 1$$

which ensures that $S_{xn}$ is a similitude and the transformed surface does not vanish or flip over. We will also restrict $S_n$ to be a set of non-overlapping transformations:

$$\bigcap_n S_n(A) = \phi, \ \forall A, \ \text{where } A \text{ is an arbitrary compact set}.$$

Additional constraints can be specified for the purpose of generating a fractal interpolation surface ensuring joining up of the transformed surfaces at the interpolated points.[a]

## 3. TWO-DIMENSIONAL WAVELET TRANSFORM ON FRACTAL FUNCTIONS

### 3.1 The Two-Dimensional Continuous Wavelet Transform

The (real-valued) wavelet transform $W$ decomposes the function $f(\mathbf{x}) \in L^2(\mathbf{R}^2)$ in the base of elementary wavelets created by the action of the affine group on a single function $\psi(\mathbf{x})$.

$$(Wf)(\overline{\overline{s}}, \mathbf{b}) = \frac{1}{|\overline{\overline{s}}|} \int d\mathbf{x} f(\mathbf{x}) U(\overline{\overline{s}}, \mathbf{b}) \psi(\mathbf{x}) \tag{6}$$
$$\text{where} \quad U(\overline{\overline{s}}, \mathbf{b})\psi(\mathbf{x}) = \psi((\overline{\overline{s}})^{-1}(\mathbf{x} - \mathbf{b})).$$

Formally, the choice of the wavelet $\psi(\mathbf{x})$ is subject to one restriction: $\psi(\mathbf{x})$ must be *admissible*, i.e., of zero mean, if the function is to be reconstructed from its wavelet transform. Since we investigate only invariance properties of the functions' wavelet transform, we lift this requirement. In practice, the wavelet $\psi(\mathbf{x})$ is chosen to be well localized both in scale and space, and if no directional sensitivity is required uniform scaling in all directions is a natural choice. The matrix $\overline{\overline{s}}$ can then be reduced to two equal scaling factors $s$ on the diagonal and $|\overline{\overline{s}}| = s^2$.[b]

Further, the choice of the analyzing wavelet $\psi(\mathbf{x})$ is dependent on the application. For our purpose, we will need the wavelet with $m$ vanishing moments, i.e., orthogonal to polynomials of order $l < m$:

$$\int_{-\infty}^{+\infty} \mathbf{x}^l \psi(\mathbf{x}) d\mathbf{x} = 0 \qquad \forall l, \ 0 \leq l < m. \tag{7}$$

Arguments for the choice of $m$ will be given in Sec. 4 of this paper.

In this work we use continuous parametrization of the affine operator $U(\overline{\overline{s}}, \mathbf{b})$ known as continuous wavelet transform (CWT). However, restricting $\overline{\overline{s}}$, $\mathbf{b}$ to discrete values gives the discrete wavelet transform which, for the particular choice of a wavelet satisfying the orthogonality criteria, constitutes an important class of orthogonal wavelet transforms.

### 3.2 The Multi-Resolution Scheme

Since the self-affine attractor is by definition invariant with respect to some affine operator $T(\overline{\overline{u}}, \mathbf{v})$:

$$f = T(\overline{\overline{u}}, \mathbf{v})f; \quad T(\overline{\overline{u}}, \mathbf{v})(\mathbf{x}) = \overline{\overline{u}}\mathbf{x} + \mathbf{v} \tag{8}$$

we expect this to be reflected in the invariance of the wavelet transform. Indeed,

$$\begin{aligned}(Wf)(\overline{\overline{s}}, \mathbf{b}) &= \frac{1}{|\overline{\overline{s}}|} \int d\mathbf{x} f(\mathbf{x})(U(\overline{\overline{s}}, \mathbf{b})\psi(\mathbf{x})) \\ &= \frac{1}{|\overline{\overline{s}}|} \int d\mathbf{x}(T(\overline{\overline{u}}, v) f(\mathbf{x}))(U(\overline{\overline{s}}, \mathbf{b})\psi(\mathbf{x})) \\ &= \frac{1}{|\overline{\overline{s}}|} \int d\mathbf{x} f(\mathbf{x})(T^\dagger(\overline{\overline{u}}, \mathbf{v}) U(\overline{\overline{s}}, \mathbf{b}))(\psi(\mathbf{x}))\end{aligned} \tag{9}$$

which, using an easily verifiable relation $T^\dagger(\overline{\overline{u}}, \mathbf{v}) = \frac{1}{|\overline{\overline{u}}|} T^{-1}(\overline{\overline{u}}, \mathbf{v}) = \frac{1}{|\overline{\overline{u}}|} U(\overline{\overline{u}}, \mathbf{v})$, where $T^\dagger$ denotes an operator adjoint to $T$, can be written as:

$$\begin{aligned}(Wf)(\overline{\overline{s}}, \mathbf{b}) &= \frac{1}{|\overline{\overline{u}}|} \frac{1}{|\overline{\overline{s}}|} \int d\mathbf{x} f(\mathbf{x}) \\ &\quad \times (U(\overline{\overline{u}}, \mathbf{v}) U(\overline{\overline{s}}, \mathbf{b}))(\psi(\mathbf{x})).\end{aligned} \tag{10}$$

[a] These conditions are however irrelevant to this work. In fact, for the sake of simplicity, we will limit ourselves to examples without additional components in the function value which translates to $\gamma_n = 0$ and $\delta_n = 0$.

[b] The normalizing factor $s^2$ is taken per default if working in two dimensions. This may need 'correcting' in Eq. (15) using the fractal dimension of the support of the attractor recovered from the solution to the IFP.

One can easily show that the action of $[U(\overline{\overline{u}}, \mathbf{v})]^k$ $U(\overline{\overline{s}}, \mathbf{b})$ is equivalent to applying the operator $U(\overline{\overline{s}}^{(k)}, \mathbf{b}^{(k)})$, where $k$ indexes the action of the operator $U(\overline{\overline{u}}, \mathbf{v})$:

$$\begin{cases} \overline{\overline{s}}^{(k)} = \overline{\overline{u}}^k \overline{\overline{s}} \\ \mathbf{b}^{(k)} = \overline{\overline{u}}^k \mathbf{b} + \mathbf{v} \sum_{i=0}^{k-1} \overline{\overline{u}}^i \quad k \in \mathbf{N}\,. \end{cases} \tag{11}$$

Therefore, the invariance relation for the wavelet transform of the self-affine function is:

$$(Wf)(\overline{\overline{s}}, \mathbf{b}) = (Wf)(\overline{\overline{s}}^{(k)}, \mathbf{b}^{(k)})\,. \tag{12}$$

The action of the new operator $U(\overline{\overline{s}}^{(k)}, \mathbf{b}^{(k)})$ with respect to index $k$, defines a "multi-resolution representation" of the function $f$, where all the nested subspaces are translated and scaled versions of each other. The construction of such a multi-resolution scheme for an arbitrary self-affine attractor, as defined in Sec. 3.2, is possible and is equivalent to finding the operator $T(\overline{\overline{u}}, \mathbf{v})$. A demonstration of such a construction was given in Refs. 1 and 2, where a special kind of wavelet transform representation — the wavelet maxima bifurcation representation, was introduced to accomplish this.

## 4. RECOVERY OF THE SCALING PARAMETERS FROM THE WAVELET TRANSFORM BIFURCATION REPRESENTATION

We will use wavelets with $m$ vanishing moments [see Eq. (7)]. This will allow us to perform filtering off of the polynomial behavior of the $m-1$ order according to:

$$(W)(D^{(m)}f(x)) = |\overline{\overline{s}}|^{-m}(W^{(m)}(f(x)))\,, \tag{13}$$

where the operator $D^{(m)} = d^m/dx^m$ and $W^{(m)}f$ the wavelet transform using a wavelet with $m$ vanishing moments.[c]

[c] In this communication we used an example with no additional terms, which is equivalent to both $\gamma_n = 0$ and $\delta_n = 0$. Therefore, orthogonality of the wavelet to polynomials would not be essential and the wavelet with $m = 0$ could be used. Nevertheless, in the formal considerations we will, however, retain generality.

The self-affine function can generally be described by a finite set of $n$ affine transformations [compare Sec. 2, Eq. (4)] of the generic form:

$$\begin{cases} \mathbf{x}' = S_{xn}(\mathbf{x}) \\ y'(\mathbf{x}') = \alpha_n y(\mathbf{x}) + \gamma_{n1}x_1 + \gamma_{n2}x_2 + \delta_n\,. \end{cases} \tag{14}$$

If by taking the wavelet transform of $y$ with the wavelet with the appropriate number (in this case at least 2) of vanishing moments, additional components in Eq. (14) are removed, the following invariance relation holds with respect to the affine operator $S_{xn}$:

$$\begin{aligned} D_x^{(m)} y(\mathbf{x}') &= D_x^{(m)} y(S_{xn}(\mathbf{x})) \\ &= \sigma_n^{-m} \alpha_n D_x^{(m)} y(\mathbf{x}), \quad m >= 2\,. \end{aligned} \tag{15}$$

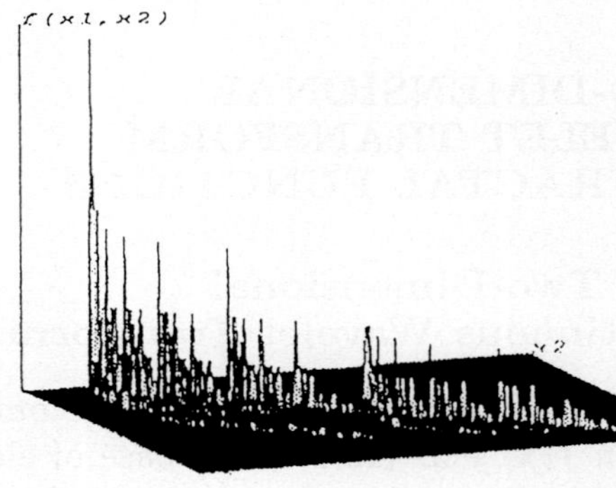


**Fig. 1** The test case: Sierpiński triangle with non-uniformly distributed measure.

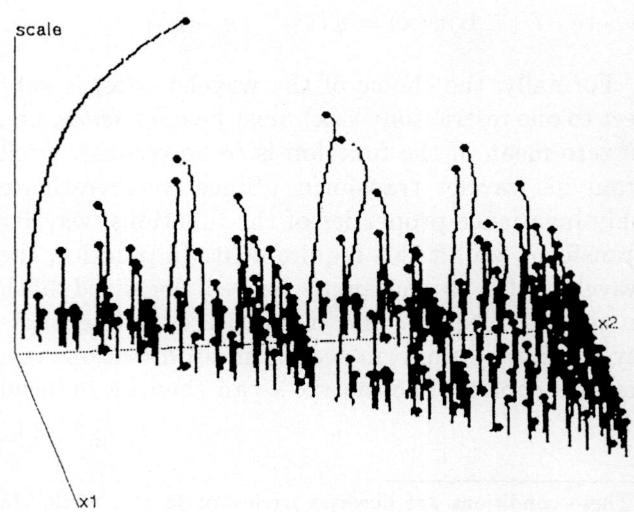


**Fig. 2** Local maxima with associated bifurcations for the example from Fig. 1.

As already shown in the previous section, Eq. (12), for $T(\overline{\overline{u}}_n, \mathbf{v}_n) = S_{xn}$ this invariance reflects (up to the multiplicative factor) in the wavelet transform coefficients:

$$w_D(\sigma_n)(W^{(m)}y)(T(\overline{\overline{u}}_n, \mathbf{v}_n)(,)) \\ = \sigma_n^{-m}\alpha_n(W^{(m)}y)(,), \quad m >= 2\,. \qquad (16)$$

The factor $w_D(\sigma_n) = \sigma_n^{2-D}$ corrects for the possibly fractional support of the attractor; the fractal dimension $D$ of the support can be estimated in the usual way: $\sum_n \sigma_n^D = 1$.

It is therefore the second (or higher) derivative of the wavelet transform $W^{(2)}(F)$, which, for the generic self-affine attractor defined by Eq. (4), involves the invariance with respect to the transformation $S_{xn}$ on the coordinates $x_1$ and $x_2$, and the multiplicative dependence on the parameters $\alpha_n$, and will allow the recovery of the parameters of $S_{xn}$ and the $\alpha_n$'s.

What is needed for this purpose is the representation capturing the invariance present in the wavelet transform of the function and allowing the solution of the Eq. (15) through the comparison of related points in $W^{(m)}(y(,))$, Eq. (16). This is achieved in the wavelet transform maxima bifurcation representation.

Once the invariant representation is established, the remaining parameters ($\gamma_{n1,2}$ and the constant terms $e_n$, $f_n$, $\delta_n$) can be recovered from $W^{(1)}_{x_1,x_2}(y(,))$ and $W^{(0)}(y(,))$: the wavelet transform taken with lower order wavelets $D^{(1)}_{x_1,x_2}\psi$ and $\psi$ respectively, sampled with the invariant grid of bifurcations.[1]

### 4.1 Wavelet Transform Local Maxima Bifurcation Representation

We will not go beyond being very brief on the arguments for the construction of the bifurcation representation, referring the reader to earlier publications[1,2] for more details.

Indeed, neither the continuous wavelet transform nor the orthogonal discrete version can achieve the kind of multi-resolution analysis we require. While the continuous version maps all the scale-space relationship originally present in the analyzed function onto the scale-space domain in the most redundant way, leaving the searched-for affine invariance undiscriminated, the discrete one represents it in an orthogonal fashion on an *a priori* defined grid, which in general is not 'compatible' with the affine invariance of the function.

The wavelet transform maxima bifurcation representation was therefore derived from the (modulus) maxima representation of Mallat.[5] The particularly convenient feature of Mallat's maxima representation is its translation invariance, which is of great importance in pattern recognition problems. Since fractal functions are a special class of functions where the translation invariance is accompanied by scale invariance, we used unique 'landmarks' in the form of bifurcations of maxima lines to derive scale invariant representation from the maxima lines.

The necessary condition for the local extrema of the wavelet transform $Wf(s, x_1, x_2)$ of the function of two variables $f(x_1, x_2)$ is zero of the partial derivatives:

$$\begin{cases} \dfrac{d(Wf)(s, x_1, x_2)}{dx_1} = 0 \\ \dfrac{d(Wf)(s, x_1, x_2)}{dx_2} = 0 \end{cases} \qquad (17)$$

which points are further classified according to the sufficient condition for local extrema:

$$\begin{cases} \mathcal{H}(s, x_1, x_2) > 0 \\ \text{either } \dfrac{d^2(Wf)(s, x_1, x_2)}{dx_1^2} < 0 \text{ in the case of a local maximum} \\ \text{or } \dfrac{d^2(Wf)(s, x_1, x_2)}{dx_1^2} > 0 \text{ in the case of a local minimum}, \end{cases} \qquad (18)$$

where $\mathcal{H}(,,)$ is the Hessian $\mathcal{H}(,,) = D^{(2)}_{x_1^2}(Wf)(,,)\, D^{(2)}_{x_2^2}(Wf)(,,) - (D^{(2)}_{x_1,x_2}(Wf)(,,))^2$.

The bifurcation case is defined by:

$$\begin{cases} \text{Eq. (19)} \\ \mathcal{H}(s, x_1, x_2) = 0\,. \end{cases} \qquad (19)$$

Further, we will differentiate bifurcations depending on the sign of the gradient of the second derivative of the function, which is related to the direction of the development of the branches in the tree structure apparent in the maxima representation.

This tree structure serves as a basis for the parameter recovery algorithm utilized in the solution to the IFP. The tree construction is similar to that described in our communication on the one-dimensional case.[1] The expansion of the attractor in scale thus achieved, gives direct insight into the action of the transformation, in this case IFS, used

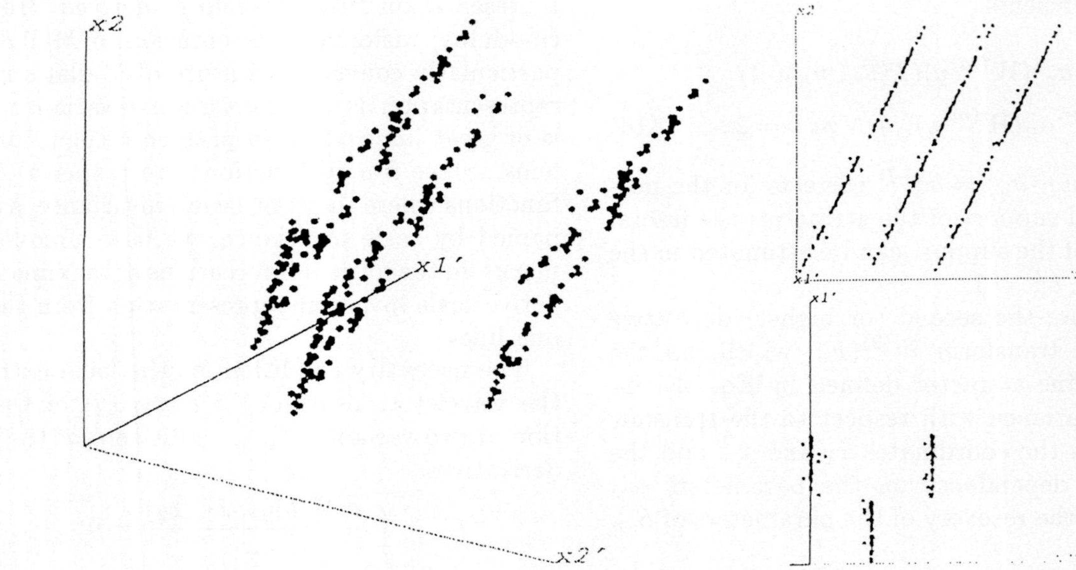


**Fig. 3** The maps recovered from the bifurcation representation in Fig. 2. Three of the total of six two-dimensional maps $x_2 = f(x_1', x_2')$ are visible in the figure on the leftmost. The Sierpiński triangle was rotated in order to expose the maps. The view on the maps showing the dependence on one coordinate $x_1'$ in the upper right and on $x_2'$ on the bottom right. Note that, as expected, the linear dependence shows in only one of the coordinates.

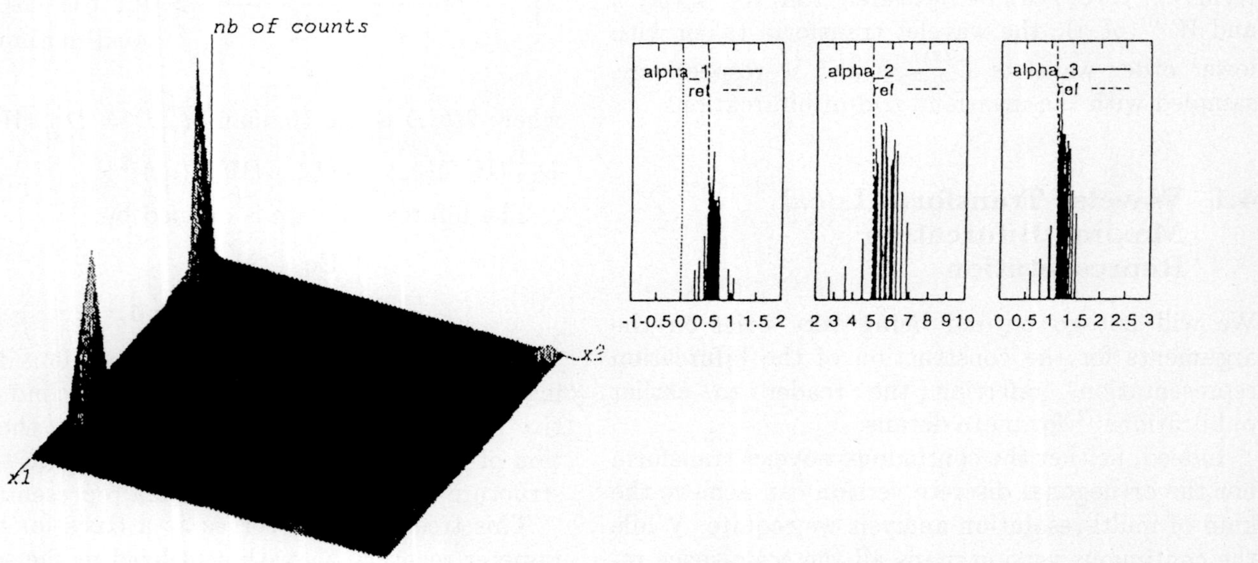


**Fig. 4** The histogram of the parameters $e_n$ and $f_n$ shows the location of the peaks on a triangle corresponding to the translation vectors of the Sierpiński triangle (left). The modes of the parameter $\alpha_n^{-1}$ responsible for the measure distributed on the triangle (right).

to create the attractor. The 'stages of construction' are therefore revealed and linked with one another in the bifurcation representation.

The next crucial step comprises the finding of the invariance of the bifurcation representation. This can be done by means of tree matching (see Ref. 1), where the invariance of the representation is sought in the optimal match of the tree to its branches. From the match of pairs of bifurcations found invariant, the parameters of Eq. (16) are estimated.

Space limits prevent elaboration on the procedure and we will restrict ourselves to presenting and interpreting the results for the given example. The affine maps $x_{\{1,2\}} = f'(x'_1, x'_2)$ representing scaling and translation parameters of $S_{xn}$ as shown in Fig. 3. As expected from the diagonal character of $S_{xn}$ for the Sierpiński triangle the linear dependence with consistent slope $1/\sigma_n = 1/0.5$ shows in only one of the coordinates. The translation components are directly visible in the histogram in Fig. 4 and the repartition of measure can be verified from the modes of parameter $\alpha_n$ in the upper right of this figure.

## ACKNOWLEDGMENTS

We are grateful to Dr E. H. Dooijes for careful reading and suggestions for improvements to the manuscript. This project is supported by the Dutch Foundation for Scientific Research (NWO-SION).

## REFERENCES

1. Z. R. Struzik, *Fractals* **3**(2), 329–350 (1995).
2. Z. R. Struzik, E. H. Dooijes and F. C. A. Groen, in *Fractal Reviews in the Natural and Applied Sciences*, ed. M. M. Novak (Chapman & Hall, 1995).
3. M. F. Barnsley, *Fractals Everywhere* (Academic Press, NY, 1988).
4. K. Falconer, *Fractal Geometry — Mathematical Foundations and Applications* (John Wiley, 1990).
5. S. G. Mallat and S. Zhong, *IEEE Trans. PAMI* **14**, 710–732 (1992).

# Author Index